Advanced Practice
in Human Service Agencies

About the Authors

Lupe Alle-Corliss, formerly Lupe Alley, is a bilingual/bicultural licensed clinical social worker who has worked in many areas of human services. She has practiced in a variety of settings—a rehabilitation hospital, a family services agency, a county-contracted mental health clinic, a county department of mental health, and the psychiatry department of Kaiser Permanente—and remains active in clinical work. Lupe's professional positions have included medical social worker, minority outreach coordinator, lead clinician, graduate fieldwork supervisor, private practitioner/EAP, managed care provider, outreach/crisis intervention counselor, and lecturer. Lupe also teaches and supervises fledgling human services practitioners; she is an instructor in the Human Services program at California State University, Fullerton, where she has served for more than 12 years. Most recently, she is faculty liaison and professor in the Graduate School of Social Work at California State University, San Bernardino. Lupe has also taught at the University of Southern California. She attended the University of Southern California from 1974 to 1980, earning a bachelor's degree in social gerontology and a master's degree in social work.

Randy Alle-Corliss, formerly Randy Corliss, is a licensed clinical social worker with a broad range of professional experience. He has worked in a variety of community agencies and with many client populations. He is a therapist for adult clients in the psychiatry department at Kaiser Permanente and has worked at both county-contracted and state mental health centers. Randy also has a small private practice, in which he uses a range of therapies to help children, adults, and families. Randy is committed to human service education. For the last 13 years, he has taught undergraduate courses in the Human Services program at California State University, Fullerton, where he has also served as fieldwork coordinator. Randy was a featured speaker on men's issues and men's groups at a recent annual conference of the National Organization of Human Service Providers. His professional experience includes clinical therapist/psychiatric social worker, community outreach coordinator, outreach/crisis intervention counselor, group counselor, fieldwork coordinator, and field instructor. Randy earned a bachelor's degree in psychology, with a minor in human services, at California State University, Fullerton. He entered the Graduate School of Social Work at the University of Southern California in 1978, concentrating in mental health.

Advanced Practice in Human Service Agencies

Issues, Trends, and Treatment Perspectives

Lupe Alle-Corliss

California State University, Fullerton
California State University, San Bernardino

Randy Alle-Corliss

California State University, Fullerton

BROOKS/COLE
CENGAGE Learning™

Australia • Brazil • Japan • Korea • Mexico • Singapore • Spain • United Kingdom • United States

BROOKS/COLE
CENGAGE Learning

Advanced Practice in Human Service Agencies: Issues, Trends, and Treatment Perspectives
Lupe Alle-Corliss, and Randy Alle-Corliss

Sponsoring Editor: Lisa I. Gebo

Marketing Team: Steve Catalano, Aaron Eden

Editorial Assistant: Susan Wilson

Advertising Communications: Margaret Parks

Production Editor: Kelsey McGee

Manuscript Editor: Patterson Lamb

Permissions Editor: May Clark

Interior Design: E. Kelly Shoemaker

Cover Design/Illustration: Roger Knox

Art Editor: Lisa Torri

Indexer: Steve Rath

Typesetting: Omegatype Typography

For product information and technology assistance, contact us at
Cengage Learning Customer & Sales Support, 1-800-354-9706

For permission to use material from this text or product,
submit all requests online at www.cengage.com/permissions
Further permissions questions can be emailed to
permissionrequest@cengage.com

Library of Congress Control Number: 98-30151

ISBN-13: 978-0-534-34811-3

ISBN-10: 0-534-34811-4

Brooks/Cole
20 Davis Drive
Belmont, CA 94002
USA

Cengage Learning is a leading provider of customized learning solutions with office locations around the globe, including Singapore, the United Kingdom, Australia, Mexico, Brazil, and Japan. Locate your local office at
www.cengage.com/global

Cengage Learning products are represented in Canada by Nelson Education, Ltd.

To learn more about Brooks/Cole, visit **www.cengage.com/brookscole**

Purchase any of our products at your local college store or at our preferred online store **www.cengagebrain.com**

Printed in the United States of America
7 8 9 10 11   19 18 17 16 15

In recognition of our mothers,
Lily and Violet
for their enduring love, dedication,
and profound impact
on our growth and development.
Despite their absence, we know they
are always with us in spirit and
share in the joy of our accomplishments.

Contents

CHAPTER 3

Micro Level Practice: Working with Individuals   88

CHAPTER 4

Mezzo Level Practice: Working with Families and Groups   135

CHAPTER 5

Macro Level Practice: Working with Organizations and Communities

CHAPTER 6

Practice with Diverse Populations

CHAPTER 7

Ethical and Legal Issues in Advanced Practice   307

Preface

This advanced practice text was born out of a need to provide upper-division undergraduate, graduate, and post-graduate students within the helping professions with valuable information about the issues, trends, and treatment perspectives commonly found in human service agencies. This, our second text, is a natural outgrowth from our first, *Human Service Agencies: An Orientation to Fieldwork*. We intended the previous book to help beginning professionals deal with agency life by introducing them to the most important concepts that human service providers face. In this text, our intent is to provide helpers with a more *advanced* look at the many intricacies of *direct practice* with clients in a variety of agency settings. We take a more clinical approach in discussing an array of issues most helping professionals are likely to encounter. Those embarking in a career in human services, counseling, psychology, social work, or any other related profession will find the material we present to be both informative and relevant to current-day practice concerns.

Our purpose is to provide practitioners with the tools that will help them in maintaining a healthy balance between self-awareness, knowledge development, and skill attainment. Throughout the text, we encourage healthy introspection coupled with the acquisition of knowledge about a variety of practice issues that range from working with individuals, families, groups, organizations, communities, and public policy, to working with diverse populations and with advanced ethical and legal considerations. At each juncture, we also invite helpers to begin to cultivate the specific skills necessary for effective work with individual clients and specific problem situations. Each chapter is organized similarly; knowledge regarding specific issues is presented first, followed by suggestions for treatment considerations that are then applied to a case scenario designed to test know-how and skill. This "hands-on" approach to learning is useful given its practical applications, but perhaps even more significant in alerting helpers to the many complexities of "real-life" agency work.

Our text is intended to be very comprehensive and practice-based. In Chapter 1 we begin by discussing common practice issues experienced in agency settings and the importance of supervision, peer review, and case consultation in the pursuit of ongoing learning and career development.

In Chapter 2, we discuss key elements and challenges of practice. These include developing sound assessment and interviewing skills, understanding transference and countertransference issues, recognizing one's own limitations, learning to work with resistant and difficult clients, and being able to face successfully the many organizational and environmental challenges that are typical of work in agency settings. The emphasis is on learning assertion and effective communication skills in working with clients, co-workers, administrators, other professionals, and in understanding the ever-present dangers of encapsulation and burnout.

Chapter 3 centers on *micro* level practice with individuals. We present an overview of some common theoretical orientations including psychodynamic, person-centered, cognitive-behavioral, and developmental approaches. We then review treatment modalities—crisis intervention models (including critical incident debriefing); brief therapies, such as the popular solution-focused therapy; traditional and contemporary long-term treatment approaches—and versatility in case management.

Chapter 4 is a logical transition to working with families and small groups on a *mezzo* level. Here we focus on working with families by presenting a historical focus of the family, the systems approach, the developmental life cycle, and family assessment. We follow this by reviewing various treatment orientations to family therapy and specific practice models that fall within each of these categories. From the psychodynamic school, we feature Bowen's Family Systems Theory; from the humanistic school, the Satir Model; from the cognitive-behavioral school, Structural Family Therapy; and from the developmental school, Haley's Life Cycle Model. We focus on the changing family life cycle and its implications for practice by touching on role changes, diverse family life styles (given the increase in single parents, divorce, blended families, and two-provider families, gay and lesbian parents, and so forth), the lack of economic stability, and the rise in family violence. In addition to family work, we highlight learning about small group work that is becoming a common treatment modality in many agency settings. Besides reviewing the basics of group work, we focus on the personhood of the group leader, the need to appreciate group dynamics and group process, and the important ethical and diversity issues involved. We survey varied group treatment modalities, discuss current trends, and provide insight on developing one's own approach to group within an agency system.

Our focus in Chapter 5 changes to a *macro* emphasis in which we discuss the many subtleties of working within organizations, the community, and in the social policy arena. We present the broad spectrum of macro level practice with historical views as well as theories and models of practice for each of the three areas. Social action, social planning, and locality development are discussed in depth. Macro practice administration, research, and the importance of empowerment practice are also included.

In Chapter 6, our focus zeros in on practice with diverse populations. Our emphasis is on becoming "diversity-sensitive" by considering the multidimensional aspects of people's lives. We closely examine issues related to race, culture, ethnicity, age, gender, socioeconomic status, and sexual orientation with the hope of encouraging a more open and accepting climate for working with clients who may be different from us. For each of these categories of diversity, we present historical perspectives, current trends, and treatment issues. In discussing race, culture, and ethnicity, we feature the importance of a culturally relevant assessment and treatment and we touch upon the four major ethnic groups: African Americans, Asian Americans, American Indians, and the Hispanic/ Latino populations. In discussing *age*, we take a developmental slant and feature working with all age groups, including children, adolescents, young adults, middle-aged adults, and older adults. In considering *gender issues*, we discuss the individual challenges faced by both men and women and consider effective treatment modalities with each. We discuss working with differences of varied socioeconomic status, with emphasis on the poor or homeless. We address the importance of understanding and developing the ability to work effectively with gay and lesbian clients in our discussion about *sexual orientation*.

We end with Chapter 7, where we address advanced treatment considerations of ethical and legal issues. We discuss such topics as informed consent, confidentiality, child abuse, elder and dependent adult care abuse, suicide, homicide, spousal abuse, AIDS, and dual relationships. We also consider the reality of ethical dilemmas and the need to learn sound ethical decision making.

Because we have attempted to provide a wealth of relevant, current-day information about "practice," we believe our text has the potential to serve various functions. As a textbook, it provides a sound knowledge base about theory and practice, and as a guidebook or manual, it offers useful treatment perspectives and the all-important opportunity to engage in skill attainment.

Acknowledgments

Many individuals have contributed to our professional development and influenced us, in some way, in the writing of this text. Many experiences and interactions with supervisors, colleagues, and clients have helped shape us into the clinicians we now are and have provided us with the impetus to share our knowledge and skills. We thank our past supervisors Brian Conlan, Bruce Hume, Leo Juarez, Hal Platts, and David Rodriguez for their keen insight, support, and encouragement that have allowed us to persevere and continue to cultivate our clinical skills. In similar fashion, our colleagues throughout the years have also influenced us through peer review, consultation, and education. Their dedicated relationships have been a source of ongoing strength and validation. We also wish to acknowledge our clients, who have inspired us with their courage, honesty, and resilience. They have taught us more than we can ever express and are a major reason for our continued commitment to clinical practice and teaching within the human services arena.

Faculty and students of various universities have also been instrumental in the undertaking of this project. Special thanks go out to those within the Graduate Department of Social Work at California State University, San Bernardino. They include: chair, Teresa Morris; professors Lucy Cardona, Mel Hawkins, Marshall Jung, Nancy Mary, Ira Neighbors; field work director, Steve Petty; and students Narda Judd, Gawad Hajawad, and Glenda Boatner. We also appreciate the guidance and assistance of several individuals in the Human Services department of California State University, Fullerton: professors Gerald Corey, Mikel Garcia and Kristi Kanel, and students Debbie Stout and Patricia Huerta. We also thank Terrence Forrester from Loma Linda University and Ray Liles from the Public Child Welfare Academy of Southern California for their interest, support, and encouragement.

We cannot fail to recognize the reviewers from a variety of academic institutions who have so generously provided us with positive reinforcement during the rigors of the review process. Their specific suggestions, constructive feedback, and insightful comments served to help us refine our ideas and consider additional areas of interest. They include Vickie Gardine-Williams, Ira Neighbors, and Cynthia Vaske.

Additionally, we cannot say enough about the dedicated members of the Brooks/Cole team who guided us throughout this project. We are especially indebted to Lisa Gebo, editor of human services and social work, for her ongoing support and inspiration; Kelsey McGee, production editor, who so enthusiastically helped us through the editing and production process; Patterson Lamb, manuscript editor, for her fine editorial assistance; May Clark, permissions editor, for her contribution; Martha Wood Simmons, for her sharp-eyed proofreading; and Steve Rath, for his careful indexing. Without any one of these, the publication of this text would not have become a reality.

Finally, we thank our children Jessica and Justin for their patience, love, and playfulness. Their sense of humor and ability to take things in stride and to enjoy the simple pleasures of life has helped us on many occasions to stop, relax and reevaluate our own priorities. We are forever grateful to them for keeping us on the right track and teaching us that the most important thing in life is our love for each other.

Lupe Alle-Corliss
Randy Alle-Corliss

Advanced Practice in Human Service Agencies

Issues, Trends, and Treatment Perspectives

1

Practice Issues in Human Service Agencies

Professional development in human services is an ongoing, continuous process that typically begins in the classroom and extends into the field, where theoretical knowledge, concepts, and ideas are applied. The value of and need for fieldwork cannot be overstated, for this is often where an individual's professional practice actually begins. Through fieldwork opportunities, a beginning practitioner gains additional knowledge, cultivates skills, and acquires invaluable hands-on experience. Without appropriate fieldwork opportunities, professional development is limited. Without an emphasis on training and ongoing learning, the potential for professional growth and self-awareness could be diminished. In this text, we provide the advanced student and beginning human service practitioner with a framework that allows them to maximize their learning and guides them through their fieldwork and other work experiences. To this end, we examine current trends, issues, and treatment perspectives of human service practice, with emphasis on advanced practical fieldwork applications.

MAINTAINING A BALANCE: AWARENESS, KNOWLEDGE, AND SKILLS

Effective practice requires a balance among the three key elements of self-awareness, knowledge, and skill development. The absence of any one of these will impact some aspect of practice negatively and could possibly have an adverse effect on clinical treatment.

The Need for *Self*-Awareness

To be sound, clinically and ethically, practitioners need to be self aware, possess the capacity for introspection, and be motivated for growth. We believe that together with a personal commitment to service, self-awareness is essential for successful helping and for becoming an effective professional.

Self-awareness allows you to use yourself in helping others. To do so, you must be knowledgeable about what makes you think, feel, and act as you do. Hutchins and Cole (1992, p. 4) believe that "to be an effective helper it is first necessary to understand major aspects of your own behavior and how you interact with others." When you have become attuned to your

1

own feelings, you will be more sensitive to those of others. Goldman (1995, p. 43) noted that "the ability to monitor feelings from moment to moment is crucial to psychological insight and self-understanding. People with greater certainty about their feelings are better pilots of their own lives, having a surer sense of how they really feel about personal decisions" (p. 43).

Working in agency settings requires you to relate to many different people—from clients to agency administrators. Your ability to communicate effectively with these individuals depends on how well you can read your own feelings and thoughts; you must be able to choose wise behaviors if you are to help others and yourself gain the most from your work together.

In embarking on a career in human services, you are setting out on a journey of self-awareness that will last the rest of your professional life. Be aware of how much you have to learn. Overconfidence at this stage could undermine your ability to help others and ultimately impede your own ongoing learning. If you are not serious about the importance of self-evaluation and introspection, expect reminders that you are human and have room for personal growth. We stress using all available resources to help you toward self-awareness, including supervision, consultation with co-workers and instructors, group sessions, and your own counseling if needed.

Working with clients can trigger your own issues that may directly or indirectly impact the work you do. Self-awareness is crucial at this juncture. As practitioners, you should "be aware of your own needs, areas of unfinished business, personal conflicts, defenses, and vulnerabilities" (Corey, Corey, & Callanan, 1993, p. 29). Included in this introspective process is the willingness to look at your own past history, including conflicts related to your family of origin. Some of these childhood experiences will undoubtedly influence your perceptions and reactions as an adult. When we, as helpers, are unaware of our own unresolved issues, we can be hindered in our effectiveness to help others. If you experience emotional turmoil from uncovering unexplored territory in your past, these unresolved areas may precipitate a crisis. If this happens, we advise you to seek guidance from your supervisor, a field instructor, or a counselor. It is essential that you be aware of anything in your past that may enhance or hinder your work with clients.

Self-acceptance is another area often triggered by the self-awareness process. Because acceptance of oneself is a prerequisite to accepting others, it is crucial for helpers. All of us are continuously learning who we are and how we need to change to become more self-actualized. If you have difficulty accepting yourself, competency issues are likely to surface. Helpers who are not comfortable with who they are may be ineffective in helping others battle their own negativity.

Self-acceptance starts with exploring your own limitations and sources of low self-esteem. Helpers who are willing to work on themselves are more likely to develop a positive self-concept, to be genuine, and to inspire others to work on themselves too.

Last, self-awareness is vital as you learn to work with clients of different backgrounds. Understanding how you feel about your social class; your racial, ethnic, and religious characteristics; and your similarities or dissimilarities from your clients is critical to your effectiveness as a helper. We believe that the first step in becoming diversity sensitive is to look at yourself—to understand where you come from and to recognize that your roots were indeed important in shaping you and your values. To appreciate your clients and their history, you must first be willing to appreciate your own past. It is also important to explore your own family dynamics and generational components. Tracing family lines through a family tree or a genogram can help you understand your family of origin more clearly. It also may uncover some of the biases and prejudicial beliefs of your family that you may not have been aware of. This may be an unpleasant discovery, but it can be an opportunity for you to explore your own values and biases and determine how they might impact your clients. By recognizing your prejudicial views, you may learn when you might have difficulty remaining objective, and thus become more sensitive in your work with certain types of clients. If you find that you cannot remain objective despite your efforts, it may be an appropriate time for a referral.

The Importance of a Solid Knowledge Base

Without a solid knowledge base to guide you in your work with clients and the many issues they present, you are likely to be ineffective in your practice. You must have knowledge before you can develop skills. Your knowledge must also be current. Keep up with the literature in the field. Staying abreast of new theory and practice allows you to adopt novel approaches to treatment and to assess current trends and future directions in practice.

In addition to learning from the classroom and textbook readings, we advocate other types of knowledge building: reading journal articles and literary works, watching movies and documentaries, engaging in self-help activities, being involved in friendships with others, and having some type of exposure to diversity and diverse populations. These are only a sampling of the many ways you can learn.

Because you will likely work with diverse populations as a human service practitioner, you must try to understand your clients and their life situations. By being knowledgeable and well informed about the population groups you will work with, you will not only foster trust and respect in your working relationships but also be able to tailor treatment to more accurately meet your clients' individual needs. This will necessitate your remaining open to continued educational opportunities to enhance your awareness and knowledge of your clientele. We strongly recommend building a specialized knowledge base regarding the needs of certain client groups in addition to seeking to understand their worldview.

Sue and Sue (1990) stipulate two main ingredients for developing this understanding. First, you must know the sociopolitical systems in operation in the United States with respect to the treatment of minorities in this country; second, you must be conscious of the institutional barriers that obstruct service provision for many population groups. This will lead you to explore the forces of oppression and the role of racism, sexism, heterosexism, ageism, and other biases that affect certain groups.

To gain the knowledge necessary for effective work with certain population groups, you will have to begin by assessing your own knowledge base. If you plan to work with clients who are different from you, you will need additional knowledge. However, many of you will be working with a population you believe you know and are comfortable with; therefore, you may not feel the need for additional education in the area. Even though you will already have a sound knowledge base to draw from, there are always new things to learn. We encourage you to remain open to learning and to recognizing your limitations. Be very alert to generalizing or stereotyping based on your experience. Remember that the most effective way to get accurate information about your clients and their issues is to ask them. They can often provide you with new insights, and asking them for their opinions tells them that their views are important and valued. The principles behind ethnographic interviewing that have long been applied in anthropology are useful here. Important knowledge about a person's cultural values, family structure, religious practices, child-rearing customs, experience of oppression and racism, and level of acculturation can be obtained by direct ethnographic client interviewing and by informal observation of clients in their environment. When direct interviewing is not possible, you can still try to understand your clients and their point of view. Doing so will tell them that they are the "experts" and you are the "learner." You can observe your clients while they are in the agency or by visiting their community. Observing their style of dress, their interactions with others, the language or mode of communication they use primarily, and the general environment in which they live can tell you much about their culture and lifestyle. Being open, flexible, and willing to understand clients in their world, not yours, is ultimately what should be emphasized.

Skill Development and Application

Skills are developed through practice. Fieldwork settings allow you to apply practically the knowledge you have gained.

To build skills, you must begin to apply the knowledge and awareness you have acquired. Often, this is an anxiety-ridden step, as you will have to face some of your insecurities and fears. Remembering that learning is ongoing will encourage you to take these risks and to accept yourself as someone who is continually evolving, both personally and professionally.

Developing appropriate intervention strategies is important when you are working in human services with diverse client groups. You can enhance your effectiveness when you use methods that are consistent with the life experiences and cultural values of the specific client you are treating. Sue and Sue (1990) cite the following considerations in developing sound practice skills:

- Economically and educationally disadvantaged clients may not be oriented to engaging in advanced treatment modalities, such as journaling, logging, charting, reading self-help books, or using cognitive restructuring techniques. Instead, you can use techniques on the clients' level where they may feel more comfortable and be more responsive.
- Encouraging self-disclosure may be futile or even detrimental to the helping process, as it may be incompatible with the client's cultural values. In some cultures self-disclosure is considered a violation, and talking about personal issues may be perceived as exposing the family's private affairs.
- The process-oriented, time-consuming, and ambiguous nature of the helping process may be contrary to the life values of your clients. In order to avoid conflict, it is important to understand your clients' views and be aware of both your and their values.
- Some clients prefer a more active and directive approach that may differ from the manner in which some helpers practice. Sometimes clients will want specific directives on how to cope or concrete feedback, rather than wanting to develop awareness and insight.

Competence develops when you can successfully translate knowledge into action. This requires ongoing self-evaluation and a willingness to seek client feedback about your effectiveness. You should be open to using a variety of strategies in working with diverse clients. Lum (1996), Soloman (1976), and Goldstein (1993) advocate using joint intervention strategies, which involve mutual involvement of the client and the worker. You must start where the client is, which means that you must be guided by the client's perception. Also, the strategies you use must fit within the client's culture. The enthnographic principles discussed earlier can be applied here. We believe that clients should be viewed as the experts in their lives; the helpers are those who are knowledgeable about problem solving and the helping process. When client and helper work together, the client's needs can be met in ways that are both therapeutic and culturally acceptable.

Developing good communication is also essential to becoming an effective helper. Practitioners must maintain clear communication with clients, colleagues, and the community at large. Brill (1990, pp. 59–75) cites several aspects of communication that we have found especially useful in skill development:

- Understanding the messages being sent and received by being cognizant of the attitudes and beliefs of both client and helper
- Recognizing similarities and differences among people
- Understanding a client's capacity to use both verbal and nonverbal communication
- Using varied means of communication
- Being able to seek feedback
- Ensuring that gestures and symbolism are understood
- Communicating at the client's pace
- Interfacing and keeping distractions to a minimum or at least being aware of their impact on the client's attention and ability to interact with the helper

Another consideration in skill development is for helpers to be open to alternative treatment options or out-of-office strategies in their work with some clients. Traditional methods may not always be the most conducive to helping clients resolve their issues. Engaging in outreach work, consulting, taking an ombudsman's role, and acting as a facilitator and advocate for your clients are examples of alternative approaches.

A last step in skill development is to seek appropriate opportunities for practice. Practicum, fieldwork, and internship placements are excellent ways to begin applying what you have learned. We also encourage you to explore additional practice opportunities through volunteer work and involvement in self-help ventures.

Integration and Balance

A competent practitioner will integrate and balance self-awareness, knowledge, and skills. Possessing any one of these qualities without the others can seriously impair your ability to engage in effective practice. Below, we illustrate this point with several scenarios showing how the absence of any one of these elements can impair the helping process.

SCENARIOS

AWARENESS WITHOUT KNOWLEDGE OR SKILLS

Joel is a clinical social work intern who has been involved in his own therapy to deal with depression and loss. He is now beginning to see clients in his fieldwork setting. Although he may be in tune to what it is like to be a client, given his own experience in this role, he is perplexed about how to treat a client who enters therapy complaining of excessive anxiety. Joel has heard about systematic desensitization, which can be helpful with this type of client. However, because he has no experience with this technique, he does not know how to begin. He realizes that although he is self-aware, his beginning knowledge and skill level limit his current ability to help the client. He will have to study systematic desensitization and develop the skills in that technique before he can work effectively with anxiety conditions.

KNOWLEDGE AND SKILL WITHOUT SELF-AWARENESS

A counseling intern, Judy, is being trained in Gestalt techniques and is very eager to apply her knowledge and skill in working with her new clients in the mental health center where she will be interning in this year. One of her first clients enters treatment hesitantly and presents with many unresolved issues regarding childhood abuse and neglect. Judy, excited by the possibilities of using Gestalt techniques in this case, insists that the client engage in the empty-chair exercise whereby she can have an imaginary talk with her mother about her hurt and anger. The client becomes very anxious and leaves the session prematurely indicating she does not feel well. When she calls to cancel her next appointment, Judy decides to discuss the case with her supervisor. After a review of the case, the supervisor concludes that the client's anxiety and reluctance to return to treatment could be related to the helper's lack of awareness that the client may not have been ready for Gestalt work. Judy learns that it is important to be self-aware to be effective; she realizes that she may have been too insistent or domineering in her push to apply the newly learned Gestalt techniques.

AWARENESS AND KNOWLEDGE WITHOUT SKILLS

Steve is a clinical social work intern who has been trained in cognitive therapy. However, he is having difficulty applying his knowledge and seems to lack certain skills necessary to assist a client in dealing with her angry feelings toward her spouse. This client enters treatment upset by a succession of losses: the death of her year-old daughter, the stillbirth of her second child, and a recent divorce by her husband who announced that he didn't love her any more and was going out with a younger woman who could successfully give him the child he wanted. The client was devastated by these events and enters treatment hoping to feel less depressed. As Steve begins to probe into the client's past, she becomes angry with him and he responds by becoming overly defensive and irritated. In retrospect, Steve realizes that the client's anger toward him was really a projection of her anger toward her ex-husband, but he was not skillful enough to help her see this. His defensiveness caused the client to shut off the expression of her feelings, which was an important factor in allowing her to deal with her grief and depression.

The ideal to strive for is a balance among awareness, knowledge, and skill. We know that a perfect balance is not always possible but we encourage you to remain open to recognizing your limitations and committed to continuing growth in pursuit of this ideal.

PRACTICE ISSUES

Practice is a concept that can be defined in many different ways. It can occur in a variety of settings and is dependent on many factors. In child welfare, practice might be defined as working with families and the court to protect children from abuse and neglect. In clinical social work, practice may mean advocacy, case management, community organization, counseling, social policy, psychotherapy, and research. In counseling and psychology, practice may be more narrowly defined as providing counseling or psychotherapy. Our working definition of practice includes all these views with an added emphasis on the clinical capacity of helping that we believe is instrumental for *advanced* practice.

The Helper's Role in a Clinical Capacity

The term *clinical* can be viewed on a continuum encompassing a broad range of services. From a more conservative view, clinical refers to providing psychotherapeutic interventions to clients in mental health, psychiatric, or counseling agencies. This might include using psychodynamic, cognitive, or behavioral approaches that are based on specific theoretical frameworks that guide treatment and must be employed by licensed clinicians (i.e., licensed clinical social workers, marriage and child counselors, psychologists, and psychiatrists).

From a more liberal standpoint, *clinical* may describe such services as professional case management in which clinical assessment, intervention, evaluation, and follow-up take place. In contrast to counseling or psychotherapy, case management may be provided to clients in a wide variety of settings including public child welfare agencies, medical and chemical dependency units, rehabilitation and mental health centers, senior centers, and so on. Also, unlike the first, more conservative definition, this definition allows both licensed and unlicensed human service professionals to provide these types of services—dependent, of course, on their education, experience, and specific agency roles. Certainly, the more clinical the role, the greater the need for licensed professionals who have more extensive training and experience.

The term *clinical* is well defined by the National Association of Social Workers (NASW). This group describes clinical social work as "the professional application of social work theory and methods to the treatment and prevention of psychosocial dysfunction, disability, or impairment, including emotional and mental disorders" with intervention directed toward "interpersonal interactions, intrapsychic dynamics, and [or] life-support and management issues" (NASW Provisional Council on Clinical Social Work, 1987, pp. 956–966). This definition can be broadened to include theoretical perspectives and methods additional to those specific only to social work. These might include developmental, behavioral, cognitive, humanistic, and family systems, and psychoanalytic theories and interventions that are also common to human services, counseling, and psychology.

Professional Diversity

The terms *clinician* and *therapist* are commonly used to describe professionals in the fields of counseling, psychology, and clinical social work who most typically engage in *clinical practice*. Others who also may use clinical skills in some aspects of their work include human service professionals such as case managers, child development specialists, community organizers, career and occupational counselors, human service administrators, managers, outreach workers, psychiatric nurses, psychiatrists, and residential treatment workers. Our view of clinical practice fosters the notion of professional diversity that reflects the real world; in this world there are many different kinds of human service professionals with varying levels of education, training, and experience. Despite our specific roles, we believe we are all part of the larger family of human service professionals. We all deserve respect and validation for our individual contributions to the helping profession.

Profiles of Human Service Professionals

Below are profiles of various human service professionals who are often engaged in some form of clinical practice as described above. We offer these brief sketches to broaden your own view of clinical practice and increase your awareness of the professional diversity in the human service field. Remember that many of these professionals will employ clinical and nonclinical skills, both of which are equally important in human service delivery and in the role of helping. We focus here and throughout the text primarily on the clinical aspects of these positions.

Case Manager   Susan works as a case manager at a large rehabilitation facility. Her job description is multidimensional given the many roles she often performs on a daily basis. Typically, when patients are first admitted to the facility, Susan conducts a psychosocial evaluation that will assist her in determining the most important needs of each patient and the patient's immediate family. Because of the critical situations of many of her patients, she often must engage in crisis intervention services with both patients and their families. Susan needs sound communication skills that enhance her ability to form strong relationships with her co-workers, patients, and the community. Frequently she is involved in information and referral services that require her to act as a broker, an advocate, and at times, a community organizer. She also provides education to patients and their families on various aspects of the rehabilitation process. Similarly, she offers inservice training to staff on the psychosocial considerations of patient care. Last, she plays a pivotal role in successful discharge planning for patients that requires sensitivity and skill.

Child Protective Service Worker   Rick, who specialized in child welfare studies in graduate school, is now employed by the county department of social services where his primary function is to investigate child abuse allegations and assess service needs. He must assist individuals and families by recognizing dysfunctional behaviors and help clients take corrective actions to ameliorate these. He must be skilled at conducting ongoing assessments of his clients' needs and be able to provide continuous case management. He is knowledgeable of community resources so that he may provide appropriate client and family referrals to other programs or agencies. He is often called on to monitor minors' in out-of-home placements as well as to prepare periodic reports for the courts. Although his role is not that of a clinician, he often utilizes clinical skills in child abuse risk assessment and family interventions. Because he is the primary professional managing the case, he may be the first one on the scene when problems erupt and immediate action must be taken. Skill in emergency response is critical as is the ability to work with families who have been in the system for an extended period of time and may be resistant to following the stipulated case plan. Overall, Rick has found that his prior skills in interviewing, referring, counseling, maintaining records, practicing crisis intervention, and providing other social services has served him well in his current position.

Clinic Administrator   Mark works as a clinic administrator for a small psychiatric facility that is part of a larger health maintenance organization. Although each day is an adventure in ensuring that the clinic runs smoothly, Mark's most consistent roles are supervisor, mediator, and manager. In each of these functions, Mark must use sound clinical skills to perform his job effectively. He is often called on to resolve conflicts between staff and clients or among co-workers, so his assessment skills are essential. His knowledge of human nature, problem solving, and conflict resolution are all crucial in his daily efforts to help the facility run smoothly. Additionally, Mark acts as a buffer between upper management and his staff, a role that requires keen insight, sensitivity, and effective communication skills. Other administrative functions he performs, such as budget preparation, personnel evaluation and hiring, and program design, are equally important in his position.

Clinical Social Worker   Jane is a clinical social worker whose primary role is to provide therapeutic services to clients seeking counseling for mental health concerns. Because of the broad range of clients she treats, Jane's job requires a great deal of flexibility and creativity; she may engage in traditional psychotherapy with some clients, brief counseling with others, or crisis intervention with those requiring immediate services. Because of the high demand for rapid access to services and the cost-effective nature of group therapy, her skills as a group therapist are also valuable in Jane's daily professional activities. Her knowledge of assessment and treatment planning is essential to her success in working with her clients. Jane must also be adept at working with staff in her agency as well as with other professionals and agencies in the community. In addition to counseling or therapy, case management services are also a part of her job description. Sound writing skills—needed to record a detailed psychosocial evaluation, document symptoms in a behavioral manner, or correspond with numerous community agents—are also important to her advocacy role in helping her clients. Other activities Jane engages in on a weekly basis include peer review, consultation, and inservice training for interns and newly hired clinicians.

Career Counselor   Joe is a career counselor at a local community college. His primary role is to provide students with information about educational programs and career opportunities, including available academic and financial resources. Joe is often involved in crisis intervention with students who are struggling with personal, academic, or social issues. He has found that his assessment skills and ability to relate to his students in a compassionate way are vital to helping them. He has also realized the importance of maintaining positive community contacts and knowledge of existing community referrals. Because he does not provide ongoing counseling, he often helps students find additional counseling at a local agency or in private practice. He also may refer students to other agencies that provide health services, chemical dependency treatment, legal aid, and so on.

Chemical Dependency Counselor   Sally is a counselor at a chemical dependency unit at a major community hospital. She provides support, education, and other services through individual, group, and family therapy to clients in her caseload and their families. Her knowledge about addiction and recovery is central to her ability to assess and treat her clients appropriately. She must also be aware of differential and dual diagnosis. She often leads psychoeducational groups with clients and their families. In her work with clients, she emphasizes relapse prevention by educating them about the various aspects of their addictions, including the triggers that commonly lead to relapse. In her work with families, she teaches about the family dynamics often in place and the nature of co-dependency. Sally has found that crisis intervention skills are crucial in managing volatile and suicidal clients.

Marriage, Child, and Family Counselor   Tom is a marriage, family and child counselor who works at a family service agency where he treats children, adolescents, and adults for a variety of behavioral and emotional problems. Depending on the particular case, Tom may use one of several modalities, including individual, marital, or family therapy. Regardless of the type of treatment, Tom's initial and ongoing assessment skills are an essential aspect of his role. Knowledge of child development, child abuse, systems theory, and family therapy techniques is important in his work with his clients. Additionally, Tom provides counseling and consultation in local schools and often works with students, teachers, and school administrators directly on the school site where he can observe problem behaviors firsthand.

Crisis Worker   Michelle is a crisis worker in an outpatient psychiatric facility. In addition to knowledge of crisis theory, Michelle needs excellent assessment skills, especially in regard to child abuse, suicide, or homicide risk. Given the nature of emergency services, Michelle has learned that she must not only be quick but also assertive with the many individuals she often comes into contact with, including family, police, child protective service workers, chemical dependency unit staff, and inpatient psychiatric facility personnel. Crisis intervention skills and the ability to defuse potentially dangerous situations are critical to her job performance. Equally important is her ability to practice triage with others both within her facility (i.e., with psychiatrists, mental health specialists, clinical social workers, and psychiatric nurses) and in the community (with police, inpatient facility staff, outpatient counseling or mental health center staff). Knowledge of community services and available resources is also necessary for her to ensure that clients follow through with appropriate referrals and treatment planning.

Psychiatric Nurse   Lisa is a psychiatric nurse who works closely with others on the psychiatric unit of a large county facility. She is responsible for monitoring clients and ensuring that they are following their medication management and ongoing clinical treatment protocols. She is often involved in crisis work, some ongoing counseling, and much patient education. She conducts weekly medication groups with the psychiatrist and is available to provide case management services to individual clients as well. Lisa is an integral part of the therapeutic treatment team whose members work together to treat clients in a holistic manner.

Psychologist   Brian is a clinical psychologist in a community mental health center. His work activities include testing, counseling, psychotherapy, research, supervising, teaching, and consulting. He is especially valuable to the treatment team as he administers and interprets the various psychological tests that are often key to better understanding client dynamics and personality traits. He collaborates with other treatment team members and, at times, with other professionals within the community.

Psychiatrist   Kathryn is a psychiatrist who heads up the clinical treatment team at a major behavioral health corporation. Of the many functions she performs, a most important one is investigating, diagnosing, and treating mental, emotional, and behavioral disorders. As a medical doctor, Kathryn is the only person on the treatment team who is legally authorized to prescribe psychotropic medication; consequently, she spends a major part of her day in medication evaluations and medication management. She also consults with other workers often on cases for which a psychiatric opinion is considered necessary.

Because of the variety of roles these professionals practice in the human services field, we refer to them in various ways; terms like *helper, human service worker/provider, practitioner, clinician,* and *therapist* are all used interchangeably and in an appropriate manner throughout the text.

Practice in a Variety of Agency Settings

General practice in human services occurs in a wide variety of settings and, as noted previously, can be performed by many different types of helpers who may differ in their role, degree status, and orientation. Clinical practice, although more narrow in focus, can also be provided by various human service professionals working in diverse practice settings. Human service agency settings in which both general and clinical practice occurs are numerous. Some of the better known types of agencies that often serve as internship and fieldwork sites for students include adoption agencies, chemical dependency facilities, child welfare agencies, career counseling agencies, community centers, foster home agencies, occupational and vocational centers, medical hospitals, human resource departments, mental health agencies, inpatient and outpatient psychiatric facilities, rehabilitation hospitals, residential treatment facilities, school programs, and senior service agencies. Services provided range from career and interpersonal growth to crisis counseling.

Types of Agency Settings

Agency practice will expose you to a variety of community agencies where you can acquire invaluable hands-on experience. Exposure to various types of agencies will broaden your perspectives about human services and allow you to see firsthand the type of work that occurs in agency settings and the many differences that exist. This exposure can also be advantageous in your future career choices and areas of specialization. We encourage you to explore some of these different types of agencies through internships, fieldwork, volunteer work, and of course, direct work experiences. The literature offers many views on how best to categorize human service agencies. Netting. Kettner, and McMurtry (1993) identify *nonprofit*, *public*, and *for-profit* as the main types of human service organizations that exist today.

Nonprofit (Voluntary)   The nonprofit agency is essentially bureaucratic in structure and is governed by an elected, volunteer board of directors. Professional and/or volunteer staff are employed to provide a continuing service to a clientele in the community (Kramer, 1981). Many local agencies would not exist were it not for this type of agency. Funding may be from private donations, fund-raisers, grants, and perhaps some governmental sources. Examples include family service agencies that can provide counseling, crisis intervention, parenting classes, respite care for the elderly, teen programs, anger management groups, and other services. In such agencies, professional, paraprofessional, nonprofessional, and volunteer staff all provide services to the community.

Working in such family service and mental health agencies that contract with state, county, and federal entities for additional funding will reinforce your need for negotiation, flexibility, and compromise. It may also teach you to be receptive and able to adapt to the ongoing changes inherent in the political and economically changing environment that directly and indirectly impacts the operation of human services.

Public Agencies (Governmental)

Government entities include federal, state, regional, county, and city agencies. Their main purpose is to provide services to designated target populations; their funding is primarily from government. An example would be a county mental health center that provides crisis and short-term counseling to clients in the community who are suffering from emotional disorders. Individual, conjoint, family, and group treatment can be offered as well as parenting and anger management classes. Psychotropic medication is sometimes provided along with day treatment services for those who are chronic mentally ill. The center, which is probably funded from a combination of federal, state, and county funds, is governed by government policies and procedures. Staff could be county employees who range from clerical, paraprofessional, to professional status. These settings will expose you to the challenges of working within the parameters of government policies and regulations, which can be quite frustrating. However, we have found that workers can develop successful coping skills and that public agencies can offer rewarding experiences. Personally, we have found that by maintaining a positive attitude and seeking out support from colleagues and friends, we have been able to survive the more trying aspects of working in bureaucratic settings. We believe that the positive elements of public agency work far outweigh the negatives; hence, our focus should be on these.

For-profit (Commercial/Private)

There has been a notable increase in for-profit agencies in recent years. Some of the reasons for their emergence are reduced funding of public agencies, limited resources for voluntary agencies, and changing political and economic times. Known as proprietary agencies, they have a dual function. The first is to provide a service; the second is to make a profit in doing so. Because they are in competition with public and nonprofit agencies, for-profit agencies are often the center of controversy (Netting et al., 1993). Critics often charge that these agencies can become so focused on profit motives that their concern for their clientele is compromised.

In the 1990s, the issues of privatization and managed care took center stage (Kettner & Martin, cited in Austin & Lowe, 1994; Goodman, Brown, & Dietz, 1992). The premise of privatization is that governmental responsibilities are shifted to nongovernment entities with the expectation that market forces will improve productivity, efficiency, and effectiveness in the provision of services (Kettner et al., 1994). Some services that are often privatized are needs assessments, funding, policy making, program development, service provision, monitoring, evaluation, or any combination of these. Examples include employee assistance programs (EAPs), managed care, health maintenance organizations, private health care organizations and businesses, and service agencies.

Our most recent experiences in working in large health maintenance organizations and with managed care systems in our own private practice have provided us with new insights and frustrations concerning the changing world of health care in this country. Although we aspired to work in the private sector because we thought this would let us be more autonomous and free from agency restrictions, we have discovered this to be an erroneous perception. Because of the economic state of flux of our health care system, the private, for-profit arena can be equally as demanding and restrictive as community agencies; therefore the private agencies require the human service provider to be just as flexible and open-minded as if he or she were working in public facilities.

Despite the differences in agency types, effective practice is possible within each. We believe that knowledge of your agency system can guide you in developing your skills as a helper. Furthermore, understanding your agency setting is crucial if you are to work cooperatively with others and fit in with the agency's mission and goals. Developing a good sense of your agency will help you develop adaptive coping strategies to deal with various facets of agency life, including rules and regulations, clientele, and personnel.

KEY INGREDIENTS FOR A SUCCESSFUL FIELDWORK EXPERIENCE

Values

As important as self-awareness is value awareness. Helpers cannot be effective if they are not aware of their own values and the impact these can have on the helping process. Many decisions we make as helpers are based on the values we hold. Your fieldwork placement may be an experience in which your values become apparent.

Understanding our own values and those of others is important for various reasons. Values held by legislators, board members, and directors of agencies and nonprofit organizations can influence the formulation of the social programs we work in. Values impact our work at the community and policy level. Decisions about the how, when, and why of service delivery are often colored by the values of administrators, funding sources, and practitioners. The way we as helpers feel about different issues, the beliefs we hold, and our overall attitudes make up our personal value system that consciously and unconsciously enters our work with clients. Recognizing that we have such values is a first step in learning how to keep from imposing these on our clients or allowing them to negatively color our work with certain individuals or specific population groups.

Value Awareness and Congruency

We cannot be aware of values if we don't understand them. Values are what ought to be, not what is; they are "standards or ethical guidelines that influence an individual or group's behavior, attitudes, and decisions" (Nugent, 1990, p. 262). They are fundamental to decision making.

Some of our values are obvious. We can see how they reflect a part of us. However, many of them are not so evident. These hidden values can be the most harmful—biases and prejudices that we manifest without intending to do so. Striving to evaluate our values continually is mandatory to remaining objective in our work.

Recognition of possible value conflicts often results from this awareness process. Because helpers and clients are not identical, there will be differences in the values each will hold. Some of these differences may be slight; others may be great. Common value conflicts that engender controversy in human services include issues related to family, gender roles, religion, abortion, sexuality, sexual orientation, AIDS, and cultural and racial identity. You may be able to identify other areas where value conflicts could occur.

Because conflicts often arise when divergent values exist among helpers and clients, it will become important for you to decide whether you are comfortable enough with these differences to remain objective and effective in your work. Many helpers believe that even though their values may differ greatly from those of their clients, co-workers, or agencies, they can still provide effective treatment. If value conflicts are too great, referrals or transfers are recommended.

If we have difficulty recognizing our own values, or if we try to impose some of our more strongly held values on our clients, we can be guilty of unethical behavior. This can harm clients, especially those who are already feeling vulnerable and receptive to the helper's influence. We can't keep our values separate from our work as they are part of who we are, but we can learn to expose them in ways that are not intrusive on clients and co-workers (Corey & Corey, 1993). We must be open, aware, and receptive to realizing when our own biases do not allow us to be objective with certain clients.

SCENARIOS

VALUE AWARENESS: HELPER IS AWARE OF OWN VALUES

Susan is a student placed in a Catholic charities agency that provides social work services to the surrounding community. She is expected to provide assessment, case management, and brief counseling services under the supervision of a staff social worker. Although Susan was raised as a Catholic, she is clear about her open beliefs and values relating to the issue of abortion. The agency informed her, as a part of her initial training, that no one on the staff is supposed to discuss the option of abortion as it directly conflicts with the teachings of the church. Susan is performing an intake one day with a young woman who reveals that she is pregnant and wishes to discuss the options available to her, including abortion. Susan informs the client about her own personal beliefs about the right of a woman to choose whether to have an abortion and about the agency's policy about not discussing this option with women who possibly want this option. She tells the client she will have to discuss the client's request further with her supervisor to decide on the best course of action for her.

VALUE CONGRUENCY: HELPER'S VALUES MATCH HIS OR HER BEHAVIOR

Peter was an intern placed in a school setting working with elementary age children. These children were referred to him by classroom teachers when the children were having problems that interfered with their learning in the classroom. Peter prided himself on his developing communication skills, and although he knew he was still learning, he realized that he could be direct and assertive if he needed to be. In the middle of his internship, the principal of his school complained to him that he was not working with the children enough and had failed to assist in the classroom of one of the teachers as he had promised. Peter was initially shocked by this information as he had been meticulous in keeping his appointments with the children and teachers alike. He informed the principal of his busy schedule along with his perception that he had not missed any appointments. He then asked the principal to identify the source of the complaints against him so he could discuss the problem directly with those individuals. At first the principal was reluctant to give that information, but finally she gave in and told him of one teacher, in particular, who had complained about him. After thinking through what he might want to say to this individual, Peter went to the teacher and stated in a firm but kind way that he was aware she had some complaints about him and that he would like to discuss them with her. The teacher reluctantly agreed and told him that he was not fulfilling her expectation that he would come into her classroom like the education intern before him had done. Peter explained that he was a counseling intern and that his role was to talk individually to the children who were referred to him. Although he could see that the teacher was quite uncomfortable with the direct discussion of this matter, he was also proud of himself for dealing with this matter directly and assertively. He was aware that his behavior matched what he so often attempted to teach others about the importance of direct, honest communication.

VALUE CONFLICTS: HELPER'S VALUES CONFLICT WITH CLIENT'S VALUES

Tracy is an intern working in a community mental health agency that provides an array of counseling services to the surrounding community. She is expected to do assessments as well as to provide brief counseling and case management services to the clients the agency serves. In one case, she is to work with the parents of a rebellious teenage girl who has been having severe problems in school. In discussing the parenting skills and disciplinary techniques of the parents, Tracy learns of their strong beliefs in the importance of respect for elders and the use of corporal punishment to reinforce this belief. Tracy realizes that she too believes in respect for others, but that she does not believe in using corporal punishment as a means of discipline. She also realizes that there are laws about the excessive use of corporal punishment by a parent. Despite her fears about revealing her beliefs and the laws pertaining to child abuse, she decides that doing so would be the wisest step to take with these parents from the beginning of their therapeutic relationship. She reveals her belief about the negative consequences of corporal punishment and the possible consequences of reporting if she believes that their use of discipline constitutes child abuse. The parents become quite angry about her telling them how to raise their daughter and ask to be transferred to another worker. In discussing the outcome with her supervisor, Tracy realizes that despite the clients' anger, she did what was most consistent with her own values and beliefs; her behavior was ethically sound according to the code of ethics of the profession, which spells out the need for clients' informed consent.

The Importance of Supervision

Supervision is a fundamental part of your educational experience. It enables you to begin reflecting critically on your performance within your fieldwork setting. For both beginning and advanced students, supervision is critical in their professional development. According to Thomlison, Rogers, Collins, and Grinnell (1996, p. 135), "Educational supervision in your practicum setting is essential to your learning: It is where the being, knowing, thinking, and doing parts of becoming a social worker can be explored, developed, and enhanced." Although its purpose is didactic, supervision involves dynamic and emotional factors. The supervisory relationship is therefore a complex one. A unique link between preparation and skilled practice, supervision is critical to your career development.

Chiaferi and Griffin (1997, p. 24) describe supervision as "a method of training and teaching in which experienced professionals interact with students and interns to provide guidance, on-site education, skill development, and general support." Becoming aware of the many aspects of a supervisory relationship is essential in learning how to use supervision effectively. However, before we concentrate on specific supervision issues, we should first explore general facets of learning that we need to understand and incorporate into the supervisory process.

Facets of Learning

Different individuals learn in different ways; some learn best from reading and research whereas others need practical application in real-life situations. Each of us is responsible for cultivating our own learning style and exploring innovative ways to learn. We encourage you to be creative and open to new pathways to learning. Thomlison et al. (1996, p. 40) cite eight myths about learning and explain why these are invalid. We believe that being familiar with these myths will help you understand the overall nature of learning. Each of them is briefly discussed below with illustrative scenarios that relate them to the supervisory relationship.

Myths About Learning

Myth 1: *Learning should be an enjoyable experience.* Of course, this would be the ideal. However, learning can also create anxiety, frustration, and self-doubt. Sometimes, struggling to learn can actually lead to more powerful and memorable learning.

SCENARIO

A student who was older and thought of herself as experienced encountered some difficulty in her first placement in a battered women's shelter. She believed the agency was not respecting her level of experience and had her performing tasks that were too easy for her. She had not considered that she might have to be assertive with her supervisor to help the supervisor recognize her level of experience and skills. The student decided to discuss her frustration in her practicum class and received feedback about the level of her expectations and how she might explain this to her supervisor. In the following week, she reported that she had discussed her concerns with her supervisor and was received well. The supervisor decided to have her co-lead a group instead of folding brochures and answering the phone as they had initially agreed. The student spoke of the importance of learning that being experienced does not mean that this experience will automatically be recognized and acknowledged. You may need to be assertive in letting others know what you want. Her difficulty had led to a significant learning experience for her.

Myth 2: *Adults are self-directed learners.* Some adults may be self-starters, but many are not. At times, even the most independent and self-directed person will need guidance and structure in learning new and perhaps difficult concepts. "The degree to which adults can or want to be self-directed depends on what is to be learned, how familiar they are with the content, the subject, their skills and the circumstances" (Thomlison et al., 1996, p. 41).

> ### SCENARIO
> An older returning student who had years of experience in the business world was placed in the fieldwork office of a major university's fieldwork program. Although he was a hard worker who generally needed little guidance and worked very independently, he experienced problems when he encountered an angry female student who had a history of dissatisfaction in her placements. When she called the office very upset about a recent difficulty and came across as very angry, the student intern himself became angry and defensive. He then was not able to hear her concerns and responded by cutting the conversation off, therefore creating more anger in the student. He later talked about this experience in group supervision and began to understand that his reaction had not been helpful. He was able to see that, although he had years of experience in dealing with people in the business world, he still needed some guidance on how to respond better to upset students in a fieldwork office setting.

Myth 3: *Adults bring a lifetime of prior experiences that can be drawn on and used productively in the present learning situation.* Past experience can be helpful, but not all of it will be useful in the new learning situation. Sometimes, prior ways of thinking and viewing the world must be challenged as they no longer apply to current trends, and unlearning old ways may be more difficult for some than starting out fresh.

> ### SCENARIO
> Julie's prior placement had been in a college counseling center. She had successfully co-led several growth groups and had learned a substantial amount during her placement there. One of the principles she had learned in group dynamics was to limit the questions she asked members, as this might discourage individuals' thoughts or feelings from surfacing. In her new placement at a child guidance center, she was asked to do intakes and assessments and to work with families and children. When first encountering the assessment forms and procedures, she initially was uncomfortable because of her prior training that asking too many questions was inappropriate. After a discussion with her supervisor and her practicum instructor about her concerns, she realized that the training she had received in her prior placement was not applicable in her new setting and that asking questions was an important skill to have in assessing a client's situation.

Myth 4: *Adults know what they need to learn.* Many adults are very sure of the direction they wish to pursue; others, however, may need guidance in finding the specific path they should take.

> ### SCENARIO
> Sally was certain before she began her placement in a group home setting that she wanted to work with children and adolescents. She believed that exposure to this population alone would help her learn what she needed to know. On one of her first outings to pick up some children from their school, a young boy who had been diagnosed with attention deficit disorder was extremely agitated and began to hit the other children and the house parent who was in the van. When the house parent decided to

restrain him, he opened the door and began to run down the street. Sally was quite confused and began to run after him, asking him to stay on the curb and imploring him to stop. The boy finally did stop and was disciplined for his behavior when he arrived back at the group home. Sally later stated in her practicum class that she had never thought something like that could happen and had never before realized that she might need specific training on managing the behavior of a child who was acting out.

Myth 5: *There is one adult learning style.* As mentioned earlier, there are many learning styles. Both the student and the supervisor need to explore learning styles to find the one that best suits their situation and their supervisory relationship.

SCENARIO

In becoming more aware of myself through the years of my internships and work in agency settings, I (Randy) began to notice that I enjoy learning by doing. If I can actually try out something new with a client or do a role play with my supervisor, I feel much more satisfied and assured that I have grasped the concept. After this realization, I began to capitalize on these kinds of experiences with my supervisors and in my work.

Myth 6: *Satisfaction is the measurement of successful learning.* Although satisfaction is pleasant, it alone is not a sufficient measure of success. Everyone has his or her own definitions of both satisfaction and success. Some may be satisfied with learning just enough to get by, whereas others may not feel satisfied or successful until they have exhausted all their capabilities or until they have completely mastered a specific skill.

SCENARIO

Alan was placed in a family service agency close to his home. There had been a large turnover in the staff during the summer and the former supervisor, who was familiar with the school's fieldwork policies, had left the agency. As Alan began to ask for time with the new supervisor, it became clear that this person was unwilling to commit to the time necessary to meet the needs of the student or the requirements of the school. Alan reported his disappointment to his practicum class and was ready to drop out of the class altogether. After receiving constructive feedback from his classmates, he decided to become assertive with the supervisor and strongly request the time he needed. If the agency was not willing to provide him the supervisory hours required, he had decided that he would pursue a more appropriate agency setting after clearing the change with his practicum instructor and his program's fieldwork coordinator. Appropriate supervisory hours eventually were provided. Alan was later able to say that his disappointment, when communicated effectively, led him into taking better care of himself and encouraged him to be more assertive in his own behalf.

Myth 7: *Learning is a sequential and progressive building process.* Learning can and does often occur sequentially, but not for everyone nor in all situations. Sometimes individuals need time to digest what they have learned before they can move on to the next step. Also, some of us may have to take two steps forward, and one step back to master a certain situation.

SCENARIO

Jessica was entering an internship in which she was expected to work with women who had been abused as children. Although she had a strong desire and eagerness to learn, she realized early in the internship that she had neither the personal experience nor the knowledge to understand completely the reactions and choices these women had made in their lives. She decided, in collaboration with her supervisor, to do some research and reading about the effects of abuse on adult survivors before attempting to co-lead a group for this population. The time, distance, and knowledge that she gained during this period allowed her to reenter the helping relationship with a renewed sense of her role and a better understanding of the clients she was expected to serve. Jessica thanked her supervisor for supporting her efforts to learn in a graduated and unforced manner. She was later able to see how this experience expanded her knowledge and added to her skills in her agency setting.

Myth 8: *Making mistakes means you cannot learn.* Making mistakes is a part of life and of being human. Certainly, we can all attest to this. Some of our most potent learning experiences have resulted from mistakes. The key is to acknowledge your error and to scrutinize the situation honestly in order to learn from it.

SCENARIO

Jim was an intern at a local mental health center. He began to be frustrated as the agency did not have enough office space for the interns and often required them to move from one space to another every time they would see a client. He believed that this was disruptive to himself and the clients he served. He did little in the beginning to voice his complaints because of his fear of being evaluated negatively by his supervisor. One day in a staff meeting, he became very angry when the topic of office space came up. He blurted out his frustrations in a hostile manner; the staff, surprised, reacted defensively and were uncooperative with his request. Discussing this incident later in his practicum class, Jim realized that he had made a mistake in the way he had dealt with his feelings about the lack of office space. He saw that he could have approached the issue by being assertive from the beginning and discussing his concerns with his supervisor and the agency staff in a more constructive manner. He was able to learn a valuable lesson from this incident: to speak up earlier and more clearly about his thoughts, feelings, and needs.

We hope these myths about learning have given you some insight into your own views and the possible misperceptions you may have been harboring. It is most important for your own learning that you become aware of your particular learning style and how you take in new information and experiences. You may need to review your past learning experiences to see whether you can find a theme or pattern to how you learn best. Learning can take place in many different ways: It can be intentional, incidental, serendipitous, by trial and error, spontaneous, accidental, or, as mentioned earlier, from mistakes (Thomlison et al., 1996, p. 42). Take time to discover which ways best fit your personality, your supervisory relationship, and your particular clients and agency setting. Each of these will help you determine the pathway that is best suited to your learning style.

Levels, Stages, and Styles of Learning

Nelson (1994, p. 26) has identified the following progression from awareness to skill mastery:

1. Awareness: Develops appreciation, sensitivity, the beginnings of understanding
2. Recognition: Identifies a person, fact, or concept as something seen before
3. Knowledge: Knows about the problem; can articulate its causes and possible solutions
4. Application: Uses the knowledge with varying skill and frequency
5. Mastery: Uses the knowledge in a skillful, disciplined manner most of the time

You will always be in the process of learning something new and therefore will not be at level 5 all the time. Realize that you will probably learn in stages. Reynolds (1965) has identified five stages of learning—from stage fright to teacher of others. Although you will probably not accomplish every stage during your internship, you will eventually experience them all. Here are the learning stages identified by Reynolds.

Stage 1 is acute consciousness of self, of which stage fright is typical. Each of you will probably react differently at this stage. Some withdraw, others may talk excessively, tell jokes, or become overly active or assertive. The fight-or-flight response, typical in crisis situations when anxiety is high, often occurs at this stage. It is important to seek support and security from your supervisor as well as from others. Focus on your areas of adequacy, strength, and successful past experiences as a foundation on which to build future successes.

Stage 2 is the sink-or-swim adaptation. You will have overcome the anxiety and fright experienced earlier, and are now beginning to work with clients and on projects. Although you are more relaxed and are beginning to adapt to your surroundings, you may still feel overwhelmed but less apt to acknowledge it. We encourage you to be honest with yourself about your anxieties and fears; to focus on your past successes and seek support and guidance.

At stage 3, you understand the situation without being able to control your own activity in it. At this point, you will find that you are more conscious and aware of yourself and your role, yet your practice skills are still developing. You, as well as your supervisor, may feel disappointed and frustrated that you are not yet able to master the skills you are working on. Instead of blaming yourself or your supervisor, use this stage to sharpen your newly gained knowledge and explore various ways of applying it in your daily practice.

Stage 4 is the stage of relative mastery in which you can both understand and control your activity. At this stage, you may feel comfortable with yourself and your growth and have faith in your ability to help others. You may also find that you are now able to engage in self-evaluation without always being prodded to it by your supervisor. You may find that you can consciously evaluate your interventions and alter your approach to new situations. You are now functioning as a competent professional, utilizing yourself as an instrument. When you have reached this point, you may feel the need to be further challenged by more advanced learning activities. Once again, it will be important for you to inform your supervisor of your needs and directly request additional challenges.

In stage 5, you learn to teach what you have mastered. This may not occur until years after you've completed your fieldwork and are an experienced and skilled helper. Along the way, however, you may find that you have mastered certain skills that you can now teach to someone else.

Remember that various levels and stages of learning will occur within your own preferred style of learning. Years ago, three distinct learning patterns among social work students were identified in the literature: the experiential-empathetic learner, the doer, and the intellectual-empathic learner. Today, these styles are still good guidelines for assisting student learning. It is important for you to learn which style best fits you so that your learning can be individualized to your specific needs. The approach your supervisor takes can be geared to the specific ways you learn best. It is primarily up to you, however, to inform your supervisor of your particular style and past experience with learning. The information in Table 1.1 should help you do this.

You may identify with one of the learning styles in Table 1.1. You also may be able to determine the particular level or stage of learning where you are at the present. If this information about learning styles and stages is useful, please incorporate it into your learning and discuss it with your supervisor. Allow yourself to learn at your own pace and to individualize how you evaluate your progress. Most important is for you to remain true to yourself and to be open to continued growth.

Influence of the Fieldwork Setting

Learning to work within the agency system can be challenging, yet doing so is important in your learning experience. The specific character of your particular fieldwork placement

TABLE 1.1   Learning Patterns

Learning Style	Learning Activities
The Experiential-Empathetic Learner: Characterized by dependence on feelings and intuition. Self-focused, intuitive, reflective. Sensitive to early feelings reactivated. Fearful of confrontation and authority issues; slow starter due to initial anxiety.	Learns from reflection on repetitive experiences over time. Encourage your supervisor to be supportive, to allow you to ventilate feelings and to explore your reactions.
The Doer: Action-oriented and eager to conform to agency ways of doing things. Initially dependent on supervisor's directives, seeking help with procedure. Steady progress.	Learns from well-planned opportunities, repetitive experiences, careful case selection; concrete services reinforce sense of adequacy. Encourage your supervisor to be supportive and directive and to use active teaching; to be positive and reinforce your feelings of adequacy through accomplishment; to encourage identification.
The Intellectual-Empathetic: Characterized by initiative, self-mobilization, and self-critique. Shows initiative, is self-critical and self-mobilizing; able to conceptualize, imaginative. Readily integrates theory and practice, anticipates.	Learns from a range of activities and levels of involvement, from exploring issues and theories with supervisor, and from testing and self-evaluation. Encourage your supervisor to provide you with the opportunity to test theory before accepting it. Ask for reading assignments that let you explore issues. Encourage supervisor to direct your motivation through client need.

can directly impact the learning pathways you choose. For instance, you may be the intentional type of learner, yet the agency you work in may be a crisis center where serendipitous and spontaneous learning prevails. Knowing how to blend your learning preference with the available situation is preferable either to insisting that your style be accommodated or succumbing totally to the placement situation.

Supervisory Roles, Functions, and Responsibilities

Typical supervisory roles identified in the literature include teacher, model, evaluator, mentor, counselor, and adviser (Alle-Corliss & Alle-Corliss, 1998; Corey et al., 1993; Faiver, Eisengart, & Colonna, 1995; Martin & Moore, 1995; Travers, 1993). The main functions of a supervisor can be narrowed down to three: administrative, supportive, and educational. The ideal supervisor would incorporate all three of these functions well. Real-life supervisors will, of course, be stronger in one of these areas than in the others.

Supervisor roles and functions are numerous and will vary. In general, however, all human service supervisors must maintain numerous areas of accountability. Specific supervisory responsibilities involve: ethical and legal issues, evaluative and educational skills, protection, and compliance. Consider which of these most closely fit the responsibilities of your current supervisor.

Qualities of an Effective Supervisor

To be able to successfully partake in a supervisory role, you must know the qualities necessary for effective supervision. The following are characteristics we consider essential to supervisory excellence.

The supervisor is supportive and understanding.   We believe the supportive component is the most valuable and important of all the qualities in effective clinical supervision. People learn best when they feel valued and respected. "Students and workers will persevere even under the most difficult circumstances, alter behavior and modify attitudes, when they perceive positive support from their supervisor" (Nelson, 1994, p. 7). Furthermore, supervisors who offer reassurance, encouragement, and positive reinforcement foster trust and rapport with their supervisees, just as a counselor does with a client. Such trust can be furthered by providing approval, constructive criticism, catharsis, and attentive listening.

Practical ways supervisors can show support include coaching, modeling skills, and offering suggestions about practice issues. They can offer emotional support through reassurance, encouragement, affirmation, and acknowledging the beginning helper's uncertainties. (Martin & Moore, 1995).

SCENARIO

Paul is a bit anxious going into his first internship at a child guidance agency. He has had little hands-on experience and believes that he should be more experienced and knowledgeable at this point in his development. When he is greeted by his supervisor, he notices the warmth and recognition that she gives him and his fellow interns. His supervisor has gone out of her way to introduce him to all the staff and has given him lots of material to review about the policies and procedures of the agency. She takes the interns out to lunch with several other staff members and seems genuinely interested in getting to know each one. Paul is still somewhat anxious about his level of development, but is much more relaxed than before he met his supervisor. Her greeting has helped him feel supported and open to the idea of revealing himself and his fears in supervision.

The supervisor is knowledgeable and experienced. A primary supervisory role is to provide education. Therefore, the supervisor must be knowledgeable as well as experienced in the areas in which he or she will be supervising. Such knowledge and experience will certainly enhance the trust the supervisees will have in the supervisory process. Of course, when a supervisor is knowledgeable and experienced yet is also rigid and expects others to be so without exception, there can be problems. This will be especially true of students who need permission to make mistakes and already feel quite anxious about their competency level.

SCENARIO

As an intern working in a child guidance setting, you receive supervision from two different social workers. One day after you have dealt with a particularly difficult child who is quite angry and cursed at you in the session, you feel overwhelmed and unsure about what to do if this happens again. One of your supervisors takes the time to discuss her knowledge and experience in working with anger in sessions with these children. She shares her own struggles and insecurities as well as some of the strategies that she has developed through the years to deal with excessively angry children who act out inappropriately in a session. She tells you about some of the limit-setting strategies she has used and how she typically tells the client about them. You feel relieved and grateful that your supervisor has shared her knowledge and experience with you just when you needed it.

The supervisor provides structure and organization. Beginning your fieldwork experience in an environment that is structured and well organized can enhance your sense of security, especially if you are just beginning in the field. It would be ideal if your supervisor were able to direct your learning in ways that compliment your style and are relevant to your learning goals and objectives. However, either too much or not enough direction can create obstacles that you may need to confront. It is good to be challenged, but not to the point that you become overwhelmed. Guidance from your supervisor can make a tremendous difference in your learning experience. Ideally, a good supervisor is one who will select cases, tasks, and projects that are appropriate in level of difficulty to your current skill level. He or she will also recognize and acknowledge when these learning activities are either too narrow and limiting or too difficult and overwhelming. Because of the integral role your supervisor will play in your learning, we suggest that you maintain open, direct, and assertive lines of communication with him or her. It will be up to you to let your supervisor know when you feel stifled or overburdened as well as when things are going well. To ensure a more successful fieldwork experience, it is important to ask certain questions early in the development of your supervisory relationship:

1. Will my learning activities be designed for direct learning? Indirect learning? Both?
2. Will theory and practice be linked to my fieldwork assignments?
3. Will assignments be adapted to my particular needs?
4. Will you help me understand myself better by identifying patterns and themes in my work with clients?
5. Will you help me mold my natural abilities into conscious techniques and assist me in developing the skills I need to work effectively with my clients?
6. How much do you expect or want me to talk about myself in my work?
7. What do you want the focus of our supervisory time together to be?

SCENARIO

As an intern at a college counseling center, you are assigned a caseload of over 25 students by the executive director when you first enter the agency for your internship. You are surprised and a bit overwhelmed by the high number of cases you have been given. After you discuss this with your supervisor, he realizes that giving you that many cases right at the start is unrealistic and potentially frustrating to your learning. The supervisor tells you that he will go through the cases and give you a small percentage that are appropriate to your particular level of experience and learning. He also says he will discuss the intern workload issue with the executive director so this type of thing does not happen to future interns. You are thankful for your supervisor's willingness to work within the system and for his ability to set appropriate boundaries to protect you and aid your learning process.

The supervisor is realistic in expectations of students.   Thomlison et al. (1996, p. 138) believe that your supervisor or "practicum instructor's ability to help you integrate 'classroom theoretical material' and 'practicum practice' is a very important aspect of your total learning experience." We agree, although in real life this may not always be the case. It is highly unlikely that your supervisor will know exactly what your strengths and weaknesses are. Sometimes, a supervisor will expect too much and therefore assign you cases or projects that are beyond your capacity. Conversely, he or she may have lower expectations about your abilities than is realistic for you. In either case, it will be up to you to discuss these issues early in your relationship. We recommend that you continually clarify your needs and specify your goals, as they will surely change as you grow. Striving to develop a strong, positive working relationship with your supervisor will be key in helping you deal with the various concerns and conflicts you are apt to experience.

SCENARIO

You are a new intern working in an administrative position in a large human resource office within a larger organization. Initially you are told that the last intern who worked in your position had developed and taken full responsibility for a company newsletter that was published out of the office. Having little to no publishing experience, you are glad that your supervisor tells you she does not expect you to carry out the same assignments as your predecessor. In fact, the supervisor engages you in a discussion about your level of experience and asks you what your expectations are for working in this type of setting. This discussion lets you tell her what you would most like to learn while you are in your internship.

The supervisor is flexible.   An important quality for any human service professional is flexibility. Change is a constant when you work with people and in agencies, and a good supervisor will work with change, not against it. He or she will be able to accept unforeseen shifts in your needs and circumstances and adapt accordingly. If your supervisor is overly rigid or inflexible, it would be wise for you to consider this and decide early how you will handle this relationship.

SCENARIO
Your internship consists of assessing families who apply for financial assistance for food, shelter, and/or emergency loans. You are a few weeks away from finishing your hours when your grandfather dies unexpectedly of a heart attack. You know your taking time off would be inconvenient for your agency because you are the only intern working there. Although you hesitate to discuss the importance of taking time off to go to the funeral, your relationship with your Grandfather was important to you and you feel strongly about attending his funeral. You hesitantly ask your supervisor about having a few days off to attend the ceremony and, to your surprise, he is very understanding and flexible. He inquires about how you are doing and tells you that having a few days off is no problem. You are grateful for the flexibility that both your supervisor and your agency were able to show to you in this matter.

The supervisor provides boundaries.   An ideal supervisor will have a sound knowledge of appropriate boundaries in all relationships within the agency, including the supervisory one. He or she will see when boundaries are being crossed and take the necessary precautions to ensure that limits are maintained. Chiaferi and Griffin (1997, p. 31) speak directly to this issue: "Training in the field of human services encourages self-awareness in the areas of personal and professional development. One result of this dual emphasis is the confusion students experience as they attempt to make an appropriate selection of topics to bring to the supervision meetings." When boundaries are crossed, dual relationships can develop between the helper and the client or the student and the supervisor. Also, a boundary breakdown results when a supervisor becomes a therapist to the supervisee. Supervision may have therapy-like qualities, but it is *not* therapy. For instance, a supervisor may need to help a student deal with transference or countertransference issues that are triggering painful emotions. The degree to which the supervisor delves into these issues will determine their therapeutic nature. Usually, when a supervisor realizes that he or she is beginning to assume a therapeutic role with the student or when the student's issues begin to impact the working relationship with clients, the supervisor should refer the student for ongoing counseling. "An experienced supervisor will help you discover how a specific issue is becoming an interference, and will make a recommendation for personal therapy when it might be necessary or helpful" (Chiaferi & Griffin, 1997, p. 31). Sometimes, students become defensive when they are referred. If you are ever in this situation, we encourage you to consider the underlying issues and view a referral as another way you can learn more about yourself and ultimately become a more effective helper.

SCENARIO
As an intern in a runaway shelter for teens, you are expected to co-lead various groups and sit in on family sessions when appropriate. One day, a female resident who has a

good relationship with you tells you that she was sexually abused by a relative when she was younger. She further states that she trusts you with this information and does not want you to tell anyone else in the shelter about it. The client says that she wants to work with you only because you are so nice and understanding. You are ambivalent about keeping this information to yourself and realize that there are reporting laws that may affect the situation. Therefore, you tell the client that you will have to discuss her revelation and request with your supervisor. The supervisor listens attentively and compliments your strengths as a worker in engendering such trust in a client. But he also expresses concern about you as a learner being asked to treat a client who has possibly been sexually abused, and he has doubts about being able to keep this information confidential as it may meet the criteria for the exception to confidentiality due to possible child abuse. You and the supervisor decide to deal with the client together and then have the supervisor take over the case because of the sexual abuse that may be involved.

The supervisor is able to provide constructive feedback.   A supervisor who is skillful at providing students with ongoing feedback about their performance is much more desirable than one who offers little input. Students and beginning helpers need their supervisor's frequent assessments to learn and progress. To be effective, feedback must be several things, as listed here:

- Timely: It should be provided shortly after the event.
- Specific and clear: It should be stated directly in ways that make sense.
- Balanced: Emphasis should be on a students' strengths as well as limitations.
- Useful: It should help you examine alternative approaches and options, or help increase your self-awareness.
- Relevant: It should be related to a specific event or action and be applicable to your learning objectives and roles.
- Reciprocal: Your comments and individual perspectives on the feedback should be both elicited and accepted by the supervisor providing the feedback. (Thomlison et al., 1996)

SCENARIO
You are ending an internship at a family service agency where you had a caseload of adult clients. You are thankful for your internship at this agency and for the ongoing feedback that you received while you were here. Your supervisor is now going over your evaluation, and as she had done before, she is able to point out both your strengths and weaknesses as they relate to the work you have performed with the clients. In this evaluation, she tells you about your great depth of caring as well as your difficulty confronting clients when they might need it. Although you are mildly annoyed with your supervisor for writing this in your evaluation, you know that she is being accurate and fair in her appraisal of your performance. You discuss her perception of your strengths and limitations and she gives you ideas for how to work on those skills in your next placement.

The supervisor is accessible.   For supervisors to perform all the functions and roles mentioned earlier, they need to be available and accessible to students on an ongoing basis. They need to set a regular supervisory time that is diligently maintained. In addition, good supervisors will encourage students to seek them out whenever necessary. When supervisors limit their interactions with students or discourage involvement, students often feel uncomfortable approaching them and in more extreme cases may feel abandoned or rejected. The element of approachability is just as important as availability and can greatly enhance the supervisory relationship.

> **SCENARIO**
> As a new intern at a children's protective services agency, you feel extremely over-whelmed by the many responsibilities and demands of your role. Besides learning to assess child abuse risk factors and to intervene appropriately in emergency response situations, you are also confused by the court reports and legal ramifications. You find that with each new task you have many questions. Luckily, your supervisor is readily available and willing to answer your questions, or has left other seasoned workers in charge of providing you with the structure you need. You feel especially lucky when you hear how other interns are struggling and seem to have little or no guidance, resulting in their frequent feelings of abandonment by inaccessible supervisors.

The supervisor is self-aware and open to learning.   One of the main functions of the supervisor is to help you learn more about yourself so you can effectively help others face their own personal crises. Good supervisors encourage introspection and self-reflection in their students. They are also open to self-awareness and ongoing learning. These supervisors are receptive to constructive feedback, cognizant that no one is immune to making mistakes, and open to learning from their students.

> **SCENARIO**
> John is a human services student who is placed in an AIDS service foundation agency. His supervisor has had many years of experience working with AIDS patients and is very knowledgeable about them. On this particular day, John is observing his supervisor work with a client when the client suddenly becomes quite angry. The client states that he is upset because the helper did not call him back when he left two messages recently. His attack is met with defensiveness by the supervisor. After this session is over, the supervisor encourages the intern to discuss his behavior with the client in that situation. He can look at his mistakes without being overly self-critical and encourages the student to report his own feelings and reactions.

The supervisor is a role model and mentor.   During the course of your fieldwork experiences you will encounter many styles of supervision. As you reflect on your learning, you may find that some of your supervisors taught you more than you could have imagined. You may have even attempted to emulate them in some way. These particular supervisors may have become role models and perhaps mentors, further enhancing the supervisory relationship. This type of relationship between a supervisor and student is indeed special, although not without its problems. Sometimes when students admire their supervisors without exception, they are disappointed when the supervisors' human flaws appear. Because of this possibility, we encourage you to be realistic in your expectations and accept your supervisors with both their strengths and limitations.

An important concept known as "parallel process" illustrates how the supervisor becomes a direct role model for students and an indirect one for clients. In the supervisory process, the supervisor becomes the model for behavior, attitudes, and expectations that the student transmits to the client. If the supervisor tends to be inflexible with agency regulations with the student, the student may tend to be rigid in interpreting regulations to the client. Similarly, if the supervisor gives ready answers and advice rather than encouraging the student to think through and select alternatives, the student may do the same with his or her clients. However, when the supervisor helps the student think through case problems and supports independent decision making, the student is more likely to integrate this process and utilize it in client interactions.

SCENARIO

Jerry had worked with his supervisor, Randy, for two semesters and realized that he really looked up to Randy and appreciated the way he worked with people. Jerry became aware over time that he was trying to emulate Randy and his particular skills and approach. He was further aware that he admired Randy's lifestyle and relationships. Feeling so close was good for Jerry as he believed that he had opened himself up in the relationship with Randy, which had allowed him to learn a tremendous amount. He felt respected and supported by his supervisor, and this encouragement made him willing to try new things. He was aware, as well, that there was a downside to this relationship as he had real difficulty saying no to his supervisor when he was to take on more tasks at the agency. At one point, Jerry became very overwhelmed with his work load and decided he had to be assertive with his supervisor. This encounter went very well and helped Jerry appreciate his relationship with his supervisor even more.

The Supervisory Relationship

The supervisory relationship is an evolving one that is dependent on many factors. Both the student's and supervisor's professional and life experiences as well as their personal qualities are important in determining the nature of their relationship. When both make efforts to blend their teaching and learning styles and to remain active in their communication, a strong working relationship can develop.

Developmental Nature of the Relationship

The progressive nature of learning has been alluded to earlier, and although learning is rarely a linear process, it does generally seem to occur progressively in steps or levels. Likewise, the relationship between student and supervisor develops over time and is a process. It is commonplace for students to move from high levels of dependence on their supervisor toward greater autonomy in reaching conclusions and taking action. For example, in a typical academic year beginning in the fall, students look for specific instruction and are more passive in their learning. By the spring, however, the students have usually become more active in their learning. Although supervision continues to be important, it is used more for support, action, approval, or consultation than for instruction and direction (Nelson, 1994, p. 26).

Much like the working relationship with a client, the supervisory relationship can be divided into a beginning, a middle, and an ending. Initially, students may need more direction and guidance to develop learning goals and objectives and specific helping skills. They will need information on policies and procedures of the agency and may require help in navigating their way their first few weeks at their placement. Also, students may need help in beginning to understand themselves and the introspective nature of becoming a helper.

As in most relationships, first impressions and early interactions often set a tone that may be long-lasting. When student and supervisor are willing to develop clear channels of communication from the start, they can create the type of cooperative environment that leads to a positive working relationship. For a student or beginning practitioner, being supervised will raise questions about competence, authority, and specific concerns related to seeking help. Having an open and positive attitude will help you wrestle with these issues constructively.

The middle stage of the supervisory relationship is typically marked by increased trust and rapport. The student is less overwhelmed than in the first few weeks, and he or she can begin actually working with the supervisor. Agency ground rules have been set and a routine is most likely in place. "Each individual is becoming acquainted with the other's personality, work habits, communication style, and areas of interest" (Chiaferi & Griffin, 1997, p. 26). Your supervisor will continue to evaluate your progress and perhaps begin to challenge you with more complex tasks. Your responsibility is to seek a balance: to encourage additional work when warranted as well as to assert yourself when you believe the load is too heavy. Compromise and negotiation are crucial; both you and your supervisor may find that earlier expectations and circumstances may have changed, requiring reevaluation and adjustment.

During this middle stage, you may have developed relationships with others including interns, co-workers, and preceptors with whom you have begun to share your highs and lows. You may seek support from or consult with them. This type of involvement may lead you to compare yourself with them and perhaps to be overly self-critical or negative about your supervision. Don't succumb to this temptation. We encourage you to deal directly with your competency issues and/or the supervisory relationship with your supervisor. Certainly, if your supervisor is not responsive and problems evolve to a serious level, you should ask for guidance from the university-based field instructor who oversees your fieldwork experience.

The ending stages of the supervisory relationship can be both exciting and melancholy. Both you and your supervisor will be excited about your success as you approach the final days of your field placement experience. You may also feel saddened by the pending closure of your relationship and the termination process that is about to take place. Because of these feelings, some students and supervisors may want to rush through this period. They may want to avoid dealing with the painful emotions or difficult issues that are inherent during this phase. Some may experience separation issues and anxiety. Also, because the student role is comfortable, you may find moving on especially difficult. The transition from student to professional is a time of anxiety for most. If you experience any of these feelings, we caution you not to surrender to your fears. In actuality, termination can be a very important time when the supervisory process is reviewed, the student's progress is evaluated, and anticipatory planning takes place. It can be an opportunity for honest review of your strengths and limitations, allowing you to explore areas where you need further growth.

The Learning Agreement or Contract

The elements of learning discussed thus far can be formally incorporated into a learning contract or learning agreement. The purpose of the learning plan is to establish the learning experiences the student is expected to have and to plot the objectives for the duration of the fieldwork experience. The tasks involved in developing this learning contract include defining learning objectives, identifying activities for achieving these objectives, and delineating the procedure for evaluating the student's progress. Specific learning objectives may cover the basic areas of practice: individual, family, group, administrative, and community. The learning plan may also include development of certain competencies:

- Awareness of and ability to use individual learning patterns
- Effective interaction with the agency
- Demonstration of the beliefs and ethics of the profession
- Skill as a practitioner
- Sensitivity to cultural diversity

Having an agreement in writing about your learning plan from the outset of your fieldwork will give you structure and direction. The written plan can help prevent misunderstandings or confusion about your role and expectations. It can also help you begin a solid working relationship with your supervisor as you assess your areas of strength and limitations. Often, the very act of meeting with your supervisor to explore your learning goals and objectives can alert you to such issues as what is realistic within the agency and what its expectations are for student interns. Early interaction with supervisors gives students insight about their styles and approaches that will be useful as they work together.

Be serious about your role in developing this learning plan. Remember that it is *your* learning that is at stake and that you need to be integrally involved. When students do not take this first task seriously, they often feel frustrated later when their needs have not been met or the learning goals and objectives delineated are not truly theirs. Be open about asking your supervisor to renegotiate the plan as needed. According to Chiaferi and Griffin (1997), "A working contract defines the expectations of both parties. Objectives are specific and arranged in a progressive manner, reflecting activities that are geared to an appropriate level of skill development and are designed to promote learning. The contract describes the frequency, length, and time of scheduled supervisory meetings" (p. 29). This is typical of most learning contracts, although every agency will be slightly different from the others in how it completes the learning plan. In most cases, the university will provide a standardized form that each agency must follow.

The challenge of any learning agreement is to assign to students the tasks that let them accomplish the specific objectives in the plan. Problems occur most often when the learning assignments do not match the specific goals identified. Sometimes the activities are too narrow or too broad; at other times the goals and objectives themselves are too specific or too vague. These types of problems may not surface until the activities or objectives are put into practice. The difficulties that can arise again reinforce the need for open communication with your supervisor and for renegotiation when necessary.

Another barrier may appear when you and your supervisor meet to complete your learning agreement if either of you is resistant. Authority issues may be challenged and assertive communication put to the test. The outcome of this first interaction with your supervisor may set the tone for future supervisory conferences. Your professionalism will show if you are willing to take this aspect of your learning seriously. However, your relationship will be complicated if your supervisor does not value the learning agreement and minimizes its importance. This attitude may require early interventions from you in using assertive techniques and perhaps in consulting with your university-based instructor or liaison.

A last benefit of the learning agreement is its usefulness during the evaluation and termination process. Reviewing the goals that were initially set and evaluating whether they were or were not met can promote important dialogue about your performance and growth. In essence, the learning plan can be helpful throughout your entire fieldwork experience. At the beginning, it sets the tone for the work that is to be accomplished; in the middle, it is a guide that keeps learning focused and continual; at the end, it is an essential evaluation tool.

Understanding and Managing Differences

The ideal supervisory relationship fosters trust and an open atmosphere. The relationship is easier when the supervisor and student share common values, views, and goals and have similar backgrounds and styles. This is not always the case. However, if the supervisory relationship is to be open and productive, both the supervisor and supervisee must be able to acknowledge their differences and similarities. Making an effort to understand one another fosters the development of a plan for working together in a cooperative fashion. Some of the areas where differences may exist are discussed next.

Differences in Power   By virtue of their status and role as evaluators, supervisors possess more power than students. If either of you has strong authority issues, these could get in the way. However, if you accept the difference in power and consider your supervisor an ally in your learning process, you are less likely to have problems. If the supervisor misuses power, then you must point this out in supervisory meetings; if you can't resolve the issue, you may need to involve your university field instructor. Power issues can surface with a line supervisor, administrator, or even a student that you may someday be supervising. Learning to manage conflicts related to power will not only be helpful during your fieldwork but can also be useful later in your career.

Differences in Gender   The impact of differences in gender may be minimal or they can be great. Some believe that gender differences may lead to reduced or heightened expectations from supervisors about student performance, given the transitional state of male and female roles (Thomlison et al., 1996, p. 149). For instance, a female might expect the male to be courteous and gentlemanly whereas a male might expect a female student to be submissive or caretaking. In reality, a female may be very independent and the male might be very sensitive and understanding. Certainly, when there is no consensus on expectations based on gender, conflicts are more likely. You must be aware of your feelings on gender differences and how your experience of gender issues may affect your practice. If you feel some discomfort or are aware of conflicts, bring this up early on with your supervisor.

Differences in Culture   There are many types of cultures—some related to ethnicity and race, others to geographical location, and still others related to religious and spiritual practices. These differences can enhance the supervisory relationship when both individuals are diversity sensitive and open to learning from one another. Difficulties arise when differences are ignored or there is an underlying message that one person is better than the other. Perhaps the most commonsense approach is the best: Be aware of and discuss these differences from the start. A direct dialogue will open an avenue for discussing differences in culture, and this may be sufficient in dealing with them. Nonetheless, there will be times when racist attitudes and discriminatory beliefs exist in you, your supervisor, or both of you. This can become a delicate matter requiring careful intervention. We recommend that you consult with your field instructor or faculty liaison if you believe there is a serious problem between your supervisor and you in this regard.

Differences in Age   Age differences are common and may not present any conflict for you. Today it is not unusual to find a wide age range in college and graduate schools as more and more individuals are returning to school in pursuit of a second career. The supervisory relationship can be affected in various ways by a difference in age. When students are very young and inexperienced, the supervisor will need to take this into consideration in making assignments and developing an appropriate learning plan. Also, the supervisor may need to be more actively involved and structured from the start with this population. With more mature students, the supervisor may be comfortable in allowing greater independence. Problems often arise when mature students who believe they are very knowledgeable and experienced meet supervisors who think otherwise. Also, when supervisors are much younger than the students they supervise, the age discrepancy can pose its own set of conflicts. In any of these cases, it is important to have open communication and an effort from both students and supervisors to be direct about their own troubling issues. Respecting one another and recognizing that everyone has something unique to offer, independent of age, is an essential first step.

Differences in Physical Abilities   Differences in physical abilities may not pose any problems at all. However, problems sometimes arise if there is a physical disability in either the student or the supervisor. Perhaps the nondisabled individual feels uncomfortable or thinks the disability hinders the relationship in some way. In certain situations, some may even believe that the disability is accompanied by intellectual deficits that can seriously affect supervision. This is usually not so, but this belief can remain strong unless it is challenged. Again, we encourage both students and supervisors to be open and aware of their own concerns as these relate to disabilities.

Learning to manage differences requires open minds, attitudes, and communication styles. Sometimes, similarities have the same requirements. Perhaps because two people are much alike, one person will expect a specific response from the other; when this response does not occur, the first individual may feel frustrated or upset. Transference and countertransference issues may develop, and unless the situation is discussed there may be higher and more unrealistic expectations because of the similarities.

The Challenges of Supervision within Agency Settings

Even though you have identified your learning style, are working on becoming more self-aware, and are prepared to have an open, productive relationship with your supervisor, things can still go awry. When this happens, you can handle the situation more easily if you know in advance some of the difficulties that might arise. We have identified here several of the possible challenges you may encounter in a supervisory relationship. Accompanying each one is action you can take to lessen its impact.

• Poor match between supervisor and supervisee: Accept the fact that you will have to learn to work with different personalities in a cooperative fashion.

• Frequent change in supervisors: Realize that it is normal to be disappointed and upset at a change in supervisors. Remain open and assertive about having your needs met. Use your interpersonal skills to adapt as well as you can to the changes.

• The absent supervisor: Try to deal with your concern as soon as you notice a potential problem. Be clear about your need for direct and regular supervision. Use assertiveness skills. If appropriate, use your academic requirements to substantiate your requests and needs.

• The inadequate supervisor: It is most important to share your concerns about the supervisor's inadequacy as soon as they arise and to be assertive if your needs are not being met. If this does not work, you may need to seek out others whom you can learn from as mentors and role models.

• The critical or difficult supervisor: Having a critical supervisor can be a trying experience. Make an effort to understand the person's perspective. Remaining assertive is crucial, but try not to personalize the situation. Develop a support system where you can vent your frustration safely and where you can receive needed validation and support.

• The unethical supervisor: Depending on the nature of the supervisor's unethical behavior, you may have to report it, or at least discuss it with others—fieldwork instructors, faculty members, or advisers. This is a difficult situation, especially for beginning human service workers, who may be surprised that unethical behavior exists at this level.

• The overloaded, overworked, or overwhelmed supervisor: Make an effort to be sensitive and supportive, yet remain assertive about having your needs met. You may ask if there are others who could help with your supervision. Informally, you can seek out others who can provide you with appropriate feedback.

• Dual relationships: Attempt to keep the line between supervisor and supervisee clear. Be aware of the complications and potential negative consequences of dual relationships and avoid them.

• The involuntary supervisor: At times, someone is made a supervisor without wanting the role. This may be due to a shortage in staff or budget constraints, but it can create animosity in the supervisor who may already feel overwhelmed and exploited. Also, there may be times when a supervisor has had a negative experience with a student and may not want to supervise again, yet is told he or she must do so. Last, there are situations when a person is hired for a specific position and due to unforeseen circumstances, their role changes to include supervision, an activity they did not agree to when hired. In any of these instances, there will likely be some reluctance on the part of the supervisor. If students in this situation are not careful, they can see the supervisor's attitude as a personal affront and allow it to affect their entire fieldwork experience. We encourage you to ask your supervisor directly if he or she is comfortable with the supervisory role if you note a problem in the person's motivation.

• The unqualified supervisor: Sometimes a person is not qualified to supervise but for various reasons may be asked to do so. This can create problems for the student who may believe that he or she is not receiving adequate supervision as a result. If you find yourself in this situation, seek appropriate guidance and consult a preceptor, university field instructor, or liaison with whom to discuss your concerns. The school may need to address this issue on a policy level.

- Time limitations: Some supervisors are often so overwhelmed with the many responsibilities they have that they cannot do an adequate job in all areas. The student is then given very little supervisory attention or is left alone most of the time. Attempting to understand the reasons for these time limitations is important in clarifying matters. Also, it will be essential that you discuss your concerns directly with your supervisor. You may be surprised to find that he or she is not even aware of a problem or did not realize the impact it was having on you.

- Value differences: Differences in values between supervisee and supervisor are to be expected. Each should be able to accept and respect the other. However, when values are too different or when a conflict arises, the supervisory relationship can be affected. Be sensitive to these differences and explore ways to deal with them appropriately.

- Fiscal restraints: Value conflicts can occur over a supervisor's focus on productivity. Beginning students who want to ease into working with clients may be overwhelmed if they are expected to carry a large caseload. When the focus is primarily on numbers and not on clients or the learning process of the student, ethical practice may be at question.

- Environmental issues: There could also be times when such environmental issues as limited office space and/or unsafe surroundings may negatively impact the learning process. This situation may also reflect a difference in values. The supervisor may not see a problem whereas the student may find it difficult to feel comfortable and safe. If these issues are to be resolved, they need to be discussed as they arise. You must let your supervisor know of any problems you have in this area.

Group Supervision

In addition to individual supervision, advanced practice often involves group supervision. University programs, particularly graduate ones, will frequently require both individual and group supervision on a weekly basis. Although group and individual supervision are different, group work can be just as important as the other in your overall learning. Group supervision allows you to meet with your peers and share common issues and concerns as well as learn from one another. You will be able to see firsthand how other students solve problems as you compare your style with theirs. Learning can be fun when there is open sharing and constructive feedback. However, fierce competition can thwart the learning process.

Group supervision can give you a wider selection of learning opportunities, including films, videos, lectures, presentations, and role playing. We especially favor the use of experiential exercises in small groups. Group supervision can also help you develop a professional identity because it involves you with other professionals. When students can experience firsthand the way group process actually works and learn the dynamics of group sessions through direct participation, many will become more comfortable in their own group leading.

The Role of the Preceptor, Faculty Liaison, and Fieldwork Instructor

In addition to your supervisor or practicum instructor, several other individuals will also play key roles in your learning. Understanding their roles and functions can help you make the most of your relationship with them.

The Role of Preceptor

The preceptor role may be formally designated or informally developed. Either way, the preceptor is typically an assistant to the supervisor and may be a secondary source of information and an auxiliary resource for the student. Responsibilities of a preceptor may include the following:

- Providing on-site operational supervision
- Acting as a resource to the student in absence of the supervisor or for special assignments
- Being a role model to the student, observing and providing feedback
- Assisting the student in learning paperwork requirements and agency procedures and policies
- Assisting in orienting the student to the agency and its clients
- Sharing experiences from his or her own caseload
- Offering the student an opportunity to work with him or her on a particular case, in co-leading a group, or in treating family members or offering case management services to certain clients in the preceptor's caseload.
- Providing process-oriented, clinical supervision to augment the supervisor's expertise (Nelson, 1994, p. 4)

If you find yourself working directly with a preceptor, you may find that you learn just as much from this person as from your supervisor. You may also find that the relationship with a preceptor is less anxiety producing as he or she is not usually in a position to directly evaluate your performance. Because the pressure to perform is lessened for both you and the preceptor, you may feel more comfortable and at ease in this relationship. As in the supervisory relationship, we encourage open, direct, and assertive communication that is accompanied with respect and appreciation.

The Role of Faculty Liaison

In many instances, you will interact with a faculty liaison regularly. His or her role is to represent the university program and meet with you and your supervisor frequently to monitor your progress and identify potential problems. In general, the liaison's responsibility is to coordinate the individual student's educational program in the field. Specific functions of the faculty field liaison include (1) facilitating field teaching, student learning, and the integration of theory and application; (2) monitoring and assessing learning opportunities offered by the placement agency; and (3) evaluating student achievement in the placement. If conflicts arise between the student and the supervisor or agency, the liaison can act as a mediator as well as an advocate for the student. You should maintain open lines of communication with this individual also. He or she may be very involved in your midterm and year-end evaluations and may have a major say in which assignments are most relevant to your learning. It is typically this person who will review your learning agreement and who will be responsible for the formal paperwork requirements of fieldwork. Often, the liaison will also be your fieldwork instructor.

The Role of Fieldwork Instructor

In many programs, internship, practicum, and fieldwork are accompanied by a seminar class that meets regularly in the university setting. The person who leads this seminar will be your fieldwork instructor. His or her role is to help you integrate practice skills with theory and knowledge, to show you how to bridge the gap between what you learn in the classroom and how it can be applied in the field. The seminar can let students share their experiences in a larger group setting than in their placement. Students can learn a great deal about other placements and helping roles by listening to their seminar classmates. Also, the seminar is an excellent place to network. We have found that many students learn of placement and job opportunities through discussion in seminar courses. Bonding among students can also be high in seminars.

Seminars can be conducted in numerous ways. We prefer to teach our seminars in an informal yet structured manner. We provide group supervision, didactic exercises, and experiential learning through small group experiences and role-play assignments. In these seminars, our goals are to help enhance self-awareness, impart knowledge, and encourage skill development. Students may be graded on their academic and participatory performance, although some required seminars are designated credit/no credit or are not given any university units. Regardless of these particulars, we encourage you to make the most of your seminar experiences. You should be willing to risk sharing and to seek out information and support you may need.

Guidelines for an Effective Supervisory Relationship

The following guidelines summarize the actions you can take to get the greatest benefit from your supervisory relationship:

- Be clear about what you expect.
- Be assertive about your learning.
- Take advantage of your role as learner.
- Recognize that a certain amount of anxiety is appropriate.
- Take an active stance.

The Value of Various Learning Methods

As noted, learning can take place in a variety of ways. Traditional textbook learning remains essential in providing a basic foundation of knowledge that you will need to become an effective helper. Hands-on learning typically occurs when there is an opportunity for practice, either experientially in the classroom, during group supervision, or in actual practice situations involving a helping relationship. Providing actual practice is the purpose of internships and fieldwork placements. These are the settings in which the beginning helper is given the chance to apply knowledge gained in the classroom and through textbooks to real-life situations. This is when skills are likely to be cultivated and developed.

Above, we outlined the importance of supervision in a student's overall learning. Here, we elaborate on ways additional learning can be gained and the supervisory process enhanced. These include process recordings, audiotaping, videotaping, and journaling.

Process Recordings

Wilson (1980) has written extensively on recording in general. About process recordings specifically she writes: "Process [narrative] recording is a specialized and highly detailed form of recording that is a very valuable tool for enhancing student learning if it is properly used" (p. 18). Any type of learning has limitations. In our experience, process recordings can be misused or disregarded; they can also create additional busywork for the student who is already overwhelmed by school and fieldwork responsibilities, thereby creating undue resistance toward this type of learning. We certainly advocate the constructive use of process recordings, and further encourage that they be used as creatively and dynamically as possible. Before discussing specifics, we begin by orienting you to the purpose for process recordings.

Purpose of Process Recording   The primary purpose of this type of recording is to help students process what took place during some type of human interaction in their fieldwork setting. This is true regardless of whether the interaction is individual, group, administrative, or community based. Typically, a process recording contains the following information:

| Supervisory comments | Content-dialogue | Student's feelings | Analysis of the situation |

According to Nelson (1994, p. 13), "Process recording is the chief learning vehicle of field instruction, sorting out feelings, making inferences, developing critical thinking, and selecting alternatives for action." Process recordings were developed by social work educators who envisioned this type of learning to be a catalyst for professional growth whereby individual experiences, feelings and values could be integrated into a professional self (Wilson, 1990). Today, process recordings are used primarily during a student's internship or fieldwork experience and by field supervisors whose purpose is to help students apply theory or empirical-based knowledge to practice. In sum, the process recording serves as a method for both professional/personal growth and acquisition/integration of knowledge. Specifically, the written process recording facilitates student learning by helping students do the following:

- Rethink the interview content
- Deal with their feelings about the content and the client
- Analyze interventions
- Draw conclusions
- Understand their impact on the client system
- Demonstrate their understanding of information and theory gained in the class-room
- Build social work skills
- Participate in a growth relationship with the field instructor (Nelson, 1994).

Process recordings can have additional purposes:

- Process recordings are an effective way to help you learn interviewing skills.
- Process recordings can help your supervisor determine how you are performing and can delineate areas where you have strengths and limitations. It is especially important for your supervisor to determine your capabilities if he or she is to develop an educational training plan that is tailored to your needs and skill level.
- Unlike audio- or videotaping, process recordings often create less anxiety in the student. They allow you greater control over which verbal and nonverbal communications will be revealed to the supervisor. This selective factor, criticized by some, can be helpful to others. Students who are especially anxious about their performance seem to benefit most from starting off with process recordings rather than audiotaping or videotaping their work with clients.
- Process recordings provide added security for both the agency and the new student. By asking you to maintain a detailed account of your client interactions, your agency can stay informed of your activities and overall functioning. This helps your supervisor provide guidance when necessary, especially when a situation requires skill or intervention that may be beyond your capabilities as a student or a beginning professional.
- Process recording can enhance your self-awareness and help you separate factual data from gut-level reactions of what you observed. This lets you analyze the interaction more objectively. Wilson (1980, pp. 21–22) writes that "the act of putting personal feelings into writing is an important step toward achieving self-awareness in the new role of professional helping, and enables the student to talk more freely about his or her own feelings and how they affect service delivery." For those involved in more advanced clinical work, process recordings can be crucial to identifying and exploring transference and counter-transference issues that sometimes emerge in the therapeutic relationship and which are commonly dealt with during supervision.
- Process recordings can also be used by experienced practitioners in special situations. For instance, the advanced clinician may want additional feedback on a specific case and may find use of a process recording valuable in discussing the diagnostic issues or treatment considerations. By processing his or her interactions with a supervisor, the student may gain insight and objectivity.

• Students can sometimes compare their process recordings in a kind of informal peer consultation. This can help the entire peer group. In this sense, process recordings are used effectively in group supervision.

• Process recordings can enhance the review and evaluation process when an internship ends. Similar to what the therapist asks a client to do at the termination of a helping relationship, your supervisor may review your work with you, evaluate your progress, and explore your future goals. Examining process recordings you wrote at the beginning of your internship can be valuable in charting your progress. The recordings will help you review your most commonly experienced feelings and thoughts. The comments made most often by your supervisor can be easily identified in a review of process recordings written during the beginning, middle, and ending of your fieldwork. Comparing your beginning and end work can be a very effective way to demonstrate your growth (Wilson, 1980, pp. 21–22).

Different Types and Uses of Process Recordings   The literature describes several forms of written process recording (Nelson, 1994; Holden, 1992; Wilson, 1976). Below we sketch five types; these can be tailored to your specific learning objectives and your individual placement or agency. The actual process recording forms described here appear in the appendix.

Matrix Format   The matrix is a verbatim account of the interview with the students' observation of their own feelings and thoughts. This format emphasizes students' feelings and how these impact the students' interaction with clients. As students reach the middle or working phase of treatment, a more abbreviated version of recording may be appropriate.

The matrix format is used to emphasize student-client interaction and student feelings, analysis, and interpretation. It provides a good beginning in understanding and assessing process issues.

Analytical-Type Process Recording   The analytical recording stresses close examination of the interaction. Once the student has a good understanding of the importance of the interaction and a sense of the feeling component of the sessions, this type of recording is recommended; it includes analysis of content, purpose, and professional role as well as selected verbatim exchanges. The recording can also give much information about the student's writing ability, use of professional jargon, understanding of the core diagnosis, grasp of the worker's role, and ability to develop a treatment plan.

Short Version Process Recording   This abbreviated version or recording provides the structure for processing, but reduces repetitiveness. It is recommended for the experienced student and/or in processing later sessions. Use of this outline assumes that the student has a good grasp of interaction, feelings, and professional role.

Process Recording with Groups   Group recording is designated for use with a wide variety of units: family, educational, support, or psychotherapeutic. Many agency experiences do not include individual psychotherapy or counseling but they do expose students to some type of group involvement. The group recording allows students to document these experiences, which can be used in further discussion during supervisory conferences.

Process Recordings with Meetings   The next type is used in recording a wide variety of meetings—community-based, agency-based, information-giving, information-receiving, problem-solving, and other types of meetings. Many students will be placed in agencies whose main function is community organization. Attending various types of meetings will be a major role of the student intern. This recording allows students to analyze their roles and to determine future meeting goals and objectives while helping them become aware of the many facets of group dynamics in a community setting.

Responsibilities for Students and Supervisors   In ideal settings, we recommend use of the different types of process recordings as they represent the student's learning progress. As we have seen, process recordings can be a creative and effective teaching technique. Nevertheless, several factors must be present for students and supervisors to benefit optimally from their use. First, students must be open and honest in their process recordings; second, supervisors need to give students meaningful feedback. Wilson (1980) cites the following guidelines for both student and supervisor.

• Both students and supervisors must acknowledge that total recall of nonverbal and verbal communication in an interview is impossible. At best, only selective recording is possible. The interview must remain the primary focus, not the recording of it.

• Students must remain open and honest in recording their interactions during an interview. Although they may be tempted to polish the written record, leave out awkward statements, and change the wording just a little, they need to refrain from this. Making alterations reduces the learning opportunities and limits the ability of the supervisor to provide needed assistance.

• Be aware that each student is unique in how he or she perceives a specific interview, group, or meeting. If five different student interns were recording the same interview, their writings would differ, as perceptions and memory vary from one person to another. Also, each student is apt to focus on or avoid issues of particular relevance to him or her.

• Process recording, to be meaningful, must be done as soon as possible after the interview has occurred. As time lapses, many important details are lost.

• Accurate process recordings can be time-consuming. Use your time management skills to structure enough time for writing them.

• Due dates for each process recording are recommended. Both student and supervisor can benefit from deadlines.

• When the student is procrastinating, the supervisor is responsible for identifying the reasons for the resistance. Often students feel insecure in their skills and thus fear the supervisory evaluation of their interview.

• When supervisors neglect to provide students with feedback or are harshly critical in their evaluations, conflicts may ensue. Students must be prepared to ask supervisors about either of these issues as they are sure to impact the supervisory relationship negatively if they are not discussed.

The evaluation of process recordings is a function of the supervisor, who should give students constructive feedback that identifies both their strengths and their weaknesses where further development is needed. Students need specific and detailed comments so they will know what and how to improve. In sum, when process recordings are used appropriately, they can enhance students' self-awareness, identify gaps in their knowledge, and help them learn skills.

Audio- and Videotaping

Audio- or videotaping one-to-one interviews, family sessions, and group work with clients can be tremendously useful. Taping can give you an instant replay of your interventions. Even though audio and video recordings are much more accurate than process recording, they can also be overwhelming. Nonetheless, listening to or viewing yourself on tape can help you see how your clients perceive you. Using tape can also alert you to habits you would like to change as well as to good qualities you were not aware you possessed.

Audio- or videotapes are a direct view of your interviewing and problem-solving styles. "A supervisor viewing or listening to a tape can furnish you with specific suggestions for what you might have said or how you might have proceeded at any particular part of the session. In this way, you can get ideas about alternative and possibly more effective intervention strategies" (Kirst-Ashman & Hull, 1993, p. 544).

According to Wilson (1980), recordings can be used in a number of creative ways. The most obvious is review by the student's supervisor. Students should first have the opportunity to review the tape before the supervisor sees it. "The tape can be played back during a supervisory conference and stopped at various points for discussion and perhaps role play of alternate techniques that might have been used. The learner could be asked to identify instances when reflection, interpretation, and other basic techniques were used, or to identify certain behavior on the part of the client.... The tape itself can be used as a therapeutic tool with the client; the client can listen to or view the tape, which is then discussed in the session with the helper" (Wilson, 1980, p. 14).

Because taping always involves more than one person, certain guidelines should be followed in using this recording method:

• The helper must first get the client's explicit permission to tape. It is unethical and a violation of confidentiality to tape a client without the client's knowledge. Be prepared to explain the exact purpose of the recording, who will hear the tape, and what you plan to do with it. This information should be included in the consent form the client is asked to sign.

• Before the actual taping, help the client prepare for it. Take time to answer clients' questions and perhaps introduce the taping experimentally: "We'll try it for a few minutes, but I'll stop if it bothers you." Keep your word.

• Taping should be something the client not only agrees to readily, but is also comfortable with. Never tape an interview with a suspicious or paranoid client as this is likely to increase their suspicion and heighten their anxiety and resistance. Similarly, never attempt to hide your equipment; this could also increase the client's guarded behavior.

• It may be helpful to transcribe what happened during the taping. Due to time constraints, you may need to concentrate on the most crucial part of the interview.

• You will probably be asked to write a summary of the taped session. The structure used for this summary may vary according to your style and your supervisor's request. What is important is that you be able to identify the significant aspects of the interview, as these will help you learn the most from the process.

• Following your summarization, you and your supervisor will discuss your strengths and weaknesses.

• A last step involves "adapting what you have learned during this discussion and writing a report that complies with the conventional format your agency uses" (Kirst-Ashman & Hull, 1993, pp. 544–545).

Apart from the clinical aspects of taping, there are personal and technical points you should consider:

• Be prepared to recognize and deal with your own feelings about taping. Any anxiety, resistance, or ambivalence should be expressed and discussed with whoever is requesting the taping (probably the supervisor). Self-awareness is important here. Students often rationalize that clients will feel uncomfortable with taping, using this as a strong reason not to do it. Although some clients will not be comfortable with taping, many are amenable. Perhaps it is not the client who is the most uncomfortable.

• Be familiar with the equipment you will be using. Having to adjust the taping or video machine can distract the client and ultimately intrude on the session being conducted.

• Whenever a tape is to be used in the classroom or for training purposes, the client's permission must be obtained. Also, be careful to secure the tape in a safe place where confidentiality can be maintained (Wilson, 1980, pp. 15–16).

In addition to taping interviews with clients, you can also tape staff meetings, inservice training sessions, agency board and committee meetings, and group therapy sessions for later self-assessment or critique. Consider alternating taping with process recordings; you will be amazed at the different things you can learn from each (Wilson, 1980, pp. 15–16).

Journaling

Journaling is a form of free-form writing about various topics within your field placement setting and/or even within your life. It can be a way to access your feelings and thoughts in regard to the work and relationships you are forming within your agency. At times your instructors may ask you to write about specific topics to encourage your critical thinking on the various practice and process issues that evolve out of your work in the agency. Sometimes you may just want to keep a diary of your reactions from day to day in your work to help you become more focused and to serve as an evaluative tool in your growing self-awareness.

> **SCENARIO**
> Randy remembers that it was extremely helpful in his beginning years in the field to keep a journal and just write about what was happening in his thoughts and feelings at the time. Years later he could return to these writings to see some of the progress he had made in his ability to recognize and deal with his feelings and thoughts.

The Benefits of Peer Review and Case Conferences

Having a forum in which you can discuss your work, receive constructive feedback, and learn from others is important. To provide this, many agencies advocate peer review and/or case conferences. The benefits of each are numerous and are outlined below. Not all agencies or practitioners are amenable to these forms of learning, however. Much depends on the emphasis on learning within the agency and the orientation and style of the supervisors and/or management. In settings where these types of review are used, students generally find them very helpful.

Peer Review

Peer review is a process in which individuals receive feedback from their peers on some aspect of their work. For students, the process of having to present their work and be evaluated by their peers can be anxiety producing, but they are encouraged to keep their focus on continual growth, self-evaluation, and awareness.

Peer review is widely accepted as a method for both ongoing evaluation and increased communication between members of an organization. Sometimes the peer review meeting might have a specific focus, such as reviewing intakes and deciding on a treatment plan within an agency setting. Some agencies use peer review for management issues and development of staff cohesion.

> **SCENARIO**
> Early in Lupe's educational experiences in an agency setting, she approached a staff member to introduce her idea that discussing cases within the peer review meeting would be beneficial to all. She was told by the worker that she just needed that time "to complain about the agency" and that she did not want to utilize the time to discuss cases as she was seeing clients all day. What would you do if this happened to you as an intern in an agency setting?

> **SCENARIO**
> Andy remembers working in a community mental health center in which the majority of cases involved multiproblem families and/or abused children. Although there were administrative meetings in the agency, there were no peer review meetings. Andy talked to several of the staff members in his unit about the possibility of meeting once

a week to discuss treatment issues. After getting support from most of the staff, he presented the idea to the administrator, who agreed to let them take the meeting time out of their schedules. The staff soon found these meetings extremely helpful as they allowed the staff to discuss issues about the particular cases that were being treated within the agency. Being able to discuss concerns with peers gave staff members a chance for valuable examination and evaluation. The meetings also became a sort of think tank where new and different ideas could be shared.

Case Conferences

Case conferences offer some of the same benefits as the peer review process. In addition, they may be mandated by the agency in which you work and may thus be more formalized. Case conferences may require practitioners to engage in case presentations. This experience is invaluable for beginning workers as it teaches them to consolidate and organize their thinking. These presentations also give workers the benefit of constructive feedback and added insight from co-workers that may enhance the therapeutic outcome of a case. Finally, case conferences can be an important vehicle through which helpers can gain needed support, validation, and encouragement.

CONCLUDING EXERCISES

1. As a human service worker, you must be aware of your strengths and limitations. The following exercise is structured to help you determine these strengths and limitations for further reflection and improvement.

    List ten things you like about yourself:

    1. _____
    2. _____
    3. _____
    4. _____
    5. _____
    6. _____
    7. _____
    8. _____
    9. _____
    10. _____

    Now, list ten things you don't like about yourself or believe you would like to change:

    1. _____
    2. _____
    3. _____
    4. _____
    5. _____

6. _____

7. _____

8. _____

9. _____

10. _____

Which was easier to complete? Why? If the negative things were easier for you to get in touch with, why is this so? Looking at the list of things you like about yourself, circle the items that are solely internal characteristics (friendly, empathic, caring). Now underline those items that are solely external (handsome face, good body, nice car). Is there a balance between the two types of items? Traits that are external are vulnerable to loss and the judgment of others. It is important to balance your traits between external and internal and between likes and dislikes.

2. How do you see yourself as a learner at this point in your professional and educational development? Using the stages of learning described by Nelson, attempt to describe honestly where you see yourself now. Indicate which stage you most closely identify with:

1. Awareness: _____

2. Recognition: _____

3. Knowledge: _____

4. Application: _____

5. Mastery: _____

Now describe how you might help yourself to move along on the continuum to the next level. What do you need to do to help yourself reach the next stage of learning?

_____

_____

_____

3. Supervision will undoubtedly become an important part of your professional training if it is not already. If you have had experiences in supervision, you will have memories to draw on for the following exercise. If you haven't had any supervision, this will help you think about what type of individual you would most want to supervise you.

Write what you believe to be the top ten positive qualities of a good supervisor:

1. _____

2. _____

3. _____

4. _____

5. _____

6. _____

7. _____

8. _____

9. _____

10. _____

Now, write what you believe to be the top ten negative or destructive qualities of an ineffective supervisor:

1. _____

2. _____

3. _____

4. _____

5. _____

6. _____

7. _____

8. _____

9. _____

10. _____

Which qualities were easier to write? Why do you think that is so? Now, using the same list, circle the qualities you have experienced in a supervisory relationship and discuss the nature of this relationship. Would there be any changes you would make?

2

Key Elements and Challenges of Practice

AIM OF CHAPTER

An integral part of clinical practice involves working with individuals in some therapeutic capacity. In this chapter, we examine essential aspects of that work including the personhood of the helper, communication techniques, and interviewing skills. We discuss the importance of being able to perform a thorough assessment and develop sound treatment interventions. We believe the issues presented are essential ingredients for successful practice at any level and are therefore the foundation for any helping relationship. This is true whether you are working with individuals at the micro level, with families and groups at the mezzo level, or involved in organizational or community change at a macro level; the importance of the helper-client relationship and the helping process is at the core of any effective work.

Some of the more common challenges of practice are also identified. First, we address personal, organizational, and environmental challenges of practice such as transference, countertransference, resistance, working within organizations, and the fiscal and political realities of agency practice. Second, we discuss the importance of effective communication and conflict resolution in both interpersonal and professional relationships. Last, we discuss surviving in agency practice. We introduce the concepts of encapsulation and burnout and explore means of prevention and intervention.

ESSENTIAL ASPECTS OF PRACTICE

The Personhood of the Helper

The personhood of the helper is perhaps one of the most important aspects of helping. The values and behavior of the helper are instrumental to the success of the helping process. The quality of the client/helper relationship seems to be a crucial factor in fostering growth. Although some of these issues have been discussed earlier, an examination of the personal characteristics of helpers that contribute most to this process is fitting here.

Characteristics of Competent Practitioners

Many characteristics are considered essential for effective practice. Carl Rogers (1951), founder of client-centered therapy, emphasizes the importance of empathy, genuineness, and unconditional positive regard. We elaborate briefly on each of these, given their undisputed importance in the helping process.

Empathy relates to the helper's ability to "perceive the internal frame of reference of another with accuracy and with the emotional components and meanings which pertain thereto as if one were the person, but without ever losing the 'as is' condition" (Rogers, 1959, pp. 210–211). The two steps necessary for empathetic understanding are (1) accurately seeing the client's world—being able to see things the way he or she does, and (2) verbally sharing your understanding with the client.

Genuineness simply means being honest in the relationship with a client; being real. *Congruence* is seen in the helper's bringing to the relationship a consistent and honest openness in which he or she remains constant in thoughts, feelings, and actions.

Positive regard is considered by Rogers to be a necessary antecedent to empathy. Basically, it means "accepting other people's rights to their own unique individualities and perspectives" (Long, 1996, p. 71). This is closely tied to the condition of *acceptance*. Acceptance is manifested in the helper's ability to recognize the essence of being human; it requires helpers not to judge clients by their problems, but to seek actively to understand them. Issues of *trust* are bound to arise, and expectations of the helper are also natural. According to Compton and Galaway (1994), "trust or expectation means that workers have a belief and faith in the capacity of individuals for self-determination and self-direction—that they consider it is the right and responsibility of each individual to exercise maximum self-determination in the person's own life with due regard for the welfare of others" (p. 277). Compton and Galaway (1989) cite other elements of the helping process that are basic ingredients for a positive helper-client relationship. These include *concern for others*, seen in the helper's sincerity in caring about what happens to the client and being able to communicate this feeling openly; *commitment and obligation*, which are built on involvement and investment that allows clients to feel safe and committed to the helping process; and *appropriate use of authority and power* in that the helper uses his or her role to influence the client positively toward constructive action. Additionally, helpers are more successful if they are actively engaged in ongoing learning and growth and thus are maturing; are creative, as shown by originality, expressiveness, and imagination; have the capacity to observe themselves and seek constructive feedback throughout their careers; truly desire to help others; have the courage to engage in a helping relationship in an authentic and concerned way; and are sensitive and caring.

A list of these attributes appears in Table 2.1. You may already possess many of these traits; others you may wish to cultivate in your evolution as a professional helper. Remember that learning is an ongoing, individual process. Remain open to discovering both your strengths and your limitations.

The Importance of Awareness

In the list in Table 2.1, *awareness* is mentioned in two contexts, self-awareness and value awareness. This repetition highlights our view that self-awareness is especially important in human services. As mentioned in Chapter 1, self-awareness, with a personal commitment to helping others, lies at the center of successful helping. Helpers who are encouraged toward lifelong self-exploration are likely to be satisfied and successful in their work. We believe that all professional helpers, novice or experienced, must continue this process if they are to remain useful to others and true to themselves.

The Importance of Establishing the Helping Relationship

We are the most important tool in our work with clients. The stronger the helping relationship is, the more likely it is to produce a positive outcome. Therefore, the value of establishing a sound helping relationship becomes even more significant. Helpers must know the personal characteristics they possess or need to cultivate to nurture this process. In

TABLE 2.1   Primary Characteristics of Professional Helpers

Self-awareness	Value awareness
Intentionality	Willingness to make appropriate self-disclosure
Awareness of feelings	Acceptance
Dynamism	Desire to help
Sensitivity	Realism
Commitment	Purposefulness
Responsibility	Personal presence
Integrity	Careful use of authority
Openness	Ability to show emotions
Patience	Flexibility
Motivation	Curiosity
Willingness to accept constructive feedback	Ability to communicate

Table 2.2, we offer Brammer's (1988) inventory of personal and growth-facilitator characteristics of helpers that in large measure, determine the nature of the helping relationship. Similar qualifications have been identified by Brill (1990, pp. 89–97) and Neukrug (1994, p. 89) as basic to the helping relationship.

In addition to these traits, the development of *trust* and *rapport* is essential for a helping relationship to be effective. Words such as understanding, compatibility, and harmony have been used to describe rapport. It can also be considered an interactional process containing the three components of (1) "perceived warmth and caring," (2) "a perceived safety and acceptance," and (3) "a perceived trustworthiness and credibility" (Long, 1996, p. 81). Trust is thus an ingredient of rapport. According to Long (1996), "Trust is an affective experience and a desired outcome within the relationship. It grows from rapport and is built on acknowledged cognitive beliefs, particularly the belief that individuals have the right to be themselves and to have their own feelings, thoughts, and actions" (p. 81).

Trust is the underlying force that enables helpers and clients to negotiate the stages of helping. Trust may continually be tested and may waiver when clients are challenged by

TABLE 2.2   Brammer's Inventory of Helper Characteristics

Helper's Personal Characteristics

1. Awareness of self and values
2. Awareness of cultural experiences
3. Ability to analyze the helper's own feelings
4. Ability to serve as model or influencer
5. Altruism
6. Strong sense of ethics
7. Responsibility

Helper's Growth-Facilitator Characteristics

1. Empathy
2. Warmth and caring
3. Openness and congruence
4. Positive regard and respect
5. Concreteness and specificity
6. Communication competence
7. Intentionality

new insights or painful experiences. Without trust, a client will seldom risk opening up and enlisting help from the helper. Maintaining trust requires helpers to remain responsible, committed, and ethical in their practice.

Last, helping is a process; helpers may plant seeds that flourish and develop later. Helping is not an isolated event; it begins at your first interaction with the client and continues long after. Knowing that it evolves over time, you should approach the relationship in a patient and committed manner.

The Quality of the Helping Relationship

Corey (1996), in his work on counseling and psychotherapy, highlights the quality of the helping relationship. We have applied his concepts here more generally to a helper engaged in any form of clinical work. Corey identifies two additional personal characteristics of helpers that are likely to enhance the helping relationship: (1) authenticity of the helper, and (2) the helper as a therapeutic person.

Authenticity is essential, as the helping relationship is founded on honesty and trust and involves the interaction between the helper and the client. Although authenticity is closely tied to genuineness, various authors (Corey, 1996; Neukrug, 1994; Hepworth, Rooney, & Larsen, 1997) separate the two. Hepworth, Rooney, and Larsen (1997) define authenticity as "the sharing of self by relating in a natural, sincere, spontaneous, open, and genuine manner. Being authentic, or genuine, involves relating personally so that expressions are spontaneous rather than contrived" (p. 120). Helpers must be aware of how they present themselves to their clients. Congruency between our verbalizations and our actual thoughts and feelings is part of being authentic. Also, taking responsibility for our part in the helping relationship and accepting responsibility when we make errors is central to conveying authenticity to our clients. If we hope to encourage clients to be open and truthful in their interactions with us, we must be open with them.

Corey (1996) contends that when helpers hide behind the professional role, only the technical aspects of helping are revealed. Leaving our own reactions, values, and self out of the helping process accomplishes little. "It is through our own genuineness and our aliveness that we can significantly touch our clients. If we make life-oriented choices, radiate a zest for life, and are real in our relationships with our clients, and let ourselves be known to them, we can inspire and teach them in the best sense of the words" (Corey, 1996, p. 16). This view reinforces the integration of the humanistic ideas proposed by Rogers and many of the characteristics listed above.

Modeling and *self-disclosure* are two facets of being authentic. Perry and Furukawa (1986) view these as two kinds of behaviors that, used appropriately, can dramatically influence clients in their change process. Helpers are constantly modeling for their clients. When we are empathetic, clients also learn the importance of listening and accepting; when we are assertive and endeavor to resolve conflict in appropriate ways, we model the value of assertive behavior and conflict-resolution skills. Similarly, when we engage in unprofessional behaviors, we model these to our clients. There may be times when we are unaware of the behaviors we are modeling as they are very subtle in nature, so we need to be continually introspective and self-evaluative.

It is often through *self-disclosure* of our own life events that we model how we cope and interact with others. Self-disclosure has been defined in many ways. According to Hepworth et al. (1997, p. 121), it is "the conscious and intentional revelation of information about oneself through both verbal expressions and nonverbal behaviors (e.g., smiling, grimacing, or shaking one's head in disbelief). Viewed from a therapeutic perspective, self-disclosure encourages clients to reciprocate with trust and openness."

Self-disclosure can be defined as personal verbal revelations made by helpers to their clients. However, self-disclosure may also occur through nonverbal behavior and may be purposeful or unintended. Egan (1994) notes that "we always disclose information about ourselves though nonverbal channels and by our actions even when we don't intend to."

Cormier and Hackney (1988, p. 22) have identified several forms of self-disclosure: (1) disclosing the helper's own problems, (2) disclosing facts about the helper's role, (3) disclosing the helper's reactions to the client, and (4) disclosing the helper's reactions to the client-helper relationship. These can be categorized into the two major types identified by Danish, D'Aguelli, and Hauer (1980) as (1) self-involving statements, and (2) personal self-disclosure.

Self-involving statements include messages that express the helper's personal reaction to the client during the helping process. In contrast, personal self-disclosing messages are related to struggles or problems the helper is currently experiencing or has dealt with in the past that are similar to the client's problems. Self-involving self-statements "appear to be low risk and relevant to the helping process" (Hepworth et al., 1997, p. 121) whereas personal self-disclosure can be potentially problematic. Personal self-disclosure should be used judiciously because excessive disclosure of personal issues may undermine the client's confidence in the helper as well as divert attention from the client's issues.

You should consider both the timing and the intensity of any type of self-disclosure. Before you share personal feelings and experiences, be sure that enough trust and rapport have been established that clients are willing to engage in a more personal level with helpers. According to Hepworth et al. (1997, p. 121), "The danger in premature self-disclosure is that such responses can threaten clients and lead to emotional retreat at the very time when it is vital to reduce threat and defensiveness. The danger is especially great with clients from other cultures who are unaccustomed to relating on an intense personal basis."

Practitioners should also use caution in the amount of self-disclosure they engage in. Even when trust and rapport are present, only moderate self-disclosure should be exercised. Beyond a certain level, it is no longer helpful to the therapeutic process (Truax & Carkhuff, 1964). Also, self-disclosure is more useful when a helper is working with higher-functioning clients and may be counterproductive with mentally ill clients (Shimkunas, 1972; Doster, Surratt, & Webster, 1975).

Some of the purposes of self-disclosure cited in the literature have been summarized by Cormier and Cormier (1998, p. 43), who believe self-disclosure is useful in various ways:

1. It may generate an open [therapeutic] atmosphere
2. [It can reduce] the role distance between a counselor and a client
3. [It can increase the disclosure level of clients] to bring about changes in clients' perceptions of their behavior and to increase client expression of feelings
4. [It can help] clients develop new perspectives needed for goal setting and action

When used properly, self-disclosure is valuable. However, it must be appropriate rather than indiscriminate. A helpful way to determine the appropriateness of any self-disclosure is to ask whose needs it would meet—the client's or the helper's. If the helper's needs take precedence, the self-disclosure could be more harmful than helpful.

Being a Therapeutic Person

Just as being authentic is important to the therapeutic process, so is being a therapeutic person. Corey (1996) views the willingness of helpers to struggle to become more therapeutic persons themselves as a crucial quality. He has identified personal qualities, listed below, that helpers need to be effective. Review this list and determine which qualities you already possess and which you must continue working toward.

Personal Qualities for Becoming a Therapeutic Person

- Effective helpers have an identity.
- They respect and appreciate themselves.
- They are able to recognize and accept their own power.
- They are open to change.
- They are expanding their awareness of self and others.

- They are willing and able to tolerate ambiguity.
- They are developing their own helping style.
- They can experience and know the world of the client, yet their empathy is nonpossessive.
- They feel alive, and their choices are life oriented.
- They are authentic, sincere, and honest.
- They have a sense of humor.
- They make mistakes and are willing to admit them.
- They generally live in the present.
- They appreciate the influence of culture.
- They are able to reinvent themselves.
- They are making choices that shape their lives.
- They have a sincere interest in the welfare of others.
- They become deeply involved in their work and derive meaning from it.
- They are able to maintain health boundaries. (Adapted from Corey, 1996, pp. 16–18)

Communication Techniques

Communication is at the heart of successful human interaction and certainly crucial for anyone hoping to be an effective helper. Communication is defined as "the process of passing along information and understanding from one person to another" (Travers, 1993, p. 188) or "as an interactional process that gives, receives and checks out meaning and occurs when people interact with each other" (Compton & Galaway, 1989, p. 332). It is one of the primary instruments in the helping process.

When communication is poor, the potential for misinterpretations, misleading information, and erroneous perceptions is high. As helpers, we are responsible for ensuring that we understand our clients and that they understand us. A key to preventing poor communication is recognizing the principles of communication itself. Among these are attentive listening and nonverbal communication, which have been identified as essential to effective communication.

Attentive Listening

Attentive listening, sometimes referred to as *active listening,* involves using a variety of techniques and behaving in such a way that the client feels heard and understood. In active listening, you try diligently to understand what the other person is saying. This requires new ways of responding so that people you are listening to feel understood and open to further interaction. There is an important distinction between hearing and listening. Hearing "is a word used to describe the physiological sensory process by which auditory sensations are received by the ears and transmitted by the brain," whereas listening "refers to a more psychological procedure involving interpreting and understanding the significance of the sensory experience" (Drakeford, 1967, p. 67).

We can hear what people are saying without really listening to or understanding them. As helping professionals, we must be cautious not to fall into this common trap of hearing without listening.

To develop attending skills helpers must recognize the areas in which they might experience difficulties and be willing to improve these. Attending skills are maintained through involvement with clients, by using appropriate body language and eye contact, and by creating an environment with few distractions. Furthermore, you must attend with all the senses. "Attending is giving your physical attention to another person. I sometimes refer to it as listening with the whole body. Attending is nonverbal communication that indicates you are paying careful attention to the person who is talking" (Bolton, 1979, p. 23).

In attending, helpers listen to what clients say and how they say it. They avoid interrupting and allow clients to complete sentences and ideas. They use silence to encourage clients to talk when ready, and they use reflection to clarify the clients' meaning. Helpers ask questions that elicit important details. They help clients to explore relevant contexts rather than letting them ramble. When appropriate, helpers reflect similarities and discrepancies in what clients say (thoughts), how they say it (feelings), and what clients do (actions). Finally, helpers intermittently elicit feedback to ensure the accuracy of their perceptions of the clients' intent.

Listening carefully and attentively indicates caring, concern, and respect. Ten techniques that are helpful in sharpening your listening abilities are listed below. Review them from time to time to make sure you are continuing to be attentive to your clients.

1. Give your wholehearted attention to the person.
2. Really work at listening intently.
3. Show an interest in what the speaker is saying.
4. Resist distractions.
5. Show patience.
6. Keep an open mind.
7. Listen for ideas.
8. Judge the content, not the delivery.
9. Hold your fire.
10. Learn to listen between the lines. (Bolton, 1979)

Nonverbal Language

Nonverbal language is also important as it is the most prevalent form of communication. Mehrabian (1971) offers the following statistics about communication in general:

- 7% of messages come from *what* people say (their words)
- 38% of messages come from *vocal aspects* of their communication (tone of voice, volume, etc.), but not actual words used
- 55% of messages come from *nonverbal* aspects of their communication (i.e., body language)

Aspects of nonverbal communication include vocal or subvocal sounds; vocal tones or pitch; speed of speech; and body language, including eye contact, facial expressions, gestures, posture, and overall body movement. Nonverbal language is the most used form of communication and can provide invaluable cues to a person's feelings.

Assertiveness is necessary for effective communication. Assertiveness involves verbal and nonverbal skills that help "maintain respect for oneself and others; and defend one's rights without dominating, manipulating, abusing, or controlling others" (Bolton, 1979, p. 12). Engaging in assertive behavior with clients is beneficial in two ways: First, it models to clients the value of assertive behavior; second, it helps minimize potential problems from developing within the client-helper relationship by dealing with them directly and swiftly as they arise.

Multicultural and Gender Issues and Communication

Communication styles and patterns vary from culture to culture and between men and women in each culture. Evans, Hearn, Ulhemann, and Ivey (1993) propose that attending and listening patterns in Euro-North Americans are different from those of other cultures. Such recent research on culturally sensitive communication patterns is causing us to reexamine earlier findings. A 1972 study found that more effective interviewers are those who lean toward the client, maintain steady eye contact, and establish a conversational distance of slightly over an arm's length. We are learning through research with culturally diverse participants that this approach to interviewing is not appropriate for all cultures.

Although listening skills are foundational and important in all cultures, the following are cultural differences in basic listening skills as identified by Evans et al. (1993):

- *Eye contact:* The direct eye contact pattern of mainstream Americans is considered intrusive and rude in some cultures. In many societies (e.g., Native Americans and Latinos), less eye contact is preferred; in some, direct eye contact is avoided altogether. Among Native Americans, for instance, direct gazing is considered an intrusion of privacy. Many clients, discussing intensively private matters, may feel more comfortable if you avoid gazing at them directly. It is important for helpers to modify their use and expectations of eye contact to meet both the individual and cultural needs of their clients.

• *Body language and space:* Body language meanings vary widely from culture to culture. Russians may shake their heads up and down to indicate "No;" Vietnamese may consider motioning to a person with one's fingers to "come here" as a rude and derogatory gesture.

In regard to space, some North Americans consider an arm's length between people to be a comfortable conversational distance. Others, especially those of Arabic descent, may prefer 6 to 12 inches. Still others, such as Australian aboriginal people, are not comfortable unless there is great distance. For many, maintaining a lengthy distance is desirable until trust and rapport are established.

• *Verbal following:* A common North American style is for helpers to listen to clients by following their words as directly and literally as possible. However, a more subtle, indirect style of listening for generalities, rather than specifics, may be appropriate for the Chinese-American or Japanese-American client.

• *Listen first, then act:* Although we agree that it is important to listen before reaching a conclusion, some clients want at least a sign of direct helper action before they trust enough to share deeper emotions. For example, some Asian clients may view helping professionals as authorities who can solve their problems by direct advice. This points up the importance of the timing between listening and acting that may vary among cultures.

Similar to cultural variations are those tied to gender. Men and women clients may differ in their responses, and this may be further complicated by the clients' culture. In talking about feelings, men may intellectualize, while women may respond emotionally. Men may be culturally influenced to focus on thinking and problem solving, while women may be more influenced to focus on feelings and process. Even though the ideas of listening and acting are universal, helpers should modify their style and approach when necessary because of gender, culture/race, affectional orientation, physical issues the client may be facing, age of the client, and a multitude of other factors. Professional helpers must be committed to continued learning about gender and cultural differences in the diverse client population they will be working with.

Interviewing Skills

Positive communication skills lead to effective interviewing. "Interviewing is a specialized form of communication that is contextual, purposeful, limited, and involves specialized role relationships" (Compton & Galaway, 1989, p. 335). Competent interviewers, regardless of orientation and training, have found that communication is basic to their relationship with clients. In effect, the core communication skills we have discussed are essential to any interview.

Interviewing skills are developed through time and require constant readjustment and fine-tuning by the helper. The more seasoned the interviewer, the easier the interview process becomes. Only practice and experience will give beginners mastery of the skills necessary to becoming effective interviewers.

Working with individuals requires use of sound interviewing skills. This is true whether the helper is involved in individual counseling or other forms of service, such as case management. Interviewing is a basic process for information gathering, problem solving, and information or advice giving. Clinical interviewing, used in such advanced therapeutic interventions as counseling and therapy, is a more intense and personal process that is geared toward helping people develop adaptive coping skills and to grow.

Brammer (1988) categorizes into seven clusters the skills that help promote understanding of self and others. They provide a quick overview of many aspects of the interviewing process. Some of these have already been discussed in relation to communication but all are listed here for review:

1. *Listening skills:* attending, paraphrasing, clarifying, and perception checking
2. *Leading skills:* indirect leading, direct leading, focusing, and questioning
3. *Reflecting skills:* reflecting feelings, reflecting experience, and reflecting content
4. *Summarizing skills:* pulling themes together
5. *Confronting skills:* recognizing feelings in oneself, describing and sharing feelings, feeding back opinions, mediating, repeating, and associating
6. *Interpreting skills:* interpretive questions, fantasy, and metaphors
7. *Informing skills:* advising and informing

In addition to these, we believe that interviewing must include an interpersonal focus. Interpersonal factors that contribute to the development of sound interviewing skills include these:

- Being appropriately nurturing and supportive to help the client grow. Becoming overly nurturing or lacking in this area can result in a potentially negative outcome.
- Interpreting clients' issues from one's own perspective to help them clarify and enhance their understanding of their issues.
- Actively involving the client in the helping process from the beginning phase to termination.

Cormier and Cormier (1998) have summarized the broad areas in helping relationships that are consistently associated with positive outcomes:

1. *Facilitating Conditions:* Empathy, positive regard, and genuineness. Conditions especially empathy, which, if present in the counselor [helper] and perceived by the client, contribute a great deal to the development of the relationship
2. *A Working Alliance:* A sense in which both therapist and client work together in an active, joint effort toward particular goals and outcomes
3. *Transference and Countertransference:* Issues of emotional intensity and objectivity felt by both client and counselor. Usually related to unfinished business with one's family of origin, yet triggered by and felt as a real aspect of the therapy relationship. (p. 35)

The third factor, *transference and countertransference,* is discussed in detail later in this chapter when we examine the personal challenges of direct practice.

Building Core Counseling Skills

Regardless of your level of practice or the specific discipline and role you are involved in, having a foundation of skills is essential. Your reward for developing these key elements of practice will be your successful involvement with clients.

To summarize, we turn our attention to the basic core counseling skills described by Fenell and Weinhold (1989) that we believe are also crucial to professional development. They include rapport building, information gathering, structuring, information giving, reflecting content, reflecting feeling, summarizing, self-disclosure, confrontation, interpretation, behavior change, and closure. These are outlined and discussed below.

1. *Rapport building* is the development by practitioners of a positive connection with the clients they are working with. It requires helpers to be attentive to both verbal and nonverbal forms of communication, to trust their own reactions, and to utilize them in a therapeutic way to promote a positive working alliance with their clients.

2. *Information gathering* refers to the helper's ability to gather relevant information about the client that will aid in the development of a treatment plan and interventions. Another name for this process is assessment or systematic inquiry. The use of open-ended and closed questions can help the counselor learn more about the client. Open-ended questions invite the client to elaborate on responses. A closed question is one that can be answered yes or no. Both types can be helpful in gathering information that lets the helper see the world from the client's perspective.

3. *Structuring* lets the helper set clear boundaries and goals with the client and form realistic expectations for what can and cannot be accomplished in the therapeutic relationship. Structuring can also refer to the physical environment. We have found that actively structuring the helping process results in a more mutually satisfying outcome.

4. *Information giving* refers to the helper's ability to provide important information to the client that comes from his or her education and experience in the helping profession. Fenell and Weinhold (1989, p. 62) state three main purposes of information giving:

1. To identify alternatives
2. To evaluate alternatives,
3. To correct erroneous information.

As an example, we have learned that providing accurate information to clients about the difference between thoughts, feelings, and actions can help them accept their feelings, challenge their thoughts, and implement change in their behaviors.

5. *Reflecting content* is demonstrating to the clients that you understand them by succinctly restating the content of their messages. This skill may also be thought of as a "perception check" as it gives you, the helper, a chance to see if your perceptions match those of your clients.

6. *Reflecting feeling* is listening and accurately reflecting the feelings the client is experiencing. This skill helps clients increase their self-awareness and self-understanding. When you use this skill, pay attention not only to verbal cues, such as specific feeling words the client uses, but also to nonverbal behavior. This may be a sigh, the tone of the client's voice, a tear, or the client's arms crossing, for example. Reflecting is a particularly important skill to have as many clients seem especially confused about the nature and importance of understanding and expressing feelings accurately.

7. *Summarizing* occurs when helpers "recapitulate, condense, or crystallize the essence of what the client said and felt" (Fenell & Weinhold, 1989, p. 66). This skill requires more than merely reflecting as it is usually more comprehensive and covers larger amounts of material than dealt with in reflection. Summarizing is often useful at the beginning and ending of sessions.

8. *Self-disclosure*, to reiterate, refers to sharing significant and relevant personal information about yourself as it relates to what the client may be experiencing or working on. This sharing can demonstrate your connection with and understanding of the client. It also will instill hope and knowledge that there are options open to the client. A useful check on your self-disclosure would be to ask yourself about the motivation for your personal sharing at any given time. If the information will further the client's goals rather than your personal ones, it is likely to be appropriate.

9. *Confrontation* is pointing out effectively the discrepancies among the thoughts, feelings, and actions of the client. For example, the client's eyes may be full of tears, but she might be saying that she feels "fine" or "okay." Pointing out these discrepancies can heighten clients' awareness of self and may help them discover new avenues for problem resolution. If you are unwilling to point out these discrepancies or too invested in getting the client to like you, then you are missing opportunities to foster growth in your clients.

10. *Interpretation* is presenting "the client with a new frame of reference through which the client can view the problems and better understand the situation" (Fenell & Weinhold 1989, p. 68). In studying the various theoretical orientations, you may find you can use

these in your interpretations, as each different orientation provides a unique perspective on human behavior.

11. *Behavior change* refers to your ability as a helper in assisting clients to bring about change within their life situation. These changes may occur in the client's thoughts, feelings, or behaviors. This skill is built on a foundation of the other skills we have discussed. According to Fenell and Weinhold (1989), helpers who want to effect behavior change must be able to do the following:

1. Identify and clarify the problem presented
2. Establish workable and realistic goals
3. Establish criteria to measure the effectiveness of the action plan
4. Implement the steps needed for goal attainment

Changing behavior involves knowledge of what clients can do to alter their thinking, feeling, and actions; it also requires the ability to break down goals into small, realistic, and achievable steps that can be accomplished by clients. Both take time, experience, and patience in working with clients.

12. *Closure* is the resolution and end of the helping process. Successful closure requires the helper to know the client's feelings and issues involved in termination, and to face them in a way that brings the client to a satisfactory ending of the helping relationship. The helper may summarize or evaluate the helping relationship for the client. Knowledge of the stages of grief is extremely helpful in the use of this skill.

ROLE PLAY

In a role play with a fellow student, one person will play the client and the other will be the helper. Your task as the helper is to do everything you can do to build an effective relationship with this client through practicing the specific skill you choose from our list of core skills. Attend to the client's behaviors both verbally and nonverbally. After about 10 to 15 minutes of practice, stop the role play and discuss how each position felt. What was it like to be the helper? The client? How did the practitioner do at rapport building? What were his or her strengths and limitations?

The role play can be interchanged for practice with any of the core skills listed. Select the skill you want to practice and proceed with the role play. Another variation is to pick a role play and practice it without telling the class what you are trying to demonstrate. Then perform the role play in front of the class to see whether they can tell which skill you are highlighting.

Developing Treatment Interventions

Now that the foundation has been laid, practitioners can proceed to the next phase of helping: developing treatment interventions. Whether working in a traditional clinical setting or engaging in some other form or clinical work with a client, helpers must be able to develop both a therapeutic relationship and appropriate treatment interventions if the outcome of the helper-client interaction is to be effective. Most of the models for developing interventions consist of stage-based approaches. Almost all contain levels similar to the five major stages of the counseling process. We believe these should be incorporated in any of the treatment modalities we discuss later in the text. The five stages are these:

1. *Relationship-building stage:* developing the foundation for a sound collaborative working association
2. *Exploratory stage:* examining and understanding the client and his or her frame of reference
3. *Decision-making stage:* formulating a counseling goal and an intervention strategy

4. *Working or implementation stage:* expending effort to ameliorate the situation or solve the problem

5. *Termination stage:* concluding the counseling process (Doyle, 1992)

The Importance of Assessment

Like establishing the helping relationship, assessment is a critical and fundamental process of clinical practice. Hepworth et al., (1997, p. 194) emphasize the global importance of this skill in their view that "assessment is the basis for contracting, goal setting, and intervention planning. Indeed, the effectiveness of selected interventions, and ultimately the case outcome, depend in large measure upon the accuracy of the assessment."

A thorough assessment can help counselors resist the impulse to formulate a premature hypothesis about their clients. Without obtaining the necessary information about the presenting problem/concerns; its duration; precipitators; social, family, psychiatric, and medical history; and current stressors, helpers will not truly understand their clients or become aware of possible underlying dynamics. An inaccurate or incomplete assessment will likely result in inappropriate goals and interventions that may ultimately hinder positive changes in the client. Assessment has been defined as "the process occurring between practitioner and client in which information is gathered, analyzed, and synthesized into a multidimensional formulation" (Hepworth et al., 1997, p. 194).

In the past, assessment was referred to as *psychosocial diagnosis* (Hollis, 1972). However, because many practitioners objected to the term *diagnosis* and the negative connotations attached to it, the term *assessment* is preferred today, with its positive and comprehensive focus. In an assessment, in addition to learning what problems exist, the helper also gains insight into existing "resources, strengths, motivations, functional components, and other positive factors that can be used in resolving difficulties, in enhancing functioning, and in promoting growth" (Zastrow, 1995, p. 76). Three major components of a thorough assessment include (1) determining areas of concern, (2) assessing the client's strengths, and (3) learning about the client's history.

Determining areas of concern usually occurs during the initial phase of treatment when the helper inquires about what led the client to seek treatment. At this juncture, it is important to emphasize the positive nature of seeking professional guidance and to refrain from using the term *problem* as much as possible. In this respect, we concur with Weick, Rapp, Sullivan, and Kisthardt (1989), proponents of the strength perspective. They assume that helping can proceed successfully from the identification and enhancement of individual strengths. The emphasis is on clients' strengths and their ability to use them, together with resources from family, friends, and the community, to overcome obstacles and proceed on a positive path. This philosophy opposes diagnosing and identifying problems. Weick et al. (1989) view a "problem with problem focus" and identify several ways in which such a focus may impede change and growth. First, engaging in a problem-oriented assessment encourages individualism rather than sociocultural explanations of human difficulties. Second, talking about "problems" tends to insinuate that people are the cause of their own difficulties, which further reinforces their deficits rather than their strengths. Third, treatment is circumscribed when problems are the main focus.

Based on these considerations, we encourage helpers to focus on client *concerns* while stressing the value of clients' active participation in the helping process. Emphasis is on clients' current coping skills, past successes, and existing strengths. In addition, a concentrated effort is made to define present areas of difficulty and work toward setting goals to alleviate these.

Assessing the client's strengths is a key part of the assessment process. Helpers identify strengths by asking specific questions that elicit positive feedback from clients about their qualities. This approach not only aids in the development of a positive relationship but also enhances the client's confidence and self-esteem. Additionally, it is important to acknowledge clients' diverse backgrounds and to remain sensitive to the possible obstacles presented by cultural views on seeking professional assistance.

Learning about the client's history is also important. To understand a client well, a helper must learn as much as possible about the person's history and life as a whole. Gathering information about the client will assist the helper in forming a sound impression and developing an appropriate treatment plan. We encourage a biopsychosocial approach that allows a comprehensive picture to emerge.

Assessments conducted at different types of agencies will vary in nature. At times, helpers make independent assessments; at other times, they are part of a team working together to assess the client. In either case, the more information that can be gathered about the client, the better he or she can be served.

Cormier and Cormier (1998) cite six purposes of assessment originally proposed by Paul (1967, p. 111) that are specific to counseling and therapy. We present these here as they add insight into the value of the assessment process for anyone engaged in clinical practice.

1. To obtain information on the client's presenting problem [concern] and on other, related problems [concerns]
2. To identify the controlling or contributing variables associated with the problem
3. To determine the client's goals/expectations for counseling outcomes
4. To gather baseline data that will be compared with subsequent data to assess and evaluate client progress and the effects of treatment strategies
5. To educate and motivate the client by sharing your views of the problem with the client, increasing the client's receptivity to treatment, and contributing to therapeutic change through reactivity
6. To use the information obtained from the client to plan effective treatment interventions and strategies

Data used in an assessment are derived from a variety of sources; when combined, they allow a more complete view of the client and his or her situation to develop. These sources include the client's verbal report, self-assessment forms, collateral sources, psychological tests, nonverbal behavior, interactions with significant others, home visits, and worker's intuition from direct interactions (Zastrow, 1995, p. 78). Although all of these may not be used at any one time, efforts to compile as much information as possible will enhance the overall assessment.

Knowledge of human behavior, the social environment, and diversity issues are also key ingredients in conducting assessments. Frequently, the purpose of the interview determines the factors to be examined; it will also influence the areas of knowledge the helper needs to master. Conversely, the approach, theories, and techniques the helper applies may affect what he or she explores during the assessment phase.

Assessment is sometimes a product and at other times an ongoing process. It is a product when it is a formulation at a single time about the nature of a client's difficulties and resources (e.g., a mental status assessment completed in a psychiatric setting). In the second instance, an assessment is said to begin at the initial interview and to continue until termination of the case. As clients and their circumstances change, the assessment will change. Often, with increased trust and rapport, clients may be willing to share new information with the helper that will require a revision of the initial assessment that was made.

Last, as part of the assessment process, helpers have an ethical and legal obligation to assess the urgency of client needs as well as the priority of these needs (see Chapter 7 for details about ethical and legal issues). Cultural differences between client and helper must also be taken into consideration. In some instances, a client may show resistance or reluctance in accepting help; however, further exploration may show that this reluctance reflects a cultural norm that discourages or criticizes seeking help.

When helpers perform assessments, they often use existing forms as guides. There are so many assessment forms that it would be impossible to include even a small percentage here. We do, however, provide you with several examples (found in the appendix) that may be helpful as you familiarize yourself with the assessment process. The first is a specific format titled *Initial Evaluation* that is a compilation of many assessment forms we have used through the years while working in the mental health arena. We offer the long, detailed version followed by an outline form that is shorter. The second sample form we present (also in the appendix) is a shorter assessment format developed specifically for work in managed care and HMO settings. It is titled *Initial Intake* form.

With interviewing and assessment skills in place, the practitioner is able to begin his or her actual therapeutic work, which requires a theoretical base as a guide.

The Role of Theory in Practice

Theory is derived from practice; in turn, it directly affects practice. Webster (1984, p. 1223) defines a theory as "an exposition of the general abstract principles of any science or humanity which have been derived from practice; a plan or system suggested as a method of action." This definition illustrates the role of theory in practice. In this context, a theory can be seen as a way of assisting helpers to organize their ideas and guide them in the development of appropriate treatment plans.

Neukrug (1994, p. 57) contends that "regardless of whether you adhere to an individual or systems approach, you must have a theoretical base with which to approach your clients. Otherwise, everyone would be 'just doing their own thing' and there would be no rhyme or reason to client interventions."

On a similar note, Hansen, Stevic, and Warner (1986, p. 16) believe that trying to function without a theory is "to operate in chaos, for without placing events in some order it is impossible to function in a meaningful manner." In accord with these views, we also value the need for a theoretical base that serves as a foundation for the philosophies and techniques we use in our practice with clients. Some helpers may be particularly fond of a specific theory that directs their practice; others may prefer to draw from a variety of theories depending on the client population they are working with. Whatever your preference, we encourage you to remain open to exploring the many facets of helping that are often deeply rooted in theory.

We have found that willingness to learn about theoretical perspectives is well worthwhile. At times, learning new theories or refreshing our knowledge of those we already know has inspired and excited us to continue helping clients in new ways. In later chapters when we focus on micro, mezzo, and macro levels of practice, we will examine various classic and contemporary theories that are relevant to the specific level of practice that is being considered. We encourage you to keep in mind our views on the role of theory in practice at that time.

Inevitably practice can be quite rewarding but also quite challenging. We address some of the challenges in the following section.

CHALLENGES OF PRACTICE

Being an ethical, skillful, and caring helper requires you to learn to face a multitude of challenges—some obvious and easily dealt with, others elusive and difficult to confront. The challenges we have found most prevalent in our own professional development are grouped into three main categories: personal, organizational, and environmental. Even so, overlap among these is common. We deviate from a more traditional approach by introducing these issues at the beginning. Because of their importance and the high likelihood that you will encounter them early in your professional development, we present them now when they can be of immediate use to you.

Personal Challenges

Personal involvement is one of the greatest assets human service practitioners possess, allowing us to understand someone else's pain and struggles as well as their joys and excitement. The ability to empathize contributes to developing trust and rapport with clients. In this personal involvement, however, there is a risk of becoming overwhelmed or disillusioned. Below, we discuss the personal challenges we have experienced ourselves or have witnessed in our students and colleagues.

Beginning Fears

Beginning workers, often students in training, feel anxious about the unknown, overwhelmed by their lack of knowledge, and pressured by the demands of the agency they are placed in and the educational institution they are enrolled in. If you are in this position, you may be telling yourself that you should know it all and be able to help everyone you come into contact with. Do not be overly hard on yourself or create undue stress. Not even the most experienced professionals can help everyone. In fact, seasoned practitioners realize that they are able to help only those who are able and willing to change. We believe that the helper's desire and skill must be paired with the client's willingness if change and growth are to occur. Acknowledge your limitations and seek guidance when necessary. Experience will help you develop a broad knowledge base and allow you to refine your skills. Remember the adage: "The more you know, the more you realize how much you don't know."

Typical Difficulties and Obstacles Facing Beginning Professionals

Beginning professionals often believe the fallacy that they are responsible for their clients' progress. If you accept this idea, it will lead to much self-criticism and disappointment. One way to avoid this pitfall is to be clear from the outset about your responsibilities and those of your clients. These will vary depending on the agency, the type of clients you work with, and your overall helper role.

A transition often occurs when beginning practitioners leave the safety of their educational setting for work in the real world. Even though you are looking forward to this, be aware that the switch from a structured and somewhat sheltered life to the many challenges of the real world can be difficult. When you are truly on your own, with no educational supports to guide you, you may feel frightened, disappointed, and frustrated. The major task will be to make the transition from having an external structure provided by the educational program to having to provide your own structure from within.

The transition process can be traumatic. Caseloads can quadruple, the demands triple, and the expectations double. As an intern, you may have had time to give 100% to each case and may now find yourself focusing primarily on the most pressing cases and less on the least urgent ones. This can be overwhelming and discouraging to those who have always taken pride in giving their all to every client. Time management skills and prioritization can help you survive. Setting limits and being realistic can help you make the transition smoothly.

Woodside and McClam (1994, p. 170) highlight the stressors involved in the transition from student intern to professional:

> It is typical for entry-level professionals to be totally consumed by their jobs, spending long hours at work and worrying about their clients during their off-time. It is also common for beginning professionals to start work with high job satisfaction, which decreases when realistic assessments of the work replace earlier dreams of what the job would be like. If the transition is successful, beginning professionals emerge with a more stable sense of professionalism.

Overall, we encourage you to remain realistic about your role as a beginning helper. A career in the helping professions is a lifelong process. Allow yourself time to settle into your new role, and recognize that even as a professional, you may need guidance and support.

Balancing Family Life and Career   On the surface, balancing family life with your professional career may seem simple enough. However, this balancing act is often far more complex than it first appears. Every professional has a personal style and boundaries for how and when their career is allowed to influence their personal lives. Visualize a continuum: On one end of the spectrum, you will find professionals who keep their careers totally separate from their family life. At the other end of the spectrum are professionals who live their professional lives both at home and at work. In between are people who maintain more of a balance between the two.

Certainly, individuals differ in their comfort level and attitude in this regard. Depending on the circumstances, many human service practitioners fluctuate in how much they allow their careers to be part of their family life. Some believe it is possible to integrate career with family if this is done carefully and constructively. However, doing so is not easy. It requires ongoing self-evaluation and adjustments.

Keeping Personal and Work Issues Separate   Keeping personal issues apart from work with clients is often a true challenge. Entering work, we are still the same individuals we were earlier that morning—we might have received a speeding ticket, had a flat tire, argued with our children about their clothes, or awakened an hour late. We may subconsciously bring such frustrations to our work.

It can also be a challenge to keep from taking your professional concerns home. You may bring excessive paperwork from the office and find yourself consumed by it rather than interacting with your family or engaging in leisure activities. Agency demands may at times make it inevitable that work will affect your home life. It may be helpful to set specific times to do paperwork and give yourself a deadline.

Another common experience is to feel physically exhausted after a long day of helping others. Ongoing emotional investment can be physically draining. You may need to develop some healthy after-work activities to reenergize yourself. You may find yourself worrying about a particular case and allowing it to affect your home life. Before going home, find a supportive colleague, spouse, or significant other to talk with. This can minimize the tendency to ponder the day's events and may keep your worries or concerns from interfering with your family life. Sometimes the drive home from work can be a nice transition time, the bridge you need to help you unwind. You might listen to your favorite music or plan what to do when you get home—which can help take your mind off work.

Caretaking and Rescuing

The human service practitioner is often naturally adept at caretaking and rescuing. Caretaking is certainly a natural and healthy aspect of helping, a strength that allows the helper to be empathetic and motivated. However, it has a counterproductive side. The worker may be trying to avoid painful encounters by caring for clients in ways that minimize experiencing the full range of clients' emotions. Clients often seek help with the assumption that the helper will take care of them by giving them advice or pointing them in a specific direction. Depending on the situation, this may or may not be appropriate. Long-term gains may be sacrificed when a short-term solution is simply handed to the client. Be aware of your role and the goals of your work.

Rescuing is a form of caretaking. Some types of rescuing are obvious, such as wanting to take an abandoned child home. Subtler forms include discouraging clients from expressing sadness, pain, hurt, anger, or sexuality and avoiding discussion of such topics. The reasons for this avoidance may be the helper's own difficulties in expressing these emotions. In fact, change usually occurs when people explore and express their painful emotions and deep feelings, not by shutting them off. When painful emotions are suppressed, depression may develop. When feelings are left unexpressed, they may later surface in a pathological form.

If you find yourself experiencing difficulty when you encounter pain or anger in clients, or if you struggle with sexual issues, perhaps you need to engage in self-exploration. Seeking supervision or therapy is often advised in such cases.

Wanting to Be Liked

A most difficult challenge is the helper's need to be liked by clients and co-workers. Wanting to be liked is not a negative reaction, but it carries an inherent danger. If a helper wants too much to be liked, he or she may neglect to confront a client when appropriate or may discourage the expression of painful or difficult feelings. Certainly, alienating a client or co-worker is not recommended, but successful human service work does not mean being popular with everyone. Our work at times requires us to deal with raw emotions and people in crisis. Clients may feel vulnerable when you lead them to express strong feelings they might normally keep subdued. A crisis can be a dangerous event as well as an opportunity for growth (Parad, 1965; Hoff, 1989). The crisis state may be most conducive to change because of the client's vulnerability. Clients may be more receptive to looking at themselves and moving toward positive change than they were before the crisis. Although this process may have a positive outcome, it is likely in the short term to produce pain and other strong emotions. The helper is often the catalyst that encourages the expression of these painful feelings; thus he or she may not be seen favorably at the time by the client undergoing therapy.

We encourage practitioners not to please their clients at the risk of failing to give honest feedback. Our role is not merely to put out fires or to please everyone we come into contact with; we must act as change agents. Clients who may not respond favorably to our interventions at the time may experience benefits years later. We often plant seeds that may not flourish immediately; often, they come up later without our knowledge.

Transference and Countertransference

Transference and countertransference are issues you will undoubtedly face as you begin fieldwork or agency employment. Initially analyzed in Freudian psychoanalytic theory (Corey, Corey, & Callanan, 1993; Nugent, 1990), transference and countertransference are likely to exist in any helping relationship. Many authorities say they are universal in all relationships—social or therapeutic (Gelso & Carter, 1985; Brammer & Shostrom, 1982). "To varying degrees, transference occurs in probably all relationships. The therapeutic situation, however, with its emphasis on a kind of controlled help-giving, magnifies and intensifies this natural reaction" (Gelso & Carter, 1985, p. 169).

Transference   Transference is considered to be an unconscious process "whereby clients project onto their therapist past feelings or attitudes they had toward significant people in their life" (Corey & Corey, 1998, p. 91). Furthermore, "In transference, clients' unfinished business produces a distortion in the way they perceive and react to the therapist. They are rooted in past relationships but are not directed toward the therapist" (Corey et al., 1993, p. 41). The feelings typically involved in transference range from love, lust, high praise, and regard to anger, ambivalence, hate, and dependence. "Unrealistic perceptions of and reactions to a practitioner are known as transference reactions; that is, the client transfers to the practitioner, wishes, fears, and other feelings that are rooted in past experiences with others (usually parents, parental substitutes, and siblings)" (Hepworth et al. 1997, p. 561).

It is not uncommon for transference to occur in the helping process, no matter what theoretical orientation the practitioner assumes. According to Cormier and Cormier (1998, p. 49), transference commonly occurs "when the emotional intensity has become so great that the client loses his or her objectivity and starts to relate to the counselor as if he or she were some significant other person in the client's life." It can occur regardless of gender and may develop as a response to old unmet needs, triggered by the presence of an empathetic helper (Kahn, 1991; Kohut, 1984).

In our role as clinicians, we can certainly attest to the existence of transference in therapeutic relationship with our clients. We have also witnessed transference in nonclinical settings. By a broader definition of transference, the client's feelings and attitudes are directed toward the helper regardless of whether the feelings are rooted in early childhood. All helpers must be cognizant of the impact transference can have on the helping relationship. Some examples of transference are listed below:

Examples of Transference

- As a caseworker in an adult care center, you are faced with several seniors who treat you like a daughter and tend to infantilize you.
- You are working in a child care agency where several children are especially attached to you and often call you "Mom."
- As a probation officer at juvenile hall, you have encountered several teenagers who look up to you as a father figure. They alternately test your limits and ask you for fatherly advice.
- You are a career counselor for women reentering the workforce. You often come across women who identify closely with you because of your history of returning to school in your 40s. Some seem to idolize you and see you as a perfect person.

Transference can be positive, negative, or neutral. If not handled properly, it can impede progress in the helping process. Also, given the nature of transference, overgeneralizations and distorted perceptions are likely; when this happens, difficulties in other interpersonal relationships may result.

Hepworth et al. (1997, p. 263) list common manifestations of transference that we have ourselves experienced from clients in our practice, including these:

- Having a clinging, dependent relationship with the practitioner
- Needing excessive praise and reassurance from the helper
- Making excessive efforts to please the clinician, which may range from compliments and praise to offering personal favors or gifts
- Asking excessive personal questions of the helper
- Behaving provocatively; arguing with or baiting the therapist
- Questioning the interest of the helper
- Seeking special considerations of the practitioner
- Attempting to engage the practitioner socially through lunch, party, or other social invitations
- Dreaming or fantasizing about the therapist
- Making defensive responses to the practitioner
- Feeling rejected; considering therapist's responses unwarranted criticism or punishment
- Behaving seductively; wearing revealing clothing or making affectionate gestures
- Resorting to regressive or destructive behaviors when upset with therapist
- Being silent, inattentive, or dozing off in sessions
- Being late, or pushing for time to be extended

Corey and Corey (1998, pp. 91–93) also offer examples of transference situations: "Clients who make you into something you are not; clients who see you as a superperson; clients who make unrealistic demands on you; clients who displace anger onto you; and clients who easily fall in love with you."

Because transference is often subtle, it is important that you become adept at recognizing it in your client's behaviors. Once you have become aware of its existence, then can you proceed with appropriate interventions to deal with it. Furthermore, you must also become aware of your own needs and motivations that left unrecognized might result in your becoming tied "into your clients' projections and ... lost in their distortions" (Corey & Corey, 1998, p. 93). Paying attention to your client's feelings can give you key insights into how they interact with others in their social sphere; exploring these further in treatment can be beneficial.

Transference in a Clinical Setting   There is no consensus about appropriate responses in cases of transference. Generally, the interpretation and analysis of transference depends on the nature of the therapist's role and the settings. If the helper is in a clinical position, it may be therapeutically necessary to interpret, analyze, and work through the transference. By recognizing and working with this phenomenon, helpers can gain a better understanding of their clients' conflicts and therefore develop a tailored approach to helping.

Sometimes, transference reactions in therapeutic relationships can symbolize an opportunity for growth whereby the practitioner is challenged to "assist such clients to recognize their distorted perceptions and to develop finer interpersonal perceptual discriminations so that they can differentiate and deal with the practitioner and others as unique individuals rather than as overgeneralized projections of mental images, beliefs, or attitudes" (Hepworth et al., 1997, p. 561).

In many instances, when transference issues arise, they can be therapeutically dealt with by helping clients to recognize that what they expect of the clinician they may also expect of other people in their life. Also, practitioners can work with transference in a helpful way "by empathetically reflecting on the client's desires or wishes—for example, the wish to be loved, the wish to be important, the wish to control, and so on" (Cormier & Cormier, 1998, p. 49).

> ### SCENARIO
> Jamie is a 24-year-old female of mixed descent; her father was born in Italy and her mother is American. She enters treatment with you, a middle-aged male licensed clinical social worker from an Italian family. Her presenting concern is her passivity, especially in heterosexual relationships. Gradually as she begins to trust you, she becomes more easily irritated with you—sometimes, for no apparent reason. When you question her about her frustrations toward you, she is not even conscious of them and becomes embarrassed. In time, when the focus turns to her unresolved issues with her father who was never emotionally close to her, you are able to make the connection between her struggles with her father and your resemblance of him. Although you are much more empathetic and emotionally available to her, she seems to feel safe enough to transfer her feelings toward her father onto you. How might you proceed in working with Jamie? Would you make this connection for her? Would you interpret the transference?

Transference in a Nonclinical Setting   Transference can occur in helping relationships that are not clinical in nature. In situations where there is not a therapist-client relationship, it may be necessary only to understand the transference issues. In fact, in some settings, it is considered both unethical and inappropriate to explore the nature of transference with clients. Such interpretations may lie beyond the scope of the services offered and the competencies of the helper.

Regardless of whether you are a clinician, case manager, crisis or outreach worker, or group home caseworker, you are not immune to transference reactions by your clients. Deciding how to proceed with transference requires careful consideration.

> ### SCENARIO
> You are clinical manager in an agency where your role is to supervise staff on their casework and to oversee the agency functioning. One of your staff is easily angered by

any comment you make about his performance, no matter how constructive or positive it may be. You feel increasingly hesitant to even approach this staff person because of his overreaction to you; however, there are times when you need to discuss issues tied to his job performance or agency changes. Why do you think there may be some transference? How might you determine this? How might you proceed?

Countertransference   Countertransference is considered "the other side of transference" or "unrealistic reactions helpers have toward their clients that may interfere with their objectivity" (Corey & Corey, 1998, p. 95). Countertransference occurs when the therapist or helper projects unresolved conflicts onto the client (Nugent, 1990, p. 74). Simply, countertransference includes feelings and attitudes the practitioner has about the client. "They may be realistic or characteristic responses, responses to transference, or responses to material and content that trouble the counselor [helper]" (Cormier & Cormier, 1998, p. 49; Khan, 1991).

Some see countertransference in broad terms, as any projection that could conceivably interfere with helping a client. Examples may include "counselor anxiety, the need to be perfect, or the need to solve a client's problems" (Corey, Corey, & Callanan, 1993, p. 43). Another viewpoint by Brockett and Gleckman (1991) contends that countertransference can involve any of the helper's feelings about his or her clients. We favor this perspective, in which the term *countertransference* encompasses all the thoughts and feelings a helper has in reaction to his or her clients, whether prompted by the clients themselves or by events in the helper's own life.

Countertransference, as with transference, can take all forms, ranging from positive to negative. Because the danger of countertransference can be great if left undetected or if not dealt with appropriately, it is often viewed as potentially the more negative of the two. According to Hepworth et al. (1997, p. 565), "This phenomenon [of countertransference] involves feelings, wishes, and unconscious defensive patterns of the practitioner that derive from past relationships, interfere with objective perception, and block productive interaction with clients." It can contaminate helping relationships by "producing distorted perceptions, blind spots, wishes, and antitherapeutic emotional reactions and behavior."

"Hurtful countertransference," so named by Cormier and Cormier (1998, pp. 49–50), can occur when practitioners are (1) blinded to an important area of exploration, (2) focus on issues "that are more their own than their clients', (3) use clients for "vicarious or real gratification," (4) use "subtle cues that 'lead' the client," (5) make interventions not in the client's best interest, and most important, (6) "adopt the roles the client wants [them] to play in his or her old script."

Working with clients often sets off our own feelings. Old wounds that have not yet healed may be perfect triggers (Corey & Corey, 1993, 1998), consciously or subconsciously. If you are unable or unwilling to explore your own issues, or if you avoid certain feelings, some facet of your work is likely to be affected. Certain themes—such as separation, loss, dealing with anger or assertiveness, or denial of sexuality—may surface through your work with clients.

Chiaferi and Griffin (1997, p. 50) add that

> common reactions include a need for approval, identification with the client, sexual and/or romantic feelings toward a client, a tendency to refrain from confrontation, or a compelling need to rescue the client.... These reactions become the subject of ethical consideration when they are intense, persistent, and compelling, because their presence compromises efforts that are otherwise well founded.

Learning to recognize these themes and risking introspection is clearly advantageous for both your client's well-being and your own. Certainly, these views illustrate the importance of becoming aware of countertransference in ourselves and learning appropriate techniques to manage it. We believe, however, that countertransference is not necessarily harmful or negative in and of itself; rather, how it is handled or not handled is the element that can be potentially damaging to the therapeutic process. We concur with Corey and

Corey (1998, p. 95) that helpers can use all their responses to their clients in therapeutic ways, assuming that they become aware of the sources of their countertransference. Khan (1991, p. 127) considers countertransference to be useful in that it can be a "generator of empathy" when the helper is aware of it and can achieve "optimal distance" from the feelings creating the countertransference reaction. "Understanding countertransference can be of tremendous benefit" (Chiaferi & Griffin, 1997, p. 50).

The Identification Process   According to Watkins (1985), an identification process occurs between helper and client and influences the development of countertransference reactions. Being able to identify with clients in their experience is essential for helpers in having a positive working relationship. When there is optimal identification, the helper can understand the client yet still maintain an appropriate distance. Conversely, when such identification does not take place, the relationship may be compromised and the helper's ability to empathize is reduced. A helper's reactions or feelings toward a client are therefore affected by how well he or she is identifying with the client. Several factors help determine the extent of this identification:

- *Values*: Are your values similar to the client's?
- *Demeanor*: Is there a similarity in your attitudes and beliefs?
- *Language*: Do you share a common language?
- *Physical Appearance*: Is there a discrepancy in physical appearance between you and your client? Could this be a problem?
- *Expectations*: Do you share similar goals and expectations?

When helpers find themselves at either extreme of the identification process, destructive countertransference patterns can develop, significantly impairing the helping relationship. Harmful effects include the erosion of trust and a *unidirectional interaction* between helper and client. There are various ways helpers can minimize potential problems between themselves and clients, including self-analysis, personal counseling, supervision, genuineness, self-disclosure, and referral when appropriate.

Positive and Negative Countertransference   Both positive and negative feelings can be projected onto clients or felt by the helper (Nugent, 1990). The following list gives some examples of each type.

Positive	*Negative*
You look forward to seeing certain clients.	You feel unexplained anger and resentment toward certain clients.
You feel overly attached to certain clients.	You dislike a certain client.
You enjoy being with a specific client.	You feel annoyance toward a client.

Clearly, no one is exempt when it comes to countertransference. It is normal and expected to have feelings toward clients. Corey and Corey (1993, pp. 139–140) cite the following as possible indicators of countertransference:

"Let me help you."
"I hope he or she cancels."
"You remind me of someone I know."
"You are too much like me."
"My own reactions are getting in the way."

A more extensive list of typical ways in which countertransference is often manifested is offered by Hepworth et al. (1997, p. 565):

- Excessive concern about clients
- Persistent erotic fantasies or dreams about a client
- Dreading or looking forward to seeing certain clients
- Consistent tardiness or forgetting appointments
- Excessive protection of clients that may be expressed by avoiding certain topics in treatment
- Hostility or the inability to empathize
- Biased views seen in blaming others exclusively for client's difficulties
- Persistent boredom or drowsiness
- Ending sessions too early or letting them go beyond set time limits
- Trying to impress or being impressed by clients
- Excessive worry or concern about losing a client
- Arguing, reacting defensively, or feeling hurt by clients statements
- Being overly responsive, doing too much
- Unusual curiosity about a client's sex life
- Struggling to like certain clients

Experiencing any of the above is natural. Countertransference can occur in any type of helping relationship, whether clinical or nonclinical; thus, it is important to be aware of your feelings. Asking yourself the following questions may give you added insight into how you can best deal with your feelings.

1. How strong are these feelings? If they persist, it may be necessary to take a look at underlying dynamics and seek supervision and consultation.

2. Is your objectivity impaired? Countertransference becomes a problem when the helper's own reactions stand in the way of dealing with the client's problems.

3. How are these feelings and reactions dealt with? Are they handled in accordance with strict ethical and legal guidelines? When the feelings are not dealt with appropriately, problems are sure to arise.

We have also found it useful to have specific guidelines for coping with these feelings that are noted in the treatment considerations listed below:

Treatment Considerations for Working with Countertransference

- Work toward acceptance of your feelings. They may contain significant information about your clients or about yourself.
- Be careful to not judge or disregard your own feelings.
- Learn to utilize these feelings and reactions appropriately to enhance the helping process. Gain insight by examining your countertransference feelings. This can help you understand the client and develop more appropriate interventions.
- Consult with professional colleagues or supervisors if your feelings are interfering in your work with your clients. A more objective viewpoint is helpful and at times necessary to guide you in the right direction. Sometimes a decision to refer a client may be the most appropriate action.
- Engage in counseling or therapy of your own if extreme countertransference reactions persist.

SCENARIO

You are a new clinician at a family service agency where you work with children, adolescents, and adults utilizing individual, couple, marital, and family therapy modalities. One of your first clients is a female about your same age who presents with great ambivalence over her current relationship with a long-term boyfriend. The more she talks about her conflicting emotions and questions her decision to continue in the relationship, the more you become overwhelmed and unable to offer any constructive feedback. Later, after the client has left, you spend time questioning your own reactions toward her and suddenly realize that this client has triggered feelings in you about your own relationship struggles. Although you are not involved in a close relationship to anyone at the present time, her relationship issues are quite similar to a recent relationship you ended because of your strong ambivalence. How might you proceed with this client? Would you share this connection with your client? Why or why not? Please elaborate.

Often reactions toward clients are based on a reality and are not necessarily connected to the helper's issues. This is often the case when a helper encounters resistance in clients. In fact, resistant clients may very well be the ones who elicit the most negative countertransference in helpers. Next, we address the issue of resistance in some detail.

Dealing with Resistance

The term *resistance* conjures up negative, resentful, and anxious feelings. Brammer (1988, p. 52) considers the term as describing clients' "conscious or unconscious reluctance to begin a helping relationship as well as their covert thwarting of the goals of the interview once the process is underway." Resistance can thus occur before and during the helping relationship. However, when helpers view resistance as the deliberate refusal to cooperate, it may also reflect their own issues or excuses for not being able to help certain clients. Hepworth et al. (1997, p. 568) view this as unfortunate because some practitioners "have often exploited this definition by attributing failure of helping efforts to clients' resistance, when, in fact, failures were often caused by inappropriate or ineffectual helping methods." Further, resistance may be a result of clients' ambivalence to change, not necessarily their reluctance to change. Leader (1958, p. 22) views resistance as a natural part of living: "If we could only fully recognize that resistance in all phases of life is natural and healthy, perhaps we would be less concerned about its manifestations."

Corey (1996, p. 122) defines resistance more broadly as "anything that works against the progress of [treatment]. Resistance refers to any idea, attitudes, feelings, or action (conscious or unconscious) that fosters the status quo and gets in the way of change."

In terms of human services, resistance is defined by Meier and Davis (1993, p. 15) as an "obstacle presented by the client that blocks treatment." It can take on different forms, from open hostility to passive-resistive behaviors. Reluctant clients usually are resistant. Rooney (1992) classifies reluctant clients as follows:

- Those who don't seek help
- Those who don't respond to service offers
- Those who don't show up after accepting offers
- Those who minimally participate after showing up

Understanding the common reasons for resistance can help us be more sensitive and develop appropriate treatment interventions. We challenge you to consider ways to intervene when resistance exists for any of the following reasons.

Reasons for Resistance   *Fear of change:* "Clients are ambivalent about change, uncomfortable about the need to disclose painful or shameful things about themselves, or afraid to face what the [helper] is trying to help them face" (Nugent, 1990, p. 76). From a psychodynamic viewpoint, resistance is seen as an attempt to prevent anxiety-provoking material from entering the client's awareness (Hansen, Stevic, & Warner, 1986). The social learning viewpoint considers resistance to be related to clients' fears of change and the consequences that might result (Shertzer & Stone, 1980). Being aware of this fear can help us be more empathetic and perhaps less resistant ourselves.

Resistance as a defense mechanism: Resistance can be viewed as "defense against anxiety and as a protective coping mechanism" (Faiver et al., 1995, p. 91). Brammer (1988, p. 53) states:

Fear of confronting our feelings is a strong deterrent for seeking help. The pain must be great in most of us before we will admit that we need a helping relationship. We may try all kinds of substitutes like drugs, for example, first. It is important for helpers to recognize this ambivalent or mixed feeling in all helpees.

Resistance can also be viewed as an adaptive process. People have developed certain coping skills that have contributed to their survival. These may be internalized to such an extent that asking clients to change them may not be welcome.

Previous negative experiences: Unfortunately, some clients have valid reasons for their resistance, such as a previous negative encounter with a professional helper or agency (Corey & Corey, 1993). Resistance in this case makes sense; it is instinctive to avoid repeating a prior negative experience.

Cultural factors: Often, depending on the cultural perspective of the client, seeking help may be viewed as a sign of weakness, incompetence, disrespect, or disloyalty.

Authority issues: For various reasons, some people resist authority. They may therefore rebel or act out in a helping relationship in which the helper is seen as an authority figure. Many clients also have a need to be in total control and view the helping process as a loss of control. This may be a defense mechanism that has developed for a valid reason. Some clients may actually feel more powerful if they refuse or resist help.

Manipulation: Certain clients "are incapable of healthy relationships, and may use any relationship as a medium for manipulation and power struggles…[viewing] all other people as adversaries" (Brill, 1990, p. 242). Clients who are needy may develop a dependent relationship with a helper. They may not want help or change, but simply desire the safety and nurturance of the relationship. When asked to work on their issues, to change, or to consider termination or a referral, they may resist rigorously.

Depression: What appears to be resistance may in reality be clinical depression. Symptoms may include a negative, pessimistic outlook; apathy; lack of energy or desire; "paralyzed" behavior; a sullen attitude; self-deprecation; and inability to trust.

Involuntary clients: Clients are often resistant when they do not voluntarily seek the services of the helper but are court-ordered, coerced by significant others, or manipulated by threats of divorce by the spouse or of grounding by the parents. They are not likely to welcome your interventions with open arms. They may express their resistance through hostile behavior, through indifference or apathy, or by pretending acceptance (Brill, 1990).

It is possible to deal productively with resistant clients. The following list offers several strategies helpers can use in working with resistance.

Treatment Considerations for Dealing with Resistance

- Be accepting and strive to develop a respect for resistance, as it may serve a valuable purpose.
- Examine your own issues with regard to resistance.
- Be willing to explore the significance of the resistance and its inherent dynamics.
- Emphasize the development of a trusting relationship. Little will be accomplished without building a foundation.
- Address the resistance directly and empathetically.
- Be willing to use constructive confrontation, according to the following criteria: It must possess the elements of caring and concern, be descriptive, be specific, and be well timed. Point out resistive behaviors to the client, as well as other incongruent behaviors. Some clients may not even be aware of these incongruencies.
- Consider resistance as an opportunity for positive change and growth. Often the act of addressing and exploring the nature of the resistance can lead to important revelations.
- Respect the fact that change is not universally accepted or rigidly defined. Not everyone wishes to alter or modify his or her life as you might. Refrain from pushing your own agenda and expectations.
- Be willing to consider the resistance as possibly a result of poor communication.

- Be flexible, creative, and willing to work with the resistance.
- Settle back and listen to the client; this can open up new insights and awareness.

SCENARIO

A 25-year-old female client who has just completed a master's program in social work, specializing in child welfare, enters treatment because she feels overwhelmed about her new job as a childrens' protective service worker. She shares her frustrations about the many demands of her job, including the high caseloads, massive paperwork, court requirements, and resistant clients. You help her focus on areas where she can make an impact and to recognize her limitations. Also, you encourage her to focus on taking better care of herself. As treatment progresses, her apparent reluctance to take care of herself becomes more noticeable. You attempt to point out how she is often resistant to follow through on suggestions to minimize her stress, yet she becomes increasingly irritated and critical of your ability to help her. Based on these observations, how might you proceed in working with this client? What is the resistance about? How is it manifested? How might you help the client to recognize this?

Working with Difficult Clients

Hand in hand with resistance is the challenge of working with difficult clients. There is no universal definition of a difficult client; each of us will experience clients differently. Generally, however, certain types of clients tend to make the helping process difficult. Table 2.3 is an adapted version of Corey and Corey's (1993, 1998) analysis of difficult clients, along with suggested interventions.

SCENARIO

John is a 53-year-old divorced, Caucasian male who enters treatment because of a recent divorce and estrangement from his two teenage children. He is very angry at his ex-wife and blames her for the breakup of the marriage and the children's not wanting a relationship with him. You are very supportive of him, yet also point out the discrepancies he presents and encourage him to take a look at his own part in his problems. Despite your efforts, he refuses to consider his role and always responds with "yes, but" answers. You are beginning to wonder how to proceed with this client and seek supervision. What feelings and thoughts might you have? What suggestions might your supervisor have in working with this difficult client? What approach might you consider?

Organizational Challenges

Understanding the Organization

Most human service practitioners must work with an organization or agency in some capacity, so it is important for them to understand how these units function. When we discuss macro practice issues in Chapter 5, we examine organizations in depth; here, we only emphasize the importance of learning as much about them as you can.

Because agencies often work by informal rules and policies, it may be quite a challenge to understand the entire system. As in families, an agency system changes over time, with the loss and addition of different members. It has been our experience that key personnel in an agency tend to set the tone for the informal structure. Discovering these hidden rules and agendas can take much time, effort, and inquisitiveness, but the knowledge may be beneficial in the long run.

TABLE 2.3   Difficult Clients

Type of Client	Suggested Intervention
Involuntary clients	Be careful to not become apologetic or defensive. Place the responsibility on the client.
Silent or withdrawn clients	Be aware of the possible purposes of the silence: protection, lack of knowledge regarding the helping interaction, cultural issues, or intimidation. Explore the nature of the silence.
Talkative clients	Be aware that clients may feel intimidated, angry, or anxious. Point out body language. Discuss your own reactions in a nonjudgmental way.
Overwhelmed clients	Resist becoming overwhelmed yourself. Explore with the client the ways in which he or she becomes overwhelmed.
"Yes-but" clients	Such clients are seldom satisfied, so you may find yourself working too hard. If so, place the responsibility back on the client. Point out the behavior.
Blaming clients	Place the responsibility and focus for change back on the client.
Clients who deny needing help	Refrain from trying to convince such clients that they have problems. Encourage them, but refrain from pushing them.
Moralistic clients	Help them recognize how their judgments of others affect their relationships and may distance them from others. Ask them to indulge their tendency to lecture: Have them make up a very stern diatribe. This may reveal how they possibly have incorporated a critical parent.
Overly dependent clients	Explore how you might possibly be fostering dependence. Encourage individuation and separation in a supportive way.
Passive-aggressive clients	Explore the underlying dynamics. Be aware of how the client's behavior affects you. Share your reactions while avoiding making judgments about the behavior.
Intellectualizing clients	Recognize that intellectualization is often a defense mechanism, developed out of a need to insulate people from their deeper feelings. Refrain from insisting that these clients dig deeply into their feelings. Rather, gradually encourage their expression of feelings.
Emotionalizing clients	Ironically, these are similar to intellectualizing clients in the sense that they may also be defending against deeper emotions and have gotten stuck in their pain. Gradually encourage more cognitive reactions.

SOURCE: Based on Corey and Corey, 1998, pp. 103–110.

Fitting In

For beginning practitioners who are eager to effect change, taking time to fit into an existing agency structure may be frustrating and difficult. Allowing yourself to fit into an agency system is similar to the importance of building trust in a helping relationship. Both require helpers to be genuine, patient, and respectful of others (the client or system). This process may take more time than you anticipate; resist trying to fit in too quickly. You need time to orient yourself to the agency system, and the staff will also need to get to know you. A sense of belonging may not come immediately. Furthermore, issues you are not aware of may affect how staff members feel about hiring new personnel. Take time to learn about these nuances.

Working with Other Organizations

In addition to understanding your clients and your organization, you need to know about the community. Because the community involves new people and places outside your safety zone, exploring your larger surroundings can create anxiety and insecurity.

Agencies often have a specified catchment area, which may include a whole city or cities, rural areas, entire counties, or smaller segments of these. Learning about this catchment area can be valuable.

Knowing the community can provide important insights into the environmental and cultural aspects of the client population your agency serves. Learning about the service organizations in your community and their relation to your agency lets you build important ties that will help you network. Also, the history of the relationship between your agency and the community may provide you with useful data on how to encourage constructive affiliations. Conflicts between agencies may have previously limited any productive partnership. Knowing about this may prepare you to anticipate problems and help you devise effective intervention strategies. Clearly, you can be more helpful to your clients when you are able to call on community resources with whom you have developed a positive working relationship. Acting as an advocate for your clients requires great skill in communication. By creating positive community connections, you pave the way for such advocacy.

Among the organizations you may interact with are the local law enforcement entity, the department of social services, child protective services, counseling agencies, charitable organizations, chemical dependency programs, legal aid, shelters for battered women and children, homeless shelters, and others. Cultivating key contacts in such agencies helps you learn the most effective ways to refer or seek assistance for your clients. Minimizing the bureaucratic red tape is beneficial to everyone.

Also valuable are positive relationships with neighborhood groups, activist groups, and local political organizations. These contacts can help in fund-raising ventures, in obtaining support for your agency's continued services to the community, and possibly in formulating any new programs your agency is considering.

Organizational Limitations

According to Satir (1972), systems can be divided into two kinds: closed and open. "Closed systems are those in which every participating member must be very cautious about what he or she says. The principle rule seems to be that everyone is supposed to have the same options, feelings, and desires, whether or not this is true...An open system permits honest self-expression,... differences are viewed as natural, and open negotiation occurs..." (Satir, 1983, pp. 237–238). You may experience agencies that clearly fall into either of these categories, but most organizations have elements of both. Much or all of your professional life will probably be spent working within an agency, so learning to cope with organizational limitations will be an important asset in your development.

No organization is perfect; each organization, like each person, has its strengths and limitations. Systems that are rigid, inflexible, and bureaucratic may engender feelings of frustration in their workers. You might notice this by paying attention to how you feel when you are working at your agency. Are you frustrated and angry or depressed and down after spending a day at your placement? Then begin to pay attention to what organizational factors are influencing how you feel. Perhaps you have little chance for input within the organization. Maybe certain co-workers are antagonistic or resentful toward you. Do not ignore such warning signs: confront them directly within yourself and explore the best ways to deal with them within your agency setting.

Organizational limitations are most often discussed in terms of restrictions they place on the provision of services. For instance, many agencies now set rigid limits on the kinds of services they provide, the length of treatment, and the use of available resources. This can be exasperating. Such limits can conflict with the very values that are inherent in helping. They can be especially difficult for the beginning practitioner, who is set on making a difference and is unprepared to deal with such barriers. Do not give up; there is usually a way to help the client, remain faithful to the ethical codes of your profession, and still abide by agency policy. Flexibility and imagination are basic elements in overcoming obstacles.

Agencies often have certain role restrictions that serve as limitations as well. Some agencies may have strict policies specifying the treatments or training activities that student interns, bachelor's-level graduates, and nontenured staff may engage in. Although this may seem unjust, you would be prudent to explore the reasons behind such a policy before reacting. If, after such an inquiry, you still feel the policy is unjustifiable, you can explore appropriate channels to modify it. Again, learning the history behind such policies may give you clues on how to pursue constructive change.

Like all systems, organizations are influenced from inside and from outside. These forces have an unavoidable impact on your position within the system. Living in an organization, like living in a family, is a process of ongoing adjustment and adaptation to change. Brill (1990, p. 109) writes, "Workers are as much a part of many social systems as their clients, and as such are subject to controls and pressures of the social systems." When systems change, the worker must adapt as well. Sometimes the transition is easy; sometimes it can be an extremely challenging experience.

Considering new options and ways to cope with change is crucial to your survival in an organization. Although being upset and disillusioned by unforeseen or unwarranted changes is normal, dwelling on them can be counterproductive. Focus the majority of your time and energy on exploring ways to adapt to these changes or to modify them to meet your needs and to fit your style.

Environmental Challenges

Like individuals, agencies have a network of support systems they interrelate with. This network can be considered the agency's immediate environment. Also, the broader fiscal and political environment directly or indirectly affects the overall functioning of the agency.

Fiscal Realities

Coping with fiscal realities is an ongoing challenge in human service agencies. Many government and private agencies are being affected by budget cuts. The threat of layoffs looms over these agencies. Staff members are often demoralized when they are asked to work more and more while their pay and benefits stay the same or even diminish. According to Woodside and McClam (1994, p. 72),

> The trend of decreasing federal funding of human services has affected many state, local, and private, and non-profit agencies that previously depended on the federal government to support at least some of their services. Ideally, the community and the state would absorb costs and work with the private sector to develop ways of providing the needed services. In reality, all organizations involved in service delivery are struggling to get by.

During tough economic times, vulnerable population groups often seem to be blamed for the problems of the day. This can result in funding cuts for human services, which can greatly affect both the quality and quantity of the care offered. The client may become ineligible to receive help, and the worker's job may be ultimately eliminated. Available funding often fluctuates, depending on factors such as the nation's overall economy and the political climate.

In today's volatile economic climate, most human service agencies are being forced to implement budget cuts. This situation will probably affect you (if it hasn't already), directly or indirectly. Your agency may have had to reduce or limit services. Grant funding and new program development are also likely to diminish. Workers who specialize in community organization may be particularly demoralized by these dismal fiscal realities. How can a new human service worker, zealous to help others and effect change, cope with such economic stressors? This question has no simple answers. By keeping an open and positive mind and being creative in the use of existing funding, you can still offer effective treatment. Many helpers who have worked in the human service field for years have successfully survived repeated budget cuts at their agencies. Survival requires working together to weather the storm and creatively developing new ways to cope with demands and expectations that seem unreasonable or even impossible.

Political Realities

As human service practitioners, you may face a hostile political environment that you will have to learn to manage. At present, there seems to be a negative attitude at local, state, and federal levels of government about federal aid and funding of various social programs. As a consequence, the need for or benefit of providing certain services is questioned. Some social expenditures are labeled wasteful. This attitude breeds reluctance to fund existing programs or to create new ones.

New Technology

Even with the budget cuts felt by almost every agency, these organizations have taken part in some degree in new technology. The technology boom has been a major trend in the last decade, bringing computers, access to the Internet, copiers, fax machines, cellular phones, pagers, and modern medical equipment. Many workers are thrilled with these advancements and are naturals at learning how to use them effectively. Others are hesitant, resistant, phobic, or simply cannot afford them. By staying in tune with technology, human service practitioners can open communication channels with other disciplines and professions and become more efficient in their work. The benefits of learning new technology usually outweigh the negatives. One way to cope with such changes is to focus on how you can personally benefit. For instance, computer technology facilitates billing, payroll, and budgets; increases one's ability to plan, monitor projects, and remain accountable to others; improves statistical and tracking methods; aids in problem analysis and research endeavors; and simplifies record keeping. Moreover, the amount of time that use of technology can save is staggering.

Coping strategies for all these challenges will differ. Common to all, however, is the importance of open communication, flexibility and patience, and avoidance of negativity. Recognizing and allowing your feelings is the first step in developing useful coping strategies. A second step is being able to communicate effectively with others by maintaining open communication channels. Also, if you are flexible, patient, and willing to reevaluate your expectations and options continually, you will find it easier to face the challenges of working with others. Finally, if you hope to avoid burnout, resist being poisoned by others' negativity. Negative behavior is often contagious, so try to catch yourself as soon as you begin to engage in negative thinking.

Interpersonal and Professional Relationships

Being in human services, you must interact with many other workers at various levels; this can be as challenging as working with clients. Because these interactions often affect us directly, they can also impact our work with clients. Some interactions will be limited in time; others will be ongoing such as our relationships with clients, colleagues, agency staff, supervisors, administrators, professionals at other agencies, and the community at large. We discuss these in further depth to help you prepare for possible problems as they arise.

In working with others, your primary focus should be engaging in healthy relationships. In order for this to occur, we consider understanding and respecting different styles, the value of assertive communication, and successful conflict resolution.

Effective Communication and Conflict Resolution

Forming positive relationships is a foundation of the helping professions. Clients are more likely to benefit from the helping process if there is a strong working relationship with the helper. This also applies to relationships we have with other people at work, and with people we relate to outside the work environment. Enhancing these relationships is equally important as building strong bonds with clients. In fact, our work with clients can be greatly strengthened if we have constructive relationships at home and with our professional colleagues. Among the benefits of developing constructive relationships are these:

- Creation of a support system for venting and help with problem solving
- Increased cohesion, which decreases isolation and feelings of alienation
- Enhanced teamwork, allowing helpers to work together to benefit the client
- Reduced conflict, which counteracts tension and anxiety
- Increased creativity and a greater sense of accomplishment
- Camaraderie and closeness
- Increased ability to consider other options, viewpoints, and perceptions
- Expanded learning opportunities through increased contact with professionals within and outside your field
- Enhanced self-esteem and confidence in one's skills and ability to relate to others
- Heightened self-awareness
- Opportunities to develop skills in communicating and conflict resolution

The benefits listed last—communicating effectively and resolving conflicts constructively—are all-important. Relationships can be impaired if communication is poor, indirect, or unclear or if conflicts are not dealt with effectively. To avoid this, we emphasize the need to develop skills in both assertive communication and conflict resolution skills.

Effective Communication Skills   Earlier we discussed the importance of communication, attentive listening, and assertiveness with regard to interactions with both your supervisors and clients. These skills are also necessary in developing, maintaining, and enhancing personal and professional relationships.

You will remember that attentive listening, or active listening, occurs when the helper truly attempts to understand what the other person is saying. This requires developing new ways of responding so that people you are listening to feel understood and open to further interaction. Like attentive listening, assertiveness is also necessary for effective communication. Assertiveness involves verbal and nonverbal skills that help "maintain respect for oneself and others; satisfy one's needs; and defend one's rights, without dominating, manipulating, abusing, or controlling others" (Bolton, 1979, p. 12). Learning to be assertive in all areas of your life can benefit you personally. It may keep you from alienating others, and it can allow you to build healthy relationships. Many of the relationship problems helpers face are linked to difficulties in being assertive in their personal and professional lives.

An aspect of assertiveness is fully accepting responsibility in relationships. We define responsibility in this context as openly accepting your role in the relationships you are a part of. This includes a willingness to focus on your own thoughts, feelings, and actions, as opposed to blaming or attacking others for theirs. Maintaining a responsible stance facilitates positive relationships in which you are respected and valued.

Conflict Resolution Skills   Any two or more people in a relationship may experience conflict. No two people are exactly alike in their values, views, opinions, styles, and priorities. In the human service field, you will no doubt encounter some sort of conflict. Conflict is commonly seen in a negative light—to be avoided or quickly resolved. Surely, most of us would prefer to have little conflict in our lives, but conflict is a natural occurrence and is neither negative nor positive in and of itself. Weeks (1992, p. 7) writes,

> Conflict is an outgrowth of the diversity that characterizes our thoughts, our attitudes, our beliefs, our perceptions, and our social systems and structures. It is as much a part of our exis-

tence as is evolution. Each of us has influence and power over whether or not conflict becomes negative, and that influence and power is found in the way we handle it.

We agree that conflict, if handled properly, can actually be an opportunity for growth in certain circumstances. Managing conflict requires learning conflict resolution skills. These skills can enhance not only your work with clients, but also relations with others, both interpersonally and professionally. When conflict is delayed or avoided, the lingering effects may result in strained or failed relationships.

There are various types of conflict: conflict of emotions, value conflicts, and, conflict of needs. Conflicts arise from differences among individuals. Learning to accept such differences is much more constructive than expecting uniformity. Seven basic ingredients of conflict as described by Weeks (1992) include diversity and differences; needs; perceptions; power; values and principles; feelings and emotions; and internal conflicts.

Approaches to Conflict   The primary goal of most approaches to conflict is to work toward a resolution in a cooperative and appropriate manner. Self-awareness is the first step toward developing sound conflict-resolution skills. By looking inside ourselves, we can better understand our own part in the conflict and then work on corrective action.

One approach, developed by Bolton (1979, p. 218), is appropriate for human services and incorporates the basic elements of effective communication. It is a three-step process, intended to help people argue and disagree in a constructive manner that is "systematic, noninjurious, [and] growth-producing."

1. Treat the other person with respect.
2. Listen until you experience the other side.
3. State your views, needs, and feelings.

Conflict may engender feelings of fear and anger that must also be dealt with. A first step in most conflict resolution is to recognize the existence of these feelings. Accept fear and anger as human emotions we all experience. It is not the emotions that are negative, but rather the way they are sometimes inappropriately expressed. Understanding the underlying reasons for the feelings can help you decide how best to respond. Engaging in healthy anger management is also essential. This involves learning to work with your angry feelings in ways that lead toward appropriate resolution. Weeks (1992) lists parts of this process:

1. Learning to be in charge of your own anger
2. Learning how to receive anger
3. Learning how to express one's anger constructively

There is no way to rid ourselves completely of our angry emotions; they are a part of who we are. Potter-Efron and Potter-Efron (1995) state that anger is a natural human emotion that we need for survival. Anger may alert us that something is wrong and deserves our attention. Part of being in charge of our own anger is learning to identify our feelings when we first experience them. Expressing these feelings early, before they become unmanageable, enables you to stay calm rather than allowing your feelings to be in charge of you. Certainly, this can be a very challenging process.

Both assertive communication and conflict resolution are fundamental to healthy relationships. Learn to apply these concepts. Opportunities to practice are plentiful in daily interactions with others. Without a doubt, most of our relationships could be strengthened if we were to adopt these skills in our lives.

Personal Relationships

Human service professionals, in helping clients enhance their overall functioning and make necessary changes in their lives, are likely to witness human suffering, severe abuse, and heart-rending injustices. Involvement in such situations is bound to affect the personal lives of helpers in some fashion. The challenge lies in knowing how to cope with them appropriately.

The Effect of Helping Roles on Personal Life   Leaving at work your feelings and thoughts from a stressful day may prove to be very difficult. A certain client may have profoundly affected you; your work may have triggered an unresolved issue of your own; you may have felt overwhelmed, ineffective, and powerless to help someone. Often, you may not even be conscious of the impact until you are on your way home, or later when you are alone or interacting with family or friends.

Many helpers find that they are not suited to working with certain populations. Others can adapt to their initial difficulties, learning to work with clients effectively and managing their emotions successfully as well. Still others find that the helper's role can be most rewarding. The prospect of helping someone else can strengthen a person's self-esteem, which can contribute to improved interpersonal and professional relationships. Many helpers also report that working with disadvantaged people made them appreciate the positives in their own lives. Additionally, working with clients can temporarily take the focus off the helper's own emotional struggles.

Most of us find that our work as helpers affects us in many ways. With experience and practice, we become better at separating the emotions involved in our work from those we experience in our personal lives.

Impact on the Family   A helper's family can be greatly influenced, negatively and positively, by his or her work in human services. Intimate relationships are often those most affected. This is partly because of the deep level of trust that exists in such relationships, fostering an open environment in which sharing can readily occur. Relationships in which mates work in different professions can have both benefits and drawbacks. This is equally true when mates are both in the helping professions. In either case, the human service worker's experiences are bound to have an impact on family members, be it spouses, children, parents, or others. Family members can be sensitized to the plight of society's most vulnerable members and can vicariously enjoy the satisfaction of helping. On the other hand, the stress on the helper can lead to misunderstandings, disagreements, and heated conflicts with family members.

Helpers must be cognizant of the total impact of their work on their families. By engaging in attentive listening and careful observation, we can begin to notice if our roles as helpers are having a positive or negative influence on the significant people in our lives. If the signs are negative, we must make the necessary adjustments or changes so that our work will be less of a liability and more of an asset.

Relationships with Colleagues

Working with colleagues in a professional capacity requires sensitivity, open communication, and conflict-resolution skills. We often spend more waking hours at work than home, so our colleagues are often like an extended family. Therefore, we should strive to maintain positive relationships with our colleagues, who can offer support and professional consultation.

Benefits   Ideally, our colleagues are close friends who know us well, enjoy quality time with us, and share interests with us. However, only some colleagues are friends; others are primarily co-workers; and with others, you would prefer to have no involvement with at all. McMahon (1996, p. 181) makes an important distinction: We choose to spend time with friends and usually find these relationships fulfilling, but at work we generally have no choice who our colleagues are.

Regardless of what we might prefer, it is important for us to develop quality relationships with co-workers. This enhances our work environment and allows for increased cohesion and collaboration. In a positive environment, the agency can function more smoothly in its mission to help clients, and workers feel better overall about their roles.

Creating a supportive and cohesive environment is recommended. Although you may not choose your co-workers, you do have control over how you interact with them. When you encourage an environment of mutual support, the result can reduce stress, help prevent burnout, and enhance cohesion among workers.

Creating alliances and friendships   is also helpful. This may not always be desirable, but when it is possible, these alliances can provide you with added support and make your workday more enjoyable. Humor and playfulness in the workplace have been shown to release tension.

Nonetheless, in some instances, friendships in the workplace can lead to conflict. Differences in values and preferences involving work-related or personal issues may cloud the relationship. Healthy competition can result in jealousy. Therefore, you need to maintain healthy boundaries with friends in the workplace. Also, be open to engaging in conflict resolution early in such relationships to keep problems from developing.

Problems with Colleagues   Difficulties with colleagues create added strain in the workplace and can be difficult to cope with. Problems may result from power issues, personality conflicts, divergent views of the helping process, or agency policies and practices. When a cohesive and collaborative tone has been set, such problems are more easily resolved. When the staff is fragmented or split, however, and strong leadership is absent, problems are likely to escalate. Working can then be emotionally draining, and helpers may feel disillusioned and apathetic about their work. Clients may also be affected, indirectly: Helpers become preoccupied with their interpersonal conflicts with colleagues and neglect to focus on their clients.

Understanding, respecting, and learning to work with colleagues' different styles and personalities is certainly a challenge. Although you may not agree with your colleagues, it is important to respect them and accept their differences. An accepting atmosphere makes conflict less likely—and more easily resolved when it does occur. The principles of self-awareness, knowledge, and skill common to ethnic-sensitive practice (to be discussed in Chapter 6) can fruitfully be applied to our interactions with colleagues.

When colleagues are not amenable to our efforts to engage in appropriate conflict resolution, we must explore other options. The particular individuals may be difficult to deal with or may have personality problems that make them unresponsive to problem-solving techniques, compromise, or attempts at negotiation. In such cases, it may be more important to focus on your own needs and maintain effective involvement with your clients. McMahon (1996, p. 188) suggests focusing on taking care of ourselves and owning responsibility for what we do and who we are. We must not allow negative relationships to discourage us; rather, we can learn from them. Strive to maintain a positive attitude. If assertiveness does not work, explore ways to keep working with the difficult individuals without depending on any support, encouragement, or feedback from them. Scott (1990) advocates protecting yourself emotionally by learning to let go and not personalizing such conflicts. Others must not be allowed to hold you back from becoming the best helper you can be. Avoid negative people who won't or can't change, do not waste your energy on skeptics who will not listen, and do not allow others to manipulate you. Remain aware of your skills and capabilities, and keep your priorities in mind. Seek support and feedback from those you trust and feel comfortable with.

Assertive Communication and Conflict Resolution with Colleagues   Assertiveness with colleagues is not always easy. You may fear reprisals or increased conflict, and some colleagues would prefer that you remain passive. Assertiveness is nonetheless likely to produce the best overall outcome. Listed below are several suggestions for practicing assertiveness skills at work (Alberti & Emmons, 1995).

- Improve your decision-making skills by practicing assertive action.
- Negotiate more effectively by being assertive.

- Deal with angry co-workers assertively.
- Learn to say NO so that you do not lose yourself. Most organizations will take all you'll give, and expect more. If co-workers know you can't say no, they won't be reluctant to make unreasonable requests.
- Be persistent. Plant a seed and nurture it.
- Be assertive about goal setting. Ignore manipulation. If requests are unreasonable, say so. In response to reasonable requests you may not be able to handle, explain why.

Conflict resolution, problem solving, and negotiation with colleagues can make your workplace more enjoyable. You can also direct more of your energies to working directly with clients and their difficulties. Alberti and Emmons (1995, p. 180) suggest that conflict resolution "can be aided through an atmosphere which allows—even encourages—constructive disagreement. As different ideas are expressed, it becomes possible to work out compromises which build on strengths or everyone's contributions."

Relationships with Administrators

Working with administrators may also be a challenging experience. In some agencies, contact with administrators is minimal. Administrators may be mostly absent and delegate many of their responsibilities to assistants, department chiefs, or supervisory staff. This is common in large organizations, where administrators may not be closely involved with staff. In either case, concentrate on developing a positive relationship with administrators.

Different styles among administrators are common. Administrators' styles can be passive, distant, and withdrawn; assertive, communicative, and open; or aggressive, hostile, and noncaring. Try to understand the personality traits and management styles of the administrators of your agency. You do not have to agree with or even like your administrators, but you can learn to accept and respect them. Respond empathetically; try to put yourself in their shoes. It is important to be sensitive to the arduous tasks management often faces.

Conflict and the Role of Administrators   Conflict is inevitable in any setting, including human service agencies. What is most important is how this conflict is managed. Several possible sources of conflict within organizations are outlined in the list that follows (adapted from Alberti & Emmons, 1995; Bolton, 1979; Scott, 1990; Weiner, 1990).

- The manner in which an organization is structured may partly determine conflict. The more centralized and bureaucratic an agency is, the more likely there is to be conflict. Conversely, when agencies are flexible and encourage open communication, conflict is less likely, because of the prevalence of collaboration and teamwork.
- The personal style of the administrator is also influential. This often dictates the pattern of management and the overall effectiveness of the manager. A supportive, nondefensive leader is more effective than one who is distant, nonsupportive, and overly defensive.
- The administrator's personality and style can also contribute to the success of conflict resolution with the organization. Some administrators encourage a certain amount of conflict, which can be healthy and beneficial within a spirit of collaboration. When this is the case, there is a greater chance for sound conflict resolution.
- Agency policies and practices, set or reinforced by administrators, can have an impact on the amount of conflict generated within the agency. Policies and procedures that are overly structured, as in many rigid bureaucratic systems, tend to engender conflict. On the other hand, if agency rules and regulations are realistic and upheld by all, the risk of divisive conflict is significantly lessened.
- Policies and practices sometimes change rapidly, or new policies are introduced abruptly. The chance of conflict is then heightened. In contrast, success is more likely if such changes are presented in such a way as to allow staff involvement and negotiation.
- Having established methods of grievances, applied equally to all, is also important. When management shows favoritism and interferes with the process, conflict is certain to rise.

- Finally, effective administrators are usually well trained in assertiveness, conflict resolution, and management skills. They are also open to learning from their mistakes and receptive to constructive feedback. When such skills are undeveloped or used ineffectively, there is a greater chance for breakdown in communication, trust, cohesion, and follow-through.

Assertive communication and conflict resolution with administrators   Because of the power differential between helpers and administrators, helpers may feel that acting assertively carries the threat of being viewed negatively and the risk of losing their jobs. Nonetheless, it is sometimes a necessity. Assertiveness skills may reduce some of the anxiety you are likely to feel. You should stand up for your rights in a nonaggressive manner that respects the other person. If your feelings, thoughts, or requests are expressed assertively to administrators, there are several advantages:

1. There is a greater likelihood that you will be heard and possibly have some or all of your needs met.
2. Administrators may develop a new respect for you because of your assertiveness skills.
3. You will feel less frustrated because you will not have to keep your feelings bottled up inside.
4. You are likely to feel better about yourself and more confident in continuing to be assertive.

Conflict-resolution skills may also prove helpful when you experience conflict with an administrator. Approaching the administrator directly and attempting to resolve conflict in an open, direct, and mature manner is far better than allowing the matter to simmer to an explosive state.

Relationships with Agency Staff

Conflicts with agency staff may also arise posing additional challenges for the human service practitioner. Agency staff includes secretaries, receptionists, transcribers, housekeepers, and security personnel, to name a few. The work of these individuals makes your workday go smoothly. Because ancillary staff members have a direct impact on your daily life as a helper, it is advantageous to cultivate positive relationships with them. Accept them and be respectful of them and their role within the agency. Their jobs are often as stressful as ours, if not more so. Thus, we must be sensitive to the stressors and limitations of their jobs. We encourage taking time to work with agency staff, not against them. This has clear benefits and is in alignment with the human service principle of working collaboratively with others.

Assertive communication and conflict resolution with agency staff   It is necessary to take staff relationships seriously and respect all staff members, regardless of their status in the agency hierarchy. This sets a solid foundation from which assertiveness and conflict resolution skills can be effectively implemented. Deal with problems or issues early, when they are small and resolvable. Also, regard conflict in a positive light—an opportunity for growth or change. We have found this attitude helpful when working with staff.

Professional Relationships in the Community

In addition to all the interpersonal and professional relationships practitioners are involved with, relationships with the community are also important and sometimes the most challenging.

The role of professional helpers may involve helping not only individual clients, families, or small groups but also working within the community—another area where effective helping can take place. We can work on a macro level to alleviate broad social problems in the community. Macro practice roles are numerous and are detailed in Chapter 5. Some apply directly to work within an agency setting, and others relate to community change and mobilization. Netting, Kettner, and McMurtry (1993, p. 4), who emphasize macro practice, enumerate community activities that human service practitioners can perform.

- Negotiating and bargaining with diverse groups
- Encouraging consumer participation in decision making
- Establishing and carrying out interagency agreements
- Conducting needs assessments
- Acting as advocates for client needs in various community systems
- Engaging in policy-related activities: working as coalition builders, lobbyists, and overseers of legislative developments

Four major areas of macro practice that pose additional challenges are highlighted next.

Collaborating with professionals from other disciplines: There will be occasions when you will work with specialists from other disciplines. These may be physicians, nurses, occupational or physical therapists, speech pathologists, attorneys, court representatives, law enforcement officers, educators, and school administrators. Developing positive working relationships with these individuals is advantageous. Working together as a team can help clients receive more comprehensive treatment. Important information and referrals as well as additional support can result from such interactions. Once a solid foundation is set, these relationships should be maintained and reinforced. To be effective in our work with clients, we must strive to build reliable, open, and mutually respectful relationships with professionals from other disciplines.

Representing the agency and the profession: As a representative of both the agency and of the helping professions in general, we must present a professional image. This image facilitates negotiation and collaboration with others and enhances positive feelings in ourselves. There are three major goals: (1) increasing community awareness and sensitivity to social services issues, (2) increasing community awareness of available human resources, and (3) acting as an advocate for clients by connecting them with community resources and by promoting their involvement in community activities that can enhance their functioning.

You can develop goodwill for your agency when you strive to build strong and positive community contacts for yourself. Likewise, strong community relations can help you develop a positive and sound reputation that can enhance your career for years to come.

Assisting community change: Our work with clients often sensitizes us to problems in the community. Because of our commitment to helping, we may want to expand our role to become an agent of community change. A community change agent must be open to community involvement. He or she will come to understand the community, its pressing problems, and the resources available to support change. Homan (1994, pp. 25–26) describes five targets for community change: (1) neighborhood empowerment; (2) community problem-solving systems; (3) developing community support systems, (4) community education; and (5) developing broad-based community organization.

The targeted changes highlighted here show that community work can be helpful to clients on a large scale as well as affording real challenges to helpers who want to expand their scope.

Consulting in the community: Consultation is defined here as connecting with others in the community to share information about client needs or general issues. For effective consultation, one must have positive interpersonal skills, a genuine caring for clients, and respect for the knowledge and resources available in the community. Persistence, patience, and planning are all crucial if consulting is to be productive and mutually beneficial.

In all of these, the importance of cultivating positive relationships in the community is a given. To build relationships in the community, you must seek opportunities to take an active part in community affairs. We suggest that you participate in community activities; present a genuine, professional, and respectful image; and represent your agency with integrity and sincerity.

Assertive and conflict resolution skills   are necessary here also. Conflicts with the community often result from community perceptions of a client's behavior, needs, or motivational level, or from the scarcity of a needed resource. It is critical to act assertively on behalf of your clients and be an advocate for them whenever possible. Some conflicts may have no apparent solution; you may need strong conflict-resolution skills to reach a compromise or resolution.

In sum, we hope that you will come to value the many relationships in your life and view the challenge of cultivating positive relationships as an opportunity for growth. It is especially important for helpers to develop and maintain positive relationships with everyone they interact with, both personally and professionally.

Surviving as a Human Service Practitioner

A last major challenge we confront is that of "staying alive" as a human service practitioner. It is important to consider the major sources of stress and to be aware of the ever-present danger of encapsulation and burnout that plague many human service practitioners at some point in their careers.

Major Sources of Stress

No matter what your role in the human service profession, stress is inevitable. To deal with the stressors inherent in this profession, you need to be aware of the ones that most commonly arise. We explore these briefly below.

Coping with Overwhelming Feelings   Interacting with people in a helping capacity can be quite overwhelming, and thus a major challenge for both beginning and seasoned workers. The intensity of human emotions and dysfunction you encounter may arouse emotional reactions that you have not anticipated. You may not be prepared to deal with the impact of such pain and trauma. Having these feelings is quite normal and appropriate. If you were so out of touch with your own feelings that you were not affected by your clients' suffering, you would most likely not be an effective helper.

A key to helping others without being overwhelmed is to maintain a balance between your emotions and your focus on the necessary interventions. Learn to recognize when you are becoming too emotionally involved. We recommend that you talk freely with your supervisors, colleagues, field instructors, and fellow students. Perhaps you will discover that your feelings are connected to some unfinished business or unresolved issues in your own life. Such awareness could become the impetus for you to work through these issues and enhance your own coping skills. However, if you discover that the feelings are so overwhelming to you that discussing them with others is not sufficient, you might need to seek your own counseling.

Although some beginning helpers may regard seeking counseling as a sign of weakness, we view it as an opportunity for growth. Certainly, a willingness to seek help is a strength. If you are able to recognize and express your deeper emotions, you will more likely succeed in eliciting the same for your clients.

Working with Difficult Clients   The challenge of working with difficult clients has already been discussed. Indeed, clients can be resistant, demanding, and potentially abusive. Dealing with them can take a toll on you emotionally, physically, and intellectually. An agency demand for increased productivity can compound this situation, especially for the beginning practitioner. As we have noted, because of continued reductions in funding, many agencies must provide the same services—or more—with fewer staff. This overload creates stress, especially for those who are not accustomed to the high level of stress inherent in agency work. Learning to set limits, prioritize, reframe, and seek support are crucial to your emotional survival.

Hostile Co-workers   Another potentially stressful and challenging endeavor many practitioners must confront is having to work with hostile or difficult staff members as discussed above. Because you may have no choice about who your co-workers will be, you may have to work with people you do not like. Learning to accept others while maintaining your own identity is all important.

Self-Imposed Limitations   The many self-imposed limitations common to working in the helping professions is another potential source of stress many will experience. Being afraid to take risks is a common human experience; sometimes safety and comfort win out over trying something new. Nonetheless, too conservative a style may lead to stagnation, burnout, and ineffective helping. Taking the risk of stretching yourself professionally requires courage and willingness to endure the resulting anxieties. Allow yourself to make mistakes, and accept your limitations. This is necessary if you hope to be successful in any new venture.

Documentation and Paperwork   One of the most stressful and challenging aspects of human service practice involves documentation and paperwork. Proper documentation and paperwork are not self-imposed but are dictated by the agency. Helpers often detest these aspects of agency life. If not managed properly, the time spent documenting and keeping records will have to be deducted from your time providing direct services. If you understand the value of documentation, however, you may be able to see it from a new perspective.

Besides being required by both public and private agencies, documentation is also done for ethical and legal reasons. Many people who enter human services have the primary motivation of working with people. These eager, enthusiastic, beginning professionals are sometimes disappointed to learn that a portion of their time must go to writing evaluations, assessments, progress notes, and incident reports; making and recording telephone contacts; filling out forms; and other writing tasks. According to Moore and Hiebert-Murphy (1995, p. 433), "Paperwork and writing reports is an integral part of your work as a helping professional. Although we all, from time to time, feel as though we are about to be drowned in a sea of paper, keeping accurate records and developing your skills as a writer are important." Others view paperwork as a matter of accountability. Filip, Schene, and McDaniel (1991, p. 81) state, "It is important to recognize that the problems of your work are not tangible except as they are documented." Proving your high level of responsibility and commitment to your clients may be difficult without paperwork to verify your work and interventions.

In today's economic and political turmoil, many agencies are forced to justify their existence. This can generate even more paperwork; workers are often required to document their clients' need for continued services. For many helpers, this is a time-consuming and stressful endeavor.

Certainly, not every ounce of paperwork is necessary. However, there are numerous reasons that documentation is a required component of professional helping in any agency; some of the more important reasons for documentation include these:

- Record keeping functions
- Statistical purposes
- Ethical reasons
- Legal purposes

A positive attitude about doing paperwork is a great help. If you accept this requirement as part of your work and recognize its value rather than resisting and complaining about it, life is apt to be less stressful. Learn to make paperwork your own: Tailor it to meet your own needs when possible. This will help you develop a positive attitude. We have compiled a list of tips that may help you keep up with paperwork:

- Keep a positive mind-set; minimize resistance.
- Prioritize your work daily; make an effort to eliminate unnecessary paperwork.
- Maintain an organized system to maximize efficiency and accuracy.
- Learn from previous efforts; use previous reports, evaluations, and memos as guides for current work.
- Schedule a block of time, even a half-hour at a time, to complete paperwork.
- Develop a routine that is appropriate to your personal style and work demands.
- Consider giving the agency advice regarding streamlining certain aspects of paperwork, such as with repetitive or unnecessarily long forms.
- Recall that paperwork has many positive aspects. It does not have to be a waste of time. Recognize its value and tailor it to be useful and beneficial in your work with clients.

Budget Cuts, Realignment, and Layoffs   A last source of stress for many practitioners today is the ominous threat of budget cuts, realignment, and layoffs. Today, society is questioning its commitment to human services. In the ongoing political and philosophical debates regarding the need for social programs, human service programs are being eliminated. Working in this climate of uncertainty can be very stressful for all concerned. Not knowing whether you will have a position next week can affect you both emotionally and professionally. Your mind may be less on your clients' needs than on your own worries about the future.

Morale suffers in agencies when there are cuts, realignment of staff, or the threat of any of these. Frustration, resentment, and anger mount. Staff may feel caught in a trap: "We need our jobs, so we will have to tolerate these changes regardless of how unfair they are." The message they receive loud and clear from their employers may be "You should be appreciative that you have a job; many out there would love to take your place."

Working under these conditions can turn you off to working in agencies. We hope you can find a way to balance the negatives with the positives and strive to maintain a positive outlook toward helping.

Preventing Encapsulation and Burnout

Encapsulation and burnout are real threats for practitioners working within human services. To help you from becoming a victim of either of these, we highlight some key issues you should be aware of followed by some preventive and interventive strategies that might be helpful.

Encapsulation

Woodside and McClam (1994, p. 274) illustrate *encapsulation* in the following manner: "New workers faced with the difficulties of becoming professionals may become static in their work. They may feel so threatened and frightened by the difficult tasks they are asked to perform that they quit learning and growing—thus becoming encapsulated."

The term *encapsulated counselor* has been used at least since 1962, suggesting that this problem is not a new one to helpers. The following list of characteristics of encapsulated counselors is provided by Corey, Corey, and Callanan (1993, p. 241):

- Defining reality according to one set of cultural assumptions
- Showing insensitivity to cultural variations among individuals
- Accepting unreasoned assumptions without proof and without regard to rationality

- Failing to evaluate other viewpoints and making little attempt to accommodate the behavior in others

Others have interpreted *encapsulation* in slightly different ways: (1) substitution of model stereotypes for the real world; (2) disregarding cultural variations, in dogmatic adherence to some universal notion of truth; and (3) the use of a technique-oriented definition of the counseling process (Sue & Sue, 1990, p. 9). "Encapsulation results in helper's roles that are rigid and impart an implicit belief that the concepts of 'healthy' and 'normal' are universal" (Sue & Sue, 1990, p. 9). Furthermore the encapsulated helper, by thinking in such a narrow manner, is trapped, likely to resist adaptation, and unlikely to consider alternative approaches to working with different populations (Pederson, 1991).

Not only counselors or majority-culture helpers are at risk of becoming encapsulated; no one in the human services field is automatically immune to encapsulation. Various authors (Ho, 1991; Pedersen, 1991; Ridley, 1989) indicate that helpers with minority backgrounds are just as likely to become so narrowly focused in their work that they too become encapsulated. Terms such as *unintentional racist* and *reverse discrimination* have been used to describe the process whereby people from minority cultures begin to discriminate against those who are different from them. The following list offers some illustrations of encapsulation.

Examples of Encapsulation

- Working with a Latino family where machismo is strong, a majority-culture helper may try to change this cultural value.
- The helper believes in independence, yet the client is from an Asian-American cultural background that fosters interdependence. As this is not the favored norm, the helper is not only critical but imposes his values on the client, believing they are best.
- A recent immigrant says, "All Americans are the same. They only think of themselves without regard to their family or community." This is a case of reverse discrimination.
- The helper does not believe homosexuality is healthy and hence attempts to lecture a gay client to change his sexual orientation.
- Because of the helper's deep religious views, he attempts to sway his clients to believe and behave as he does, regardless of the client's position.

Regardless of our cultural background, we are all potential victims of encapsulation. Self-awareness and cultural reevaluation can help us avoid falling into this trap. Suggestions to minimize the risk of encapsulation are offered by Woodside and McClam (1994, pp. 272–275):

- Remain committed to life-long learning. "Part of learning is making changes in behaviors and attitudes: change and growth can be very exciting parts of the learning process, though difficult at times."
- Be willing to unlearn maladaptive patterns and views. "Part of the change process is unlearning, which can counteract the encapsulated worker's tendencies to cling to truths and to accept the values of the traditional culture."
- Strive constantly to develop professionally by "revising updating knowledge, skills and values of the profession."
- In essence, human services workers are challenged to unlearn old ways of working with clients and agency settings in response to change.

SCENARIO

Tom is a seasoned clinician working in a state mental hospital with the chronically mentally ill. He has been out of graduate school for ten years. When he first graduated, he had aspirations of becoming licensed and working in an agency setting as well as developing his own private practice where he could be more creative in his treatment approach and also work with clients of all ages and various levels of functioning. Now, ten years later, he has never left the inpatient hospital setting he was hired to work in after his graduation. He has become stagnant in his treatment approaches and tends to view all the mentally ill the same. He seems more hopeless about their progress or his efforts to make a real difference in their recovery. He finds it futile to go to any outside training seminars and attends only those he is required to attend by his unit manager. If you were newly hired to supervise Tom, what might you notice? Is Tom suffering from encapsulation? Are there any ethical issues to consider? Would you make any efforts to point out his behavior? If yes, how would you proceed?

Burnout

Most would agree that preventing *burnout* and remaining professionally vital are among the greatest challenges of all for the human service practitioner. Burnout has been defined in a myriad of ways. For example, Riggar (1985, pp. xvi–xvii) lists 18 different definitions from the literature. All seem to convey the basic message, but there are variations in the overall themes. Here is a sampling:

> A state of mind…accompanied by an array of symptoms that include a general malaise: emotional, physical, and psychological fatigue; feelings of helplessness, hopelessness, and a lack of enthusiasm about work and even about life in general. (Pines & Aronson, 1988, p. 3)

> A progressive loss of energy, idealism, and purpose experienced by helping professionals as a result of their work conditions. Given this state of exhaustion, irritability, and fatigue, a professional's effectiveness is markedly reduced with depersonalization often occurring. More broadly speaking, burnout is an internal psychological experience involving feelings, attitudes, motives, and expectations. (Maslack, 1982, p. 29)

> Burnout is characterized by physical depletion and feelings of hopelessness and helplessness. Those suffering from burnout tend to develop negative attitudes toward themselves, others, work, and life. (Pines & Aronson, 1981)

These definitions reflect not only the potentially damaging effect of burnout but also the fact that no one is immune to it. If you believe you are impervious to stress, you are among those who are most susceptible to the phenomenon of burnout. The nature of burnout and the importance of early recognition are well documented by Guilliard and James (1990, p. 572):

> Burnout is not simply a sympathy-eliciting term to use when one has had a hard day at the office. It is a very real malady.… It is prevalent in human service professionals because of the kinds of clients, environments, working conditions, and resultant stresses that are operational there. No one particular individual is more prone to experience burnout than another. However, by their very nature, most human service workers, public or private, tend to be unable to identify the problem when it is their own. No one is immune to its effects.

This statement not only reinforces the seriousness of burnout but shows that it is increasingly being recognized as a significant problem for professionals in human services (Zastrow, 1995). Farber (1983, p. 1) noted that burnout "is a complex individual-societal phenomenon that affects the welfare of not only millions of human service workers but also tens of millions of those workers' clients." Paine (1982) and Maslack (1982) emphasize the tremendous impact stress and burnout have on individuals, society, and organizations. Impairments in productivity, interpersonal relationships, and emotional and physical health are likely outcomes of job stress. The costs are certain to be high and cannot be ignored. In sum, burnout must be taken seriously by all human service practitioners; it has "major ramifications for both individuals and institutions" (Maslack, 1982, p. 39).

Ultimately, the key to minimizing burnout is learning to deal with stress more effectively. Being able to recognize the signs of stress early can help you take appropriate steps to reduce it. Knowing the causes, symptoms, and stages of burnout is most useful.

Causes of Burnout   Burnout is a reaction to high levels of stress (Zastrow, 1985). Stress contributes to many physical illnesses as well as to emotional and behavioral difficulties (McQuade & Aikman, 1974; Greenberg, 1980).

Stress as defined by Zastrow (1985, p. 281) is "The emotional and physiological reactions to stressors. A stressor is a demand, situation, or circumstance which disrupts a person's equilibrium and initiates the stress response." Corey and Corey (1993) have separated stressors into several categories:

- *Individual* (illness, allergies, obsessiveness)
- *Interpersonal* (death of a loved one, relationship conflicts)
- *Organizational* (deadlines, pressures to produce, bureaucratic constraints)

Others label these categories slightly differently, using such words as *biogenic* (coffee, amphetamines, and exercise), *psychosocial* (relationship based), and *environmental* (organizational, societal, or man-made stressors).

Symptoms of Burnout   Core symptoms of burnout, listed below, have been identified by numerous authors:

- *Behavioral*: "A marked departure from the worker's former behavioral norm" is typical (Forney, Wallace-Schutzman, & Wiggers, 1982, p. 436). An individual may become easily irritated, act irresponsibly, show little empathy, or simply withdraw and remain in isolation.
- *Physical*: Loss of energy, depletion, debilitation, and fatigue are symptoms of burnout. Poor eating habits, disturbed sleep, little exercise, and a range of chronic health problems may accompany this fatigued state (Maslack, 1982; Pines & Aronson, 1988).
- *Interpersonal Relations/Social*: Relationships at all levels are affected. An individual who is burned out may withdraw or relate to others in an apathetic or negative manner (Corey & Corey, 1993, 1998; Watkins, 1985).
- *Emotional/Attitudinal*: Burnout is characterized by a significant loss of commitment and moral purpose in one's work (Cherniss & Krantz, 1983; Pines & Aronson, 1988). Symptoms of depression may appear: hopelessness, apathy, lack of motivation, and low self-esteem. Anxiety may also be manifested as excessive worry, preoccupation, fear, and ability.
- *Intellectual*: Burnout involves a gradual loss of curiosity and desire to learn. In extreme cases, this can be seen as a loss in mental abilities (Corey & Corey, 1993, 1998). A negative attitude can also accompany such a decrease in intellectual functioning.
- *Spiritual*: Burnout-out individual may suffer a decline in their spiritual beliefs. They question the meaning of life and they may develop cynicism.

These symptoms are often manifested as follows: celebration when clients cancel; daydreaming and escapist fantasies; tendency to abuse alcohol or drugs or to indulge in any type of addictive behavior; loss of excitement and spontaneity in work with clients; delays in paperwork; deteriorating social life; and reluctance to explore the causes and cures of the condition (Kottler, 1986, pp. 92–95).

Stages of Burnout   We are all susceptible to some degree of burnout. Burnout is most often characterized by certain developmental stages; knowing them can help in early recognition. As cited by Edelwich and Brodsky (1982, pp. 133–154) the stages are as follows:

1. *Enthusiasm*: Early in their careers, helpers are extremely motivated to set the world on fire. This stage is filled with high hopes, unrealistic expectations, and a zealous drive to make a difference.

2. *Stagnation*: When personal, financial, and career needs are not being met, stagnation may begin to set in. Opportunities for promotion may be absent or seem impossible to attain. The person may feel stuck, trapped, or stagnant and therefore begin to lose the will to work in a helping capacity.

3. *Frustration*: This stage is reached when helpers begin to question their effectiveness, value, and impact. They may feel so overwhelmed by obstacles that they lose sight of their

original motivations for becoming helpers. A domino effect occurs in some organizations: Burnout can be quite contagious.

4. *Apathy:* When a person feels indifferent, unresponsive, and listless, he or she is suffering from the most serious stage of burnout: apathy. This is truly a crisis situation. There is disequilibrium, immobility, denial, and loss of objectivity, which make the condition much more difficult to treat.

Prevention and Intervention   The impact of chronic or excessive stress can be great. In the last stages of burnout, an individual may exhibit fragmentation, impaired professionalism, and erosion of the helper-client relationship. Agencies can also be affected and become increasingly dysfunctional as a result. Most critical, however, is the damage such stress can do to the helper and client. When helpers become incapable of fulfilling their roles adequately, services to their clients are sure to suffer. Regardless of your professional position, prevention and intervention must remain a high priority.

Because stress and burnout are natural risks faced by human service professionals, awareness of the symptoms and causes listed above, can be helpful in developing appropriate interventions to prevent or cope with stressors.

To conclude our discussion of burnout, we offer a list of interventions we believe are appropriate and practical. These are compiled from our review of the literature as well as from our own professional experience. Our list is by no means exhaustive; we encourage you always to be open to learning new ways to reduce stress and the potential for burnout. Remember that a person's reaction to stress is an individual matter, and stress-reduction techniques alone do not offer total protection. Work concerns, unresolved personal issues, organizational problems, and conditions outside of work must all be taken into account in preventive and interventive efforts.

Recommendations for Prevention of and Intervention for Burnout

- *Personal Awareness:* Personal awareness is a must for the human service professional. Be aware of yourself and both the joys and frustrations of working in this field. By modifying an unrealistic attitude, you can avoid disappointment and eventual apathy, thus helping to prevent burnout.
- *Realistic Expectations and Limit Setting:* Be realistic, not idealistic. Set limits; beware of overinvolvement; keep an eye on your motives (watch for countertransference) and the limitations of your professional role.
- *Professional Support:* Seek peer, supervisory, administrative, community, and faculty support. You can gain validation, consultation, and education this way.
- *Social Support Systems:* Social supports can buffer our intense involvement with our clients and colleagues. If clients are depressed and burned out, don't you encourage them to seek outside support through social channels?
- *Goal Setting:* Setting appropriate goals can enhance self-esteem, increase efficiency, and help you deal with organizational issues. Also, self-esteem and confidence are elevated when goals are successfully met, so goals must be attainable.
- *Time Management:* Time management helps people set short- and long-term goals and use their time effectively in reaching these goals. It limits feelings of being overwhelmed. We recommend "to-do" lists and prioritizing as a part of time management. Be flexible enough to reevaluate your goals, plans, and priorities.
- *Positive Thinking:* The more negative one's attitude is, the higher will be the person's risk of burnout. Positive thinking is essential in this field, where so much pain, sorrow, and tragedy are encountered. If you have a positive attitude, others will appreciate and like to be around you; this is likely to help you feel good about yourself as well.
- *Cognitive Restructuring:* Challenge yourself to change any negative and irrational thoughts such as "I'm only going to fail!" or "It's not worth it, I'm giving up," or "I always have to be perfect."

- *Relaxation Techniques:* The most common are techniques related to relaxation response: deep breathing, imagery, progressive muscle relaxation, meditation, self-hypnosis, and biofeedback.
- *Maintaining a Balance:* Being devoted solely to your work is admirable but not very wise. In order to give to others, you must be able to give to yourself as well. Get involved in outside activities that can help you relax, take your mind off your work, and rejuvenate you so that you can be most effective in your work.
- *Taking Care of the Basics:* Eat sensibly, get enough sleep, and exercise. Proper nutrition is proven to help you think more clearly and feel energized. Regular, restful sleep is also a must; sleep deprivation can cause fatigue, depression, and an unpleasant disposition. Regular exercise—working out, dancing, or taking part in a sport—also reduces stress.
- *"Personal Goodies":* Reinforce yourself for your diligence and hard work. Providing yourself with "goodies" is a way to reward yourself for your efforts and commitment to helping. Personal goodies might include traveling, listening to music, laughing, being hugged, window shopping, or sitting in a tub of warm water. These are essential in replenishing and rewarding you for your hard work.
- *Humor and Play:* Laughter is often the best medicine for stress. Humor helps us relax and enjoy our work more fully, and it also takes the edge off intense emotional situations. Some consider laughter a form of "internal jogging," saying that it serves a physiological purpose as well. Humor also allows us to perceive the paradoxes of life from an emotional distance. We can separate ourselves from worrisome incidents and perhaps gain a more realistic perspective.
- *Acceptance, Adaptation, or Change:* Assess the nature and scope of the stressors in your life. Determine whether they can be reduced, changed, or eliminated. This may involve healthy risk taking and assertiveness. At times, stressors cannot be changed, so acceptance and adaptation are necessary. Sometimes energy spent fighting an inevitable situation only compounds the stress already being experienced.
- *Active Participation in Workshops and Continuing Education:* Strive to keep learning new insights and to remain intellectually stimulated. This adds a change of pace to our daily routines and allows us to interact with others in the professional community.
- *Personal Stress Management Plan:* Develop a personalized plan for how to deal with the stress in your life. Like a diet, results will occur only if the diet is followed properly.

SCENARIO

Laura is a 40-year-old Latina who has worked in children's services for over 15 years now. She is well respected in her agency. Although not a supervisor by her own choice, she often helps the unit supervisor by training new workers. Recently, however, Laura has refused to participate in any training, indicating that her caseload is just too high and she doesn't have any extra time. She is more moody, often arrives late to meetings, and simply has become more withdrawn and isolated. Physically, she seems exhausted and often complains of not sleeping well. On several occasions, she has fallen asleep in staff meetings or nodded off in court. Her previously close relationship with co-workers has dramatically changed, with many finding her distant and difficult to talk to. She is increasingly negative about her role and the purpose of her work in reuniting families, something she used to feel great joy in. Intellectually, she seems more narrow in her thinking, with little curiosity in the intricacies of her cases that before would have interested her. Perhaps most upsetting are her frequent comments to others about the uselessness of the agency services. She no longer feels that she can make a positive impact nor has faith in her clients' abilities to overcome their difficulties. Her co-workers are so concerned that they approach their supervisor. If you were the unit supervisor what might you deduce from the information staff is giving you about Laura's behavior? How might you proceed?

Advantages of Working in Agencies

We end our chapter on a positive note by addressing the major advantages of agency work; we ask that you keep these in mind as you face the challenges of being a human service practitioner.

Working in an agency, you are at the heart of helping. The satisfaction of knowing that you are instrumental in assisting others can make up for all the difficulties inherent in working in this field. Knowing that you have made a positive impact on another person's life is often a reward in and of itself. Most of us entered this profession because we wanted to help people in need; seeing our efforts bear fruit reinforces our desire to continue working with others.

Interaction with Others and Opportunity for Teamwork

Allowing helping to occur at multiple levels is one definite aspect of agency work. The constant interaction with clients and co-workers can be a benefit. If you are a "people person" and enjoy being at the center of activity, then agency work is for you. The collaboration involved in working as a team can be beneficial to you as well as the client. You can share your fears and frustrations; gain validation, support, and feedback; and learn from other team members. Working in a supportive environment can also create cohesion and a sense of unity. For clients, teamwork, in most instances, leads to the delivery of more complete and higher-quality services. By comparison, working alone can be isolating and overwhelming because you will lack support or the opportunity to consult with other professionals.

Sharing the helping roles with others can also involve creative ways of working together; this can keep you motivated, inquisitive, and emotionally and intellectually sharp. Just realizing that you are not alone can be reassuring. In many agencies, you may be able to choose how much or how little involvement you have with others. In some agencies, staff tend to keep to themselves; in others, they work together constantly. The nature of your job, your personal style, and those of your co-workers all help determine how much teamwork there will be. If you take an active stance, as we have advocated throughout, you are likely to create a working environment that suits your needs well.

Agency work is seldom boring or repetitive. Each new day presents a new situation, helping to keep workers stimulated and eager to learn. The variety involved in working with others is intriguing although sometimes also overwhelming. Learn to appreciate the uniqueness of your work, and you will cope better with the difficult aspects of agency life.

Last, a great advantage of agency life is the opportunity to develop strong relationships with co-workers. Many will share some of your values and interests, affording you the chance to build a strong support system for yourself.

An Introspective Profession

In the helping professions, we can learn about ourselves and develop a greater capacity for growth and self-awareness. Of course, for some people, this introspection can be a stressful aspect of helping.

Dealing with clients, co-workers, and administrative personnel will no doubt challenge you to look at yourself. Such self-analysis is often difficult. Helping others involves close engagement with other people. If you enter this field without a willingness to examine your own feelings and beliefs, you will ultimately be hindered in your ability to help others. Introspection can encourage growth in both you and your clients.

We believe that the benefits of introspection far outweigh the negatives. You can learn about yourself, recognize your strengths and limitations, and develop a greater capacity for growth and self-actualization that can be beneficial to you, both personally and professionally.

CONCLUDING EXERCISES

1. Think about some aspect of your life in which there may be unfinished business with a significant individual. Briefly and succinctly, write down the situation as you can best recall on the lines below: _____

   _____

   _____

   _____

   _____

   Now write down a list of all the feelings involved: _____

   _____

   _____

   _____

   _____

   Helpers are expected to provide honest feedback and encourage clients to face difficult feelings in the process of treatment. List the reasons you believe you avoided dealing with the above situation (ex: "Wanted to avoid painful feelings") _____

   _____

   _____

   _____

   Describe how you can make a commitment by taking a small step in working on your ability to communicate and face your feelings more fully_____

   _____

   _____

   _____

2. Ask professionals in your environment, either in your educational setting or in your fieldwork agency, about their definition of transference and how they typically deal with transference reactions. Ask them for a specific example of transference in working with others and bring this to your class to illustrate how some professionals deal with these reactions in treatment. This can then be discussed or role played in the class to see how others might work with the situation differently. The emphasis here is on supporting differences and not on attacking one person's style or approach.

3. Think about what would be the most difficult type of client for you to see. List the types of situations most challenging for you below:_____

   _____

   _____

   _____

Now think about how this difficulty may be tied to experiences in your own life or strong values that you hold. Discuss how these situations are connected to experiences in your own life. _____

_____

_____

_____

If you believe these issues may be detrimental for you in working with others (unfinished business, countertransference), write a plan with beginning steps for how you might take responsibility to begin to work on these issues. _____

_____

_____

_____

3

Micro Level Practice: Working with Individuals

AIM OF CHAPTER

Although clinical work can occur at many different levels, such as family, group, and community practice, working with the individual remains an essential aspect of any therapeutic endeavor. In this chapter, we highlight working with individuals by taking a "micro" focus. We examine several common theoretical orientations and treatment modalities used with individuals, as knowledge of these can help you consider ways to work with your clients most effectively. The types of treatment examined include psychodynamic theory, person-centered therapy, the cognitive behavioral approach, and the developmental theories. The specific treatment modalities reviewed are crisis intervention models, including critical incident debriefing; brief therapies, featuring solution-focused therapy; traditional and contemporary long-term treatment approaches; and case management.

Throughout the chapter we emphasize the importance of cultivating your own style in working with individuals. Treatment considerations and vignettes are provided to further illustrate the application of theory to practice.

VARIED THEORETICAL ORIENTATIONS WITHIN A MICRO FOCUS

There are many theoretical orientations to choose from when you embark on a professional helping career. Regardless of what specific helping role you choose (e.g., counselor, social worker, psychologist), you need to know the main theoretical orientations. However, even the determination of what constitutes these main orientations varies in the literature, as shown in Table 3.1.

Each of the perspectives in Table 3.1 reflects the comprehensive nature of the general theoretical orientations commonly used in human services. Although each deserves merit, we feature Corey's view. It is extensive and seems to fit best with our comprehensive approach that can be broadly applied to working with clients in diverse settings by clinicians from various disciplines. A more detailed examination of Corey's views reveals four major categories of therapeutic approaches, each with their own set of theoretical orientations.

1. Analytic Approaches (Psychoanalytic, **Psychodynamic,** Adlerian)
2. Experiential and Relationship-Oriented Therapies (**Humanistic/Person-Centered,** Existential, Gestalt)

TABLE 3.1   Major Theoretical Orientations

Neukrug	Doyle	Ivey, Ivey & Simek-Morgan	Corey
Major Theoretical Orientations	Contemporary Theoretical Developments	Major Forces in Counseling and Psychotherapy	General Categories of Therapeutic Approaches
(1) Psychodynamic Approaches	(1) Cognitively-Focused	(1) Psychodynamic	(1) Analytic Approaches
(2) Behavioral Approaches	(2) Affectively-Focused	(2) Cognitive-Behavioral	(2) Experiential/ Relationship Oriented
(3) Humanistic Approaches	(3) Performance-Focused	(3) Existential/ Humanistic	(3) Action Therapies
(4) Cognitive Approaches		(4) Multicultural Counseling and Therapy	(4) Systems Perspectives

SOURCE: Adapted from Corey, 1996; Doyle, 1992; Ivey, Ivey, & Simek-Morgan, 1993; and Neukrug, 1994.

3. Action Therapies (Reality, **Behavioral, Rational Emotive, Cognitive**)
4. Systems Perspective (Family Therapy, **Developmental**)

We will elaborate on at least one approach (in boldface) from each of these major categories that we believe will give you a more holistic view of the range of existing therapeutic approaches. We believe these theoretical orientations can be used in a broad range of settings and helping roles. Furthermore, the approaches chosen are those we are familiar with in our own professional experiences. Even so, they are not the only ones available; we encourage you to engage in your own exploration to determine those that best fit your own individual style and philosophical base.

Choosing a Theoretical Orientation

Various factors contribute to a helper's choice of a theoretical orientation. Developing and utilizing a theoretical orientation is an ongoing process; it evolves from continued practical experience and constant self-awareness. We caution you not to seek the "perfect" theoretical base for your practice. Take time to learn about the many theories that exist, give yourself permission to experiment with some, and remain open to changing your views and using more than one orientation depending on your clients' needs and your own views.

Choosing an orientation is based on the helper's personality and style as well as the needs and type of population served. Some helpers believe in being confrontive and directive; others take a more supportive/collaborative stance; still others are more passive in their helping approach. A particular population may dictate the use of a specific approach. For instance, clients with substance abuse problems, those involved in domestic violence, or those engaged in criminal behavior typically need a more directive approach. Conversely, clients who are anxious about the therapeutic process and struggle with issues of low self-esteem may need the helper to be more supportive and humanistic.

The helper's views on human nature will also influence which helping strategy he or she is most likely to use. Neukrug (1994, p. 57) states that "our view of human nature describes how we understand the reasons individuals are motivated to do the things they do." In this respect, the way we perceive people ultimately dictates the approaches we will use in working with them. If we see people as basically good and capable we will likely trust their ability to take responsibility for the direction of their lives and will use techniques that reinforce their abilities and capacities. In contrast, if we see people as unreliable, unworthy, and irresponsible, our approach will probably be more directive and critical in nature.

A helper's beliefs about what best facilitates change can influence the therapeutic approaches he or she considers. For instance, the helper's focus on the past, present, or future may be related to how he or she views the change process and will thereby have a direct bearing on the particular strategies used. Additionally, a helper's focus on thoughts, feelings, and actions in the helping process can affect the choice of treatment modalities. If you believe change occurs when you focus on feelings, you may use a humanistic approach. Cognitive or rational approaches are often used when helpers focus on changing thoughts and when attention is placed on helping clients examine beliefs about themselves and the world. Logically, behavioral techniques are typically considered when the client's actions are the focus of change. A behavioral approach often requires the helper to be more active and directive in the change process. Last, emphasis on gaining insight may be viewed as a necessity for some practitioners who apply psychodynamic techniques. Smith (1982) and Ward (1983) believe that the integration of these various approaches is one of the most noteworthy trends in human services.

A helper's views regarding short-term versus long-term strategies can impact the approach utilized. Sometimes the agency setting dictates the length of treatment; this will prompt the helper to focus on only those approaches that accommodate the time limits permitted.

Finally, current research in the field typically influences approaches used in treatment. Today, for instance, use of cognitive approaches in the treatment of depression and anxiety have been shown to be most effective; a behavioral approach is the treatment of choice for those diagnosed with attention deficit disorders; and psychoeducational groups are typically used in work with batterers.

Analytic Approaches

Psychodynamic Interventions

Psychodynamic interventions are rooted in Freudian psychoanalytic theory. Late in the 1800s, Sigmund Freud began to practice hypnosis and discovered psychological origins for physical problems. Up to that time, symptoms were thought to be primarily organic in origin. Freud, the originator of psychoanalysis, is known today as the "father of psychoanalysis." He was an intellectual giant who made many advances in the field of psychiatry and in human development.

> He pioneered new techniques for understanding human behavior, and his efforts resulted in the most comprehensive theory of personality and psychotherapy ever developed. . . . Freud's psychoanalytic system is a model of personality development, a philosophy of human nature, and a method of psychotherapy. (Corey, 1996, pp. 91–92)

Compton and Galaway (1994, p. 134) concur that "The most complete theory of human development is psychoanalysis theory as first developed by Sigmund Freud and further refined and expanded by countless disciplines." In fact, many of the techniques proposed by Freud himself continue to influence contemporary practice. Many theorists who followed him borrowed concepts from Freud as a base for their own theoretical views and approaches. Although not all agreed completely with Freudian theory, they nonetheless were influenced in some way by his views. Various new theories and advances about human nature developed directly from psychoanalytic theory; in some cases they are primarily rooted in a traditional psychoanalytic base; in other cases, new views emerged as a direct opposition to psychoanalysis.

Freud emphasized "understanding the dynamics of behavior, the functioning of the ego defenses, the workings of the unconscious, and the ways in which present personality functioning reflects past events" (Corey & Corey, 1993, p. 64).

In his work, four key areas of psychoanalytic theory include (1) structure of the personality, (2) role of the unconscious, (3) anxiety and ego-defense mechanisms, and (4) personality development. These are central to understanding the core of Freud's work and the reasons it flourished as it did. They laid the groundwork for the many theoretical views and approaches that followed and for the basis on which psychodynamic theory rests.

Structure of the Personality   Freud developed views on personality structure by proposing the terms *id*, *ego*, and *superego*. He believed the personality consisted of these three systems that work together. Table 3.2 explains the role of these three systems.

Role of the Unconscious   Freud focused on the psychodynamic forces that motivate behavior and emphasized the role of the unconscious. In this respect, he believed that the reasons we do things may be beyond even our own comprehension and thus resulting from unconscious factors. Freud believed that much of what goes on in the mind or psyche exists below the surface of awareness and thus at an unconscious level. Theoretically, because much of our behavior is influenced by the unconscious, this level must be explored if a person's psychological functioning is to improve. By helping clients gain access to their unconscious conflicts, Freud hoped to increase their level of insight and desire to make necessary changes in their behaviors. Compton and Galaway (1994, p. 134) elaborate on this point:

> Individuals are pushed from birth by these largely unconscious and irrational drives.... Because of the operation of an unconscious defense system and the structure of the mind, people go through life largely unaware of these irrational forces, which have a tremendous effect on their behavior and on the way they relate to others.... Most of the motivating forces of personality are thus beneath the surface and are available to our conscious and rational understanding and direction only through careful exploration of these buried regions.

Based on this belief, Freud developed a set of therapeutic techniques specifically designed to help clients tap into their unconscious. Interpretation, dream analysis, free association, analysis of resistance, and analysis of transference are strategies most commonly used.

TABLE 3.2   Psychoanalytic View of the Personality

Id	Ego	Superego
The id is the biological component that is ruled by the *pleasure principle* whose role it is to reduce tension, avoid pain, and gain pleasure. The id is illogical and immoral and concerned only about satisfying instinctual needs. It is largely unconscious, out of awareness. According to the Freudian viewpoint, we are all id at birth; "it is considered the primary source of psychic energy and the seat of the instincts" (p. 93).	The ego is the psychological component termed the *executor* as it is said to govern, control, and regulate the personality by mediating between the instincts and the surrounding environment. Unlike the id, the ego is said to be ruled by the *reality principle* indicating its role in logical and reality-based thinking. "The ego, as the seat of intelligence and rationality checks and controls the blind impulses of the id. Whereas the id knows only subjective reality, the ego distinguishes between mental images and things in the external world" (p. 94).	The superego has been considered the *judicial branch* of personality, given its role in ensuring that strict moral codes are maintained. It works with both the id and ego by inhibiting id impulses and persuading the ego to substitute moralistic goals for realistic. "The superego, then, as the internalization of the standard of parents, and society, is related to psychological rewards and punishments" (p. 94).

(Based on Corey, 1996)

Anxiety and Ego-Defense Mechanisms   When the id, ego, and superego are in conflict, anxiety often develops, according to the psychoanalytic perspective. Anxiety results from repressing basic conflicts. Because anxiety is often a warning sign of impending danger, many people are motivated to act when experiencing a certain level of it. Ideally, the ego is suppose to control anxiety through rational and direct means; however, when this is not possible, ego-defense mechanisms are developed to take over this function.

Ego-defense mechanisms arise to help the individual cope with anxiety and to minimize the emergence of overwhelming feelings; either they deny or distort reality or operate at an unconscious level. Ego defenses are normal and adaptive behaviors when they help us cope in crisis situations, but they can become pathological when a person begins using them to avoid facing reality altogether. Some of the more common ego defenses you might be familiar with include repression, denial, reaction formation, projection, displacement, rationalization, sublimation, and regression.

Personality Development   A basic philosophy of psychoanalytic theory held that people are basically molded by psychic energy and by early experiences. Individuals are subject to irrational forces and greatly influenced by sexual and aggressive impulses. Freud viewed early development as critical to an individual's maturation, and he formulated a view on psychosexual development that continues to be used in some form by many therapists today. He proposed that individuals progress through stages (e.g., oral, anal, phallic, latency, and genital) that ultimately determine their personality in adulthood. According to Freud, the first six years are especially important as many later personality problems are rooted in repressed childhood conflicts. For normal development to occur, an individual must successfully progress through each stage by adequately mastering the psychological and physical tasks identified and thereby resolving the conflicts that may be present at each of these stages. When there is an inadequate resolution at a particular stage, the client is said to become "fixated" and thus be less likely to master later stages successfully because of the unresolved issues at this earlier stage. When successful progression through the identified stages does not occur, faulty personality development is said to result.

Therapists should help clients understand the nature and functioning of these internal psychological forces. In clinical work with adults, helpers must explore the ways early childhood experiences have helped shape adult personality. Proponents of this theory believe that by understanding the origins and functions of symptoms, helpers can understand their clients and help them overcome past conflicts; they can help clients resolve areas in which unfinished business continues to impact their psychological functioning.

The Psychodynamic Approach   The psychodynamic approach developed from psychoanalytic theory and is practiced by both those who adhere closely to Freud's views and those who embrace his ideas only partly. Freud believed that the drives motivating human behavior are the instinctual drives of sex and aggression (Freud, 1947); yet other psychodynamic theorists—namely, Adler (1964), Erikson (1982), and Kohut (1984)—felt otherwise. Adler and Kohut believed that people are motivated more by social drives; Erikson's work centered on life span development; and Kohut seemed to focus more on separation and attachment issues than on sex, aggression, or social forces. All shared a basic belief that drives motivate human behavior and that these drives are at least somewhat unconscious. Furthermore, they all upheld Freud's view that our perceptions of our childhood and the actual events that occurred in our early years, together with our drives, greatly affect our development and who we become as adults. According to Neukrug (1994, p. 60), "the purpose of psychodynamic therapy is to help the individual understand his or her childhood experiences, and how those experiences, in combination with the individual's drives, motivate the person today."

Psychodynamic therapy emphasizes the importance of carefully exploring an individual's past history, for this will most likely have a bearing on the person's present-day issues and functioning. Although centered in psychoanalytic theory, a psychodynamic approach is less concerned with uncovering the unconscious or dealing with issues tied to psycho-sexual development. In fact, many who now use the psychodynamic approach are focusing more on the conscious process, social causes of behavior, and individuals' ability to change and less on the long-term effects of early childhood (Dinkmeyer, Dinkmeyer, & Sperry, 1987).

Overall, we tend to favor this approach, given our successful experience with clients who have benefited from connecting present concerns to past issues. The extent to which helpers can be psychodynamic in their approach is dependent on numerous factors: the nature of the problem, the role of the helper, the time limits and constraints of the therapeutic relationship, and the willingness of the client to explore the past. In some instances, it may be counterproductive to dwell on the past; therefore, helping clients understand the connection of their problems with past issues may be all that is necessary. In other instances, in-depth exploration and working through past issues is thought to be the most therapeutic intervention. We encourage you to consider the client and his or her needs before applying an intervention that may not be the most conducive to positive change. In any event, whether you explore the past directly or simply emphasize making a connection to the past, we encourage you to be aware of how the client's past is related to his or her present functioning. In many instances, recognizing this is much like finding the missing piece of the puzzle that will allow you to understand and empathize with your clients more completely.

Treatment Considerations

In synthesizing and adapting different treatment strategies and tailoring them to your client, there are several aspects of treatment that you should consider:

- Develop thorough clinical assessment skills if you hope to understand the client's past and its impact on the present.
- Understand how history can play a significant part in a client's current issues.
- Be sensitive to the client's readiness to explore unresolved issues. Although you may be correct in your assumptions about the past, take care not to push a client to deal with issues he or she may not be psychologically ready to face (e.g., sexual abuse).
- Take into consideration the client's capacity for insight and the cultural dictates that may discourage him or her from delving into the past or uncovering private family issues.
- Consider time limits and financial constraints before beginning long-term psychodynamic therapy. Many helpers may find it valuable to learn to incorporate psychodynamic theory in crisis, brief, and short-term therapy modalities.

SCENARIOS

Helga was a 52-year-old professor at a major university. She had sought treatment after the death of her mother from cancer earlier in the year. She had noticed a decline in her motivation and level of concentration at work. She had had frequent crying spells and general tiredness since her mother's death and saw these as signs of depression. Although she had dealt with loss in her life in the past, she stated that she seemed unable to deal with the death of her mother, despite her general understanding of the grief process. While you are gathering her history, the client tells you that she was sexually abused by her father for several years and that her mother did little to protect her from that abuse. She stated that her mother was emotionally distant and Helga had been passive about telling her what was happening out of fear of her reaction. Helga had, therefore, failed to discuss the feelings and ramifications of this abuse throughout much of her adult life. She knew that she had wanted to tell her mother how she felt, which was angry and hurt, but that she always found reasons for not discussing it openly. Now that her mother had died, she felt particularly hopeless, sad, and angry that she had never taken the opportunity to discuss her feelings openly. If you were Helga's helper in this situation, how would you utilize the concepts from the psychoanalytic approach to help her deal with the unfinished business in her life?

Jim was a 25-year-old who had recently married and was having problems in his new relationship. His wife had demanded that he seek counseling as he was progressively becoming more and more angry and was yelling at her frequently. He stated at the time of the intake that he was somewhat reluctant to seek help because he didn't believe his problems were severe enough to warrant help at this point, but that he agreed to come in because his wife wanted him to. When you asked him about his family history, he told you that his father was an alcoholic who frequently would yell, verbally attack, and even physically hit him with a belt when he was younger. He stated that he always believed he deserved to be treated like this because he had gotten into a lot of trouble when he was younger. He was not sure of any effects of this abuse as he had done fairly well in school despite his problems at home. He stated that his mother had been quite passive and that she had been yelled at and hit by his father as well. Although Jim could say that he did not like for his father to have treated him this way, he did not believe that he needed to discuss the matter any further. He changed the subject back to his relationship with his wife when you questioned him further about his feelings toward his father. He said that he does not want to explore his relationship with his father; he wants to learn techniques for dealing with his anger and feelings toward his wife. If you were a psychoanalytically oriented helper, how would you deal with this client situation when he clearly does not want to explore the past and fails to make any connection between his past and current functioning?

Experiential and Relationship-Oriented Approaches

Historically, the humanistic perspective emerged in the 1940s as a reaction to both the psychoanalytic (psychodynamic) and behavioral approaches discussed here. In these first two approaches, the focus of the treatment is on the use of well-defined therapeutic skills and techniques utilized by a trained expert to bring about determined changes in the client. In contrast, the humanistic viewpoint highlights the value of a positive helping relationship that is believed to be the most instrumental factor in promoting positive client change. The deterministic views inherent in the psychodynamic therapies, as well as the scientific reductionistic notions of the behavioral approach, were both challenged by the introduction of this more humanistic view whose origins lie in existential philosophy and phenomenology. The humanistic approach "highlights the strengths and the positive aspects of the individual and rejects the concepts that people are determined by early childhood experiences or reinforces in the environment" (Neukrug, 1994, p. 66).

The view that we all have a choice in creating our existence is integral in the humanistic approaches that emphasize the growth and potential existent within the human spirit. Theorists from this perspective believe that everyone has an actualizing tendency, meaning that each of us has the ability to transcend our current existence by moving toward a more fulfilling and harmonious one. A major goal is for helpers to attempt to understand the subjective reality of their clients in an effort to help them become self-actualized.

The introduction of this nontraditional, humanistic perspective by such individuals as Carl Rogers, Rollo May, and Abraham Maslow created a revolution within the mental health field at the time. Today, humanistic approaches are highly respected and strongly adhered to by many within the helping professions. Many of today's approaches to therapy and counseling specifically and professional helping in general adhere to the central philosophy of the humanistic approach. Principles of the humanistic perspective can now be found in both educational and therapeutic environments that emphasize a nondirective and optimistic view to helping.

Person-Centered Approach

Because of its wide applicability and acceptance within human services, the person-centered approach developed by Carl Rogers is outlined here, beginning with a brief look at its development.

In the evolution of Roger's approach, four periods have been identified (Zimring & Ruskin, 1992). The time line below offers a brief overview:

1940s: Nondirective counseling theory emerged in which Rogers emphasized "the counselor's creation of a permissive and nondirective climate" (Corey, 1996, p. 198). The main focus for helpers was on reflecting and clarifying the clients' verbal and nonverbal communications in which acceptance by the helper was foremost. The main emphasis was for helpers to assist clients to clarify their feelings in the hope that this would increase their insight into these feelings.

1950s: Client-centered therapy was the new name Rogers gave his approach to reflect the change to focus on the client rather than on nondirective methods. The humanistic principles of phenomenology were central to this therapy. The major assumption was that an individual's own internal frame of reference offers the best insight to understanding the person's behavior. Also, the phenomologically based concept of the *actualizing tendency* took preeminence in Rogers's contention that "individuals have the ability to transcend their current existence and move toward a more fulfilling and harmonious existence" (Neukrug, 1994, p. 66). Additionally, it was at this time that Rogers formulated his hypothesis that helpers must possess a set of necessary and sufficient conditions for change (congruence, acceptance, and empathy, for example). The emergence of this therapy opened the way for practitioners to use humanistic skills not previously condoned in both the psychoanalytic and behavioral approaches that had preceded.

1960s: On Becoming a Person was published, which focused on "becoming the self that one truly is" (Corey, 1966, p. 199). According to Corey (1996), "the process of 'becoming one's experience' is characterized by an openness to experience, a trust in one's experience, and internal locus of evaluation, and the willingness to be a process" (p. 199).

In addition to this view, Rogers and his colleagues continued to test their hypotheses of the client-centered approach by conducting extensive research and applying central concepts of this approach to education through student-centered teaching and encounter groups led by lay persons.

1970s and 1980s: The person-centered approach developed out of Rogers's client-centered therapy. No longer were his central views applicable to only the counseling and therapeutic arenas. Client-centered concepts flourished and became applicable in many other areas: education, family life, leadership and administration, organizational development, health care, cross-cultural and interracial activity, international relationships, and finally politics, mainly in regard to world peace.

Simply, the person-centered approach is a broad perspective within which the more specific form of client-therapy can be found. A common theme of the person-centered approach is that there is "a basic sense of trust in the client's ability to move forward in a constructive manner if the appropriate conditions fostering growth are present" (Corey, 1996, p. 210).

The central hypothesis of this approach is that individuals have within them vast resources for self-understanding, for altering their self-concept, attitudes, and behavior. These resources can be tapped most thoroughly in a positive atmosphere that facilitates the change process (Brodley, 1986). The three conditions that constitute this growth-promoting climate are these:

1. Genuineness, realness, or congruence: When a helper is real, the client has a much greater likelihood of changing and growing in a constructive manner.
2. The presence of acceptance, caring, and unconditional positive regard: When a helper has a positive, nonjudgmental, accepting attitude toward the client, there is greater probability of client change.

3. The presence of empathetic understanding: The helper senses accurately the feelings and personal meanings that are being experienced by the client and communicates this acceptance and understanding to the client (Rogers, 1986).

Additional assumptions of this approach, outlined here by Brodley (1996), may also be regarded as *treatment considerations:*

- Belief that human nature is basically constructive
- Belief that human nature is basically social
- Belief that self-regard is a basic human need and that self-regard, autonomy, and individual sensitivity are to be protected in helping relationships
- Belief that people are basically motivated to perceive realistically and to pursue the truth of situations
- Belief that perceptions are a major determinant of personal experience and behavior, and thus, to understand a person one must attempt to understand him or her empathetically
- Belief that the individual person is the basic unit and that the individual should be addressed (not groups, families, organizations, etc.) in situations intended to foster growth
- Belief in the concept of the whole person
- Belief that people are realizing and protecting themselves as well as they can at any given time and under the existing internal and external circumstances
- Belief that people are realizing and protecting themselves as well as they can at any given time and under the internal and external circumstances that exist at the time
- A commitment to open-mindedness and humility in respect to theory and practice

SCENARIO

Grace, a Latina, age 27, enters counseling because of marital difficulties. She tells you that she separated from her husband about one month ago and is considering returning to him now. As you ask her to describe her difficulties, she tells you that she knows that her husband sleeps with other women and even has had a child with another woman. She states that he drinks, and when he drinks excessively, he often comes home angry and belligerent. He sometimes yells at her and has even hit her on several occasions. She has thought about leaving him permanently, but when she went to talk with her mother about this, her mother told her, "It is a woman's duty to stay with her husband no matter what" and encouraged her to work it out with him. Grace says that she has told her husband how it hurts her for him to hit her and to be unfaithful, but he has shown no indication of changing. Grace says that he wants her to come back and promises that things will get better if she returns. After she reveals her dilemma, she tells you that she thinks she will return, as it is best to work things out like her mother said. How would you react if you were the helper in this situation? How easy or hard would it be to support the client in her desire to return to her husband? What particular techniques might you use from the humanistic perspective?

Client-centered therapy is a specific therapeutic form that falls under the broad umbrella of the person-centered approach. Several terms are commonly associated with this therapy; they are summarized in Table 3.3.

The salient feature of client-centered therapy is the "empathetic understanding response process" (Bradley, 1986) in which observable responses are used by the helper to communicate empathetic understanding to the client. These vary and often include the following:

- literal responses
- restatements

TABLE 3.3   Client-Centered Therapy Terms

Accurate empathetic understanding: The act of perceiving the internal frame of reference of another, of grasping the person's subjective world, without losing one's own identity.

Congruence: The state in which self-experiences are accurately symbolized in the self-concept. As applied to the therapist, congruence is a matching of one's inner experiencing with external expressions.

Humanistic psychology: A movement, often referred to as the "third force," that emphasizes freedom, choice, values, growth, self-actualization, becoming, spontaneity, creativity, play, humor, peak experiences, and psychological health.

Self-actualizing tendency: A growth force within us; an actualizing tendency leading to the full development of one's potential; the basis on which people can be trusted to identify and resolve their own problems in a therapeutic relationship.

Unconditional positive regard: The nonjudgmental expression of a fundamental respect for the person as a human, acceptance of a person's right to his or her feelings.

Therapeutic conditions: The necessary and sufficient characteristics of the therapeutic relationship for client change to occur. These core conditions include therapist congruence (or genuineness), unconditional positive regard (acceptance and respect), and accurate empathetic understanding.

SOURCE: Corey, 1996, p. 126

- summaries; statements that point toward the felt experience of the client but do not name or describe the experience
- interpretive or inferential guesses concerning what the client is attempting to express
- metaphors
- questions that strive to express understanding of ambiguous experience of the client
- gestures of the therapist's face, hands, body
- vocal and other gestures

According to client-centered therapy, the sole intended function of these responses must be to facilitate the helper's understanding of the client's subjective reality, or internal frame of reference. The manner in which individual helpers apply these forms of empathetic understanding responses will differ, but this is not important. What is important is the outcome whereby the client feels understood by the helper.

Another trait that separates client-centered therapy from the more general person-centered approach is the extreme emphasis that practice places on the nondirectiveness of the helper. Clients are viewed as the experts on their lives whereas the helper is primarily responsible for respecting and protecting the autonomy and self-direction of the client.

Although the client-helper relationship is often viewed as unequal, client-centered therapists attempt to minimize this unequal aspect by sharing the authority as much as possible. Because of this underlying philosophy, helpers allow clients to determine the nature of the therapeutic relationship.

To summarize, the nondirective approach inherent in client-centered therapy helps protect the client's autonomy; enhances the client's self-direction; and facilitates the fostering of freedom, a positive sense of self, and of empowerment. Embedded in the growth principle, client-centered therapy accentuates promoting positive change in clients at their pace and direction. Respect for clients coupled with unconditional caring is believed to be useful in promoting positive change.

Treatment Considerations:

As you apply client-centered therapy, bear in mind these treatment considerations:

- Listen and attempt to understand the client's subjective reality (how things are from this person's point of view).

- Consistently check to make sure that your understanding of the client is accurate.
- Treat the client with the utmost respect and regard.
- Be accepting of the client no matter what.
- Be congruent or transparent by remaining self-aware, self-accepting, and not wearing a mask that could come between you and the client.
- Allow the client to dictate the pace and nature of the therapeutic issues; minimize the development of an unequal relationship.
- Encourage positive growth that the client wishes to pursue.

SCENARIO

Paul is a 37-year-old who enters treatment due to severe depression and loss. He recently lost his father who died in an automobile accident. Soon after his father's death, Paul's marriage began to experience difficulties as his wife said she could not take all his sadness and seemed angry at him often. She left him soon afterward and now did not want their two children even to visit him. Paul was devastated and wants desperately to continue to be active in the lives of his two children. He says that he has been losing sleep and not eating well for the last two weeks. He is rather isolated and has no one to talk with about what has happened. He requests that you listen and support him in dealing with the feelings he has about the death of his father and the loss of his marriage. As he says this, his voice becomes very angry and he starts to yell as if he were talking to his wife about how she abandoned him when he needed her most. How would you respond to this situation if you were using a client-centered approach? Would you respond differently if you were just following your own instincts in handling this situation?

Action Approaches

Cognitive and Behavioral Interventions

Cognitive-Behavioral Therapy   Cognitive-behavioral therapy has its roots in behavioralism and cognitive therapy. Most cognitive-behavioral practitioners today integrate principles from both these schools of thought. Taken together, the cognitive-behavioral strategies create a balanced approach to understanding and treating common problems of life. This approach is unique in that it allows two important realizations: (1) being able to examine the manner in which individuals view themselves and their environment (cognition) and (2) the way in which they act on that environment (behavior). The ultimate goal of a cognitive-behavioral therapy is to effect positive and lasting change by working with clients to modify their maladaptive thoughts and/or behaviors. Below we look at each of these theories individually. We begin with behavior therapy as it originated first.

Behavior Therapy   Behavioral techniques are central to cognitive-behavioral therapy. Views of behavior therapy lie in classical conditioning and operant techniques proposed by Pavlov and Skinner. Sometime at the turn of the century, Ivan Pavlov, a Russian scientist, discovered what was later termed classical conditioning. He found that "a hungry dog that salivated when shown food would learn to salivate to a tone if that tone had been repeatedly paired, or associated, with food" (Neukrug, 1994, p. 62). This observation implies that in classical conditioning, as a result of experience or associative learning, individuals often respond in predictable ways to certain stimuli or life events. Significantly, some of these learned responses may not always be adaptive or effective in present day living.

B. F. Skinner, a psychologist in the 1920s, later demonstrated that animals would learn specific behaviors if the target behavior was reinforced. From Skinner's work, operant conditioning was born—a technique that emphasized behavioral change through positive and negative reinforcements. Skinner found that when an individual is presented with a stimulus that produces an increase in behavior (a positive reinforcement) or the removal of a stimulus that results in an increase in behavior (a negative reinforcement), the individual's behavior can be changed successfully. Essentially, then, an individual's behavior is thought to be governed by contingencies. Behavior that is rewarded or reinforced will be repeated and the individual will learn from his or her experience.

The central premise of these behavioral techniques is the belief that maladaptive behaviors are learned and therefore can be unlearned as well. In his pioneering work in the behavioral branch of psychology, Skinner emphasized the cause-and-effect relationships between the environmental conditions present and an individual's behavior. A most significant contribution was his discovery that the physical and social environments are critically important in determining human behavior.

Various behavioral techniques are commonly used; they include training in both assertiveness and relaxation, and systematic desensitization typically employed in the treatment of phobias. Treatment and assessment are interrelated, given the fact that they occur simultaneously. According to Corey and Corey (1993, pp. 65–66), there are various characteristics that apply widely to the behavioral approaches commonly used today. These are listed below, along with others we feel are important treatment considerations.

Treatment Considerations

- The focus of treatment should be on the current influences of a client's behavior, not on the historical determinants.
- Overt measurable behavior changes are the main criteria by which to evaluate treatment.
- Treatment goals should be specified in concrete and objective terms.
- Research is incorporated in the therapeutic process as a way to evaluate therapeutic effectiveness.
- Because specific behavioral procedures have proven effective with certain target populations, clinicians working with these groups could benefit from learning such approaches.
- The therapist assumes an active, directive role in the helping relationship.
- Clinicians should encourage clients to be actively involved in selecting personal goals that guide the helping process. Also, they should enlist client cooperation in the therapeutic activities, both during the sessions in the office and in their everyday life.

SCENARIO

Julie, a 40-year-old woman, seeks treatment because of rising anxiety, which has increased to the point that she is now afraid to leave her home by herself. She tells you that she has to have someone with her whenever she goes outside and that she can't understand why this has happened to her. She knows that she feels uncomfortable and is afraid she might have a heart attack or "go crazy." She has never been in any counseling before and is unsure of the process of helping. She most wants to get over her fear of being alone as this limits what she can do. She says that she feels faint at times, has sweaty palms a lot, and is keyed up and tense. She has heard that there are some approaches that work well with her type of problem and she wants you to explain what those are and how they work! If you were utilizing a behavior-oriented approach to helping, how might you assist this client?

Cognitive Theory   The cognitive approach, which was developed much later, is now being effectively used with the behavioral approaches discussed above. Initially, when cognitive therapy emerged, many found the shift in clinical practice from a strong behavioral perspective to a cognitive one to be quite revolutionary (Baars, 1986; Dember, 1974). Not too many years later, this trend has resulted in the pairing of these two approaches, seen in the cognitive-behavioral approaches commonly used in clinical practice today. Key proponents of cognitive therapy include Albert Ellis, who developed the rational-emotive behavior therapy (REBT) (1960s) and Aaron T. Beck (1976), the founder of cognitive therapy. Ellis, who popularized this approach in the 1960s, has been widely recognized as the pioneer in cognitive therapy; he is from the discipline of clinical psychology. According to Corey (1996), Ellis's REBT assumes that individuals are born with the potential for rational thinking but tend to fall victim to the uncritical acceptance of irrational beliefs. The assumption is that thinking, evaluating, analyzing, questioning, doing, practicing, and redeciding are at the base of behavior change. REBT is therefore a didactic and directive model. Therapy is a process of reeducation based on the assumption that a reorganization of one's self-statements will result in a corresponding reorganization of one's behavior (Corey, 1996).

Beck, from the field of psychiatry, also became a prominent figure through his research and treatment applications in the cognitive arena. Beck, like Ellis, considers cognitions to be the major determinants of how we feel and act. Fundamentally, cognitive therapy assumes that clients' internal dialogue plays a major role in their behavior.

In recent years, the use of cognitive therapy has increased, largely because of its broad applicability to brief and short-term therapies. With its frequent use and proven effectiveness in the treatment of severe psychiatric illness, notably depression and anxiety, it is one of the most commonly researched forms of psychotherapy. Cognitive therapy is based on the premise the people often have maladaptive patterns of thinking that must be understood and altered for positive changes to result. Primarily, the therapy focuses on changing dysfunctional cognitions (thoughts), emotions, and behaviors. Because certain people have maladaptive patterns of information processing and related behavioral difficulties, one of the primary goals of cognitive therapy is the identification of negative and distorted automatic thoughts. These cognitions are the relatively autonomous thoughts that occur rapidly while an individual is in the midst of a particular situation or is recalling significant events from his or her past. People who suffer from anxiety and depression tend to have more negative and fearful automatic thoughts that control them and stimulate painful emotional reactions; consequently, cognitive therapy has been used with this population a great deal. Furthermore, because negative automatic thoughts can be associated with certain behaviors (such as helplessness, withdrawal, or avoidance), the problem can be intensified, creating a vicious cycle of dysfunctional cognitions, emotions, and behaviors.

Cognitive therapy aims to uncover the irrational and problematic thinking styles that often accompany emotional distress. These strategies are based on the established finding that one's feelings are a direct extension of one's thoughts. In other words, how you think determines how you feel. Central to this theory is the notion that our feelings are influenced to a large extent by the way we view life events. By changing our thoughts or thinking processes, we can change our resulting feelings as well. The goal of cognitive interventions is to challenge and ultimately change maladaptive, self-defeating cognitions, allowing the client to be more productive and satisfied in his or her life. Clients are taught how to detect cognitive errors and to use this skill in developing a more rational style of thinking. Cognitive therapy is thus directed, in part, at helping patients recognize and change these cognitive errors or distortions. Some of the typical cognitive distortions many of us use include all-or-nothing thinking, personalization, ignoring the evidence, and overgeneralizations.

Granvold (1994) describes three main cognitive intervention methods: cognitive restructuring, problem solving, and self-instruction training. In cognitive restructuring, clients are encouraged to identify dysfunctional schemata and thinking patterns that are considered "ill formed and interfering with the pursuit and attainment of goals" (Granvold,

1994, p. 20); they are then helped to modify these patterns. Cognitive restructuring can be best used in conjunction with other cognitive methods (i.e., thought stopping, covert sensitization, guided imagery) and behavioral methods (i.e., behavior rehearsal, skill training, deep muscle relaxation) (p. 23).

Problem-solving methods, depending on how they are employed, can also be considered a cognitive-behavioral procedure. D'Zurilla (1988, p. 86) defines problem solving as "a cognitive-affective-behavioral process through which an individual (or group) attempts to identify, discover, or invent effective or adaptive means of coping with problems encountered in everyday living." The problem-solving process is often enhanced when such methods as cognitive restructuring, social skills and assertiveness training, self-control techniques, and coping skills are used with it.

Self-instruction training is the third cognitive method we describe. Expanded by Meichenbaum (1975), self-instruction training involves the use of self-verbalizations to guide the performance of a task, skill, or problem-solving process. Through step-by-step self-statements, clients help themselves become competent in performing specific tasks and skills that have been targeted. The premise behind this method is that when adaptive self-statements are inserted between a stimulus and response to prevent maladaptive behavior, the stimulus/response chain is broken, thereby inhibiting the maladaptive response from occurring (Meichenbaum, 1975). This technique has been effective with a wide variety of client populations, including hyperactive-impulsive and aggressive children, those who are socially isolated, college students struggling with test anxiety, the elderly, and the mentally ill.

Cognitive therapy also includes a number of behavioral interventions such as activity scheduling and graded task assignments. Use of these procedures, as well as those discussed above, encourages a reversal of behavioral pathology and influences cognitive functioning.

Treatment Considerations

In using cognitive theory with clients, be aware of these considerations:

- Clients are born with the capacity for rational thinking, but often they have fallen victim to irrational thoughts. When this occurs, it is essential to assist them in understanding their maladaptive patterns of thinking, learning to dispute them, and eventually substituting more realistic thoughts for the irrational ones.
- Assume that thinking, evaluating, analyzing, questioning, doing, practicing, and reeducating are at the base of behavioral change.
- Be cognizant that a person's internal dialogue has a major impact on his or her behaviors and feelings.
- Help clients identify their distorted and automatic thoughts by guiding them in uncovering the irrational and problematic thinking styles that often accompany emotional distress.
- Help clients see that by changing their thoughts or thinking processes, they can change their feelings.
- Maintain the goal of helping clients to challenge and ultimately change maladaptive, self-defeating cognition so they can lead more productive and satisfying lives.
- Consider incorporating such cognitive methods as cognitive restructuring, problem-solving, and self-instruction training into your therapy.
- Consider the use of cognitive therapy when working with clients who present with anxiety and depression, as it is common for such clients to be controlled by negative and fearful automatic thoughts that can stimulate painful emotional reactions.
- Utilize a didactic approach that is also more directive.
- Maintain a reeducational focus in the therapeutic process.

SCENARIO

David is a middle-age man who comes in for help because of depression, isolation and extremely low self esteem. He says that he was married once, but that he is now single and does not date much if at all. He works as a parking attendant until late in the evening and doesn't get out too much to meet new people. He remembers his brother yelling at him a lot when he was younger and now finds himself repeating some of the same things his brother would say about him, such as "You will never amount to anything," and "You are stupid." He says that he puts himself down frequently and does not believe that he is very intelligent. He wants to feel better about himself but doesn't know how to do that on his own. How might you approach working with David if you were to use a cognitive-behavioral method?

Systems Perspective

The systems perspective relates to a systems theory approach; here the focus is on understanding individuals in the context of the surroundings that influence their development. Proponents believe that individual change can come about only when efforts are made to assess the way the individual's personality has been affected by his or her family as well as other systems. In our next chapter on mezzo practice, we examine various forms of family therapy and detail aspects of general systems theory and its importance in work with family systems. Here we focus our attention on developmental interventions that also entail taking a systems perspective.

Developmental Interventions

According to Ivey (1991, p. 18), "The purpose of counseling, therapy, and all helping interventions, is the facilitation of human development over the life span." Proponents of this view believe that considering an individual's developmental stage will lead to a more thorough understanding of the client. Furthermore, use of developmental strategies can help individuals reach their fullest developmental potential. Four main theoretical views are embodied in the developmental strategies identified by Ivey (1991). They are developmental theory, minority identity developmental theory, life span developmental theory, and developmental counseling and therapy theory. Below, we briefly discuss the most salient facets of each of these.

Developmental theory focuses on the attachments and separations that repeat throughout the life span. Emphasis is on understanding the separation and attachment theories of developmentalist John Bowlby (1969, 1973). He believed that to understand and appreciate the attachment-separation issues of our clients, we as helpers must be able to identify our own developmental connections and separations. Also, helpers will need to connect with clients to be able to understand their struggles and to encourage them to make the changes necessary to improve their lives. Without connecting with others, change is essentially impossible. According to Ivey (1991),

> Connection is as vital as separation. . . . We have twin tasks in our developmental progression. To survive, we must be simultaneously attached to others but also separate. Relationship, connection, and attachment provide a foundation for a sense of trust and intimacy. As we move through the developmental tasks of autonomy and identity, we must define our separate boundaries from our family and others. (p. 158)

Based on this viewpoint, helpers must be aware of their own attachment and separation issues to be able to form a therapeutic alliance with their clients, where trust and rapport are developed. Once helpers are self-aware of their own developmental issues, they can more easily facilitate this process for their clients.

Minority identity developmental theory (MIDT) explores the way members of minority groups view the world (Ivey, 1991, p. 18). Today's emphasis on multicultural issues highlights the importance of understanding an individual's culture and development, given his or her minority status. MIDT takes into consideration the developmental stages of various cultural minorities and of women. These five stages are shown in Table 3.4. When working with aspects such as diversity, ethnicity, socioeconomic status, and gender issues, helpers must understand the unique life situation of the particular client and the impact of diversity on the person's development.

Developmental counseling and therapy theory (DCT) provides specific guidelines for the helping interview and for implementing treatment plans. DCT is essentially a "new interpretation that illustrates how cognitive and affective development are manifested in the clinical and counseling interview" (Ivey, 1991, p. 4). In simpler terms, the therapy can be described as an integrative learning-based theory about human growth and change. Underlying this theory is the premise that there are often variations in the developmental levels of our clients. As professional helpers, we need to have a repertoire of skills and theories at our disposal that will allow us to adapt to the ever-changing needs of those we hope to help. Two central dimensions of DCT are instrumental in facilitating clients' growth and helping them expand their developmental potential.

First, specific questions and interventions can be used to facilitate client development in the interview. Basic to this counseling and therapy model is the

> premise that the actions and thoughts of the therapist affect how the client thinks about and discusses issues in the interviews. Given this, for each developmental level identified, specific developmental strategies have been generated that lead clients to talk about their issues at each developmental level. (Ivey, 1991, p. 54)

Second, traditional therapy and counseling theory can be organized in a developmental perspective. From this, helpers can use an array of skills ranging from crisis intervention to long-term treatment in an effort to help their clients reach their fullest potential. The specific skills, theoretical orientations, and intervention techniques are individualized to meet the specific needs of the clients; they are also based on the individual helpers' experience, role, and treatment preference.

Life span developmental theory includes both individual and family development and is centered on the individual developmental theories of Erik Erikson (1963) and the family life-cycle theories of Jay Haley (1973). Although criticized recently for being insufficiently oriented to women and lacking a multicultural emphasis, life span developmental theory has nonetheless contributed greatly to our understanding of human nature and thus is basic to any developmentally oriented practice. We focus here only on psychosocial theory and introduce family life cycle theory when we discuss working with families in Chapter 4.

Psychosocial Theory   Erikson (1904–1994) was a neo-Freudian theorist who, like Freud, concerned himself with the inner dynamics of the personality. Erikson proposed that the personality evolves through systematic changes. He focused on the individual through the life span and identified stages from birth to death that are common to most of us. In his work in 1963, he listed eight key stages of development, with accompanying developmental tasks at each stage. (These stages are elaborated on in Chapter 6 when we discuss diversity in the context of age issues). Erikson believed that human beings everywhere face

TABLE 3.4   Minority Identity Development Stages

Stage 1: Conformity
Stage 2: Dissonance
Stage 3: Resistance and Immersion
Stage 4: Introspection
Stage 5: Synergetic and Articulation and Awareness

SOURCE: Atkinson, Morten, & Sue, 1989, p. 44.

these eight major psychosocial crisis or conflicts during their lives. He postulated that if these developmental tasks were not completed, achieving maturity at the next stage would be difficult and perhaps even impossible. However, regardless of whether an individual successfully resolves the conflicts or crises of a particular stage, he or she is still pushed by both biological maturation and social demands into the next stage. In essence, according to psychosocial theory, if conflicts are not met or coping skills are not successful, individuals are likely to continue to struggle at each succeeding stage and perhaps throughout their lives. Conversely, "as individuals successfully resolve the central conflict of each stage of psychosocial development, they gain new personality strengths (or ego virtues)" that can help them at future crisis points (Sigelman & Shaffer, 1995, p. 270).

In considering Erikson's views about life span development, helpers need to be aware of the specific stage of development at which the client is and how he or she is coping with the particular psychosocial crisis or conflicts typical of that stage. Therapeutic interventions may focus on helping individuals achieve the tasks of their particular stage of development or in determining what factors may have interfered with their ability to successfully continue to the next stage. As helpers, we must be prepared to intervene in crisis situations, as well as help clients explore alternative coping skills that might be more effective.

Current life span perspectives consider all the stages from infancy to old age, with developmental issues of early and middle adulthood also emphasized. According to Baltes (1987), seven assumptions underlie the life span perspective. Because these assumptions are in line with our views about development, we have included a brief summary of them below:

1. Development is a lifelong process; it occurs throughout the life span rather than just in childhood.
2. Development is multidirectional in that it can take many different directions, positive or negative, depending on the aspect of functioning being considered and on the individual.
3. Development always involves both gains and losses at every age.
4. There is much plasticity (capacity to change in response to positive or negative experiences) in human development.
5. Development is shaped by its historical/cultural context.
6. Development is multiply influenced.
7. Understanding development requires multiple disciplines.

Using a developmental framework can be both simplistic and complex. Assessment is essential in determining the client's developmental level and the best course of action. Perhaps the most beneficial aspect of a developmental focus in practice is its versatility; a developmental perspective can be readily incorporated into most therapeutic approaches. For instance, elements of an individual's developmental milestones can be easily included in psychodynamic, humanistic, cognitive, and behavioral treatment strategies without interfering with the specific approach being employed.

Treatment Considerations

As you think about incorporating developmental aspects into your practice, consider the following:

- In your assessment of a client, determine the person's current developmental stage. This can help you in creating a sound treatment plan that is tailored to the client's functioning level.
- It is important to understand psychosocial and family life-cycle theory to know whether the client is on track or struggling with a particular developmental crisis or conflict. This knowledge can alert the helper to the importance of exploring underlying issues to understand the particular crisis or conflict that is presented.

- Helpers must be skilled at determining whether the presenting problem is symbolic of a delay in development or of an unrealistic expectation the client has for himself or herself or an expectation others have of the client.
- Helpers must be aware of the many diversity issues—gender, culture, race, and ethnicity, age, socioeconomic status—that may have a direct bearing on an individual's success in achieving prescribed developmental milestones.
- Helpers should be advocates for their clients who for various reasons have not achieved the developmental level where they should be ideally, given their chronological age. Also, helpers may need to educate those within the client's surroundings about what is most realistic for the clients.
- Helpers must be sensitive to internal and outside forces (physical, interpersonal, and social) that can affect a client's ability to cope with or overcome a certain crisis.
- A positive focus is crucial in helping clients achieve their fullest potential. Considering elements of the *strength's perspective* is essential in assisting clients to identify their strengths and feel hopeful about the future. (The strengths perspective is discussed later in the chapter.)

SCENARIOS

Bill and Violet, who are a married couple, come in to work on their "issues." Bill complains that his wife is extremely critical of him, especially when she is anxious about something in her life. Violet complains that he is unresponsive to her and does not support her when she needs him most. Bill says that he would support her more, but that her frequent complaining and verbal attacking pushes him further away and that he then feels more hurt and less like supporting her as she wishes. Violet says that she is very hurt and very much wants his support, especially when she is feeling anxious. You observe that this couple appears "stuck" in their interactions with each other and that both appear to have legitimate complaints about the other. In exploring the clients' histories, you discover that Violet was raised by a single mother; her father died when she was only 4. She therefore has little experience with men and had no brothers or other male figures to relate to when she was younger. Violet also states that she had difficulty separating from her mother, whom she was expected to take care of in many ways. Bill had an overprotective mother and a father who was largely absent when he was younger. He says that his mother did not let him go easily and that she had made him into her "little husband" when he was younger. Do you see any potential developmental issues that may be contributing to the marital difficulties in this situation? What might you have done if you had been using a developmental approach to help this couple change the patterns they were in?

A young Vietnamese woman enters the school counseling center where you work asking for help with her classes in school. She says that she has been in America for only a few years and that her grades have begun to fall because of her performance in the classroom. As you delve further, she tells you that she is the oldest girl in the home and because her parents speak no English, she is expected to take care of them often. She has to be at home to translate and help care for her seven brothers and sisters. The client reports that her friends are frustrated with her as she is always busy at home and can never go out to study or have fun. She feels pressure because her father expects her to make high grades in school yet also expects her to help at home a lot. When she has tried to be assertive with her parents by asking for time to study or to go out with friends, they become very upset and tell her she must respect them. Considering the developmental perspective, how would you approach working with this client? In what stage is this client stuck? What cultural and diversity issues make this a more complex client situation?

> Bobby is an 11-year-old boy who is brought in by his grandparents for treatment. The grandparents state that Bobby lost his mother in an apartment fire about a year ago. His mother and father were separated at the time, and Bobby had little contact with his father. He has since come to live with his maternal grandparents who began noticing that he was behaving in ways that seemed younger than his real age. For example, the grandparents state that he has begun wetting the bed almost every night, behavior that has not been a problem for many years. He also is very afraid to be alone and never wants to be away from his grandparents. He has been refusing to go to school, which has the grandparents very worried and upset. They state that he seems to be very insecure, afraid, and needing frequent reassurance. They report that he wakes up often in the middle of the night with nightmares and sometimes has difficulty going back to sleep. What environmental forces do you see at work in this situation? How are these changes impacting the client? How might you proceed to work with this child using the developmental perspective?

Use of Multiple Interventions: An "Integrative" Approach

In the reality of the present-day managed care system, in which treatment services are often limited to crisis and brief treatment, human service practitioners must be able to integrate their approach to helping. According to Corey (1996, p. 447), "There are clear indications that since the early 1980's psychotherapy has been characterized by a rapidly developing movement toward integration and eccletism."

Various authors (Corey, 1996; Goldfried & Castonguay, 1992; Neukrug, 1994) highlight the need to incorporate the best of many orientations into an integrated approach that embodies the most applicable features of a number of theoretical models. An integrative focus involves selecting concepts and methods from a variety of systems to create a model that is most suitable for working with specific clients in a way that meets agency demands for brief treatment.

The move toward this eclectic or integrated approach is strong, with "practitioners of all persuasions increasingly seeking rapproachement among various systems and an integration of therapeutic techniques" (Corey, 1996, p. 448). Norcross and Newman (1992) cite eight motives that are probably responsible for promoting this trend toward psychotherapy integration:

1. The mere expansion of therapies
2. The reality that no one theoretical model is adequate to address the needs of all clients and all problems
3. The restrictions by insurance companies and health care companies that mandate short-term treatment
4. The increased popularity of short-term, prescriptive, and problem-focused therapies
5. The opportunity this climate affords clinicians to experiment with a variety of therapies
6. The deficiency of differential effectiveness among existing therapies
7. Increased awareness that therapeutic commonalities play a major role in determining therapy outcome
8. The development of professional groups who foster this integrative movement

Corey (1996) has summarized well what is perhaps the central cause of this new direction in treatment: "One reason for the trend toward psychotherapy integration is the recognition that no single theory is comprehensive enough to account for the complexities of human behavior, especially when the range of client types and their problems are taken into consideration" (p. 449).

Corey continues, "No one theory has a patent on the truth, and because no single set of counseling techniques is always effective in work with diverse populations" (p. 449), there are those who do not believe it makes sense to follow a single theory. We concur, and also add that an integrated approach seems the most plausible given the vast diversity of practitioners who also have different styles and preferences for practice. This approach allows helpers to adopt the facets of various theories and treatment modalities they have found most effective with their particular clientele. This approach affords practitioners more opportunities to tailor their treatment to meet specific client needs and to use the modalities they are most comfortable with and confident in.

However, like any new endeavor, the integrated approach has both positives and negatives. The main criticism is tied to the element of choice offered to clinicians. Neukrug (1994, p. 71) cautions that helpers who use this approach must not do so haphazardly. They should "carefully reflect on their views of human nature and draw techniques that fit their ways of viewing the world." He adds that from his experience he has "seen many individuals who call themselves eclectic use a hodgepodge of techniques, which may end up confusing the client." A similar view is echoed by others (Corey, 1996; Lambert, 1992; Norcross & Newman, 1992) who believe that developing an integrated approach is an intricate process requiring much effort by the practitioner. He or she must be committed to learning about various theories and be open to deciding which key concepts of each approach best fit his or her own personality. Perhaps most important, the helper must always put the needs of the clients first in selecting treatment techniques.

Incorporating Diversity-Sensitive Issues into Practice

The need to be diversity-sensitive in one's practice cannot be overstated. Working in the human services field means working with diversity at all levels, so practitioners must be aware, knowledgeable, and skilled in treating a variety of clients. Helpers are bound to work with clients who will differ from them in age, gender, ethnicity, race, culture, socio-economic status, sexual orientation, lifestyle, life circumstances, or basic values. Learning to appreciate these differences while attempting to understand the client's needs is an ongoing challenge for the human service professional. Because of the significance of this issue in practice, we dedicate an entire chapter (Chapter 6) to practice with diverse populations. At this point, we want you to begin to consider how you might integrate your knowledge about theory with real-life practice that encompasses a wide variety of clients.

DIFFERENT TREATMENT MODALITIES

Theoretical orientations are the foundation for practice and treatment modalities, the building blocks. In this section, we present an overview of various treatment modalities commonly used in practice today, including crisis intervention, brief therapies, long-term treatment, and case management. The selection of these specific areas of focus is based on the experiences of our students and colleagues, as well as our own practice in a variety of human service agency settings. Our purpose is to give you a comprehensive view from which to begin cultivating your own style and approach to helping.

Crisis Intervention

Crisis intervention is a process to assist individuals in finding adaptive solutions to crises in their lives. Crises are a natural part of living; yet for some, the ability to cope effectively in certain crisis situations can be seriously compromised. In such circumstances, professional assistance is warranted. Because of the short-term nature of a crisis, crisis intervention is typically also considered a brief treatment modality.

Crises are commonplace in today's fast-paced life where stress is a familiar experience. Although many of us have learned to adapt, this does not mean that we are immune to needing or desiring professional assistance at some point in our lives. Seeking crisis treatment is further accentuated by the growing prevalence of brief, short treatment within the mental health arena. Historically, there has been a financial impetus for the growth in crisis intervention services, as traced by Kanel (1996):

> Crisis intervention first became valued in the mental health community in the 1970's as economic conditions led to the use of new and developing community resources. Managed care, which grew stronger in the 1980's and flourishes in the 1990's through the establishment of HMO's and PPO's, and EAP health care companies, is believed to be the most cost effective way for insurance companies and places of employment to provide mental health services for [their] employees. (p. 1)

In addition to the growing, private managed care industry that is influenced by financial pressures, a short-term treatment focus is also emphasized in most public and non-profit agencies. These organizations struggle to accommodate a high volume of clients in the most cost-effective way possible. For helping professionals, the trend toward brief therapy heightens their need to be familiar with crisis intervention and brief therapy models, as they will most likely be using such models in their work with clients. This situation is true for both the public and the private sectors.

The Nature of a Crisis

The term *crisis* conjures up many images, with each representing a crisis in a different way. In general, a crisis is an internal reaction to stressors for which normal coping mechanisms have either broken down or are no longer effective. Guilliland and James (1993) summarize crisis as "a perception of an event or situation as an intolerable difficulty that exceeds the resources and coping mechanisms of the person. Unless the person obtains relief, the crisis has the potential to cause severe affective, cognitive, and behavioral malfunctioning" (p. 3). Caplan's definition of a crisis is similar and also central to current-day crisis intervention models. He sees "a crisis [as] an obstacle that is, for a time, insurmountable by the use of customary methods of problem-solving. A period of disorganization ensues, a period of upset, during which many abortive attempts at a solution are made" (Caplan, 1961, p. 18).

Many who write about crisis intervention cite the Chinese character that means both danger and opportunity (Aguilera & Messick, 1982; Hoff, 1989; Guilliland & James, 1993; see Figure 3.1).

"Opportunity in the midst of crisis" is the literal translation of this Chinese phrase. The ancient form for danger is (危), depicting a person on the edge of a precipice; the ancient form for opportunity is (机), which is believed to depict a cocoon—a symbol of transformation. In essence, then, a crisis can be seen as a turning point in a person's life: It is an opportunity for growth in that it may require decisions that can be life changing. A crisis is typically time-limited with the resolution either being positive (opportunity) or negative (danger). Often, the chance for a positive outcome is heightened when professional services are sought and crisis intervention techniques are employed. "Successful crisis intervention involves helping people take advantage of the opportunity and avoid the danger inherent in crisis" (Hoff, 1989, p. 8).

Characteristics of Crisis

Guilliland and James (1993) elaborate on various characteristics of a crisis that we outline briefly here. These characteristics are typical in most crisis situations, but may not be present in all.

- *Presence of both danger and opportunity:* Crises often overwhelm the individual, resulting in serious pathology including homicide and suicide. However, a crisis is also an opportunity because the pain it induces motivates the person to seek help.

FIGURE 3.1 Chinese Character for *Danger* and *Opportunity*.

• *Seeds of growth and change:* If the individual takes advantage of the opportunity brought on by a crisis, the intervention can help plant the seeds of growth and self-realization. Janosik (1984) contends that "in the disequilibrium that accompanies crisis, anxiety is always present, and the discomfort of anxiety provides an impetus for change" (p. 39). This point can be seen as a threshold for change, whether reached early or as a last-ditch effort, it can still be viewed as positive.

• *Complicated symptomatology* is often presented. In many instances, a crisis is not simply based on cause and effect. Crises are often complex and difficult to understand. They may exist at various levels: individually, interpersonally, and socially. Intervention therefore must be directed at any or all of these levels for positive change to result.

• *The absence of panaceas or quick fixes* indicates that these interventions usually don't bring lasting relief, even though people often try them. In fact, clients who are suffering from severe crisis are often in such deep distress because they sought quick fixes in the first place. (Guilliland & James, 1993, p. 5).

• *Necessity of choices* exists at a very basic level in crisis situations. For instance, not to choose is a choice, which may usually turn out to be negative and destructive. In contrast, choosing to do something at least affords the person the opportunity to make changes in his or her life.

• *Universality and idiosyncrasy:* Janosik (1984, p. 13) contends that disequilibrium or disorganization accompanies every crisis, whether universally or idiosyncratically: universal because no one is immune to crisis, and idiosyncratic because each of us will experience and cope with a similar crisis in different ways.

Theoretical Foundation of Crisis Theory

Although professional crisis intervention is relatively new, in a broad sense, crisis and crisis intervention are as old as humankind. According to Hoff (1989),

> Helping other people in crisis is intrinsic to the nurturing side of human character. The capacity for creating a culture of caring and concern for others in distress is implicit in the social nature of humans. In a sense, then, crisis intervention is human action embedded in culture and in the

process of learning how to live successfully through stressful life events among one's fellow human beings. (p. 9)

The practice of crisis intervention is based on humanistic foundations. Its roots, however, are interdisciplinary in nature. Janosik (1984) conceptualizes crisis theory on three different levels:

Basic crisis theory, illustrated by Lindemann's focus on immediate resolution of grief after loss, and Caplan's expansion of crisis to encompass the total field of traumatic events.

Expanded crisis theory, which addresses the social, environmental, and situational factors that make an event a crisis. Included in this expansion are the following major theoretical components: psychoanalytic theory, systems theory, adaptational theory, and interpersonal theories.

Applied crisis theory, which emphasizes the need for a flexible approach because each person and each crisis situation are different.

To individualize each crisis situation, three domains must be considered: normal developmental crisis, situational crisis, and existential crisis. Below we take a closer look at the historical contributors to crisis theory that continue to govern crisis intervention models currently in use today.

Historical Views   Historically, psychoanalytic theory has been influential in crisis theory. Freud contributed to the study of human behavior and the treatment of emotional conflict; and despite its limitations, Freud's work laid the foundation for seeing people as complex beings capable of self-discovery and change. His emphasis on listening and catharsis (expression of feelings about a traumatic event) remain essential to helping and are especially necessary for effective crisis intervention. Furthermore, psychoanalytic theory maintains that some early childhood fixation is the primary explanation of why an event becomes a crisis. This theory may be used to help clients develop insight into the dynamics and causes of their behavior as the crisis situation influences them.

Ego psychologists such as Fromm, Erikson, and Maslow, followed. They contributed greatly to the philosophical base for crisis theory in their views that people have the ability to grow throughout life. Also, what has been labeled *military psychiatry* is said to have greatly influenced crisis theory as well. During World War II and the Korean War, crisis intervention techniques were first used when members of the military who felt distressed were treated at the front lines whenever possible, rather than being sent back home to psychiatric hospitals. By receiving immediate help, they were able to return to combat duty rapidly and seemed able to function appropriately afterward (Hoff, 1989, pp. 10–11).

Not until the 1940s, however, were individuals suffering from grief reactions actually presented with this alternative approach that differed so significantly from traditional psychoanalytic therapy. Kanel (1998) cites three instances during this time when crisis intervention techniques were instrumental in helping individuals cope with crisis. First, the Coconut Grove fire in Florida in 1942, which killed approximately 493 people, left many families and friends facing issues of loss. Second, after World War II, many women became pregnant and the number of miscarriages and stillbirths resulted in grief and loss. Third, many veterans returning from war were suffering from what we now call posttraumatic stress disorder, previously known as "shell shock." These veterans were in desperate need of a short-term, direct approach to mental health that could ease their symptoms more rapidly than long-term psychodynamic work.

The two best-known founders of crisis intervention as we know it today are Erick Lindemann and Eric Caplan. Lindemann studied bereavement following the Coconut Grove fire and defined the grieving process people went through after the sudden death of a relative. He found that survivors who developed serious psychopathologies had failed to go through the normal process of grieving. Lindemann's work on grief evolved through time.

He stipulated that grief work involves the process of mourning the loss, experiencing the pain of such loss, and eventually accepting the loss and adjusting to life without the loved person or object. Moreover, Lindemann felt that "encouraging people to allow themselves to go through the normal process of grieving can prevent negative outcomes of crisis due to loss" (Hoff, 1989, p. 11). Although Lindemann's work centered on grief, the theme of loss that is common to most crisis situations has made the greatest impact on contemporary crisis theory.

Caplan, working with Lindemann, called his new, brief approach *preventive psychiatry* (Lindemann, 1944). In 1964, he developed a conceptual framework for understanding crisis. Lindemann viewed people in crisis as being unable to take charge of their lives. He believed that when people accept this view of themselves in crisis they will be less likely to participate actively in the resolution of their crisis and more apt to foreclose the possibility of growth. He also believed that if a crisis could be resolved during the intense and vulnerable four to six weeks after the trauma, the victims would be less likely to develop psychiatric disorders. (Slaikeu, 1990, p. 7).

Caplan contributed to the development of a communitywide approach to crisis intervention. He encouraged prevention by emphasizing public education programs and consultation with such professionals as teachers, police officers, and public health nurses. Together with Lindemann, Caplan created a training clinic in Massachusetts called the Wellesley Project that incorporated the use of both professional and paraprofessional mental health workers. He noted that with brief training in the specific crisis intervention model, community volunteers and clergy could be quite effective in helping patients return to precrisis levels of functioning.

By the 1960s, the emphasis on crisis intervention had resulted in government passage of several bills funding community mental health centers that were to focus primarily on crisis intervention and suicide prevention. This trend carried over into the 1970s when many nonprofit agencies continued to use paraprofessionals and short-term crisis intervention in their work with specialized populations. Also in the 1970s the focus on systems theory affected how crises were viewed. The traditional approaches, which focus only on what is going on with the client, were replaced by a more interpersonal-systems way of thinking. The emphasis on looking at crises in their total social and environmental settings began to take precedence.

Interpersonal theory proposed by Carl Rogers also influenced crisis theory. According to interpersonal theory, people will not be able to sustain a personal state of crisis for very long if they believe in themselves and others and have confidence that they can become self-actualized and overcome the crisis (Hoff, 1989, p. 18).

In the 1980s, managed care proliferated throughout the mental health system. A cognitive-behavioral emphasis became prominent during this time. This approach is tied to adaptational theory, which characterizes a person's crisis as being sustained through the person's maladaptive behaviors, negative thoughts, and destructive defense mechanisms. The premise behind adaptational crisis theory is that the person's crisis will subside when these maladaptive coping behaviors are changed to adaptive behaviors. The trend toward short-term and crisis intervention practice that is interdisciplinary in focus continues today in the 1990s and is predicted to last well beyond the year 2000.

Crisis Intervention Models

Through the years, many different crisis intervention models have been developed in response to the changing climate of the mental health system and to the types of crises that seem most prevalent at the time. In all these models, the emphasis is on providing prompt and skillful intervention. Theorists believe that immediate involvement can help facilitate a restoration of the client's previous state of equilibrium, can help prevent the development of serious long-term problems, and most optimistically, may allow new coping patterns to emerge that can help the individual function at a higher level of equilibrium than before the crisis.

Goals   Crisis intervention begins at the first moment of contact with clients. Goals are limited from one to four that are chosen by the client. A focus on limited goals and objectives is essential for crisis intervention because clients in crisis are often disorganized and require a clear, direct, and gentle approach. Identifying and encouraging too many goals may further contribute to their disorganized state or perpetuate feelings of further chaos.

Intervention is time limited, usually from four to six sessions, although it can range from one to twelve. Often concrete services, along with counseling and referral to community resources, are provided by the helper.

As proposed by Lydia Raport (1996), crisis intervention is guided by six primary goals, all of which are aimed at stabilizing and strengthening the client and his or her family system. Below we present an adapted version that focuses primarily on the individual client.

- Relieve the acute stress symptomatology
- Restore the client to optimal precrisis level of functioning
- Identify and understand the relevant precipitating event(s)
- Identify remedial measures that the family can take or that community resources can provide to remedy the crisis situation
- Establish a connection between the current stressful situation and past experiences
- Initiate the client's development of new ways of perceiving, thinking, and feeling, and adaptive coping resources for future use.

Steps in Crisis Intervention   Steps involved in crisis intervention vary greatly and depend on a variety of factors: the theoretical orientation and model being used, the specific population and nature of the problem, the time constraints and available resources, and the client's motivation and abilities. We present an outline of a comprehensive nine-step approach that includes features of most effective crisis interventions.

Step 1: *Rapidly establish a constructive relationship:* Essential here are sincerity, respect, and sensitivity to clients' feelings and circumstances, active listening, and developing rapport that is fostered through trust and unconditional positive regard.

Step 2: *Elicit and encourage expression of painful feelings and emotions:* Anger, frustration, and feelings related to the current crisis are the primary focus, with linkages to the past made later.

Step 3: *Discuss the precipitating event:* Focus is on examining when and how the crisis occurred, the contributing circumstances, and how the clients attempted to deal with it.

Step 4: *Assess strengths and needs:* Assessment begins immediately and continues throughout; these strengths and needs are related to the current crisis and tapped to improve self-esteem and enhance motivation.

Step 5: *Formulate a dynamic explanation:* The meaning of the crisis and its antecedents as seen by the clients are explored, with emphasis on the meaning and perceptions of the crisis.

Step 6: *Restore cognitive functioning:* Clients are helped to identify alternatives for resolving the crisis that center on reasonable solutions toward which they are motivated to work.

Step 7: *Plan and implement treatment:* Clients are assisted in the formulation of short- and long-term goals, objectives, and action steps based on what they choose as priorities; the clients' active participation is emphasized.

Step 8: *Terminate:* Termination ideally occurs when clients have achieved their precrisis level of functioning. The helper reviews with clients the precipitating event(s) and response(s) and the newly learned coping skills that can be applied in the future.

Step 9: *Follow-up:* There may be continued contact with clients and/or referral sources on predetermined dates: this action places appropriate pressure on clients to continue working on the issues in a positive way and also provides them with reassurance that they are not abandoned.

Basic Crisis Intervention Models   A majority of the crisis intervention models used today incorporate some if not all these steps. In terms of specific models, Leitner (1974) and Belkin (1984) have identified three basic crisis intervention models, described next, that seem to lay the basic groundwork for many of the other models in existence.

The equilibrium model, which, as we have seen in Caplan's work, relates to the fact that people in crisis are in a state of psychological or emotional disequilibrium; their typical coping mechanisms and problem-solving abilities are ineffective. The goal of this model is to help clients reattain their previous state of equilibrium (precrisis state). The model is most effective when used early at the onset of a crisis and focuses solely on helping clients return to their previous level of coping; emphasis is not on exploring underlying factors that may be related to the crisis.

The cognitive model focuses on helping people become more aware of and to change their views and beliefs about the crisis events or situations in their lives. It is based on Ellis's premise about irrational, faulty thinking. According to Guilliard and James (1993), Ellis believed that "Crises are rooted in faulty thinking about the events or situations that surround the crisis—not in the events themselves or the facts about the events or situations." This model has proven very effective for many who tend to function at a cognitive level. However, because of the unstable nature of a crisis, use of cognitive techniques may need to follow efforts to first stabilize the client.

The psychosocial transition model takes a more holistic, biopsychosocial approach. This model

> assumes that people are products of their hereditary endowments plus the learning they have absorbed from their particular social environments. Since people are continually changing, developing, and growing, and their social environments and social influence are continuously evolving, crisis may be related to internal or external (psychological, social, or environmental) difficulties. (Guilliland & James, 1993, p. 20)

This view of crisis is perhaps the most comprehensive. It does not perceive crisis as simply an internal state; rather, it explores the role of interpersonal, social, and environmental systems in the crisis. This model affords a more collaborative focus as well. Helpers are encouraged to collaborate with clients in their assessment of the many possible factors (internal and external) that may have contributed to the crisis. Furthermore, clients are invited to explore and choose workable alternatives to their current behaviors, attitudes, and use of environmental resources. Outside systems are examined and often earmarked as areas where change must also be targeted. Because of the in-depth focus of this model, it is recommended to follow stabilizing efforts common to the equilibrium model.

ABC Model of Crisis Intervention   A fourth model, the *ABC model* proposed by Kristi Kanel (1999), is one we have chosen to include because of its versatility and propensity to incorporate facets of all the three models described above. According to Kanel (1999, p. 1),

> the ABC Model was implemented at a variety of community human service agencies with most populations experiencing both situational and developmental crises. It is a model that can be appropriately used by both the beginning counseling student and the highly trained and experienced counselor.

Kanel emphasizes the *perception* component in crisis definition. She believes that it is the client's perception of the crisis that is the most crucial to identify. Taking a cognitive view, Kanel contends that precipitating events usually cannot be changed; painful emotions responding to these events cannot be altered without first altering the person's cognitions or behaviors. The focus must therefore be in altering the client's perceptions of the stressors that will allow the distressing emotions to ease and effective coping to be restored. Kanel outlines two formulas for understanding the process of crisis intervention and increasing lowered functioning.

Formula for understanding the process of crisis intervention:
Precipitating Event → Perception → Subjective Distress → Lowered Functioning

Formula to increase functioning:
Alter Perception → Decrease Subjective Distress → Increase in Functioning and Coping
(Kanel, 1999, p. 6–7)

These two formulas are integral to the ABC model, explained next.

A: Developing and Maintaining Contact   As in most helping approaches, the purpose of the contact phase is to develop rapport. Because of the short-term nature of crisis treatment, however, rapport must be developed quickly and maintained throughout the process. Humanistic skills are essential in creating a climate of safety that is crucial to most crisis clients, who want fast relief. Because of their vulnerable state, they need emotional support and validation to help them maintain a sense of optimism and hope. Similarly, clients are typically more responsive when the helper possesses the Rogerian traits of empathy, genuineness, and unconditional positive regard.

In addition to setting the tone, basic attending skills are essential here. They include attending behavior (eye contact, warmth, body posture, vocal style, verbal following, and empathy/focus on the client); questioning (open-ended questions); clarifying (paraphrasing, accurate reflection of information); reflection of feelings; and summarizations that tie together feelings, facts, events, distress, and meanings.

B: Identifying the Problem and Therapeutic Interaction   In the second phase, the crisis is specifically identified and helpers provide clients with feedback on how they can change their perceptions of the precipitating event. The helper must explore the client's existing distortions and misconceptions and challenge these with more realistic views about the situation through education and reframing statements. Support and validation continue to be necessary in assisting clients to overcome their upset feelings.

During this phase, helpers must also assess for various ethical issues such as suicide, child abuse, substance abuse, and organic or medical conditions that may require outside referrals and immediate follow-up. Identifying the precipitating events that may have triggered the crisis at hand is essential in helping the client understand the situation differently. Also, an assessment of the client's current level of functioning is important in providing a baseline for knowing when treatment has been effective; it can also be helpful in giving the client hope in the treatment process when a higher level of functioning is sought.

C: Coping   In the final phase, the client's current coping is assessed and the client is helped to develop new coping strategies and mechanisms. It is important for the helper to end a session with the general sense that some follow-up work will be done by the client. Because of the immediacy issues relevant to crisis, it is crucial for coping strategies to be attempted without delay. If too much time passes, clients may become accustomed to their lowered level of functioning or they may adapt by using maladaptive coping methods. In either case, when clients lose their motivation to pursue alternative coping, the chances drop that they will be able to grow.

Active participation by clients in exploring solutions is integral to this process. Clients are gently pushed and encouraged. Time constraints are important to consider, especially because the window of opportunity for crisis intervention is from four to six weeks. After that, the client's typical ego defense mechanisms are likely to take over and perhaps block therapeutic movement and efforts to try out new coping strategies.

Coping attempts by the client are encouraged first before the helper offers advice or referrals. Similarly, clients are motivated to consider alternative coping methods to deal with their problems. Client participation is instrumental; clients are more likely to follow through on the suggestions they have provided than on those offered by the helper. Once a mutually agreed-on coping plan is made between client and helper, a commitment and follow-up should come next. This may be in various forms: a telephone call, further appointments, or a contract termination follows when appropriate.

Treatment Considerations

Short-term crisis intervention has unique aspects that you should consider.

• Remember that crisis intervention is a time-limited, active, and directive intervention process.

• Establish a relationship by reinforcing the client's help-seeking behavior, acknowledge the client's willingness to seek support and change. Invite open discussion and convey your involvement, commitment, and acceptance of the client (humanistic characteristics). Safety and trust issues are of utmost importance. Instill a climate of emotional safety.

• Identify the problem by assessing the situation, understanding past events as they relate to the crisis (precipitating event), determining the strengths and needs of the client, and examining previous coping skills.

• Ensure client safety by assessing the severity of the crisis and the risk for danger (e.g., suicide, homicide, child abuse, substance abuse, violent behaviors) you must be prepared to deal with—directly or through providing appropriate referrals. Helpers must be prepared to assess lethality, criticality, immobility, or seriousness of threats to clients' physical and psychological safety.

• Once the problem is identified, you may need to ask the following questions:

1. Is it short or long term?
2. Is it constructive? Is it good for the client's growth and development?
3. Is it a new problem or a chronic situation?
4. What impact does this really have on the client's life?
5. How does the client feel?

• Clients must take ownership and control of their feelings. Helpers need to communicate a nonjudgmental acceptance of the person's feelings by allowing the client to express them and by responding empathetically. Focus on the here and now; never minimize the client's feelings; instead, respond to the source of the feelings. Summarize, focus, and clarify; attend to possible conflicting feelings; and recognize possible misperceptions of events leading to feelings (cognitive distortions or beliefs resulting from irrational, faulty thinking).

• Explore alternatives by examining how the client has handled similar situations in the past; determine the person's strengths and potential to provide self-support. Place responsibility on the client. Assist by exploring alternatives, identifying advantages and disadvantages of possible options. Develop a concrete plan of action, suggest referrals and other sources of help if appropriate, and review the entire process in a final summary.

SCENARIO

Tanya, a middle-age, separated African American female, enters treatment at an outpatient mental health center in crisis over the sudden loss of her adult daughter. Tanya was very close to her only daughter, and although they lived separately, they saw each other frequently and shared many pleasant activities. Her daughter was attending college and working part time. She had been doing well until recently when she developed a breathing condition she was having treated. One evening, she became quite ill with a virus. A close friend who happened to be visiting her became worried and called Tanya to inform her of her daughter's condition. Tanya was not home and listened to the message later that evening on her arrival home. Because the message did not sound urgent, Tanya did not want to disturb her daughter so late, so she called early the next morning. When no one answered, Tanya became concerned and went to her daughter's apartment. She found her daughter dead in bed. She entered treatment shortly after her daughter's funeral with extreme feelings of sadness, anger, and guilt. If you were on crisis duty and were scheduled to work with Tanya, what crisis intervention strategies might you consider? How would you approach her? What specific issues would you focus on? What might your treatment goals be?

Critical Incident Debriefing

A crisis overrides an individual's normal psychological and biological coping mechanisms. Critical incidents, by their very definition, are serious crisis situations that almost anyone would have difficulty with. Examples of a critical incident include violent, armed robberies; assaults, or home invasions; a "near miss" that almost costs someone his or her life; an accident that injures, maims, or kills someone; a natural disaster such as earthquake, fire, flooding, tornado, or hurricane; or any other event that is sudden and emotionally overwhelming. The tragic results of these incidents can have consequences that reach beyond those directly injured. Those who witnessed the incident or who know the people involved may suffer extreme emotional upsets. Common reactions to a highly stressful or traumatic critical incident include these:

- The person feels jumpy, anxious, moody, or irritable.
- Concentration may be impaired; decision making may be difficult, as is thinking clearly.
- Individuals may have trouble going near the scene where the critical incident took place or to places that trigger memories of the incident.
- Often, if the incident occurred at work, workers may avoid work and may feel differently about their jobs and the workplace.
- They may experience trouble being around people or being alone. Some may withdraw or turn to drugs and alcohol; others may suffer from nightmares or have flashbacks to the event, often fearing the event will recur.
- People may also have feelings of anger, fear, or guilt. Some may feel responsible for the incident or blame themselves for not doing more to prevent the incident or help the victims.
- People's home life and personal relationships may suffer; those involved may displace their feelings of guilt, powerlessness, or anger on their families.
- They may feel overwhelmed and unable to cope with daily events. Often people seem numb, shocked, and helpless.
- Some may have difficulties sleeping, eating, and experiencing pleasurable activities again.

Because critical incidents may seriously affect the emotional well-being of those involved, directly or indirectly, it is important to respond to these incidents appropriately and affectively. Elements of crisis intervention have been incorporated into what is commonly known today as critical incident stress debriefing (CISD). Specific guidelines, identified below, must be followed in succession to ensure successful debriefing. The type of event is not as important as the impact it has had on those involved. Also, the mental health professionals and peer support personnel who make up the debriefing team must be well trained and experienced in the debriefing process.

As described by Davis (1995), the **CISD** is structured as follows:

Introduction, when ground rules are set.

Fact Phase, when participants are encouraged to tell their story about the event.

Thought Phase, when participants state their first thoughts on exposure to the worst part of the incident.

Reaction Phase, when participants tell their feelings and emotional reaction to the incident.

Symptom Phase, when the group discusses the physical, emotional, and behavior changes that have occurred in their lives since the incident.

Teaching Phase, when the team provides reassurance that what the group is experiencing is a normal reaction to an overwhelming, tragic incident.

Reentry Phase, the concluding phase that involves the group in asking questions and then summarizing incidents and responses.

As with crisis models, critical incident intervention follows basic steps that help individuals return eventually to their previous level of coping. Treatment is facilitated when both the helpers and the clients are amenable to the process and are working together toward the same end result. Both interventions need experienced helping professionals who understand crisis theory.

Brief Treatment Perspectives

As noted in our discussion of crisis intervention, there is a growing demand for brief therapy models, given the managed care mandates of today's mental health system. Although brief treatment modalities have been utilized before, only within the past ten years have clinicians begun to take this treatment approach seriously. With a limited number of allowable sessions becoming the norm in both public and private clinical settings, many therapists are considering the actual benefits of brief therapy. They are finding that a number of problems can indeed be resolved by brief therapy. Moreover, some who have questioned the relevance and effectiveness of traditional long-term treatment interventions are finding increased success using a brief therapy framework.

Brief therapy is defined not only by the length of treatment but also as a way of solving human problems (De Shazer, Kim Berg, Lipchik, Nunnally, Molnar, Gingerich, & Weiner-Davis, 1986). Originating in Milton Ericson's 1954 paper "Special Techniques of Brief Hypnotherapy," the key feature of brief therapy highlights "utilizing what clients bring with them to help them meet their needs in such a way that they can make satisfactory lives for themselves" (De Shazer et al., 1986, p. 208). Despite the recognition that underlying issues may be directly related to the presenting problem, in brief therapy, no attempt is made to correct any of these "causative underlying maladjustments" (Ericson, 1954, p. 393).

The growth of family therapy in the late 1960s and early 1970s directly affected developments in brief therapy. A Brief Therapy Center was established in 1968; numerous papers and articles on brief therapy models appeared in the 1970s. Different types of effective techniques using brief therapy were introduced. In all, however, the same focus prevailed: how to deal with problems, how they are maintained, and how to solve them. In the 1980s, therapists at the Brief Focused Treatment Center (BFTC) became more concerned with solutions and how they worked (De Shazer et al., 1986, p. 208).

Appreciating the main principles of the clinicians at BFTC is helpful in understanding the underlying premise behind today's brief therapy theory. Seven principles were stipulated:

1. Most complaints develop and are maintained in the context of human interaction.
2. The task of brief therapy is to help clients do something different, by changing their interactive behavior and/or their interpretation of behavior and situations so that a solution (a resolution of their complaint) can be achieved.
3. Rather than seeing clients as being resistant, this apparent "resistance" is viewed more as the clients' way of letting us know how to help them.

4. New and beneficial meanings can be constructed for at least some aspect of the client's complaint.
5. Only a small change is necessary; thus, only a small reasonable goal is needed.
6. Change in one part of the system leads to change in the system as a whole.
7. Effective therapy can be done even when the therapist cannot describe what the client is complaining about. (De Shazer et al., 1986, pp. 208–210)

In addition to these principles, BFTC defined the three basic terms, *difficulties, complaints,* and *solutions,* that they felt were integral to their approach.

Difficulties are the one damn thing after another of everyday life.... Complaints consist of a difficulty and a recurring, ineffective attempt to overcome that difficulty, and/or a difficulty plus the perception on the part of the client that the situation is static and nothing is changing.... Solutions are the behavioral and/or perceptual change that the therapist and client construct to alter the difficulty, and/or are the construction of an acceptable, alternative perspective that enables the client to experience the complaint situation differently. (p. 210)

In helping clients construct solutions, helpers may need to make some assumptions about the complaints and which solutions best fit them. The clinicians at BFTC found it useful to focus on helping clients describe their "favorite" factors, those they chose to emphasize in their complaint. They found that "those aspects of the situation that are excluded from the client's description of the complaint are potentially useful for designing interventions and prompting solutions" (p. 211). The example they present relates to someone complaining about feeling depressed and having "down days." The solution was to ask what happened on "up days."

Overall, BFTC therapists contend that clients "already know what to do to solve the complaints they bring to therapy. They just don't know that they know" (De Shazer et al., 1986, p. 220). The job of brief therapists is to help clients construct for themselves a new use for knowledge they already have.

Solution-Focused Therapy: Use of Strengths Perspective

A current-day brief therapy that has become very popular in the mental health field is solution-focused brief therapy (SFBT) or solution-focused therapy (SFT) that combines a careful utilization of the client's strengths and resources with a strong therapeutic relationship to facilitate the efficient resolution of client difficulties. SFT is theorized to be almost immediately effective in the long term, and on average requires only six to twelve sessions to complete.

Solution-focused therapy is based on an innovative theory of change. This model is changing the way psychotherapy is practiced, extracting what works from existing therapy. This brief therapy is based on a shift in thinking about therapeutic interventions. Previously, clinicians attempted to understand and locate pathology in individuals; now they are helping clients construct solutions to the problems presented. This shift is related to our earlier discussion that people's solutions to problems often seemed to have no direct relation to the problems presented. From this perspective, clients are first encouraged to consider times when their problem did not exist. Second, they are asked how these times contributed to the absence of the problem. Last, they are asked to recreate such circumstances in their present situations.

Focus is on the client's strengths and abilities rather than on weaknesses. Helpers ask questions and prescribe tasks that assist clients to focus on their own definitions of therapeutic goals and emphasize to clients how they can use their own existing and potential resources. Taking a strengths perspective is highly recommended.

A strengths assessment is necessary to practice according to the strengths perspective. This assessment focuses exclusively on the client's capacity and aspirations in all life domains. In making the assessment, both the client and the worker seek to discover the individual and communal resources from which the client can draw in shaping an agenda. (Weick, Rapp, Sullivan, & Kisthardt, 1989, p. 353)

Others, such as Hepworth and Larsen (1990, p. 195), also highlight the importance of conducting a strengths assessment. They focus on the possible negative outcomes when strengths are not assessed. For instance, they believe that concentrating on clients' deficits can impede a worker's "ability to discern clients' potential for growth," can restore "client self-doubts and feelings of inadequacy," and can influence workers "to believe that clients should continue to receive service longer than is necessary."

Cowger (1992, pp. 139–147) proposes 12 practice guidelines that foster a strengths perspective during the assessment process. Although many may seem commonsensical or simplistic in nature, we present them to reinforce some of the salient features of solution-focused therapy.

Guidelines for Strengths Assessment

- Give preeminence to the client's understanding of the facts.
- Believe the client.
- Discover what the client wants.
- Move the assessment toward personal and environmental strengths.
- Make assessment of strengths multidimensional.
- Use the assessment to discover uniqueness.
- Use language the client can understand.
- Make assessment a joint activity between worker and client.
- Reach a mutual agreement on the assessment.
- Avoid blame and blaming.
- Avoid cause-and-effect thinking.
- Assess; do not diagnose.

Additionally, solution-focused interviewing described by DeJong and Miller (1995) helps focus on client strengths by encouraging use of various types of questions. They include these types:

Exception-finding questions that are used to discover clients' present and past successes in relation to their goals

Scaling questions—clever ways to make complex features of a client's life more concrete and accessible for both the client and the helper. With each progressive step, the client is positively reinforced for existing strengths

Coping questions that help clients recognize how they have been able to cope with overwhelming circumstances and feelings. It is important to be sensitive to clients' need to talk about their problems.

Once a client's strengths are assessed, goal setting and intervention can follow effectively. Berg and Miller (1992, p. 731) have identified seven characteristics of well-formed goals that we also support.

1. The goals are important to the client.
2. The goals are small.
3. The goals are concrete, specific, and behavioral.
4. The goals seek presence rather than absence.
5. The goals have beginnings rather than endings (e.g., visualize first steps toward a happy marriage).
6. The goals are realistic in the context of the client's life.
7. The goals are perceived by the client as involving hard work.

According to DeJong and Miller (1995, p. 731), in interviewing for well-formed goals, helpers must "listen to clients' concerns, assess to determine that there is not an emergency, and then turn the focus on developing goals." They advocate using the "miracle" question at this point to facilitate the goal-setting process. A client is asked the following in an effort to elicit goals: "Suppose that while you were sleeping tonight, a miracle happens. The miracle is that the problem that has you here talking to me is somewhat solved. Only you don't know that because you are asleep. What will you notice tomorrow morning that will tell you that a miracle has happened?" (p. 731).

As stipulated by those practicing at the BFTC, SFT also emphasizes that solutions must be derived by the clients themselves. According to the solution-focused model, answers to clients' problems will eventually come from their own repertoire of coping strategies. This direct and active involvement of clients in the helping process seems to have contributed greatly to their success, and it has also allowed clients to specify solutions that are more apt to fit their unique lifestyles. When clients are encouraged to seek their own workable solutions, their self-esteem is also likely to be enhanced.

Treatment Considerations

- Remember that this model is brief and that the client is responsible for constructing the solution.
- Emphasize forming a strong/cooperative therapeutic relationship with your client.
- Early in treatment, emphasize the client's strengths and use these in the construction of solutions.
- Encourage the development of well-formed goals by using solution-focused interviewing questions (e.g., exception-finding, scaling, and coping). Also, be receptive to incorporating the "miracle question" to facilitate appropriate goal setting.
- Pay attention to the patterns of the solutions rather than the origins of the problems.
- Pay attention to what clients are doing that is good for them rather than only to the complaint.
- Become adept at using specialized interviewing techniques that invite change in the first interview.
- To keep updated, learn the latest developments in the theory such as focus on process and goaling.
- To encourage clients to work between sessions, prescribe tasks and homework assignments.

SCENARIO

Angie, a 22-year-old married female, comes to the counseling center with marital difficulties. She states that she and her husband of two years have been fighting frequently. Her husband got so angry last evening that he hit her across the face several times and this scared her a lot. She says that she feels suddenly quite numb and can't concentrate or decide how she should handle the current situation. She describes herself as usually very quiet and says she believes that she still wants their marriage to work. She says that she has not spoken to her husband about her feelings because of being numb and scared that he might hit her again. How might you work with Angie considering the information given you on brief therapy? How would you develop a sound working relationship with her? How would you use scaling questions? Exception finding questions? How might you encourage this client to work between sessions on the resolution of her concerns?

Long-Term Treatment

Long-term treatment encompasses many types of theoretical orientations and treatment modalities. When psychotherapy, counseling, and even intermittent brief therapy is provided, these treatments may be long term in duration. This long-term focus, once the norm, is rapidly being replaced by brief, short-term treatment that fits more readily with the mandates of managed care, rapid access, and cost-effective measures. Despite these changes, the value and need for long-term approaches remains integral to the effective treatment of various populations and continues to be a form many clinicians utilize.

Factors in Long-Term Treatment

Theoretical Orientation   Long-term treatment can occur for a multitude of reasons and is practiced by an array of helpers whose theoretical orientations and treatment modalities are as diverse as their individual styles. Many clinicians who engage in long-term therapy adhere to a specific theoretical framework; the concepts and techniques they employ are built from a sound theoretical base rooted in a distinct theory. Some of the more common theoretical orientations we have presented that could be used in long-term treatment include psychodynamic, humanistic or person-centered, behavioral and cognitive, and developmental approaches. These specific theories are only a few; many more exist. Corey (1996) discusses most of these as well as various other contemporary therapeutic systems he believes are instrumental in providing beginning helpers with an overview of the many directions they can take in working with people. Additional theories he surveys include psychoanalytic psychotherapy, Adlerian therapy, existential therapy, Gestalt therapy, reality therapy, and family systems therapy. In addition to specific theories that require long-term treatment, other factors may contribute to its use also. These are discussed next.

The Type of Setting   The setting can dictate whether a long-term approach is preferable over a short-term one. Many public agencies that are governed in part by budgetary constraints encourage the use of short-term treatment. This trend is also apparent in the private sector. As we have noted, the domination of managed care in the last few decades has dramatically changed the way mental health treatment is practiced in the United States. Through the establishment of health maintenance organizations (HMOs), private provider organizations (PPOs), and employee assistance programs (EAPs), insurance companies and employers have been able to provide cost-effective mental health services to their employees by dictating the specific use of brief, short-term interventions. Frequently, the typical managed care program allows only a limited number of counseling sessions. In some cases, the fine print of an individual's insurance policy may often stipulate "for crisis intervention only." This emphasis on length of treatment has greatly influenced therapists in private practice who rely on insurance referrals or who are providers within the managed care system. Some who may have previously chosen to use a longer term approach in their treatment of certain clients may no longer be able to do so given current restrictions.

Regardless of these restrictions, many clinicians continue to engage in long-term practice. In public agencies, they may have to resort to seeing clients intermittently to maintain a long-term relationship. Group therapy modalities often allow clients to be seen longer; if the group is open-ended, this method is especially conducive to long-term treatment. Also, clients who are eligible for various groups may find themselves receiving long-term treatment by ongoing participation in different groups. In certain instances, helpers may have to be advocates for some clients who may require longer treatment than the stipulated or recommended time frame.

In private practice, many practitioners have found creative ways to work around the insurance issue; more and more clinicians are setting a lower fee for services or limiting their fees to the co-payment their clients would pay if using their insurance. Still other private practitioners have joined colleagues in a group practice that is less costly to maintain and has a greater potential for marketing and referral possibilities. Last, the increased provision of group therapy in private practice can be seen as an alternate response to participating in the managed care system.

Sabin (1995, p. 163) who edits a column on how to increase the likelihood that managed care organizations will authorize coverage for extended psychotherapy, believes that there are instances when the clinician, whether working in either a public or private agency, needs to take an active role in pursuing approval for longer term treatment. He tends to focus on patients with severe problems. He describes a "clinically and ethically admirable approach to practice" that we have found quite useful. He offers "practical suggestions for clinicians on how to communicate effectively with third-party care managers and utilization reviewers," listed here:

- Make the connection of "medical necessity" work for the treatment you are proposing; show very clearly how the dysfunction and suffering that the treatment targets is a direct result of a condition recognized in the *Diagnostic and Statistical Manual of Mental Disorders* (DSM-IV; American Psychiatric Association, 1994).
- Describe your patient's history of using psychiatric, medical, and surgical resources and show that the treatment you propose may reasonably be expected to reduce overall health care costs.
- Distinguish carefully among the importance of different periods of time: the length of time you believe is required to bring about the therapeutic results, how frequently you believe the patient needs to be seen, and how long each session should be.
- Show the reviewer that your patients and, when pertinent, their families are doing their part in seeking the treatment outcomes and that you alone are not carrying the major burden of producing the therapeutic results.
- Make your treatment proposal in modular terms so the reviewer can see what your objectives are for the next segment of therapy—about ten sessions or three months—and how you will assess progress.
- In as nondefensive a way as possible, prepare to deal with the question of whether you are the right clinician for the patient.
- Do not "game the system" by exaggerating suicidal or other symptoms or by using an Axis I diagnosis instead of an Axis II diagnosis. Gaming may elicit approval of benefits in a particular situation, but as a pattern it contributes to a spiral of increasing distrust among clinicians, managed care organizations, and ultimately patients themselves.

Specific Needs and Preferences of Clients   In addition to setting, the decision to engage in long-term treatment is often influenced by the specific needs of clients as well as their preference. Clients who suffer from severe problems may require longer term therapy to treat their condition adequately or to maintain them at the highest functioning level possible.

Also, some clients prefer longer term involvement; they can work more easily with a helper they have a long-term relationship with and with whom they have established trust and rapport. Sometimes, deeper issues surface when the presenting problems have been resolved and therefore long-term treatment may also be desirable for working through these.

Therapeutic Styles and the Helper's View of Human Nature   Therapeutic styles differ greatly from clinician to clinician and can also have a bearing on whether a short- or a long-term focus will be taken. Those who consider a client's past as an important focus for exploration may choose interventions designed to assist them in understanding their past. This will take longer than if the primary focus were on the here and now or in the future. Basically, the theoretical orientation used by the helper largely determines the time frame to be emphasized.

According to Corey and Corey (1998, p. 57), the helper's views and beliefs about human nature are also very much related to the helping strategies that are employed:

> If you see people as basically good, for example, you will trust that your clients can assume responsibility for the direction of their lives. If you see human nature as basically evil, you will adopt a role as a helper who attempts to correct people's flawed nature. Your interventions are likely to be aimed at "straightening people out."

The decision to use time-limited versus longer term strategies will depend, in part, on your views of human nature. Helping people with the direction of their lives might lead to a longer term psychodynamic approach, whereas focusing on helping people "correct their mistakes" may require a shorter cognitive-behavioral approach.

Long-Term Versus Short-Term Treatment

Gustafson (1998), in his work on the use of brief versus long psychotherapy, discusses ways to determine when someone needs short-term or long-term treatment.

> Short term treatment can be successful when there is a problem in the patient's life that can be resolved by a choice. The problem is a crisis when the two paths that are open both seem disastrous.... A brief psychotherapy can often help clients arrive at a third alternative.... Long-term psychotherapy becomes advisable when a patient has the same kind of difficulty in many sectors of her [or his] life. In effect, it is a sequence of brief psychotherapies applied to different areas.... Long-term help becomes necessary when the patient cannot apply the learning from one sector to the next version in another sector in her [or his] life. Someone has to go through it with her [or him]. (p. 1)

Negatives and Positives   Engaging in long-term practice has been criticized extensively and these criticisms plague current views of long-term treatment. As a result, we have found that those who use long-term treatment are often stigmatized. Typical criticism seems to be centered on financial and dependency issues. Many believe that it is unethical to see a client for an extended period of time if the client is unable to afford the treatments. Also, because there have been instances when clinicians have taken advantage of their clients financially by encouraging long-term treatment, many now question the ethical nature of this type of treatment. In a similar fashion, there are those who believe that seeing a client for a long period is actually harmful as it fosters a dependency on the helper. Although this may be true in some instances, it is not so in many. In fact, according to a survey by Consumer Reports, long-term therapy was preferable over short-term therapy. In this report, Seligman (1996) delineates the various advantages of some form of long-term therapy:

> Long term treatment produced more improvement than short-term therapy. These results were robust, and held up over all statistical models. The group treated more than two years improved the most on every outcome measure. The advantages of long term treatment by a mental health professional held not only for the specific problems that led to treatment, but for a variety of quality of life scores as well: ability to relate to others, coping with everyday stress, enjoying life more, personal and growth and understanding, self-esteem and confidence." (p. 1)

An additional benefit of long-term treatment that we have experienced includes being able to work with clients who need more in-depth interventions, along with those who prefer to engage in ongoing treatment throughout their lives. Below we offer two perspectives on long-term treatment that we have found to be common in our work with clients at both the public and private levels. These include providing (1) traditional long-term treatment when clients are seen for extended periods of time, and (2) intermittent brief treatment in a long-term therapeutic relationship.

Traditional Long-Term Therapy

Working with clients for an extended period of time is often necessary with certain types of client populations. Sabin (1996) advocates longer term psychotherapy for patients with personality disorders, dissociative conditions, and delayed-onset posttraumatic stress disorders. We concur with these assessments and also have found long-term treatment helpful with clients who were severely abused in childhood, whether the abuse was sexual, physical, or emotional in nature. Work with incest survivors and those who suffered satanic abuse may require even further treatment, depending on the seriousness and impact of their abuse. Those with chronic and severe depression and anxiety-type problems may require extended therapy for their conditions. Similarly, those who seem to suffer from long-term relationship problems may benefit from long-term treatment. Such treatment may be curative in terms of modeling to clients through the therapeutic relationship itself the positive aspects of open communication, trust, and encouragement that can exist within a relationship.

Treatment Considerations

- Note that a strong therapeutic relationship has been shown in the research to be the most important nonspecific factor in positive outcomes in long-term treatment.
- Consider use of traditional long-term treatment when working with clients who have severe childhood abuse issues, those suffering from anxiety and depression, and those with chronic relationship problems and certain personality disorders.
- Remain flexible in your use of modalities, yet take care to remain focused on the client's issues and goals for treatment.
- As the clinician, remain aware of your potential role in modeling adaptive relationship skills such as open communication, trust, and support.

SCENARIO

Victoria came to treatment due to her concerns about severe depression and difficulties in interpersonal relationships that she had been experiencing on and off through the years. She was aware that she had been severely physically and sexually abused as a child for years and had even seen a therapist years ago for a brief period of time. She had never been married and was fairly isolated, with difficulties in trusting and forming relationships with others. Her continued isolation, fearfulness, and loneliness caused her to struggle with depression and low self-esteem. She would, at times, cut her arms to relieve a certain emptiness and discomfort she would feel. Due to her current worsening depression and suicidal ideation, she decided to seek help from her local community mental health center. As her helper, you realize that her difficulties are deep rooted in nature and may require intensive and long-term treatment. Your research and reading on treatment of severe physical and sexual abuse validates your position that longer term treatment is the treatment of choice in many of these situations. If able to engage in long-term treatment with Victoria, what might your approach be? What might be some of your initial fears and concerns? How would you proceed?

Intermittent Brief Treatment in a Long-Term Therapeutic Relationship

Different from engaging in treatment for an extended period of time is the increasingly common approach to long-term treatment in which clients engage in intermittent brief treatment with the same clinician, thereby developing a long-term therapeutic relationship. Event though this may not be the initially planned intervention, it seems to be occurring with more frequency in both public and private arenas.

Several situations tend to produce this type of long-term treatment. First, managed care clients who can attend treatment for only a brief period because of insurance limitations may continue to seek treatment for various issues they are facing. Second, clients who seek crisis, brief, or short-term treatment may continue to be involved in the therapeutic relationship over time. They may reenter treatment when there is a crisis, a new developmental task they are concerned with, resurfacing of old issues, or for additional support and validation. Clients often return to a former helper because they already have an established therapeutic alliance with that person, who knows their history and personality style and has succeeded in helping them in the past. This type of long-term therapy is a promising alternative for those who prefer longer term treatment; however, it does require clinicians and clients alike to be flexible in adapting to today's views on treatment.

Treatment Considerations

- Because of the limited number of visits allowable under many insurance plans, it is important to recognize the value in seeing clients on a short-term basis that may become intermittent through time, resulting in a long-term therapeutic relationship.
- Recognize the value of intermittent brief treatment versus becoming critical that traditional long-term treatment was not pursued. One is not necessarily superior to the other; they are different approaches that are sometimes voluntarily chosen and sometimes dictated by circumstances.
- Be aware that clients may reenter treatment when faced with a crisis or a new developmental task they are attempting to master.
- For clients to be open to reentering treatment with a prior clinician, they most likely have experienced a strong therapeutic alliance and some success in previous treatment attempts—thus the importance of the therapeutic relationship.
- Respect for the ongoing bonding and boundaries of the therapeutic relationship is essential in encouraging clients to continue to return to the same helper for additional assistance as they need it.
- Keep the door open for further treatment.

SCENARIO

Ray had initially come in to the department of psychiatry in a large health maintenance organization due to mild depression and marital communication problems. He had been seen initially to deal with his increasing sense of depression a few years back and was later placed in a 16 week men's group to explore his issues further. He had finished his work with the group and was doing much better after his treatment. Now, two years later, Ray has requested you again and says that he needs to speak to you about his depression. He is suffering from severe depression again and says that he has fallen into his old patterns of self-condemnation and shame that got him so depressed before. He wants to see you again to review what he learned before and to evaluate his current condition. Do you see a request by a client like this to be legitimate, even if he comes in several times for a brief period of time throughout his life span? How would you deal with his request for continued treatment?

Case Management

Case management is an increasingly important direct practice modality being used in human service agencies today. Rothman (1991, p. 520) finds that "case management is a practice modality that cuts across such human services fields as mental health, aging, child welfare, health, and developmental disabilities." Application of this specific modality is therefore essential to the work of various disciplines within the helping professions; some include caseworkers, human service workers, nurses, social workers, and others. Because of our emphasis on clinical practice, we focus here primarily on counseling and social work viewpoints.

Numerous definitions of case management have been proposed. According to Moxley (1989, p. 21), case management is defined as "a designated person or team who organizes, coordinates, and sustains a network of formal and informal supports and activities designed to optimize the functioning and well-being of people with multiple needs."

Barker's (1991, p. 29) detailed description of case management provides a more in-depth look at the numerous facets of intervention involved. He believes it is

a procedure to plan, seek, and monitor services from a variety of agencies and staff on behalf of a client. Usually one agency takes primary responsibility for the client and assigns a case manager, who coordinates services, advocates for the client, and sometimes controls resources and purchases services for the client. The procedure makes it possible ... to coordinate their efforts to serve a given client through professional teamwork, thus expanding the range of needed services offered. Case management may involve monitoring the progress of a client whose needs require the services of many different professionals, agencies, health care facilities, and human service programs.

Moore (1990, p. 444) presents a more general stance in that he believes "case management practice focuses on enabling individuals and primary groups to reach their full potential and on facilitating more effective interaction with the larger social environment."

In all these definitions, and in all the models of case management that have been proposed (Austin, 1990; Moore, 1990; Roberts-DeGennaro, 1987), a core premise is "the function of linking, clients with essential resources and empowering clients to function as independently as possible in securing the resources they need" (Hepworth, Rooney, & Larsen, 1997, p. 456). Three main objectives of case management have been summarized as follows:

1. Helping people to obtain resources
2. Facilitating interactions between individuals and others in their environment
3. Making organizations and society as a whole responsive to people

Although similar, the objectives specific to a case management approach to social work practice detailed by Moore (1990, p. 446) reflect the broad scope that case management embodies. These objectives are

(1) to maximize the potential of individuals to meet environmental challenges,

(2) to maximize the caring capacity of families and primary groups,

(3) to integrate formal systems of care with primary caring resources, and

(4) to maximize the capabilities within the formal system of care for meeting the needs of individuals and primary groups."

These objectives emphasize the value of self-determination and also highlight the importance of the strengths perspective we discussed earlier.

In sum, the practice of case management results when the helper, acting on behalf of the client, coordinates needed services provided by any number of agencies, organizations, or facilities. In this respect, case managers stand at the interface between the individual client and the environment, mediating between the two for the benefit of the client specifically, and the family and community in general.

Purpose and Principles of Case Management Practice   Moxley (1989, p. 21) has highlighted three purposes of case management that underscore the role of the case manager [helper]:

(1) Promotes the skills and capability of the client in using social services and social supports.

(2) Develops the abilities of social networks and relevant service providers to further the functioning of the client.

(3) Promotes effective and efficient service delivery.

Similarly, Gerhardt (1990, p. 216) has identified various principles of case management that further demonstrate its purpose:

- *Individualization of services* whereby specific services are designed to meet the identified needs of the client.

- *Comprehensiveness of services* that encompasses all areas of the client's life including housing, recreation, employment, social, financial, medical care, mental health care, and so on.
- *Parsimonious services* seen in discouraging the duplication of services and the control of costs of services.
- *Fostering autonomy* given the emphasis on helping clients become as self-sufficient as possible. Also, this allows maximum client self-determination that encourages independent decision making regarding one's own care.
- *Continuity of care* implying that case management services will ensure the continued monitoring of individual client needs that, in many client populations, may be needed throughout their lives.

Historical Views on Case Management   Even though case management has emerged as one of the most significant developments in human services this decade, Kaplan (1990) suggests that it has been in existence since the 19th century. Various reasons have been offered for why case management is now readily acknowledged as an important direct practice modality. Both federal and state laws and regulations now require case management services to be provided to specific categories of clients; for instance, approximately 75% of states require case management for the chronic mentally ill (Greene, 1987; Kirst-Ashman, & Hull, 1993). This expansion in case management is said to result from

> the expanded recognition that the needs of increasing numbers of clients with major disabilities (e.g., frail, elderly, and developmentally and disabled persons) were not being met because these people could not negotiate the complex and often uncoordinated human service delivery systems. (Hepworth et al.,1997; Austin, 1990)

Advances in both the gerontological and health fields have had a major impact in stimulating the use of case management. In gerontology, medical advancements have resulted in extended life expectancies. Also, a societal change has been noted in the responsibility of the family to attend to the elderly. These two factors have led to an increased need for professional continuing care. Likewise, advances in health care have enabled those with physical disabilities to return to natural community settings more readily than in the past. This advancement, coupled with the ever-growing managed care movement, has encouraged those with physical disabilities to continue to live at home. To prevent hospitalization or institutionalization of these individuals, they and their families need the additional support and care possible when case management services are provided. Case management services have also proliferated within the child welfare arena given the increases in family disorganization, divorce, domestic violence, and child abuse. Such negative events within the family have led to an increased number of dependent children who require sustained attention and care outside their family of origin. The movement toward deinstitutionalization in the mental health field created by economic pressures, psychotropic advancements, and the move toward more cost-effective mental health care has resulted in the need for increased case management for those suffering from mental disabilities.

Case management is also strong in the field of rehabilitation. A sample of 350 major companies surveyed in 1988 found that almost all were using case managers to help employees procure rehabilitative care for job-related injuries and to reduce the cost of that care (Institute for Rehabilitation and Disability Management and National Center for Social Policy and Practice, 1988). Kaplan (1990) has noted that in many states, there are now private case-management firms designed to help clients receive homemaker services, alternative housing, day care, transportation services, chore services, and in-home meals; this presence reflects the rapid spread of interest in this type of practice modality. Reinforcing this trend, the National Association of Social Workers (1981) identified case management as one of the 11 major functions performed by social workers. Also, the Developmental Disabilities Assistance and Bill of Rights Act (Public Law 95–602) expressly lists case management as a priority service. Current trends and issues such as concurrent planning, welfare reform, and capitation/managed care are also impacting the rise in case management services. Overall, it is clear that the use for case management in the 1980s and 1990s has been great, and this trend is destined to continue well into the 21st century.

The actual type of case management provided is determined largely by the specific population being served. In social gerontology and health care, for instance, "a greater emphasis is on the role of the case manager as planning and coordinating a package of health and social services that is individualized to meet a particular client's needs" (Moore, 1990, p. 444). The psychosocial dimension of case management is thus a main focus with both the elderly and those with physical disabilities. "Case management has been seen as a way of coordinating a fragmented service delivery" (Moore, 1990, p. 444). In the child welfare arena, the case manager has been discussed as a balance between the individual and on the environment. Both clinical and advocacy skills are integral to this position. Last, in mental health, the therapeutic nature of both the process of case management and the relationship between the case manager and the client is emphasized. Here, the clinical, advocacy, organizing, and community skills of professional case management are also necessary.

These various types reflect the diverse views of what case management is. There has been much controversy regarding the professional versus paraprofessional nature of case management. Moore (1990, p. 444) speaks to this issue:

> Some argue that it is a field of practice that requires . . . the clinical skills of a psychotherapist and the advocacy skills of a community organizer. Others assert that case management is a paraprofessional activity that is best left to those with bachelor's degrees in social work, or clerical staff. However, the most common response is that case management is just old-fashioned social work wearing new clothes.

Although we believe there is room for case management on both the professional and paraprofessional arenas, we advocate a more professional stance whereby advanced clinical and advocacy skills are both utilized. This stance will be more clearly understood as we elaborate on the tasks and roles involved in case management services.

Case Management Roles

Case managers play various roles: broker, advocate, crisis intervention specialist, counselor, teacher, therapist, community organizer or planner, motivator, outreach worker, monitor, mediator, enabler, and facilitator. Their primary responsibility is to help clients function independently (Hepworth et al., 1997; Intagliata, 1982; Kirst-Ashman & Hull, Jr., 1993; Moore, 1990; Roberts-DeGennaro, 1987). Client needs in large part will determine the helper's role. At times the case manager will fill only a few of the roles listed above. At other times, a client's needs may be so great that the case manager will have to function in almost all of them to help the client effectively.

The type of setting and the role definition within the specific agency where the client is being treated will also dictate what the helper does as case manager. For instance, the mental health setting is one where helpers typically will function as both therapists and case managers to the same client. There are settings, as in child welfare, where the role of the children's services worker/case manager will be more generalist in nature. These workers may provide limited counseling, but generally they refer clients for therapy. In this setting, both professionals and paraprofessionals may perform the case management functions. In comparison, in hospital settings (both medical and psychiatric) with interdisciplinary teams, case managers are often paraprofessionals who are supervised by the professional helpers.

Contrary to popular view, however, studies of the professional status of case managers show that more professionals than paraprofessionals engage in case management. Bernstein (1981) conducted a nationwide survey of case managers in 15 community-supported projects and found that 60% of the case managers were trained therapists. Similarly, Bagarozzi and Kurtz (1983), who surveyed directors of mental health centers, found that 75% of the directors expected case managers to perform therapy. In the 1990s and beyond, this number is expected to increase as the demand for more comprehensive and cost-effective services grows. Both helpers with advanced and bachelor's degrees will continue to be needed to provide the array of services ascribed to the case manager position.

As with all the other treatment modalities presented, case management practice requires the establishment of a sound client-helper relationship whereby the client is actively involved and treatment goals are mutually agreed on. However, different from most other modalities, those engaged in case management will likely have to assume more of an advocate role to assure that the services the client needs are provided or that untapped resources are developed. This requires case managers to be knowledgeable in assessment, community resources, and skills in connecting clients to needed resources, and thorough in ensuring that clients have received both the resources and services they require in a timely fashion. In addition, Kirst-Ashman and Hull (1993, p. 508) note another important responsibility of case managers that involves "integrating the formal support system (agencies and services) with that of the informal support system (for example, family, friends, and others directly involved with the client)."

In this respect, the client is encouraged to function independently by facilitating interpersonal relationships. This may require working at a mezzo level whereby the client's family and other existing supports may be involved in treatment to encourage them to help the client's needs.

We complete our discussion on case management roles by presenting a comprehensive table devised by one of our colleagues, Marshall Jung (see Table 3.5). It includes common roles, tasks, and functions of case management all in one diagram.

An Empirically Based Model of Case Management

Based on a review of 132 pertinent articles and a survey of 48 case managers, Rothman (1991) developed an empirically grounded model for case management that we highly support because of its comprehensiveness and versatility. We present an outline below of the various steps integral to his model as it illustrates the process of case management as well as the specific functions and tasks of the case manager.

1. *Facilitating access to agencies:* Case managers need to facilitate clients' access to the agency or system that can provide them with the services and resources they need. This may entail direct contact with the agency as well as possible outreach measures to encourage clients to enter the system.

2. *Doing intake:* Exploration of clients' problems, concerns, and needs that assist the helper in determining the client's eligibility for specified services and financial resources. Case managers must be skillful at fostering trust and rapport, and in interviewing clients.

3. *Performing assessment:* This may occur as part of the intake. Here more in-depth exploration takes place that includes interviewing family and additional support systems. The family or support system's capacity to cope with the client is assessed along with their ability to provide the client with support. Collaboration with different professionals from different disciplines may be necessary.

4. *Setting goals:* Once the client and case manager agree on the areas needing improvement, objectives are determined. Both short- and long-term goals that realistically take into account the client's capacities are then devised.

5. *Planning intervention or identifying and indexing resources:* This step consists of two functions that are dual in nature as planning interventions and linking clients with resources are clearly interrelated. Certain resources may not always be available and thus it is the case manager's job to determine which resources are accessible to the clients.

6. *Linking clients:* Linking clients with existing resources and systems is one of the major functions in case management. It can be a simple act, or it may require skill and timing; the manager may need to prepare some clients carefully by providing them with

TABLE 3.5   Roles, Tasks, and Functions of Case Management

Role	Functions	Tasks
1. Caseworker	1. Assessment 2. Care planning 3. Implementation 4. Monitoring 5. Reassessment 6. Evaluation	1. Arrange meetings 2. Provide safe place 3. Keep accurate records 4. Conduct reviews 5. Write assessments 6. Collaborate
2. Broker	1. Outreach 2. Link with resources 3. Resource acquisition	1. Make contacts 2. Plan meetings 3. Obtain directories
3. Advocate	1. Promote fairness 2. Tell clients their rights 3. Support ethics	1. Testify in court 2. Write letters 3. Change procedures
4. Educator	1. Increase client's awareness 2. Increase staff's awareness 3. Increase awareness in community	1. Conduct enrichment classes 2. Conduct inservice training 3. Conduct community meetings
5. Counselor	1. Provide crisis stabilization 2. Provide brief therapy 3. Assist in networking	1. Call clients 2. Empower clients 3. Hold family sessions
6. Consultant	1. Help client solve problems 2. Help agency solve problems 3. Help community agencies cooperate	1. Assist client is examining alternatives 2. Assist staff in examining alternatives 3. Meet with agency staff
7. Community Organizer	1. Identify community resources 2. Help mobilize community resources 3. Help acquire community resources	1. Hold community meetings 2. Research community resources 3. Write resource directory
8. Gatekeeper	1. Assess client's needs 2. Determine who is eligible for services 3. Refer clients out	1. Track cases 2. Conduct assessment interviews 3. Review case files

SOURCE: Courtesy of Marshall Jung

detailed information, helping them anticipate difficulties, using role play, and possibly accompanying them for the first visits.

7. *Monitoring and reassessing:* A vital process of case management, monitoring involves determining whether the arrangements are working adequately. Ongoing telephone contacts, agency visits, and crisis intervention may all be necessary to ensure the success of the interventions. Because change is constant, reassessment must be continuous.

8. *Evaluating outcome:* When and where appropriate, outcome evaluations of the goals should be conducted to determine the overall success of goals and means used.

As explained, Rothman's model is certainly comprehensive and captures the main tenets of case management. Below are general treatment considerations that, in many instances, reiterate some of the steps just discussed.

Treatment Considerations

• Case managers must be skilled in overall assessment that takes a biopsychosocial perspective. Assessment of the following three areas are especially important:

1. The client's ability to meet environmental challenges
2. The caring capability of the client's family and primary group
3. The resources available within the formal support system of care

• Case managers must be able to formulate service plans with those integrally involved (e.g., client, family, interdisciplinary team, or other agencies). This may require separate contacts with various others and attendance at case conferences with other agencies. Planning must result from a thorough assessment of the client's needs and capabilities as well as resources readily available to him or her.

• The case manager must be active, when possible, in supporting and enabling clients to use their personal resources in meeting environmental challenges and families and primary groups to expand their caretaking capacity.

• Case managers must be continually involved in monitoring the client's progress as well as determining the effectiveness of the plan. Managers must determine whether the designated resources are still available and whether family and others in the primary group remain interested and willing to help.

• Monitoring may necessitate reading any relevant progress notes about the client's treatment, discussing progress with other staff and family, conducting continual follow-up with all involved, monitoring medications used, assessing housing and financial needs, and in some instances, checking on client's behavior, health, grooming, and social skills.

• Case managers must be active in mediation and advocacy for their clients. Helpers must be capable of interceding in interpersonal problems and disputes in a crisis situation and when there is a problem with the client receiving the services he or she needs. For empowerment to be part of case management, helpers need to remain in an advocate role to ensure that their clients receive the services they are entitled to or that are being withheld unfairly.

• Case managers may be involved in providing crisis intervention, counseling, and even psychotherapy to their clients.

1. Crisis intervention is typically needed considering the frequency of crises in the lives of many who are vulnerable and more apt to need case management services to begin with.
2. Counseling is used at times but is usually limited to helping clients engage in problem solving and reality testing, teaching socialization skills, and providing practical help in such areas as housing, money matters, parenting, and employment.
3. Psychotherapy may also be a function of certain case managers; this is typically so in mental health settings where the therapist is also the case manager. When you do case management, focus on the present, not the past, and on helping the client learn to cope with the problems of daily living.

• Case managers must be skilled in providing information and referral services. They may need to provide clients with needed education and information on specific issues (e.g., parenting, stress management). They may need to refer clients swiftly to appropriate agencies and must work toward ensuring that both the client and the agency are receptive to the referral.

• Evaluation is important in several respects:

1. Evaluation needs to be continual to know the ongoing needs of the client.
2. The extent to which the client is adequately supported by both the family or primary group and the formal system of care must be continually evaluated as well.
3. The extent to which efforts of the family or primary care group are integrated with those of the formal system of care must also be evaluated.

• Case managers may also be required to develop individualized evaluation protocols along with administering standardized pre- and post-tests. Also, participation in agency-wide research endeavors may be necessary or encouraged.

SCENARIO

A young man named Keith, age 25, comes to the Mental Health Center where you work asking to be seen for continuation of his medication. He explains that he was working as a commodities trader in Canada when he began hearing voices and hallucinating under the pressure of his employment. He says he was diagnosed as having schizophrenia by a psychiatrist in a hospital where he was placed during a particularly difficult time when his hallucinations seemed to take over. He now wants to be followed for his medication but does not feel the need for more intensive treatment. When asked about his social life, the client tells you that he is living with his parents and that he currently does not get out much. He wishes that he could socialize more but has been somewhat afraid to get out with people since his "breakdown." Do you think this client may need case management services? How would you provide this service and what would be the areas that would need your attention as the case manager?

DEVELOPING YOUR OWN STYLE

Ultimately, practitioners all have to integrate their personhood, experience, knowledge, and skills into a style that is uniquely theirs. This can be challenging and overwhelming at times, but it can also be very rewarding.

The theoretical orientations and treatment modalities we have explored give us a framework that can be useful in guiding our assessments and interventions. As Harrow (1996) noted, "Most professional fields have a body of theory, the condensed wisdom of the ancestors, that guides the work, and the further exploration, of contemporary practitioners. This heritage of knowledge is what allows each profession to develop. Without it, each generation would have to start again from scratch" (p. 1). Our intent here has been to equip you with a basic foundation from which you can build by trying out new techniques and learning about new theoretical perspectives. Part of this developmental process is to become familiar with some of the basic theories in an attempt to find the one—or more likely, the synthesis of several—that is most suitable to you. Of equal importance is paying careful attention to what theoretical approaches and treatment methods work best with different types of clients or with specific types of issues. Always remember, however, that all the current theories work with some clients and that none of them work with all clients.

In cultivating your own style, work from a theoretical base that is congruent with your own personal values and that fits your special talents and temperament. Part of the beauty of the helping process is the use of self, so find what works best for you in helping others. Attempting to copy others can lead to a lack of genuineness and the use of ineffective interventions. Helpers are often role models; by being true to themselves, they show their clients the value of being unique. Working in a way that is congruent with who you are also allows you to feel free and uninhibited.

Three main areas will have some bearing on the style and approach you develop: self-awareness, as your style and approach will be partially based on your own personality and interests; your work setting, as your approach may depend in part on what kind of agency you work for; and your experience level, which can also dictate, to some extent, what helping approach you apply. Consider these aspects regularly in your work with clients. Be open to altering your style and approach to fit their particular needs. Also, recognize the need for additional training, supervision, or assistance in certain situations. If you remain open to learning and change, you will have a better chance at continued growth, heightened self-awareness, and the realization that becoming a helper is an ongoing, evolutionary process. As a guide to help you begin this process of integration, we invite you to review the practical and systematic approach for advanced practice in Table 3.6 that we have modified to fit the development of the human service practitioner.

TABLE 3.6   Steps to Developing Your Own Style of Helping

Step 1: Examine your personal values and views about people.

Step 2: Explore the major theories of counseling and psychotherapy and select the one that most closely resembles your own personal values and beliefs.

Step 3: Engage in an in-depth study of your chosen theory by reading everything you can find that was written by its founder and those who have developed it further. Also, become actively involved in any available workshops that can provide you with the opportunity for supervised practice with associated techniques. Determine what specifically draws you to this theory and why, as well as where you disagree with it.

Step 4: As you work with clients, begin to apply what you have learned, being cautious to keep your client's interest at heart. (If in your eagerness to apply your knowledge you neglect to consider the client's needs, you may be guilty of ineffective as well as unethical practice.)

Step 5: Once you have become comfortable and well grounded in working with your theory of choice, reexamine some of the other theories that you considered to determine whether they offer any techniques that may fit well with your chosen theoretical base. Be open to trying new methods that have the philosophical base you are most content with.

Step 6: Continue to leave yourself open to learn about ways of understanding and working with people. Keep checking theory against your own lived experience. Keep cycling through these steps. Gradually, through them, you will discover your own personal style.

(Adapted from Harrow, 1996)

Enjoy this process of learning that will ultimately enhance your ability to work with others while keeping you invigorated in your professional role.

CONCLUDING EXERCISES

1. Which theoretical orientation fits you most naturally after reading this chapter? Describe your beliefs about human beings that underpin your orientation. _____

   _____

   _____

   _____

   _____

   _____

   Practice putting your orientation into words so that if you are asked by a client or a teacher, you can explain it concisely.

2. Describe what would be the most anxiety-producing crisis situation with a client you could be in (example: child abuse, suicide, homicide). Choose one for the exercise below.

   What feelings would this situation evoke in you? _____

   _____

   _____

   _____

What knowledge would you need to handle this situation so that you would be less anxious and more comfortable? _____

_____

_____

If you knew that a referral was not an option in this case, how would you go about dealing with the situation? Consider doing a role play with another class member and take the role of the client in this situation you fear.

3.   You work for an agency that dogmatically states no client will be seen for more than three sessions. You have strong beliefs that the particular client you have seen needs more than the three allotted sessions. In fact, the client remains suicidal and very depressed despite your best attempts at helping him overcome his situation. What would you see as the options in a situation such as this? (List all options) _____

_____

_____

_____

_____

Now, pick the option that seems most ethical and comfortable for you. Tell how you would be assertive and act as an advocate for your client: _____

_____

_____

_____

_____

4

Mezzo Level Practice:
Working with Families and Groups

AIM OF CHAPTER

A second major area of human service practice involves working at the mezzo level with families and small groups. Because everyone is born into a family and interacts with a variety of groups, both these areas are prominent in human service practice. In this chapter we examine both family and group work at some length.

Our focus on working with families deals with a historical view of the family, the systems approach, the developmental life cycle, and family assessment. We follow this by reviewing various treatment orientations to family therapy and specific practice models that fall within each of these categories. Specifically, we center on Bowen's family system's theory from the psychodynamic school, the Satir model from the humanistic school, structural family therapy from the cognitive-behavioral school, and Haley's life-cycle model from the developmental school. We conclude by considering the many facets of the changing family life cycle and their implications for practice. As such, we touch on role changes, diverse family lifestyles (e.g., single parents, divorce, blended families, two-provider families, gay and lesbian parents, communal living, and extended families) as well as the lack of economic stability and the rise in family violence. Throughout, we present theoretical views, treatment considerations, and relevant vignettes.

In our discussion of working with groups, we begin by addressing the purpose of groups, key definitions of group work and group therapy, classifications of groups into treatment groups and task groups, and the personhood of the group leader. We then turn our focus to appreciating group dynamics and group process, followed by views regarding ethical and legal issues, including diversity considerations. Varied group treatment modalities are surveyed, current practice trends are addressed, and common types of groups are presented. We provide insights on developing your own approach to group therapy and profile several short-term groups for children, adolescents, and adults. We conclude by discussing specific considerations for conducting group work in agency settings.

WORKING WITH FAMILIES

In addition to working with individuals at a micro level, human service practitioners will also find it necessary to intervene at a mezzo level. Often, working with individuals leads to some form of family involvement. Some practitioners will be involved in family work to a minor degree—for instance, when members of the client's family are included in the therapeutic process in some way, be it for support, feedback, or treatment itself. Others will work with families more consistently, either through their individual work with clients or by engaging in conjoint and family therapies almost exclusively. Regardless of a helper's preference, understanding the role of the family in a person's life is critical to any therapeutic endeavor pursued. All individuals, despite their differences, have in common membership in a family system of some type. Understanding the specific aspects of each person's family history and current relationships is instrumental in both assessment and treatment. In fact, incomplete and inaccurate findings are likely when helpers minimize or neglect the role of family in an individual's life.

Family performs one of the most significant functions in a person's life. Because the family is the basic sociocultural unit from which clients grow and evolve, it has a marked impact on an individual's past development and current life situation. Also, the family plays a major role in meeting the social, educational, and health care needs of its members; therefore, it is often their greatest resource for social support, nurturance, and guidance. Hepworth, Rooney, and Larsen (1997, p. 276) have said that

> it is largely through the family that character is formed, vital roles are learned, and children are socialized for responsible participation in the larger society. Further, it is primarily through family interaction that children develop (or fail to develop) vital self-esteem, a sense of belonging, and interpersonal skills.

The family is the main institution through which culture is transmitted. The anthropologist Hoebel (1972, p. 422) speaks directly to this in his definition of the universal functions of the family:

> the institutionalization of mating and the channeling of sexual outlets; the nurture and basic enculturation of the young in an atmosphere of intimacy; the organization of a complementary division of labor between spouses; and the linkage of each spouse and the offspring within the wider network or kinsmen for the establishment of relations of descent and affinity.

Judging from these factors alone, we believe a more detailed focus on the family is warranted. We begin by considering historical factors.

A Historical Focus

As a foundation of most people's lives, the family is one of the oldest institutions in existence. Family is unique in that it can provide the security, structure, support, and intimacy many of us need and desire. In some cases, the family may also create undue stress and emotional pain. Whatever its impact, the family is an important ingredient in an individual's development. It has been called the "headquarters for human development" in that "through our families we are connected to the past—the distant times and places of our ancestors—and to the future—the hope of our children's children" (Garbarino & Abramowitz, 1992, p. 72).

Defining a "traditional" family is not a simple matter as each of us may perceive the meaning of family in a slightly different way. For some, the terms *nuclear* and *extended* family come to mind. A nuclear family is a system composed of both parents and their children. The extended family may include grandparents, aunts, uncles, cousins, and other relatives. For others, a family may be a single parent and his or her children, or a gay and lesbian couple. Given the changing norms of today's society, the definition of families is bound to be changing as well. This phenomenon of our times, in which people have discovered so many other ways to come together as "family," is no longer a rarity; rather, it is becoming more a permanent feature of American life. Practitioners who work directly with individuals and their families must remain flexible in defining families and receptive to the many forms they are likely to take. We hope you will view these evolutionary changes in a positive way that will stimulate your interest and motivate you to explore new and innovative ways to work with the many types of families that exist today.

Systems Approach

We have found a systems approach to be useful in understanding the family system. Originating in biology, systems theory has gradually been adopted in other domains such as family therapy, organizational theory, and business. As defined by Brill (1990, p. 106), a system is

> a whole made up of interrelated parts. The parts exist in a state of balance, and when changes take place within one, there is a complementary change within the others. Systems become more complex and effective by constant exchange of both energy and information with their environment. When this exchange does not take place, they tend to be ineffectual. A system is not only made up of interrelated parts, but is itself an interrelated part of a larger system.

Systems exist on various levels and in many forms: an individual, a family, a community, a society, and even the universe. Individuals themselves contain various systems—their emotion, biology, spirituality, social roles, and intellectual capacity. Fennel and Weinhold (1989, p. 24) discuss being human itself from a systems perspective:

> The human being is considered to be composed of subsystems, which in turn are composed of other subsystems. For example, the human is composed of a nervous system, a digestive system, a circulatory system, a skeletal system, a respiratory system, and a reproductive system. Each of these systems is a subsystem of the human being. Furthermore, each of these subsystems is composed of other subsystems.

When the human being is examined further within this context, each individual can be perceived as a subsystem of a larger living system: the family. The family is a part of a living system known as the extended family, which is part of a larger system called the community, and so on.

Additionally, according to general systems theory, because of these interrelationships between systems and their subsystems, change in the system will affect everyone in some way. Systems normally exist in some state of balance called *homeostasis*. When change is introduced in any one part of the system, there is likely to be a resultant change in other parts of the system. This can lead to a state of disequilibrium. Often the system is most comfortable when functioning in its original state, so systems may strive to return to the previous state of equilibrium. At times, this is not optimal; a return to the original state can serve to reinforce maladaptive behaviors. One of our roles as human service workers is to help clients and systems aspire to a healthier level and thereby form a new homeostasis.

From this viewpoint, *system behavior* as denoted by Okun and Rappaport (1980, p. 7) "attempts to regulate and control the behaviors of the different subsystems," particularly when conflict exists between the individual parts and the goals of the system. This framework from which to view change within systems is especially important in working with families. Below we elaborate on working with the family unit as a system.

The Family Unit

The family unit is a system made up of such subsystems as parents, children, and extended family. In this context, a family can be defined as a "dynamic order of people (along with their individual, emotional, and behavioral processes) standing in mutual interaction" (Okun & Rappaport, 1980, p. 7). When the family's subsystems are in agreement with the overall functioning of the family, homeostasis is maintained. However, when this harmony is disrupted and conflict arises, the system is out of balance. In other words, when members within a family deviate from the family norms, their homeostasis is disrupted. To return to their former state of balance members are likely to begin to engage in maladaptive behaviors that will ensure this. It is often at this point that professional assistance may be sought or deemed necessary. Systems therapists believe that although the presenting problem is centered on the individual, it may be maintained by behaviors of other family members. Moreover, the problem is bound to affect the behavior of other members of the family system—a critical factor supporting the use of therapeutic interventions at the family level.

A central premise of this approach is the focus on the family system in contrast to the individual system. Unlike earlier models of psychotherapy, in which the view of a problem was intrapsychic in nature and treatment of the individual was considered foremost, systems theorists prefer to view the psychological problem as a symptom of the dysfunctional pattern in the identified patient's family system. From this change in problem conceptualization new models and strategies emerged. Haley (1971) calls this movement from an individual perspective to a systems perspective of problem formation and resolution a paradigm shift. Fennel and Weinhold (1989, p. 26) state: "In order to most effectively employ system-based counseling interventions, therapists must make this paradigm shift and learn to understand how an individual's symptoms may serve to stabilize and maintain the current family system."

The concepts of *linear* and *circular* causality are important at this juncture; it is believed that when practitioners make this paradigm shift they will be able to differentiate between the two. Linear causality refers to the concept of cause and effect whereby an individual's action causes another's predictable reaction. Although accurate in some instances, this type of thinking is limited and often fails to acknowledge the multifaceted aspects of the situation. Circular causality holds that both persons maintain the problem behavior, a view that allows for a more complete understanding of the problem.

Based on our work with clients, both individually and within a family modality, we have witnessed the value of systems theory in helping us understand and work with families. Further, this view is closely tied to the ecological perspective in which the biopsychosocial facets of an individual are taken into consideration.

Developmental Life Cycle

In conjunction with taking a systems approach to working with families, we also would like to highlight the importance of considering the developmental life cycle of both the individual and the family in your work.

Individual Life Cycle   In our earlier discussion of developmental interventions, we emphasized the need to consider an individual's current developmental stage to understand him or her more thoroughly and to assist the person to reach the fullest developmental potential. Please see Chapter 3 for further details on the four main theoretical theories: developmental theory, minority-identity developmental theory, life-span developmental theory, and developmental counseling and therapy theory as cited by Ivey (1991). In addition to these, other developmental views have had a significant bearing on today's ideas about development. Jung (1971) is one of the first to delineate a theory of human development that emphasized adult development. He believed that children, although often considered the identified patient, are rarely responsible for their problems. Instead, he postulated that these problems are created by their parents, educators, or physicians. Today, many family therapists and theorists still adhere to this notion in agreeing that childhood behavior problems and symptomatology have their etiology in a larger parental system. Jung also proposed an expansion-contraction type of adult development; in early adulthood an individual expands his or her life experiences by focusing on the development of family and career; in later life, a contraction occurs. This type of development is also relevant to understanding and working with family systems.

Buhler (1968) also researched and postulated about development. She defined her views about individual development even more fully than did Jung. She grouped the experiences, attitudes, and accomplishments of the individual into five biological phases, which were then paralleled to five developmental life stages. Buhler emphasized the process of goal setting within these various phases; these goals may be personal, familial, or occupational.

According to the views of both Jung (1971) and Buhler (1968), individual development is an orderly process that includes growth, consolidation, and contraction. They see expansion and achievement as occurring in the early adult years, and reflection and introspection in the later years. Although specific to individual development, both these views are relevant to family practice in the sense that to understand the specific developmental issues of each family member is important in understanding the functioning of the overall family system.

Around the same time, Erikson (1963) introduced his views regarding individual development. He outlines these eight stages of psychological development, contending that each stage is a critical transition point in a person's development from birth through death. As with the work of Jung and Buhler, Erikson's contributions are also useful in work with families. Helping family members understand where they are developmentally, as well as where members of their families are, can be very helpful and often therapeutic.

More recent researchers have also contributed to this area of development. Newman and Newman (1995) cite studies by Levenson (1978), Gould (1972), and Lowental, Thurner, and Ciriboga (1975) who have all written further about life-span development. Unique facets of these views include their specific emphasis on mid-life development, the development of men and women, mid-life career change and extension, and the added impact of the increased life span on adolescence and all stages of adulthood.

Based on these contributions on individual development, Okun and Rappaport (1980) have identified three significant assumptions about development. We cite these here because of their relevance in both individual and family therapy and our belief that they are the foundation of any therapeutic work:

1. Development is a process of passing through a normal series of developmental stages from birth to death. Although childhood and adolescence may be past, the individual does not simply become an adult. Adulthood is a continuous "becoming" in itself.
2. Movement from one stage to the next is facilitated by the successful management of the critical challenge posed during the preceding stage. In this respect, a person cannot be fully prepared to meet the challenge of a later stage without having mastered the developmental tasks of the previous stage.
3. Each stage is different from, and no less important than, any other in the developmental scheme.

Family Life Cycle   As just noted, individuals move through predictable stages in their development. Significantly, this notion of development has more recently been applied to families, based on the observation that families also develop in predictable stages.

One of the first pioneers in family life cycle development was Duvall (1971). He outlined an eight-stage cycle through which the family passes in their normal developmental progression. These stages are (1) beginning family; (2) infant family; (3) preschool family; (4) school-age family; (5) adolescent family; (6) launching family; (7) postparental family, and (8) aging family. Kimmel (1974) proposed a slight refinement of these stages by suggesting the addition of a ninth stage, "pre-marriage," which is said to precede the other eight. Kimmel emphasizes the complex nature of pre-marriage given the personal and cultural factors that must be integrated for people to court, love, and choose to marry.

After extensive research built on these views as well as their own work with families, Carter and McGoldrick (1980) reconfirmed that families move through stages of development in the process of marriage, child rearing, and preparing for life as a couple with adult children. At first, they identified three generations of the family that were presumed to be involved in this life cycle process and presented the following developmental stages of the family:

1. The unattached young adult
2. Marriage: The joining of two families
3. The family with young children
4. The family with adolescents
5. Launching children and preparing for married life without children and parents
6. The family in later life

As with individual development theories, specific issues are expected to emerge at each of these stages and their successful resolution will allow progression to the next stage. Knowledge about these stages and the significant issues expected at each stage is essential for helpers working with families. First, with such knowledge, practitioners can guide their clients in coping appropriately with certain issues that emerge. Second, this information regarding family cycle development can allow helpers to reassure families that the events they are experiencing are normal aspects of development. Often, this "normalizing" strategy can significantly reduce stress and anxiety and enhance family functioning.

Carter and McGoldrick (1988) later revised their views to fit with current changes in the American family as well as additional features not considered previously. This newer version involved the entire three- or four-generation system as it moves through time [and] includes both predictable developmental events (i.e., birth, marriage, retirement), and those unpredictable events that may disrupt the life cycle process (e.g., untimely death, birth of a handicapped child, divorce, chronic illness, war, etc." (Hepworth, Rooney, & Larsen, 1997, p. 316)

This family life perspective considers symptoms and dysfunctions in relation to normal functioning over time. In this frame of reference, therapeutic interventions are geared to helping families restore their family's developmental momentum. Several specific aspects are integral to this perspective:

• Predictable stages of "normal" family development in traditional middle-class America, and typical fallout when families have trouble negotiating these transitions
• The changing patterns of the family life cycle in our time and the shifts in what is considered "normal"
• A clinical perspective that views therapy as helping families that have become derailed in the family life cycle to get back on their developmental track, and which encourages helpers to include themselves and their own life cycle stage in the therapeutic process. (Carter & McGoldrick, 1988, p. 4)

Another viewpoint on family life development was presented by Jay Haley (1976), which we will discuss later in our presentation of family practice modalities.

More recently, theorists such as Meyer (1990, p. 12) have presented variations in the life cycle that are even more commonplace today. He notes:

The ground rules have changed as far as the timing and sequence of events are concerned. Education, work, love, marriage, childbirth, retirement are now out of synch. There is no expectation that one phase follows another in linear fashion. In this world, life events are not pre-ordained. They are more likely atomistic, mixed and matched, responsive to self-definition and opportunity.

We examine some of the changes Meyer has alluded to in the section, "The Changing Family Life Cycle and Implications for Practice," later in the chapter.

Family Assessment

A major component of any therapeutic work with families must involve an examination of both the individual and family developmental life cycles in the family assessment. A comprehensive family assessment is necessary whether working individually with a client or collectively with the entire family system. If you are seeing a client individually, a thorough assessment of the client's concerns or problems can help determine the most appropriate treatment unit (i.e., individual, family, or group) and modality. Likewise, if you are working with the entire family, an assessment of each member as well as the entire family system is necessary to ensure a complete understanding of the various family dynamics present.

Key factors integral in assessing and interviewing families are outlined by Zastrow and Kirst-Ashman (1994, p. 521) as including the communication patterns, family norms, and problems commonly faced by families. In similar fashion, Hepworth, Rooney, and Larsen (1997) have identified a family functioning assessment from a systems framework. They stress that various factors must be considered if the family system is to be understood completely. We present these factors below, along with our own views regarding their importance.

Family homeostasis as described by Hepworth, Rooney, and Larsen (1997, p. 278) suggests that "as systems, families develop mechanisms to maintain balances and homeostasis in their structure and operations." They speak to this tendency of family to maintain their unique homeostasis as long as possible, even in the face of discomfort and dysfunction. As helpers, we need to determine and understand the specific aspects of each family's homeostatic level.

Family norms, according to Zastrow and Kirst-Ashman (1994, p. 522), "are the rules that specify what is considered proper behavior within the family group." When these rules are maintained, family homeostasis is likely to persist. The rules may be both explicit and implicit and are uniquely carried out in each family system. To be effective in working with clients, practitioners must make an effort both to identify and to understand the family norms that exist.

Both functional and dysfunctional rules (norms) exist in family systems. "Functional rules allow the family to respond flexibly to environmental stress, to individual needs, and to the needs of the family unit" (Hepworth, Rooney, & Larsen, 1997, p. 279). When norms are deemed ineffective or inappropriate, the helper can point this out and encourage alternate solutions and changes. A respect for each family system is a crucial first step in eliciting change and intervening successfully. Furthermore, recognizing both the functional and dysfunctional nature of rules allows for a more balanced family assessment to take place.

Violation of rules often leads the family to use customary means of restoring the system to its previous state of equilibrium. Often in families where rules are violated, behaviors are regulated, reinforced, or extinguished by the behaviors of other members of the system. Again, regardless of how this state of homeostasis is restored, helpers must remain respectful and cautious to not respond quickly in a critical way. Before families can begin to explore the need for change, they must have trust in the therapeutic process.

Not all families allow flexibility in rules. Of course, optimally functioning families are those who are open to discarding old rules that no longer work and developing new ones as dictated by changes in the system and in the individual members within the system. Because this is not the norm, however, we encourage practitioners to view the development of flexibility as a treatment goal itself. The family must be in agreement with the helper and the direction to be taken should be one that fosters positive growth.

Determining Patterns of Family Interactions

Communication, both verbal and nonverbal, is important in determining patterns of family interactions. When communication is effective, family functioning is typically smoother; but when ineffective communication exists, family functioning is likely to be upset. Practitioners must be observant of family communication patterns that can help them identify distortions, incongruencies, and double messages.

Also important to assess is what Hepworth et al. (1977) call content and process levels of interactions.

> Families' rules are often revealed at the process level—a level often ignored by beginning practitioners as they selectively attend to what clients are "saying." Learning to sharpen one's observational powers to attend to what people are doing as they discuss problems is crucial to assessing and intervening effectively in family systems." (p. 283)

These concepts are akin to our view of the importance of nonverbal communication, which we believe is much more telling than the actual words spoken. Similar to content and level of interactions is *sequences of interactions* that is also thought to be instrumental in understanding family systems and the way in which their members interact. By observing the sequences of interactions that occur between family members, helpers can obtain rich information about communication styles and dysfunctional behaviors of the individual members and the family as a unit. Moreover, when helpers know the ways in which the family may reinforce dysfunctional behaviors in its members, they can more easily identify and challenge these behaviors.

Use of Diagrams in the Assessment of Family Relationships   In her article "Diagrammatic Assessment of Family Relationships," Hartman (1978) writes about the value inherent in using both *eco-maps* and *genograms* in the assessment process. Embedded in systems theory and the ecological, person-in-environment perspective, these two tools have been found at both levels of practice, micro and mezzo. Further, they are useful not only in the assessment process but also as interventions. Often the mere act of asking clients to draw either an eco-map or a genogram can provide insight into their family dynamics. Additionally, clients can sometimes understand their situation more clearly and envision where change is needed by a visual representation of their lives. We encourage practitioners to be creative in using eco-maps and genograms at various stages of the therapeutic process. Let us take a closer look at each of these.

Eco-maps are specific tools used by practitioners to assess family functioning and develop treatment interventions. According to Hartman (1978, p. 3), the primary developer, "The eco-map is a simple paper-and-pencil simulation that has been developed as an assessment, planning, and interventive tool. It maps in a dynamic way the ecological system, the boundaries of which encompass the person or family in the life space."

In essence, the eco-map is a drawing of the client/family in its social environment. The eco-map is most useful when drawn jointly by the helper and the client. By providing a holistic or ecological view, the eco-map illustrates the client's family life and the nature of the family's relationships with others inside and outside the family. As a way of supplementing traditional social histories and case records, it is considered a "shorthand method for recording basic social information" that "helps users (clients and practitioner) gain insight into clients' problems and better sort out how to make constructive changes" (Zastrow & Kirst-Ashman, 1994, p. 522). In this way, the technique provides a snapshot view of the many relationships the individual and family are involved in and pertinent data regarding their interactions.

> The eco-map portrays an overview of the family in their situation; it pictures the important nurturant or conflict-laden connections between the family and the world. It demonstrates the flow of resources, or the lacks and deprivations. This mapping procedure points to conflicts to be mediated, bridges to be built, and resources to be sought and mobilized. (Hartman, 1978, p. 3)

Drawing a typical eco-map requires only a pencil and paper; it consists of drawing a family diagram surrounded by a set of circles and lines used to describe the family within an environmental context. Often, members can create their own abbreviations and symbols. In Figure 4.1 we offer a view of commonly used symbols in drawing eco-maps and then present an example of an eco-map from one of our students. We hope you come to understand the inherent value in using eco-maps. Ultimately, both clients and practitioners

(A)

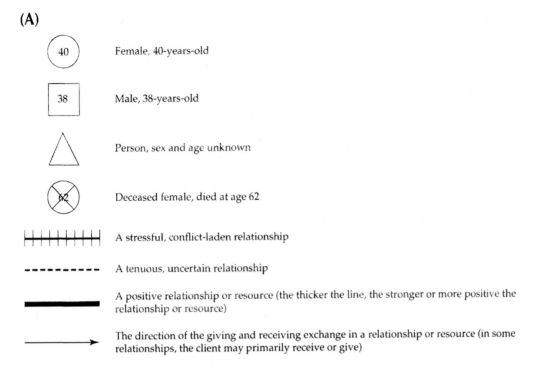

SOURCE: Zastrow & Kirst-Ashman, 1994, p. 573.

(B)

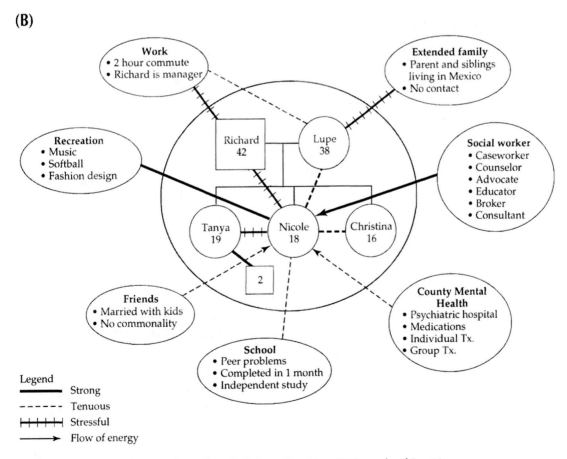

FIGURE 4.1   (A) Commonly used symbols in an Eco-Map; (B) Example of Eco-Map

can develop a greater understanding about the social dynamics of a problematic situation with the use of this type of mapping.

Genograms, used extensively by systems-oriented family therapists, were developed primarily by Murray Bowen (Kerr & Bowen, 1988) as a second important diagrammatic tool to assess family systems. As a way of depicting the family system through time, the genogram reflects at least three generations; in this regard it is much like a family tree. Zastrow and Kirst-Ashman (1994, p. 526) explain that

> the genogram is a useful tool for the worker and family members to [use in examining] problematic emotional and behavioral patterns in an intergenerational context. Emotional and behavioral patterns in families tend to repeat themselves; what happens in one generation will often occur in the next. Genograms facilitate family members in identifying and understanding family relationship patterns.

The genogram can be a simplistic diagram or may become a complicated illustration. Familial patterns can be noted more clearly and both prevention and intervention strategies can be developed once a clearer view of the family and its intergenerational context is portrayed. Often, both client and helper are surprised to learn the many facets of an individual's family history. As with eco-maps, genograms are especially useful when done in conjunction with the client. We will often give clients a homework assignment to draw their genogram, which we then examine jointly in a therapeutic manner. In Figure 4.2, we provide an example of a genogram that provided great insight and led to important treatment interventions.

Both eco-maps and genograms are useful tools in family assessment. We suggest that you explore ways to incorporate them into your own helping style.

Various Approaches to Family Assessment

Various approaches to conducting a family assessment have been proposed. We present several here as an overview of the many that can be found in the literature.

Hepworth, et al. (1997) present the following *dimensions of family assessment* they consider in working with families: family context; outer boundaries of family systems and internal boundaries of family subsystems; family power structure, decision-making process; affect and range of feelings expressed; goals, myths, and cognitive patterns; roles; communication styles; strengths; and life cycle. They encourage using these dimensions as guidelines for exploring and organizing the data gathered when working with family systems. They also recommend a specific five-step format for utilizing these dimensions in the assessment and treatment plan process. This format is briefly outlined as follows:

1. Identify the dimensions that are most relevant to one's clients.
2. Utilize the dimensions to guide one's exploration of family behavior.
3. Utilize the dimensions as guidelines for compressing new data into themes and patterns.
4. Based on the relevant dimensions, develop a written profile of functional and dysfunctional behaviors of individual members of the system.
5. Employ the dimensions to assess relevant behaviors of the entire family, developing a profile of salient functional and dysfunctional behaviors that are manifested by the system itself. (Hepworth, Rooney, & Larsen, 1997, pp. 287–288)

In a more detailed format, Tseng and Hsu (1991) consider many of the same dimensions in what they call the *scope of comprehensive family assessment*. This assessment procedure is useful when working with diverse ethnic populations and takes into account the cultural aspects of family systems as focused on by cultural anthropologists; these include marriage forms, kinship relations, marital residential choice, family authority, and family values (Tseng, 1986). We present these in great detail when we discuss race, culture, and ethnicity in Chapter 6, "Practice with Diverse Populations."

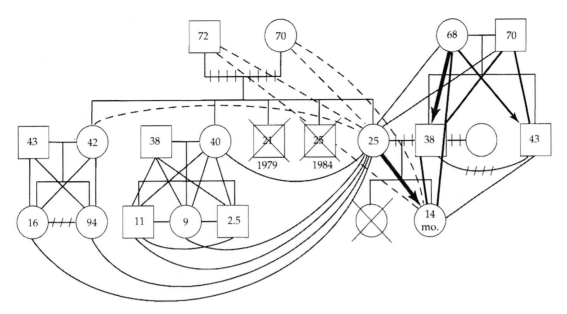

FIGURE 4.2   Example of Genogram

According to another view, *rapid assessment in family practice*, assessment can also be used for four different but related processes of family practice: (1) to define a problem area for discovering what interventions and resources are needed to help a client solve problems, (2) to evaluate the family system, leading to a formulation of a clinical diagnosis, (3) to serve as a set of interventions that simultaneously reveal relevant information about the family system and introduce information that produces change, and (4) to provide an ongoing evaluation of family's progress toward treatment goals.

This rapid approach to assessment can be beneficial, given its particular relevance to current trends in practice where brief therapy and managed behavioral health care dominate. Integral to this approach is the focus on families as whole systems. Grotevant and Carlson (1989) and Jordan and Franklin (1995) propose that a family must be assessed in terms of the interactional, interpersonal, and systems functioning of the family group. According to Goldenberg and Goldenberg (1991), focus on whole systems as an assessment modality is derived from systems theory we alluded to earlier. Some theorists (Haley, 1990; Madanes, 1984; Palazzoli, Cirillo, Selvini, & Sorrentino, 1989) believe that patterns within the family system have meaning or serve a function for the family system itself. Examples include helping a family stay "whole" or avoiding marital conflict. Others, like Cade and O'Hanlon (1993), Fisch, Weakland, and Segal (1982), and Watzlawick, Weakland, and Fisch (1974) focus very little on the meaning or function and dwell more on the behavioral aspects of the system's functioning or the self-reinforcing nature of the pattern.

Regardless of the focus taken, sound assessment methods that center on family interactions and relationships are needed. Next, we highlight seven basic methods for interviewing and questioning families during an assessment.

1. *Circular questions*, which originated with the Milan family research team (Palazzoli, Boscolo, Cecchin, & Parata, 1980), provide a structure for information gathering from all family members; its structure is nonthreatening as each member is asked to comment on the family structure and process from the view of an outside observer (Fleuridas, Nelson, & Rosenthal, 1986; O'Brien & Bruggen, 1985).

Circular questions help the practitioner assess the relationship processes and network within a family by tracing family patterns connected to complaints often made or to the presenting problem. By uncovering existing patterns, the clinician demonstrates these to the family and, when appropriate, challenges them to consider a new set of behaviors to change the pattern.

2. *Conversational/therapeutic questions* rooted in work on collaborative language systems and/or narrative perspective encourage helpers to take on a more anthropological function by seeking to understand the family from the insider's view. The premise behind these types of questions is that the helper assumes a collaborative, "not-knowing" stance that is nonthreatening. The family is considered "the expert" who can provide the most insight into the functioning of the family system (Goolishian & Anderson, 1987; Anderson & Goolishian, 1992).

3. *Hypothesizing, circularity, and neutrality* are methods also derived from work of the Milan Restack team (Tomm, 1984). These methods are used conjointly to assist clinicians in locating family patterns and in introducing change into the system as needed. By hypothesizing or introducing an explanatory narrative in the assessment, the helper explores existing patterns associated with the problems. These questions are done in a circular fashion and the helper must maintain neutrality in the interviewing process. Curiosity about, admiration for, and respect of the family's values, beliefs, and behaviors are important for a helper in being able to test his or her hypothesis (Cecchin, 1987). Clearly, if helpers are too quick to judge, the family may become suspicious and distrusting of their motives and objectivity.

4. *Tracking problems, attempted solutions, and/or exceptions to the problem* stem from work conducted by the Mental Research Institute (MRI) of Palo Alto, California (Fisch, Weakland, & Segal, 1982; Walzlawick, Weakland, & Fisch, 1974). Tracking is considered one of the best known of family systems interviewing methods in which the clinician assesses the family context and interactions associated with a behaviorally identified problem within the family. According to Weakland and Fisch (1992), the assessment process begins by learning more about the problem: "Our aim is to get as clear and specific a description as possible—who is observably, doing and saying what" (p. 308). Cade and Hanlon (1993) encourage asking such questions as "When does the problem occur?" "How often?" "Where does the problem occur?" and "What would I see and hear if I were an invisible observer?"

5. *Pretherapy change assessment* that emerged from the brief, solution-focused therapies (deShazer, 1982, 1985; Lipchick, 1986; Weiner-Davis, de Shazer, & Gingerich, 1987) is a method that assists practitioners to assess family strengths and patterns tied to their problem-solving resources. This method attempts to find out how the family is managing the problem and changing before treatment is actually sought. Many families often improve even before engaging in therapy. Also, it is important to examine times when families cope well and the identified problem is not present. This information is helpful in identifying patterns of problem resolution within the family and to refine these as needed.

6. *Scaling technique*, also derived from brief therapy models, was developed by de Shazer, Berg, and colleagues at the Brief Therapy Center in Milwaukee. It consists of having the clinician construct the client's problems along a 10-point continuum or another ordinal scale that is useful to both the helper and the client in better understanding the severity of the problem and in determining the future direction of treatment. The scaling technique is also valuable in monitoring the outcomes of practice and has been used in agency settings for this purpose.

7. *Family task observation* is a very basic technique for observing family members involved in a specific task. Such observation is considered useful in further assessing the family and its functioning. Specific tasks can include playing a game, planning a vacation or family outing, solving a problem, or making a decision. Only through such an observation can the clinician more completely recognize and understand the many aspects of how a specific family functions.

These seven assessment methods are the ones most commonly found in current family assessment models. In addition to becoming skilled in using these diverse techniques, we encourage you to remain focused on the primary, two-fold purpose of family assessment: (1) to gain a better understanding of the family as a system, and (2) to use assessment as intervention in which change in the family system is fostered.

Varied Treatment Orientations of Family Therapy

In the preceding chapter on micro-level practice, we presented key orientations and modalities for individual work with clients. In this chapter, we present the orientations and modalities most relevant to working at a mezzo level with families.

We believe that understanding the basics of working with individuals is essential to any additional work with larger systems such as the family. For practitioners to cross the bridge comfortably from individual work to systems-based treatment, they must be well grounded in their views about human nature and their purpose in helping—to assist in the alleviation of human suffering. Also, with both individuals and families, respect for the client system remains fundamental to establishing a sound therapeutic relationship that will further enhance the treatment process and assure a more positive outcome.

Family Therapies

In the last few decades, family and marital therapies have steadily increased as a viable means of treating individuals and their families. The social, economic, and political demands that affect many families today have created undue stress for some. In years past, individual treatment was the primary means of treating individuals; today, we find that more and more families are seeking treatment as a unit. Also, more professionals are recognizing the value of including certain family members or the entire family, when working with individuals. These trends have brought a proliferation of system and family therapy theories and treatment approaches. To learn new and innovative treatment modalities for working with families and to stay current in their practice, clinicians within the various disciplines of helping must become knowledgeable about the basics involved in working within a family context. We hope the overview in the preceding pages has been helpful. We now turn to the specific discussion of family therapy.

Family Therapy Defined   Many definitions of family therapy exist. In general terms, Zastrow (1995, p. 29) defines family therapy as "a type of group therapy aimed at helping families with interactional, behavioral, and emotional problems" and that can be "used with parent-child interaction problems, marital conflicts, and conflicts with grandparents." In this respect, a wide variety of problems can be dealt with when a practitioner is working with the family system as a whole or even with a subsystem. Okun and Rappaport (1980, p. 32) perceive family therapy differently.

> It is a point of view that finds its focus primarily in work with family systems. It regards problems and dysfunction as emanating from the intrapsychic problems of any one individual. Individual symptoms are seen as reflections of stress arising within a larger family system (or other societal system). This stress may occur when the family is unable to negotiate a development passage or when a nondevelopmental crisis occurs.

Others, such as Downs, Costin, and McFadden (1996), who focus on families within child welfare emphasize the value of working with the family unit in certain situations when working with a particular child or some problem that a family encounters lends itself most effectively to a form of family therapy. In this context, they consider family therapy to be "based on the view that family life is a system of relationships between people. The goals are directly related to the well-being of the children and the enhancement of their learning and growth opportunities" (p. 87).

Several advantages of family therapy are also cited:

- Attention can be given to both the individual members of the family system and the family as a whole.
- Important themes within a family can be portrayed, and a broader and more balanced diagnostic view of the strengths and weaknesses in the family will emerge.
- Pressure will be reduced on the identified family member who may have been previously singled out as the cause of the problem that he or she should strive to fix.
- Changes in the family as a whole are more likely to impact each individual member and the probability that improvement in a child's behavior and direction of growth will be sustained are increased. (Downs et al., 1996)

Origins of Family Therapy   The evolution of family therapy is quite telling. Its origins can be traced to earlier traditions of individual psychotherapy. Partly as a reaction to and partly as development of ideas proposed by such theorists as Freud, Adler, and Sullivan, later work with family systems emerged and is known today as family therapy.

Sigmund Freud, in his treatment of his most famous patient, "little Hans," demonstrated how change in one member of the family had an effect on other members of the system. Based on these early works by Freud, the Ackerman school of family therapists later emerged; they view themselves as incorporating certain Freudian concepts in a family systems approach.

Alfred Adler very much emphasized the role of social relationships; he considered personality development to be influenced by experiences and feelings about a person's social roles and the relationships among those roles. Furthermore, he believed an individual's place in the family was crucial to personality development.

Harry Stack Sullivan was heavily influenced by Adler's work emphasizing the role of interpersonal relationships in the development of personality. His view of mental illness related to family; he believed mental illness consisted of and stemmed from distorted interpersonal relationships. Sullivan's work also inspired early family therapists who went on to further develop the notion of the importance of the family system as a way to best understand the individual.

In addition to these theorists, work of both anthropologists and sociologists also greatly contributed to the development of family therapy: sociologists by focusing on the cultural influences in personality and the role structure and organizational functions of the family; anthropologists who added to this view by studying culture and ethnic systems that were believed to affect an individual's development and behavior.

In the 1950s, family therapy finally emerged as a specific and acceptable treatment modality. Family therapy developed out of the frustration many practitioners felt with their efforts to apply conventional psychiatric principles to their work with clients (specifically, adults with schizophrenia and children who were delinquent) and their families. During this time, the work of the four major research groups began: (1) the Bateson group in California, (2) the Lidz group at Johns Hopkins and later at Yale, (3) the Bowen group at the Menninger Clinic, and (4) the Wynn group in Washington, D.C. A decade later, important contributions to family therapy had been made and a new conceptualization of schizophrenia as a symptomatic reaction to stress in the family was generated.

In the 1960s, many therapists ventured into the family therapy arena, but their therapy practice was not based specifically on any of the theoretical research of the 1950s; instead, it was based on the application of existing psychodynamic techniques and concepts of family situations. In this period, Bateson and Bowen continued to develop their systems, and Haley and Jackson were constructing their own communication theories and techniques for understanding and working with families. Other pioneers in the field were Ackerman and later Minuchin; both made major contributions to the family therapy field. During this time, various family institutes were founded, each reflecting its particular orientation. Ideological conflict between analysts and system theorists and among clinicians and researchers was strong.

In the 1970s, family therapy grew tenfold. Although different viewpoints continued to be perpetuated among the different schools of thought, most seemed to adhere to the systems approach. Major new figures that emerged in the field included Whitaker, Framo, Guerin, Block, Kantor, Fogarty, Bell, Papp, Boszormenyi-Nagy, and Paul (Okun & Rappaport, 1980). Also in this decade, cultural aspects were beginning to be considered more significant and the onset of the family preservation movement took place.

In the 1980s, transpersonal therapy emerged as an important force in counseling training. Included here is Mindell's process-oriented psychology and Weinhold's transpersonal relationship therapy. Also flourishing were such systems-based cognitive-behavioral theories as Minuchen's structural family therapy and the functional family therapy developed by Barton, Alexander, and Parsons.

In the 1980s and 1990s, two trends have greatly impacted the field of family therapy. The first is the advent of managed behavioral health care; the second is the move toward a more integrated service delivery. In both, the emphasis is on improving mental health and social services and increasing the use of services, efficiency, and cost effectiveness. Demands for increased accountability on the part of the practitioner has resulted as has the massive restructuring of how clinical services are financed and delivered. As a consequence of these major changes, emphasis is now on time-limited (brief) family therapy that has a base in contemporary systems theory. The practice models considered the most conducive to brief treatment include structural (Minuchen), strategic (Haley/Madanes, Milan), cognitive-behavioral (Pavlov, Skinner, Wolpe, Bandura, Lazarus, Ellis, Beck), solution-focused (deShazer & Berg), social construction and narrative (White, Epson, & Hoffman), psycho-educational family practice (Bateson, Bowen, Laing, Lidz, Cornelison, Fleck, Terry), and family preservation.

As we approach the 21st century, the growth of family therapy is bound to continue.

Classifications   Just as there are many forms of family therapy, there are many terms for classifying these various types of family therapy. For consistency, we have categorized them the same way we did when discussing individual orientations: psychodynamic, humanistic, cognitive/behavioral, and developmental. Below we have highlighted one family therapy practice model from each of these groups as follows:

Psychodynamic: Bowen's systems theory
Humanistic systems theories: The Satir model
Cognitive/behavioral: Structural family therapy
Developmental: Haley's developmental life cycle theory

We have chosen these models as they represent a broad range of therapeutic approaches, all from a systems perspective, that you may draw from. In no way does this minimize the importance of the many others we have not included. Also, we recognize the considerable overlap among them, with certain modalities fitting the constructs of several of the main orientations.

Family Practice Modalities

Psychodynamic

Psychodynamic family therapy is an extension of psychodynamic therapy that is deeply rooted in Freudian psychoanalytic theory. Although many psychodynamic family therapists today do not adhere closely to the many principles postulated by Freud, they do strive to understand the dynamic aspects of personality and the role of early childhood. Fennel and Weinhold (1989, p. 81) indicate that

> psychodynamic family therapy can help the therapist develop insight into how the family as a social unit affects the intrapsychic development of child-parent relationships and vice versa. Using such an approach, the therapist can help each family member become aware of his or her projections and take responsibility for them, which requires each family member to develop an *observing ego* to understand and integrate this awareness.

Family therapists work toward helping each family member set appropriate boundaries in relationships with other family members. A primary goal of this approach, then, is to strengthen and preserve the family as a social unit while still respecting the individual autonomy of each member. True to the psychodynamic spirit, family therapists help individuals resolve past parent-child conflicts rooted in early childhood experiences. Most psychodynamic family therapists view treatment as a process with four main stages: (1) the therapy contact, (2) the initial phase of treatment, (3) the working through stage, and (4) termination in which the therapeutic relationship is considered a tool for change by modeling healthy relationships. By providing conflicting families and couples an opportunity to learn more about their inner workings and how these issues impact their relationships with other people, change is possible.

The work of Harry Stack Sullivan, Erick Fromm, and Erik Erikson in the late 1940s and 1950s paved the way for this analytic family therapy. Sullivan's work was especially instrumental, given his view that the growth and development of the child is often a response to the social environment of the family. In fact, several early leaders in family therapy, including Don Jackson, Virginia Satir, and Murray Bowen, were significantly influenced by Sullivan. Below we briefly examine the work of Murray Bowen and his family system's theory.

Bowen's Family Systems Theory   Murray Bowen is considered one of the leading proponents of family systems theory. Bowen, a psychiatrist, is well known for his early work with patients diagnosed with schizophrenia and their families. During this time, he developed the Bowen Group who practiced in the Menninger Clinic in Topeka, Kansas; they began to research schizophrenia from a family systems perspective. Based on his findings, Bowen eventually modified his theory to include the entire family constellation in his understanding of schizophrenia in an individual patient.

In his early work, Bowen initially focused on the mother-child dyad that led to the concept of *mother-child symbiosis*. From his work with these families, he found that because the mother and child bond could be so intense, it was often difficult for the mother to differentiate herself from the self of the child (Bowen, 1960, 1961). This led him to suggest that schizophrenia was a result of this "emotional-stuck-togetherness" of the mother and child (Kerr, 1981). He hypothesized that emotional illness in the child resulted from a less severe problem in the mother such as being overadequate or overprotective. Later, when it became apparent that the child also lived in a family with others besides the mother, Bowen began to consider the role of other family members, including the father, in his research.

By the mid-1950s, the Bowen group had moved to Washington, D.C., to work at the National Institute of Mental Health (NIMH), where Bowen began to identify families who had a member with schizophrenia and hospitalize the entire family for observation and research. The length of the hospital stay for many of these families ranged from 6 months to between 2 and 5 years. Based on their observations, Bowen and his team identified various themes that seemed to characterize these families. The three most notable family traits included "(1) their restrictive relationships; (2) their rigid, stereotyped roles; and (3) their selective use of external variables only insofar as they validated their inner projections" (Okun & Rappaport, 1980, p. 46).

Through research, Bowen's group also discovered that parents of patients with schizophrenia seemed unable to separate their feelings from their thoughts; this degree of fusion or differentiation of feelings and intellect varied among families. From these findings, the group determined that people with the greatest fusion between feelings and thinking seemed to struggle the most; those who were better able to make this distinction appeared to have more effective, adaptive, and flexible coping skills. As a result, Bowen came to believe that *emotional illness*, which he preferred over the term, *mental illness*, is more than a disorder of the mind; rather, it is a disorder of relationships. His use of the term *emotional illness* seemed to fit nicely with his view that what occurs in family relationships can cause emotional impairment in individual family members.

Bowen's work with families also led to his hypothesis that parents who have a child with schizophrenia have a weak, tense marriage. Because the marital relationship is a distant one, the child born into the system is often *triangulated*. This "concept of triangulation refers to a situation in which two people are unable to relate to each other and so use a third person, in this case the child, to reestablish contact and some kind of homeostatic balance. It is as if the third person is needed to maintain a relationship between the first and second persons" (Okun & Rappaport, 1980, p. 46).

To help you understand Bowen's approach more fully, a few of the key concepts central to his theoretical views are elaborated here.

Key Concepts   *Emotional triangle,* as explained above, leads to triangulation. Bowen found that when there is an unstable two-person system, a third person is typically pulled in to create the needed stability in the system. Often in these emotional triangles two sides are stable whereas one is conflictual. An example would be a family in which the relationship between mother and father and mother and child is conflictual, yet the father-child relationship remains stable.

Differentiation of self from family of origin, also noted above, relates to the individual's ability to develop a strong sense of self and to make choices independently from the influences of others. Bowen's work on *fusion* relates to the level of differentiation with a system. He developed a conceptual scale to describe this; on his scale, zero indicates the least amount of differentiation, and 100 represents the maximum level of fusion and differentiation. The scale shown in Figure 4.3 further illustrates this concept and demonstrates the importance of people becoming differentiated in order to function adequately in the world.

The views on differentiation illustrated in Figure 4.2 are integral to Bowen's approach in working with families. He found that a person's level of differentiation is established in childhood and is thought to be stable throughout life unless specific interventions are used to alter it. Furthermore, he believed that this level reflected how adequately children were able to separate from their family of origin to establish a sense of self of their own. Notably, when parents are fused and undifferentiated, the child will be more unlikely to develop a

---

High Differentiation

100   Individuals are aware of the difference between their thoughts and their feelings.

They are able to separate these two appropriately without allowing their feelings to cloud the rational process.

They can state their positions clearly and are open to change.

50   People are usually able to separate their thoughts and feelings; however, they may not be able to express themselves openly in a relationship if they feel threatened.

They live in a world dominated by emotions, and their ideas are usually based on others' opinions.

They are unable to separate thoughts from feelings readily; their rational processes are frequently clouded by their emotional experiences.

0   Most of their energy goes into seeking love and approval from others or punishing those who do not provide this. Little energy is left to engage in productive, goal-directed activities.

They have a greater incidence of emotional and physical problems.

Low Differentiation

FIGURE 4.3   Differentiation-of-Self Scale
SOURCE: Based on Fenell & Weinhold, 1989, p. 104.

higher level of differentiation. From a Bowen perspective, then, the goals of family therapy are directly related to understanding the differentiation process within the family and exploring ways to encourage each member to become more differentiated.

The *family projection process*, which is very much related to the emotional triangle, explains how parental problems are projected onto a child. Most typically, Bowen found this process to exist when the mother was more emotionally involved with her child than her husband; the opposite, where there was overinvolvement between the father and child is also possible. In either case, the father (or mother) senses his wife's (or her husband's) overinvolvement with the child. Usually the anxiety in the mother (father) is immediately felt by the child who is most involved with the mother (father). This situation, in turn, leads the mother (father) to place all her (his) energy in relieving the child's anxiety instead of working on her (his) own. This scenario is likely to repeat itself often and results in the child's becoming increasingly impaired and unable to function independently. Also, the child may develop significant behavior problems as he or she matures, typically in adolescence. The other parent continues to support the established behavior pattern in the child by either backing the mother (father) or by completely withdrawing from the situation when he or she is not in agreement. Often, problems become so severe that professional intervention is sought. According to Bowen, attempts to "triangle in" the helper to help stabilize the system often result.

Nuclear family emotional systems occur as a result of the family projection process. In families where there is much instability, efforts are made to attain a balanced state. However, when there is a high level of fusion (little differentiation), the attempts of one member to stabilize the system may be at the expense of another family member. In these families there is likely to be a high level of anxiety. According to Bowen, dysfunctional family systems attempt to reduce this anxiety level in one of four ways. We present these because they are often found in families who seek family therapy.

- Attempt to establish emotional distance, which is a means to avoid the nuclear family emotional process (Kerr, 1981).
- Marital conflict through which spouses can meet their needs for both emotional closeness and distance. Marital conflict is normal; however, when the children are made to feel guilty or responsible for the conflict, it can be emotionally harmful.
- Dysfunction of one spouse that occurs when one spouse becomes submissive and the other dominant. The submissive one becomes a "no-self" and is the one who adapts most (Bowen, 1971); this increases his or her susceptibility to physical illness, emotional illness, or social dysfunction.
- Projection onto children, which occurs through the family projection process discussed earlier. When intense fusion exists, the child becomes impaired. This allows the parents to focus their attentions on the child and avoid looking at their own role in the family's lack of differentiation.

Multigenerational transmission process addresses the multigenerational factors involved in individual and family development. Bowen contended that emotional illness resulting from undifferentiated or fused family systems is projected onto children over a series of generations. Unfortunately, this often results in increased emotional illness with each succeeding generation.

On a more optimistic note, this process can be slowed or altered when favorable life circumstances are present or when professional intervention is effective in calming the family's high level of anxiety and helping them become increasingly differentiated.

Personality profiles of sibling positions relates to the significance of birth order. Drawing from Toman's (1961) study on birth order, Bowen furthered his own view that birth order of children can serve as a predictor of adult behavior. The positions most typically recognized are of the oldest and youngest family members. Often, the eldest will assume an overfunctioning role, whereas the youngest may be more reluctant to take any initiative.

Although these positions may be functional and adaptive, when the family is fused, these birth order patterns also become dysfunctional.

Therapeutic Functions   Bowen's therapy is based on the belief that the identified patient is not a source of the problem but rather a result of emotional fusion within the family system. For the therapist to be effective, one of his or her first tasks is to remain as neutral as possible thereby avoiding entering the family's emotional field. For this to occur, practitioners must be reasonably undifferentiated and are encouraged to engage in their own differentiation process as part of their training. In much the same way, we encourage helpers to engage in ongoing self-awareness that alerts them to the need for possible therapy of their own.

A primary way that helpers neutralize the family's emotional system is through rational means (cognitive interventions). In effect, practitioners gradually assist families to understand their situation better by learning the basic concepts of systems theory. For instance, when a child is the identified patient, the parents are helped to understand their role in the child's problems. More specifically, it can be said that the child's problems result from their parents' dysfunctional relationship. When the parents are able to redefine and modify their relationship, the child's symptoms are likely to change as well. Based on this, Bowen preferred to work with the parents alone, without the child. This allowed them to focus more directly on their relationship and the issues regarding their own differentiation of selves from their families of origin. Also, through this process, the therapist is better able to reinforce to parents how a lack of differentiation may be what is actually affecting the child.

The role of the therapist is very important in Bowen's therapy. Not only is it necessary to remain neutral, but it is also essential to be able to teach clients about families and the way they function. To be most effective, Bowen also suggests that the spouses talk directly to the therapist rather than engage in an emotional battle with one another. He believes this strategy allows the couple to understand more clearly the underlying family dynamics and be more open to changing dysfunctional patterns. When clients begin to focus on each other or they overreact to something the other has said, therapists are encouraged to return the communication flow back to themselves; they do this by having the clients talk about their thoughts, feelings, and reactions to the therapists rather than to each other.

A last important facet of the Bowen approach centers on teaching the family the functioning of emotional systems that encourages differentiation. This is done through such techniques as "I messages," parables, metaphors, descriptions of similar family situations, and eventually through didactic measures. When couples are able to understand their family as an emotional system and how it functions, they will be able to relate to each other and others in and outside the family in a differentiated way.

Treatment Considerations

Bowen identified four main functions of the therapist that we believe make up basic *treatment considerations* of his model.

- Therapists must define and clarify the relationship between the spouses that requires a thorough assessment, effective communication, and trust in the therapeutic process.
- Practitioners must keep themselves detriangled from the family emotional system. This requires the helper to be highly differentiated and to be continually aware of situations when he or she might be tempted to side with a particular member or become emotionally involved in the family's process.
- Teaching the functioning of emotional systems is necessary in helping families learn ways they can become more differentiated and functional. Helpers must therefore be comfortable with didactic measures.

- Therapists must be open to demonstrating differentiation by taking "I position" stands during the course of therapy as well as using other skillful interventions that assist families to differentiate.

SCENARIO

A couple in their 40s apply for services for their 10-year-old daughter who they state is being harassed by the neighbor children to the point that it is affecting her grades in school and her relationships in the family. They say that the neighbors are hostile and frequently attack their daughter verbally. You notice as they talk about this that the mother complains the most and is extremely affected by the neighbors' behavior, stating that she has been very angry and anxious, almost to the point of not being able to go out of the house for fear the kids might be there. As you do a further assessment of the family you discover that this marital couple has multiple issues between them which have been going on for many years. The mother states that the husband used to be an alcoholic and that they fought terribly for years in the beginning of their relationship. Now she says that he is very distant and often sleeps on the couch in the other room; she is very frustrated. She states further that he reads pornographic magazines frequently and that they have a very limited sexual relationship. The husband admits that this is true but does not see these issues as much of a problem. You can feel the hostility and anger of the mother as you are assessing their marital relationship. The father is more distant and passive. The daughter denies that there is a problem and says that although the neighbors used to attack her, she stays away from them now. She says it is more of a problem for her mother, who gets very afraid of the neighbors and is very angry about what they did to her in the past. As you interview them further, you notice that a concentration on the marital relationship seems to make the parents anxious and they then shift back to the daughter, who continues to say that she is "okay." What questions or issues might you focus on with this family? If you were the helper in this situation with this family, what might you do to handle the current family dynamics and presenting concerns? Would you continue to see the family as a whole or see the marital couple separately?

Humanistic/Existential

Application of humanistic/existential theory in family therapy has been an evolving process. Early works of such leaders as Rollo May, Carl Rogers, Abraham Maslow, and James Bugenthal has given way to current humanistic family therapy practice. Similarly, such existential leaders as Victor Frankl, Rollo May, and philosophers Martin Buber and Jean-Paul Sarte have been influential in use of existential concepts in family treatment. The similarities of the philosophical underpinnings and their common focus on subjective states held by both humanistic and existential schools suggest a benefit in using them together in treatment of both individuals and families.

From a humanistic/existential stance, a basic goal of treatment is to increase the clients' awareness of their options and potentials and to help them make choices and decisions. Clients are encouraged to increase their autonomy and become self-actualized rather than merely adjusting to their situation. For this to occur, clients are helped to accept their own freedom to change and grow and supported in accepting responsibility for their own actions. Because anxiety is common when clients attempt to look at themselves, humanistic therapists also help clients face and deal with their anxious feelings. According to humanistic theory, all these goals can be achieved only when the therapists possess certain humanistic characteristics (e.g., genuineness, caring, unconditional positive regard, empathy) that allow clients to feel connected and trusting of the therapeutic relationship.

In family therapy, the focus is on not only the individual but also the family system as a whole and what happens between family members. Table 4.1 outlines a typical family

TABLE 4.1   Family Therapy Process Based on Humanistic/Existential Principles

1. *Milling around* relates to clients' confusion about the therapeutic process.

2. *Resistance to therapy* is noted in family members who are resistant about sharing their thoughts and feelings.

3. *Talking about the past* is often a way family members begin to talk about their feelings.

4. *Expressing negative feelings* toward each other and the therapist may signify that clients feel safe enough to test the waters within the therapy session.

5. *Expression and exploration of personally relevant issues* usually follow when clients feel safe and hopeful about the prospect of change.

6. *Full expression of feelings* in the here and now takes place within the therapy session and can result in changes in family structure.

7. *Healing capacity of family members* is evidenced when family members begin to turn their attention to helping others within the system. A climate of trust and freedom allows for acceptance and understanding to emerge

8. *Self-acceptance and change* occur as family members begin to accept themselves and take responsibility for their actions and their power to change their behavior.

9. *The facades fade away* as clients allow a sense of realness to develop. Clients who remain hidden behind a facade are encouraged to share their real feelings and thoughts in the hope that they will allow their true selves to emerge.

10. *Feedback is given* by family members to each other with the goal of helping the family grow.

11. *Confrontation often occurs* when clients feel safe enough to challenge those who remain defensive and closed.

12. *The helping family develops* as clients are successful in applying what they have learned in therapy on the outside and are more consistent in helping one another.

13. *The family encounter occurs* as the family begins to experience itself as a unit, each member independent but working together with the others.

14. *Fuller expression of closeness* results when family members possess warm and genuine feelings toward each other. They have learned to listen and accept each other's feelings and are able to resolve disputes much easier and faster.

15. *The family is changed* as the members now can experience both "oneness and separateness" in their interactions with one another. They are hopeful about their future relationships and their ability to work together to resolve any problems that may arise.

SOURCE: Based on Fenell and Weinhold, 1989, pp. 177–178.

therapy process based on humanistic/existential concepts and therapeutic frameworks. This process is comprehensive in nature and has a beginning, middle, and ending. Respect for clients is important and a positive view of human nature is necessary. This type of therapy is most effective with families who are motivated, open, and receptive to change and their role in the change process.

Three major humanistic/existential therapies that apply this therapeutic process include "person-centered therapy" proposed by Rogers, "Gestalt-oriented family therapy" influenced by Perls and Kempler, and "psychodrama-oriented family therapy" created by Moreno. Additionally, the two major systems-based theories that are humanistic/existential in nature are Satir's "communication theory" and Whitaker's "symbolic-experiential family therapy." Below we focus on the work of one of the best-known family therapists, Virgina Satir.

The Satir Model   Virginia Satir, trained as a psychiatric social worker, began her work in 1955 training family therapists at the Illinois State Psychiatric Institute in Chicago (Fennel

& Weinhold, 1989). In 1959 she became a founding member of the Mental Research Institute (MRI) in Palo Alto, California. She was joined by other leading authorities in the field, including Jackson, Haley, Riskin, and other MRI researchers. Although she did not add any new theoretical concepts to those already presented, she did provide a different focus, interpretation, and application of theory to practice. In 1964, she published her first book, *Conjoint Family Therapy*, in which she introduced her family therapy model. Several years later in 1967, she revised her work by incorporating the humanistic, existential foundation of her communication theory. In 1972, Satir wrote her second book, *Peoplemaking*, in which she focused on the adults in the family as the peoplemakers. Based on this perspective, she believed it was essential to place special responsibility on the parents for the nurturing of their children as well as the perpetuation of a loving and flourishing family environment.

Satir's work is renowned for its growth-oriented philosophy base that she so skillfully translated into practice. She believed that individuals are innately good and that they possess the capacity to develop to their fullest potential. Pathology occurs only when this development potential is blocked or hindered. Clearly, Satir's phenomenological outlook was positively influenced by the earlier works of such humanistic theorists as Rogers and Maslow.

Satir considered her treatment of families to be an ongoing process that was constantly evolving. She did not believe in adhering to a rigid set of principles, as she was constantly seeking ways to improve her model and refine her techniques. Her work with families focused on identifying the process of their relationships, with the primary intent being to help them heal and thus be able to grow and develop to their fullest. She believed that every relationship is an encounter between two people at a particular moment in time. In her therapy, she attempts to recreate these encounters between family members to be able to identify and deal with underlying issues and feelings that promote healing.

Satir's work also emphasized the need to integrate what she called the four basic parts of the self: (1) the mind, (2) the body, (3) the report of the senses, and (4) social relationships. One of her most significant views is tied to the functioning of individual family members within the family system. She believed that for the family to function at its optimum, individual family members must themselves possess strong self-concepts. Her strong conviction in the power of the individual is noted in key concepts of her theory presented below.

Key Concepts   *Maturation* is a concept that is central to Satir's (1967, p. 91) viewpoint that "the most important concept in therapy, because it is a touchstone for all the rest, is that of maturation." Mature people are those who are able to accept responsibility for their own choices and decisions and to take full charge of their lives. Inherent in this maturational process is the need for each individual to become a *differentiated self*—that is, able to separate oneself from one's family. Characteristics of a mature person include (1) the ability to be in touch with his or her own feelings, (2) the ability to communicate clearly with others, (3) the ability to accept others as different from oneself, and (4) the willingness to see differentness and differentiation as a potential for growth and learning rather than as a threat (Satir, 1972).

Self-esteem is a second core concept for Satir. She postulated that individuals cannot reach a high level of maturity until they develop a feeling of self-worth. Only when clients have a strong self-concept can the family unit progress.

The importance of communication as stipulated by Satir has evolved into one of the most salient aspects of her therapeutic work. In the context of the family, Satir believed that communications within the family system reveal how individuals within the family system view themselves. She believed that behavior is communication, which is both a process of giving and getting information and a verbal and nonverbal process of making requests of the receiver (Satir, 1983). Furthermore, she believed that *feeling*, or the *emotional system* of the family is expressed through communications. Her views about dysfunction are directly connected to her beliefs about the communication process; when communication is incongruent—that is, when communicational and metacommunicational aspects of the message do not agree—dysfunction occurs.

Thus, problems typically arise in families when individuals send conflicting information both verbally and nonverbally. In *Peoplemaking*, Satir describes four types of dysfunctional communication—blaming, placating, super-reasonable, or irrelevant—that result when people have low self-esteem and feel threatened (see Table 4.2). Whether they fear rejection, judgment, or exposure of their weaknesses, they respond by using one or more of these survival stances, which are often learned in childhood.

According to Satir, when individuals use any of the dysfunctional communication styles, they not only communicate nonacceptance of the others involved but also reflect low self-esteem. In her view, low self-esteem hinders a person in learning to communicate effectively and thus impedes the maturation process that allows that person to successfully differentiate from his or her family. Because of this view, Satir emphasized early formative relations with parents as being critical to healthy development. She felt strongly that the inability to separate oneself from the family and the inability to communicate clearly are intertwined. In this respect, she noted circular causality in the relationship between communication and self-esteem: Incongruent communication causes low self-esteem, and low self-esteem leads to incongruent communication. Conversely, when people have relatively high self-esteem they respond mostly in congruent ways, seen when their communication reflects their feelings.

Observation of the family's communication system is important in understanding the system and how it functions. Particular attention is paid to dyads, or two-person relationships in the system. One of the most important of these dyads is the marital relationship, which Satir felt was pivotal in the family's development. She believed that when the marital relationship is functioning well, children in the family are permitted to have different relationships with each parent; this allows them to develop high self-esteem and to eventually be able to separate successfully from the family system.

Through observing the family's communication, a therapist can discover the "rules" of the system, namely the "shoulds" that are either clear and explicit or vague and implicit. Often these rules reveal the family's value system; they are also helpful to clinicians in targeting areas for intervention. When rules are faulty, they cause people to interact dysfunctionally and result in system problems. Because of the importance of the rules, Satir strives to discover them by assessing the level of freedom members have to express their feelings within the family, whether others agree or disagree; identifying which family members are more amenable to open communication; and observing ways in which family members ask questions when they do not understand (Satir, 1972).

Therapeutic Functions   Satir emphasizes an open and fluid therapeutic approach that is open to many principles and ideas. For instance, she has incorporated ideas from such disciplines as dance, drama, religion, medicine, communications, education, and the behavioral sciences in her work. She encourages clinicians to develop a broad repertoire of tools to use in family therapy, not only those specific to psychotherapy. She also believes in flexibility and variability in terms of length, location, clients seen, the use of therapists, and the techniques and procedures for therapy. She encourages creativity in working with families that will allow them to succeed in overcoming their problems or level of dysfunction.

TABLE 4.2   Dysfunctional Types of Communication

1. Blaming, when an individual sends the message "I'm OK—you're not OK."
2. Placating by denying one's own feelings about oneself and sending the message "I'm not OK—you're OK."
3. Reasonable analyzing by denying one's own feelings, thereby remaining neutral, yet distanced from others.
4. Distracting to avoid dealing with the central issues at hand as well as the process.

Although Satir is less concerned about how the clinician engages in therapy, she does highlight the need to create an environment for the family that is supportive of the continued positive growth of the individuals in the families and their relationships with one another.

To be effective, family therapists who follow Satir's ideas must be willing to risk themselves in therapy, to be spontaneous, and be able to establish intimate contact with family members that supports their growth and promotes their self-esteem. Satir believes that therapists, as positive role models, can help clients become congruent, genuine, caring, and more open and capable of healthy risk taking in their relationships with others.

Treatment Considerations

• Therapeutic interventions must be designed to allow people to express their potential fully, both individually and within the family system.

• It is important for clinicians to encourage healthy risk taking by family members that will allow them to express their feelings openly and take responsibility for their own behavior.

• Because the crux of Satir's theory emphasizes the importance of both communication and self-esteem, practitioners need to help family members develop increased self-worth; clear, direct, and honest communication; and flexible and appropriate rules (Satir, 1967).

• The therapist must have strong humanistic traits that reflect a deeply caring individual whose goals are strictly to help the family heal.

• Therapists must recognize that family members often are fearful on entering treatment. Because of this, it is important to create a safe environment that is conducive to encouraging introspection and behavioral changes. This requires therapists who are confident in their skills and are also sensitive and understanding.

• Helpers elicit trust and rapport with the family when they demonstrate confidence in the therapeutic process—namely, the direction and purpose of therapy.

• Once a positive therapeutic relationship is established, helpers must continually encourage family members to look at themselves more objectively and share their growth with one another.

• Positive, nonjudgmental feedback about communication patterns within the family is encouraged between members themselves and from the therapist as well. Sensitivity and timing are important to consider when providing constructive feedback.

• A central role of therapy is to help clients develop high self-esteem. Satir cites the following as techniques to help clients have positive feelings about themselves.

1. Making "I value you" comments to the client
2. Identifying client strengths and pointing these out to the client
3. Increasing clients' confidence by asking them questions in areas in which they feel competent.
4. Encouraging clients to ask for clarification if the therapist is unclear in his or her communication.
5. Asking each family member what they can do to increase other family members' sense of happiness.

• Emphasizing the team concept so the therapist and family can work together to effect change. This requires the therapist to be perceived as a knowledgeable person who is able to work at the family's level.

• Perhaps the most important, clinicians must be integrally involved in helping family members develop effective communication skills. This entails teaching clients about dysfunctional communication styles and helping them identify these within their family along with a plan to replace them with more open, direct, and congruent communication.

• Above all, in remaining true to the spirit of Satir's work, therapists must approach treatment in a warm, humanistic way that deemphasizes technique-oriented therapeutic strategies and accentuates working with the family at their level.

> ### *S C E N A R I O*
> A family with three boys enters treatment on the urging of the family physician who had been treating the father of the boys. The father had been diagnosed with a brain tumor and had been operated on about 2 years earlier. The tumor was found to be non-cancerous at the time of his surgery. Subsequently, the parents' youngest son was also discovered to have a brain tumor which, when removed, was found to be cancerous. This necessitated a long hospitalization for the son at a time that the father was still recuperating from his own illness. The father was on antiseizure medication and had experienced a few small seizures during his recuperation process. The son was under-going chemotherapy and had had radiation treatments several months before. The other two sons were healthy but fearful that the same thing could happen to them. The mother in this family seemed very verbal, supportive, and active in caring for her hus-band and her children. She also worked outside the home. The father had been forced to stop working after his surgery. His income was now from disability payments, fur-ther stressing the family finances. The family physician had referred the father because of his worsening depression and sadness about the family circumstances. He states that he is not suicidal, but he is very sad about what has happened to him and his son. All members of the family state that they are willing to come in to help the father if nec-essary. How would you employ Satir's model if you were assigned to see this family in family treatment?

Cognitive Behavioral

As we saw in an earlier chapter, cognitive and behavioral approaches used together are often quite effective in treating individuals in brief or short-term therapy. This combination in therapeutic approaches has also had effective treatment outcomes in a variety of areas such as depression, anxiety, and phobias. It is useful at both the micro level with individu-als, and the mezzo level with families and groups. According to Beck (1976), "Cognitive therapy suggests that the individual's problems are derived from certain distortions of reality based on erroneous premises and assumptions. These incorrect conceptions origi-nated in defective learning during the person's cognitive development" (p. 3). Because of these incorrect premises and distorted images, people develop maladaptive emotional and behavioral responses to external events. The premise behind cognitive therapy is that help-ing clients alter their thought patterns, beliefs, and attitudes will lead to lasting behavioral change.

Two major contributors to cognitive therapy cited earlier are Albert Ellis and Aaron Beck. Their work emphasizes the two main goals of minimizing self-blame and placing blame on others. Proponents of cognitive therapy underscore its effectiveness, having seen the following notable outcomes:

> enlightened interest that honors the rights of others—self-direction, independence, and respon-sibility; tolerance and understanding of human fallibility; acceptance of the uncertainty of life; flexibility and openness to change; risk taking and a willingness to try new things; and self-acceptance. (Fenell & Weinhold, 1989, p. 130)

Certainly, based on these outcomes, the benefits of cognitive therapy are well worth con-sideration when you are working with clients who struggle with their thought processes.

Behavior therapy, in contrast, focuses primarily on behaviors. Its proponents believe that by targeting specific behaviors in predetermined ways and by focusing exclusively on observable events in the environment, clients can effect behavior change that is objectively measurable. The works of Pavlov, Skinner, Wolpe, and Lazarus have been well docu-mented as being major to the advancement of behavior therapy. Also, the work of Albert Bandura from the social learning realm has been influential in the development of behav-ior therapy as we know it today.

In behavior therapy, a general goal is to help clients change unwanted or dysfunctional behaviors through systematic application of experimentally established principles of learning. In addition, problems relating to family members are also targeted. Notable outcomes of behavior therapy as outlined by Fenell and Weinhold, (1989, p. 131) include these:

> learning to ask clearly and directly for what one wants; learning to give and receive both positive and negative feedback; being able to recognize and challenge self-destructive behaviors or thoughts; learning to become assertive without becoming aggressive; being able to say no without feeling guilty; developing positive methods of self-discipline, such as regular exercise, controlling eating patterns, and eliminating stress; learning communication and social skills; and, learning conflict resolution strategies to cope with a variety of family situations.

As with cognitive therapy, the positive aspects of engaging in behavior therapy are numerous. Furthermore, when these approaches are used together, the possibilities for effective treatment outcomes are magnified. To illustrate a cognitive-behavioral therapy approach we present systems theorist Salvador Minuchin and his structural family therapy.

Structural Family Therapy   Salvador Minuchin, originally from Argentina, was initially trained as a physician with a strong interest in pediatrics. He later received additional training in child psychiatry in the United States where, together with his colleagues, he developed a program in New York City that centered on delinquent minority children. Through his work at the Wiltwick School for Boys, Minuchin began to ponder on the role of the family in the children's behaviors. Because these children frequently came from very disorganized families which had little or no structure and few rules for conduct, he began to consider the correlation between patterns of delinquency and family disorganization. As Minuchin pursued his work, he was awarded a grant from NIMH, which allowed him to study closely the structure and process of disorganized, socioeconomically disadvantaged families with one or more children deemed delinquent. This specific work fostered many of the concepts of structural family therapy as it exists today. The text, *Families of the Slums* (1967), is an excellent recounting of the work of Minuchin and his colleagues with these disadvantaged families. In the text, the strategies of the Wiltwick project are summarized, profiles of the families involved are presented, and specific implications for therapeutic interventions with this type of family are identified. Today, Minuchin's advances in the treatment of symptomatic dysfunction in individuals and families continues to be relevant.

In addition to his work with delinquency, Minuchin became the director of the Philadelphia Guidance Clinic that, to his credit, emerged into one of the top family therapy service and training institutes in the world at that time. In years that followed, Minuchin (Minuchin, Rosman, & Baker, 1978) has also worked with psychosomatic families; an account of this work appears in his publication, *Psychosomatic Families: Anorexia Nervosa in Context*. Additionally, he has written extensively about structural family therapy; see *Families and Family Therapy* (1974) and *Family Therapy Techniques* (Minuchin & Fishman, 1981).

Structural family therapy is one of the most action-oriented systems theories in practice today. In addition to its base in general systems theory, it contains elements of communication theory, network therapy, and others that further add to its unique and dynamic nature. As implied in its name, structural family therapy emphasizes the importance of structure within the family system. Its directive approach is anticipated to bring about rapid changes in family functioning. According to Minuchin (1974), the organization of the family system is essential to the well-being and effective psychological functioning of the members within the family. Because a main premise of this approach maintains that there is an inherent drive in all living things toward organization, structural family therapists must understand behavior in a family as being a product of the structure of the family.

When behaviors in the family are dysfunctional, the family is helped to reorganize its structure so that it no longer supports or requires the disturbing behavior. A main goal of this type of therapy is to change the structure of the family, including its subsystems, boundaries, hierarchy, and transactions. More recently, goals of this approach include promoting changes in other social systems as well (Aponte, 1994, p. 22) as the structure of the family is often not the only dysfunctional social structure that needs to be changed.

Key Concepts   Integral to understanding and being able to apply the principles of structural family therapy to practice is knowledge of various key concepts. *Homeostasis and disequilibrium* are important concepts we have discussed earlier in the context of general systems theory. Because a family system is evolutionary, dynamic, and changing, the system is bound to experience various periods of homeostasis and disequilibrium. Periods of homeostasis are characterized by changes that occur in small and acceptable ways that do not require any major alterations with the family system. Disequilibrium occurs when the homeostatic nature is challenged, often requiring modifications in the family structure to accommodate the changes. When the family system fights to retain its original structure rather than adapting to the new situation, symptoms in one or more of the family members are apt to emerge. The role of the family therapist is thus to help the family reorganize in a more productive manner that encourages ongoing family development and change.

Subsystems and boundaries are also important to understand when working with family structure. The family is viewed as a system, and within that system are countless subsystems based on generation, gender, interests, and so on. Surrounding the subsystems are boundaries considered to be invisible barriers that regulate the amount of interaction with others.

Four subsystems considered important to understanding structural family therapy in general include the individual family system, the spousal subsystem, the parental subsystem, and the sibling subsystem. Although these are the ones most commonly found within families, a variety of other subsystems may exist that are equally as important in explaining the family system as a whole. The functions of these subsystems may be either helpful or problematic for the family. Sometimes when they are seen as beneficial, they may not be so to the entire family, and vice versa. Thus, family therapists should be equipped to identify subsystems and their functioning prior to developing techniques for family change.

In addition to boundaries, subsystems have rules that determine who is in and who is out of the subsystem. When the boundaries within the family are diffuse and family members are overinvolved with one another, they are considered to be enmeshed. Conversely, when boundaries are rigid and the family members have little contact with one another, they are said to be disengaged. Complications often arise when coalitions form within families and some members join together at the expense of others.

It is believed that families at different developmental stages will experience a certain degree of enmeshment or disengagement. Examples include the parent who is enmeshed with his or her newborn child or the adolescent who disengages from his or her parents as a way of asserting a certain amount of autonomy. Because these behaviors are natural and expected, they constitute a problem only when the family or individual member continues to function in this manner beyond the expected time frame.

Communication theory and circular causality are concepts we have discussed before. In terms of structural family therapy, they are related to the reality that family members, individually and as a part of subsystems, are affected by and affect all other members. According to communication theory, *how* a family interacts over a long period of time defines the relationships between members. Notwithstanding the importance of verbal and nonverbal communication, metacommunication is also important. Considered a third level of communication, metacommunication conveys to others additional information about the actual communication being made. Each message that a person sends has a "report" (the actual content) and a "command" (a statement defining the relationship). Relationships between the communicators can either be symmetrical, which is based on equality, or complementary, which is based on hierarchy. Overt (from the report) and covert (from the command) "rules" become known to the family, and members act accordingly. In terms of homeostasis, these rules must be abided by in order to maintain family equilibrium. In this regard, families operate according to what communication theorists term the *redundancy principle*, suggesting that although families are aware of alternatives, their interactions are based on a limited number of behaviors.

In tying all these concepts together, the main premise of structural family therapy can be better understood. Simply stated, the family has structure that is based on the patterns of behavior that emerge from a family's overt and covert rules, which stem from the boundaries between family members. From the structural family perspective, dysfunction lies in the operation of the family system. Whereas normal families tend to modify their response to change, pathological families tend to become more rigid and dysfunctional.

Therapeutic Functions   Assumptions made by structural family therapists stress that family dysfunction results when the organization of the family system has been disrupted to the extent that the family is unsuccessful in reestablishing its homeostatic state. Therefore, it is not unusual to expect families seeking treatment to feel confused, angry, and perhaps fearful. They may be confused that their typical problem-solving abilities are not working or allowing them to fix the problem. Anger may develop because at least one family member may be seen as the cause of the dysfunction. Last, fear may exist because families do not know what to expect from the treatment process. To help reduce these aversive conditions, Minuchin has identified various techniques, often referred to as "change techniques" that are often useful.

Joining is the process of establishing a solid, working relationship with each family member. This technique is especially critical to structural family therapy because the therapist and clients may not agree on the sources of the problem or what has to be done to resolve it. Only through engendering trust can the therapist join with all the members effort to help them understand the nature of the dysfunction. Without this level of trust, family members may not feel confident enough to share their feelings and problems with the therapist. Successful joining requires the therapist to become a significant source of self-esteem and validation for each person in the family. It is also important for the therapist to be able to demonstrate to each family member his or her ability to understand their individual situation. When the therapist successfully joins all the members of the family, there is a greater likelihood that they will be receptive to new ideas, suggestions, or challenges that are provided as part of the treatment. In contrast, when joining is minimal or absent, the level of resistance and mistrust is sure to rise.

Elements of joining include tracking, maintenance, and mimesis. *Tracking* occurs when the therapist uses the words, symbols, history, and values of the family in communicating with them. *Maintenance accommodation* results when the therapist demonstrates respect for the family by acknowledging their current rules and roles. The goal here is to help the family feel respected and therefore be more open to the therapeutic process and to change. *Mimesis* happens when the therapist imitates the family members' moods, tone of voice, postures, communication, and behavior. The goal of this technique is for the family to feel more accepting of and accepted by the therapist.

Activating family transaction patterns through enactment begins once joining has been successful and a positive therapeutic relationship between each member and the therapist has been established. At this point, the therapist begins to learn through the technique of enactment how the family is structured (Minuchin & Fishman, 1981). Specifically, enactment occurs when the family is asked to act out the problem; stated differently, it occurs when the family behaves in sessions the way it does at home. Enactment may occur spontaneously in family sessions; however, when it does not, the therapist may need to facilitate this process by encouraging key subsystems to discuss an acknowledged family problem that is likely to elicit strong emotions. Through enactment, the therapist is able to see exactly how a family functions and thus be able to provide alternatives for change or to model more positive behaviors.

Transformation of the family structure occurs when both joining and enactment have been successful in clarifying the family's structure. At this juncture, efforts are made toward altering the structure and the functioning of the family; this results in moving to restructure existing maladaptive patterns. There are three ways in which restructuring may be approached: system recomposition, symptom focusing, and structural modification (Van Deusen, Scott, & Todd, 1981).

System recomposition results when the therapist has chosen to add or eliminate a subsystem; this is especially necessary when the existing subsystem does not meet the needs of the family as a whole. *System focusing* happens when specific techniques are used to affect family functioning by focusing on the symptom itself. Examples of techniques used to alter symptoms are (1) relabeling the symptom (reframing) or altering the effect of the symptom, (3) expanding the symptom, (4) exaggerating the symptom, (5) deemphasizing the symptom, and (6) focusing on a new symptom (Minuchin, 1974; Minuchin & Fishman, 1981).

Structural modification occurs once emphasis on the symptoms changes and the family is encouraged to begin making structural changes. A variety of techniques that help facilitate this change are (1) challenging the current family reality, (2) creating new subsystems and boundaries, (3) blocking dysfunctional transactional patterns, (4) reinforcing patterns, (5) and educating and guiding (Fenell & Weinhold, 1989; Minuchin & Fishman, 1981). Additional techniques used by family structural family therapists include intensity and intervening isomorphically (Percy & Sprenkle, 1986).

Treatment Considerations

Engaging in structural family therapy is highly encouraged as a way to assist a family in the change process actively and swiftly. Because it is a complex as well as a delicate process, we recommend that those entering this practice realm be knowledgeable about the therapeutic concepts and theoretical underpinnings of the approach and skilled in applying the many techniques it encompasses.

- In the beginning stages of therapy when the therapist must join and accommodate the family, he or she must convey respect, understanding, and acceptance of each family member. This is done by speaking to each member, responding empathetically to what each one says, and avoiding challenges to the family structure.
- Because it is common for the family to have identified one member as the source of the problem, the therapist must reframe the family's thinking by having them understand the problem as indicative of a system structural problem.
- Throughout all stages of therapy, either as a means of assessment or intervention, the therapist needs to use enactments, whereby family members act out the problems they are talking about, or spontaneous behavior sequences, in which problematic behavior patterns emerge spontaneously.
- Enactments when used as assessment tools allow the therapist to assess the family's functioning by making a structural map on which he or she notes dysfunctional subsystems and the types of boundaries between members, keeping in mind the family rules.
- Assessment of the family system is necessary. Colapinto (1991b) has identified various steps of this assessment process that seem to overlap with the intervention process. They include (1) preplanning, (2) tracking, (3) staging enactments, (4) searching for strengths, and (5) reframing the problem.
- Enactments, when used as interventions require that the therapist highlight positive family interactions and modify dysfunctional ones. Modification takes place by a

therapist's use of intensity, when he or she controls his or her affect, repetition of a particular theme, and duration of a sequence.

- It is also important for the therapist to change, strengthen, or weaken boundaries between subsystems. Techniques used for this can be blocking a member's attempt to interrupt other members' conversations or blocking an attempt for someone to detour by changing the subject.

- Therapists need to learn additional techniques for altering boundaries such as having family members physically move closer or farther away from other members or to give directions to a member of a subsystem while purposely excluding other members from the interaction. This is tied to the notion that to change the relationships within a subsystem, as in breaking a coalition, it is necessary to unbalance the subsystem by siding with one member at the expense of others.

- Another technique structural family therapists need to learn is related to restructuring the family by using the "stroke-kick," in which the therapist compliments a member about a particular behavior, but then tells another member the downside of that behavior. Because creating paradoxes confuses and frustrates the family, this strategy should lead them to find alternative behaviors.

- Often, structural family therapists must challenge family members' views of reality. They can do this by challenging the labels they place on others, educating them about pertinent subjects, and advising them about particular behaviors.

- Recognizing the importance of the termination process helps bring proper closure to treatment. Termination may be initiated by the therapist, by the family, or by all acting in mutual agreement (Minuchin et al., 1967). If done effectively, termination should assist the family to look forward and recognize that they will continue to experience changes and normal transitions. Also, leaving the door open for future assistance conveys a positive message.

SCENARIO

The parents of a Latino family of five come in bringing a daughter as the identified patient. They state that she has been depressed and has talked about suicide recently to one of her teachers in school. She is 16 and a junior in high school and reportedly is doing very well in both her grades and friendships with her peers. The mother and father are at a loss to explain her feelings, but do know that her older sister was married about 2 months ago and that the daughter felt very sad about her leaving. Both parents in this family work outside the home to support the family financially. This meant that the oldest daughter had to take over many of the parenting tasks, which she had done faithfully until the time of her marriage and subsequent moving away with her husband. Out of the five children there were two boys and three girls. The identified client was the second oldest girl, then there were the two boys and the youngest daughter.

The parents explain that things in the home always seemed to be good when their oldest daughter was there, but now the children are fighting. The parents were unsure what to do because they are away working much of the time. They were also quite worried about this daughter when she began talking about suicide because that was "not like her." Both parents seemed genuinely interested in the daughter. The father stated that he did drink some after work at times but not excessively. He further stated that he did expect to come home and have his dinner ready, but that he didn't believe that was too much to ask as he worked really hard during the day. The mother seemed very quiet and did not talk much during the family assessment. She did express concern for her daughter and seemed mildly annoyed when her husband was expressing his desire to have his dinner ready when he comes home, but she did not say anything directly. How might you intervene in this family to bring about change if you were utilizing structural family therapy techniques?

Developmental

The developmental perspective has been examined at length earlier in this chapter with emphasis on both the individual and family life cycle. In Chapter 3, we also touched on developmental issues. Because of this earlier discussion, we will not elaborate any further on the developmental perspective, going directly to our example: Haley's developmental life cycle theory.

Haley's Life Cycle Developmental Theory   Haley defined the family life cycle as consisting of a six-stage cycle; explanation of this is known as a family life-cycle theory. Much like Erikson's psychosocial theory, Haley's family life cycle also has specific and predictable developmental sequences he viewed as common to most families. He cautions practitioners not only to assess each family individually, but also to take into consideration cultural differences, gender issues, and the family's affections orientation that may have a critical impact on how the family proceeds through life. Furthermore, individual development of each family member must be taken into account. The developmental difficulties of one family member may directly or indirectly influence the entire family's ability to cope and successfully navigate their way through the life cycle. Conversely, the family's difficulties can have a significant bearing on each individual and his or her development. All these facets must be carefully examined if treatment interventions are to fit both the individuals and the family client system.

Treatment Considerations

- Because of the specific and predictable developmental sequences that are common to most families, it is important to recognize these when working with individuals and their families.
- Assess each family individually as well as each family member.
- Cultural differences, gender issues, and the family's affections orientation should be assessed as they may have a significant bearing on both individual and family development.
- When designing treatment interventions, consider how individual family members impact the family and how the family affects its members.

SCENARIO

Jim and Laura present their teenage daughter as their "problem," stating that Jennifer has been angry and belligerent with them recently and will not listen to what they have to say or respect the limits they have laid down. They say the difficulties have gotten worse as she has begun to see more of her latest boyfriend who they believe has a "bad influence" on her. Jennifer sits quietly while they are talking to you but is obviously angry about being here and rolls her eyes every time her parents speak about her. When asked to give her side of the story, she says that she believes her parents are overprotective and that she doesn't agree with their rules for her. She is angry because she believes they favor her older brother, who she claims never had any rules and got to do whatever he wanted. The parents seem agitated as Jennifer talks and they become angry, denying her accusations about them. When you get a chance to talk to Jennifer alone, she tells you that her mother has always been overprotective of her and that she doesn't need that kind of treatment from her mom anymore. She says that she knows her mother was raped by a boy her age when she was in high school and has often told her that she would never let that happen to her daughter. Considering a developmental perspective, how would you treat this family situation? Would you continue to see them separately or see them only as a family?

The Changing Family Life Cycle and Implications for Practice

The Changing Family Life Cycle

The notion that the family is a system is not new. As we have seen, according to a systems theory perspective, the family is a social system embedded in larger social systems. As a system, the family can be viewed as a "whole consisting of interrelated parts, each of which affects, and is affected by every other part, and which contributes to the functioning of the whole" (Sigelman & Shaffer, 1995, p. 390). Inherent in this systems view is the recognition that change is ever present and is apt to upset the homeostasis, thereby creating some disequilibrium. In many such instances, the family strives to restore its state of equilibrium; this is often true regardless of how dysfunctional or unstable this state may be. As helpers, we are often in the position of helping a family return to a previous level of coping they are most comfortable and able to live with. Because of our own issues and biases, helping families do this may not always be easy. Respecting clients' decisions and working with them at their level remains an essential ingredient to effective clinical practice.

When the family itself is considered a developing entity, change is to be expected. As the family system undergoes change, so does the family life cycle. As described by Sigelman and Shaffer (1995), the family life cycle is a "sequence of changes in family composition, roles, and relationships that occur from the time people marry and die" (p. 391). Ashford, Lecroy, and Lortie (1997) view this life cycle as a "sequence of developmental stages that families typically move through" (p. 90). They caution, however, that each family is unique and may not always fit the typical patterns. We echo their concerns and advocate continually striving to individualize each family system. Likewise, we encourage viewing the family from several different perspectives: the family system as an entity in itself; the individual people making up the family system; and the impacts of the social, political, and economic environment on how the family functions.

The effect of today's changing world on families varies. In the past few decades, several dramatic social changes have altered the makeup of the typical family and the quality of family experience. Several trends have been cited:

Increased number of single adults
Postponement of marriage
Decreased childbearing
Increased female participation in the labor force
Increased divorce
Increased number of single-parent families
Increased number of children living in poverty
Increased remarriage and percentage of reconstituted families
Increased years without children
More multi-generational families (Sigelman & Shaffer, 1995, pp. 391–392)

These trends are not all positive. Furthermore, it is unfortunate and telling that many of the negative trends seem to affect some ethnic and racial groups more dramatically than others. An example can be seen in the high percentage of single-parent African-American mothers, resulting in the startling statistic that half the African-American children in our country live in poverty (Eggebeen & Lichter, 1991).

Additional changes affecting families in recent times have been identified by Sigelman and Shaffer (1991):

More adults are living as singles today than in the past.

Many young adults are delaying their marriage while they pursue educational and career goals.

After marriage, couples are having fewer children.

An increasing number of couples are deciding to remain childless.

Most women, including those who have children, work outside the home.

Up to 50% of young people are expected to divorce sometime in their lives.

Up to 50% of children born in the 1980s will spend some time in a single-parent family.

Because more couples are divorcing, more adults—about 75%—are remarrying.

Adults today are spending more of their later years as couples or as single adults without children in the home primarily because people are living longer.

For helping professionals, these changes and trends may have a significant bearing on the types of families requesting treatment. It is important for helpers to recognize the changes (and reasons for these) that many families have undergone in order to understand and help them; deciding which specific factors impact their functioning is crucial in developing an appropriate treatment plan and tailoring treatment services specifically to meet their needs.

Some of the most prevalent changes affecting families today are discussed below. We begin with the notable aspects of role changes and their effect on the family system.

Role Changes   According to Schmolling, Youkeles, and Burger (1993), today's family can well be called the "changing American family" in that "the family is no longer organized primarily around child rearing" (p. 35). They cite the lowered birth rate, the longer life expectancy, and increasing divorce and remarriage rates as some of the contributing factors. Unlike the past, the role of women is no longer primarily tied to child-rearing activities. More women are entering the labor force, and for many, their goals and identity are not solely related to their role as homemaker and mother. The 1992 U.S. Bureau of the Census highlights this trend with the following figures: In 1989, 60% of all married women with children under age 6 and 67% with children under age 18 worked outside the home. For single-parent mothers, the percentage was over 53% who worked outside the home (U.S. Bureau of the Census, 1992, p. 388).

This changing role for women not only impacts the woman and her child, but may also affect the role of the father. For dual-career families especially, the need for increased sharing of both child care and household responsibilities is heightened. In some cases, roles are reversed and the male becomes the house husband and the primary child care provider while the female becomes the primary or sole breadwinner.

Controversy still exists on the effects of the social and emotional development of children whose mothers work. Reviews of research by Kadushin and Martin (1988, p. 178) reflect that there has been no adverse effect noted on a child's development when "the mother is satisfied with her job and the provision for child care is reasonably good and suitable." We believe that attitudes of both the mother and father are important in determining whether the concept of a working mother is viewed negatively or positively. Likewise, if a father decides to stay home, it is important for both him and his wife to view this as a positive and innovative approach. Unfortunately, old stereotypic views still prevail that a woman's place is in the home and a man should be the primary breadwinner. It will be vital for helping professionals to accept these role changes and to help dispel the negative myths about the ideal or traditional view.

Treatment Considerations

- Working mothers: Support and validation of their efforts, elimination of guilt-inducing comments and views.

- Stay-at-home dads: Reinforcement of their courage to strike out in relatively untouched territory. Validation for their commitment and dedication to their family.
- Acceptance of role changes is important in working with many of today's families; undue criticism based on stereotypic views is only detrimental.

Current Challenges: Diverse Families and Lifestyles

Today's families come in all different shapes and sizes. The diversity noted in family forms and lifestyles is apparent in almost every sector of society.

The many changes affecting families of today are partly a result of industrialization and technological advances. Many of the functions of the family have been lost (e.g. the economic-productive and protective functions) or sharply reduced (e.g, educational, religious center, and status recognition functions). Nonetheless, the affectional function still prevails as this is where many individuals receive social and emotional gratification and have many of their companionship needs met. Also, as noted by Zastrow (1996), "there are still important functions that families in modern industrial societies perform to help maintain the continuity and stability of society" (p. 144). These include replacement of the population, care of the young, socialization of new members, and regulation of sexual behavior.

Single Parents   The phenomenon of single-parent or "solo-parent" families as denoted by Germain (1991, p. 98) is certainly not new; but they are more dramatically seen in today's world. Recent statistics report that more than 20% of all children in the United States are raised in homes with only one parent present. Reasons cited for this include divorce, desertion, death of spouse, unwed teenage parenthood, and births outside marriage. Most of these families are headed by women, with the rate of African-American women twice that of white females (Zastrow & Kirst-Ashman, 1994, p. 515). These statistics reinforce the need for continued efforts to understand the cultural variable and work toward addressing this in both direct practice and on a societal level. Gender issues must be addressed as well and the "feminization of poverty" must be more closely explored and understood.

Although single parents are notably women, many of an ethnic group, and many living in poverty, this does not negate the fact that there has also been a striking increase in solo parenting across most racial, ethnic, religious, and socioeconomic groups. This reality heightens the need for today's practitioner to be knowledgeable of the issues related to single parenting and to develop the necessary skills to work effectively with this population.

Many still view one-parent families as less functional than two-parent families. Often because of their single status, these women may require more assistance, guidance, and support. Their self-esteem may be low and hope for the future scarce. Those without informal supports may also experience isolation and loneliness. This may require us as helping professionals to be their primary supports while we assist them to develop healthier support networks.

Even though statistics show that a majority of single parents are women, there has been a notable increase in single fathers as well. The situation of single fathers in qualitatively different from that of single mothers (Mendes, 1976). For one, little attention is given to the plight of single fathers in the literature. This should not negate that fact that they too are vulnerable to special stressors; although these are different from women's stressors, they are nonetheless difficult and often challenging to cope with. According to Germain (1991, p. 102)

> solo fathers confront role ambiguity and the lack of norms, values, and environmental expectations for role performance. While single parents of both genders are at a disadvantage in regard to societal perceptions, solo-father families are frequently assumed to have been created by greater pathology in the mother than solo-mother families. And the solo-father is apt to be perceived as poorly prepared by his very nature for the performance of the necessarily expressive roles in child rearing" (p. 102).

In addition to support groups and educational programs for these fathers, development of further services to meet their unique needs is essential. We concur with Nieto (1982) and Germain (1991) who advocate for research on this area. Germain specifically cites the need for "research on theoretical frameworks for the analysis of single-father parenting; clarification of the unique roles, tasks, and stressors involved; and studies of the psychological, social, cultural, economic, and educational dimensions of the sole father populations" (p. 102).

Treatment Considerations

- Practitioners should recognize and appreciate the strengths and resources of single-parent families and help them build from these strengths rather than presuming weakness and isolation (*NASW News*, 1987, p. 16).
- New counseling and ancillary services are needed to reach some single parents to facilitate their meeting the life tasks of the family.
- Self-help and support groups (e.g., Single Parents, Single Moms, Single Dads, etc.) are needed.
- Parent-child play groups should be created.
- Divorce-mediation services should be promoted.
- Family life education is needed.
- Parent aide programs (parenting classes) should be developed.
- Provision should be made for the following: "Alternative counseling, physical and mental childbirth, counseling on legal issues, counseling on interpersonal relationships, alternative living arrangements, alleged-father counseling, family planning counseling, educational and employment counseling, self-development counseling, financial and money management counseling, child care, child development counseling" (Zastrow, 1996, pp. 170–171).
- It is important to understand cultural variables, gender issues, and the feminization of poverty and work in addressing single parenting, both in direct practice and on a societal level.
- Recognition of the specific issues related to single parenting is essential as is the development of the necessary skills to work effectively with this population.

Divorce

The incidence of divorce in the United States has increased steadily from a rare occurrence before World War I to the current 50% level, based on figures citing that nearly one out of every two marriages ends in divorce (U.S. Bureau of Census, 1991). This rate is not only indicative of the societal acceptance of divorce, but it also points to the probability that human service practitioners will have to deal with some aspect of divorce in their daily work with clients. Whether working directly with the grief, loss, and adjustment difficulties, custody and mediation concerns, or more indirectly with the aftermath of divorce, practitioners are increasingly having to confront the many issues encompassing divorce in their work with both individuals and families.

With divorce such a common occurrence, it is being viewed more and more as a normative event. Although this is a natural reaction, the inherent danger of this view is the tendency to overlook or minimize the impact of divorce. It is crucial for practitioners to be aware of both the short- and long-term ramifications for the individuals involved, the entire family, and even society at large. Carter and McGoldrick (1989) have found it useful to conceptualize divorce as "an interruption or dislocation of the traditional family life cycle, which produces the kind of profound disequilibrium that is associated throughout the life cycle with shifts, gains and losses in family membership" (p. 21).

The impact on the family as a whole may be great. In such a crisis, additional efforts from all may be necessary to restabilize and proceed. The extent of the crisis will vary from family to family; some will be able to overcome this difficult time successfully and learn from it as well; others may never completely recover from the trauma of loss, abandonment, or rejection.

Many have written about the divorce process. Wallerstein (1988), for instance, describes a three-stage process in which he emphasizes the impact on the children's ability to focus on their developmental tasks. Although not experienced by every family, these stages do seem to apply to many. In succession, these stages are as follows:

Stage 1: Acute Phase—Characterized by parental fighting, conflict, anger, depression, and the actual separation, which is often not mutual. The stage may range from months to years and is especially difficult for children who are witness to the strife, conflict, and overall tension within the home. Parents are often so caught up in their own emotional turmoil that they are often not emotionally available for their children.

Stage 2: Transitional Phase—In this phase, both parents and children attempt to adjust to their new life in a restructured family. The period can last several years before the family settles into the new roles and routines.

Stage 3: Stabilizing Phase—This stage occurs when the family has adjusted to the shifts in roles and allegiances. An optimal level of functioning is regained and the family has moved on to tackle new changes.

Carter and McGoldrick (1989) visualize the divorce process as analogous to a roller coaster ride, given the peaks and valleys of emotional tension during major transition points that include the following:

- When the decision to separate or divorce occurs
- When this decision is announced to family and friends
- When the money and custody/visitation arrangements are discussed
- When the physical separation takes place
- When the actual legal divorce takes place
- When separated spouses or ex-spouses have contact about money or children
- As each child graduates, marries, has children or becomes ill
- As each spouse is remarried, moves, becomes ill, or dies

In both these views, it is clear that the divorce process is often a difficult one that may require professional guidance and support. In almost every case, everyone involved is affected.

The married couple going through the divorce are apt to experience a myriad of feelings. Usually both experience emotional as well as practical difficulties. Some feel they have failed; they may doubt whether they will be able to give and receive love; they may feel a sense of loneliness. Others may feel concerned about the stigma of divorce; they may worry about the reactions of friends and relatives and may question whether they are doing the right thing by ending the marriage. The fear of making it on their own is also common and often experienced by both husband and wife. Each faces the difficult task of revising his or her identity as a single versus a married person; this requires work known as "emotional divorce" whereby there is a retrieval of self from the marriage. Also, both males and females may feel socially isolated from former friends, especially couples they were close to, and unsure of their ability to establish new romantic relationships. In cases where the woman obtains custody of the children, she is likely to be angry, depressed, moody, lonely, and otherwise distressed, although often relieved as well. Money concerns and unresolved custody issues or visitation problems may create additional stress. The husband is also likely to be distressed, particularly if he did not want the divorce, feels shut off from his children, and is caught in a legal battle over custody and child support issues.

The children can be especially stressed in the crisis of divorce. This event "disrupts one of the core relationships in their lives and dissolves the family structure they depend on for secure development" (Ashford, Lecroy, & Lortie, 1997, p. 301). They must also adapt to the newly structured family once emphasis on the symptoms changes and the family is encouraged to begin making structural changes. A variety of techniques that help facilitate this change are (1) challenging the current family reality, (2) creating new subsystems and boundaries, (3) blocking dysfunctional transactional patterns, (4) reinforcing patterns, (5) and educating and guiding (Fenell & Weinhold, 1989; Minuchin & Fishman, 1981). Children often face many adjustment difficulties during as well as after the divorce. Feelings of insecurity, loss of self-worth, loss of love and feeling lovable, anxiety, loneliness, anger, resentment, guilt, fear of abandonment, depression, and helplessness are all common (Wallerstein & Blakesless, 1989). Additionally, increased stress and turmoil is often experienced when children face issues of relocation to new homes and schools, when a stay-at-home mother has to go to work, and when parents are in crisis themselves over the divorce. Disorganization and unpredictability in daily events such as meals and bedtime often affect children, who may need more structure and routine to help them cope with the chaos of their lives. Difficulties in school performance and peer relationships are often noted; in more severe cases, delinquency can also occur. Last, the process can be seriously complicated by the suspicion or actual existence of abuse by a parent or a significant other with whom the parent is now involved. The full spectrum of stressors children experience when their parents divorce appear in Table 4.3.

Various authors (Hetherington, 1989; Kaler, Kloner, Schreier, & Okla, 1989; Wallerstein, 1988) have identified certain determinants of children's adjustment. These include

- The personality and temperament of the child
- The age of the child (younger children tend to suffer more short-term consequences whereas older children are affected more in long-term ways)
- The quality of the parent-child relationship, pre- and postdivorce
- Stability of the custodial parent's household
- Financial security
- Postdivorce relationship between the ex-spouses

According to Sigelman and Shaffer (1991), the following are five factors that can facilitate a smooth and adaptive recovery from divorce:

- Adequate financial support
- Adequate parenting by the custodial parent
- Emotional support from the noncustodial parent
- Additional social support
- A minimum of additional stressors

TABLE 4.3   Stressors That Occur with Divorce

Life Changes	Consequent Effects on Children
Hostilities between parents	Sadness, anger, loyalty conflicts
Distraught custodial parent	Anxiety, put in roles of parent, co-parent to custodial parent
Loss of relationships with noncustodial parent	Self-blame, low self-esteem, depression
Parent dating	Competitive feelings with parent's new partner, fear of loss of parent's affection, curiosity (for older children) about parent's sexuality
Remarriage	Sharing parents, accepting parents' intimacy. Relationships with stepparents and step-siblings, accepting new parent as authority, resolving issues of loyalty to new stepparent and parent of same gender
Socioeconomic decline	Downward economic mobility, emotional stresses, changes in residence, loss of peer relationships and familiar school environment, change in consistent caregivers

Although providing each of these five is not always feasible or realistic, being aware of these factors can help parents become better prepared for this difficult transition period in their lives and that of their children. The information is also useful for helpers who can begin to incorporate some of these factors in their own individual treatment plans.

Treatment Considerations

- Be prepared to deal with an array of issues that may accompany divorce. Some require direct work with grief, loss, and adjustment difficulties as well as custody and mediation concerns. Other concerns may involve more indirect issues related to the aftermath of divorce.
- Take care not to overlook or minimize the impact of divorce as it becomes more common.
- Recognize both the short- and long-term ramifications for all the individuals involved.
- Accept the variance among families in their coping and adjustment with divorce.
- Understand the various stages of divorce that are presented in the literature and consider utilizing them in your work with clients when appropriate.
- Recognize how the trauma of divorce impacts the entire family as well as the specific ways it can affect each family member. For instance, the adults and children may be affected differently.
- Whether working with the family as a whole, the marital couple, or the children, helpers may need to cultivate specialized skills to ensure successful work with each of these units.
- Incorporate factors that have been shown to facilitate a smooth and adaptive recovery process into one's interventions.

Blended Families   Blended families occur when one or both spouses have biologically produced one or more children with someone else prior to their current marriage. In many of these families, the married couple also gives birth to their own children, thereby creating the scenario that fits the "yours, mine, and ours" slogan.

This type of family, joined by the marriage of one parent to another, can take on many forms and has been referred to in many ways; the most common include remarried family, the step-family, or the reconstituted family. Until recently, blended families were not as common as they are today and thus have been referred to as nontraditional families. The high rate of divorce and remarriage along with increased childbirth out of wedlock are a few of the reasons that account for the increase in these types of families. Regardless of the reasons for the development of this type of family or the specific form, all involve complex situations and present members with additional adjustments and stressors. Zastrow (1996) states, "Blended families, in short, are burdened by much more baggage than are two childless adults marrying for the first time" (p. 39). Some of the added issues and adjustment difficulties faced by these families are listed below:

- Both the adults and children involved have to face issues of loss, grief, and rejection sustained from divorce, abandonment, or death; some may be afraid to trust and love as a result.
- Previous bonds and loyalty between children and one or both of their parents now separated may interfere in the formation of a new and positive relationship with a stepparent.
- Conflicts between stepchildren and their stepparents is likely. This is especially true when children travel back and forth from different homes where values, standards, and life-styles are different.
- When divorced parents continue to quarrel and feud—over the past or about current issues—children are at special risk of being caught in the middle, creating even further strife for the newly formed blended family.

- Holding onto past resentments, ways of doings things, and negative experiences may impact the healthy formation of new, more positive relationships in the newly established family.
- At times, marrying a new partner involves marrying a whole new family that may include an entire new set of characters. This is complicated if some of the family was opposed to the divorce of its family member and finds it difficult to accept the new spouse into the family. Also, if the married couple still maintains contact with their ex-in-laws, this relationship can be overwhelming and stressful.
- When the remarriage is a result of an affair for which much resentment and anger still exist, issues of betrayal and disloyalty may also need to be dealt with.
- Adults have to adjust to raising children who were biologically parented by someone else—perhaps someone they do not particularly feel close to or like.
- Children must learn to form new relationships with step- and half-siblings. Jealousy, resentment, and sometimes extreme bitterness may result, further complicating the family dynamics. This often happens when the couple has a child of their own whom they show preference over their step-children.
- Sharing parental attention, space, and belongings with others is often difficult for children who are not accustomed to doing so.
- Last, there are situations in which one parent is inexperienced in child rearing or who views parenting in a totally different way from the child's other parent. Children in these families have to learn ideas, values, rules, and expectations they may resent or rebel against.

In addition to helping families overcome and cope with these many adjustment difficulties, we must be careful not to buy into or perpetuate the typical myths regarding blended families, such as the "wicked stepmother," the idea that "step is less," or the notion that the minute they are joined as one family, everyone will have instant love for all new family members. The damage engendered by these myths often leads to difficult and conflicted interpersonal relationships within these families. In your work with families you will no doubt encounter this type of family system and therefore need to be aware of the various issues specific to their plight. Only then can you develop effective interventions.

Treatment Considerations

- Be alert to considering the feelings and perspectives of all family members in order to join effectively with all members of the blended family.
- Be alert to the special issues that a blended family will bring up among its members. Be prepared to address these issues with your unique style and approach.
- Recognize the complex situations that blended families involve and the additional adjustments and stressors they present to their members.
- In addition to recognizing the many stressors common to blended families, be prepared to help families overcome and cope with the adjustment difficulties they face.
- Do not perpetuate the typical myths regarding blended families, such as believing in the "wicked stepmother," accepting the idea that "step is less," or thinking that instant love is possible.

Other Emerging Types: Two-Provider, Gay and Lesbian, and Communal Families

Two-Provider Families   Even in families that have remained intact, it has become commonplace for both parents to work outside the home. Economic need is the primary motivation for many families, although for more and more, career development is also an incentive. Germain (1991, p. 118) believes it is important to make a distinction between two-worker families versus two-career families "because their motivations, benefits, and costs are markedly different." For the first group, both parents must work to make ends meet. The rising cost of living coupled with the increased cost of raising children has contributed to this financial need. For the two-career family, the many changes brought on by the women's movement have contributed, in part, to the changing perspective on the status and roles of women. Today, women are entering higher education and professional education in greater numbers, the typical family size is smaller, and women are now more aware of their increased life expectancy and of their vulnerability arising from the high rate of divorce. These changes have led women to aspire to both a career and motherhood. Not only does a career offer a new and higher social status and role for middle-class women, but the income from the career has become a necessity in many two-career families.

The effects on families in which both parents work can be great, whether parents work solely for financial reasons or to further a career. Juggling schedules, compromising needs, or just letting things slide are typical. Parents may have less time to spend with each other and may feel a certain amount of guilt for being away from the home for extended periods of time while working. A healthy marriage in which there is communication and shared responsibilities can avoid many problems. The job satisfaction of both parents may have a great bearing on whether each parent enjoys balancing family life with work. Also, satisfaction with child care arrangements and having job flexibility for meeting special parenting demands contributes greatly to how smoothly a two-provider family can function.

Treatment Considerations

- Distinguish between two-worker families and two-career families as their motivations may be different and may affect the selection of treatment interventions to be used.
- Be prepared to acknowledge the special forces that affect the two-provider family and to lend the additional support and validation they may need.
- Encourage a healthy marriage with open communication and shared responsibility that can help reduce some of the stressors many of these families face.
- Be cognizant, however, of the cultural factors that may be involved and the partners' particular views on gender as these must also be considered in treatment planning.
- Take into account how and when the decision to become a two-provider family was made. For those who have always been a two-provider family, the issues may be different from those in families in which one member recently joined his or her mate in the workforce.
- Discover whether the decision was a joint one or made solely by one parent without consulting with the other; this could be a major issue in the treatment process.

Gay and Lesbian Families   As our society has become more diverse and, one hopes, more accepting, there has been an increase in the number of gay and lesbian individuals living together as couples. Although exact statistics are not available, many of us know of a family in which both parents are of the same sex and consider themselves a married couple. This is not to say that our society has completely accepted this different type of lifestyle, nor that these families are free from criticism, rejection, or stigma. Throughout history, anyone who deviates from the norm is subject to additional scrutiny and censure, and this is true of many of these families. Therefore, in addition to performing most of the same functions and tasks that other families do (e.g., household management, child rearing, arrangement for decision making and role allocation, and meeting the emotional needs of the members and the sexual needs of the adult partners), they must also contend with many other stressors directly resulting from their different lifestyle.

Stigma and prejudice can result in discriminatory practices against them. The existence of homophobia and associated myths continue to contribute to this negativity and stigmatism. Negative attitudes are more prevalent in some geographical areas than others and are further perpetuated by the irrational fear of AIDS contagion for the children. Three prevalent myths are (1) the gay or lesbian parent or parent's lover will molest the child,

(2) gay or lesbian parents will try to convert their child to their own sexual preference, and (3) children of gay and lesbian parents will be ostracized and damaged because of society's reactions to homosexuality. In each of these cases, there is no factual evidence to justify the assumptions. The third assumption is perhaps the only one that has merit in some instances, yet there is no compelling evidence that children in gay and lesbian families suffer disproportionately because of their parents' sexual preference; however, they may suffer some of the same effects of prejudice seen in children of minority poor families, parents with physical impairments, and divorced parents (Brown, 1995; Baptiste, 1987; Germain, 1991).

These families may face additional pressures:

- For some, the inability to provide linkages to the community for their members to the same extent that heterosexual families do, except for connections to the gay or lesbian community when one exists.
- Ostracism and loss of relationships with families of origin because of the prevailing attitude that homosexuality is a sin, is wrong, and is not acceptable.
- In some cases, denial of credit, insurance, licensing, the joint purchase of a home, inheritance rights, and joint tax returns.
- Overt or subtle rejection and exclusion of the entire family, the parents, or the children resulting in additional emotional strife.
- Conflicts with ex-spouses who do not agree or accept the homosexual lifestyle for their children, often resulting in custody battles and increased criticism and rejection. In fact, in divorce cases in which one parent is gay or lesbian, the courts have tended to grant custody to the heterosexual parent on the basis of the sexual preference of the other parent and regardless of parenting ability or the wishes of the child (Germain, 1991, p. 116).

Treatment Considerations

- The exceptional child-rearing tasks for the lesbian and gay parent and her or his lover are similar to many of those of the blended family in terms of step-relationship and roles. However, they are made more difficult for same-sex families because of social intolerance. Be prepared to acknowledge these social injustices as they might affect the family members.
- Offer assistance with exceptional child-rearing tasks, which include giving special attention to open communication, coming out to the children—that is, telling them that the couple are homosexual, dealing with difference and stigma, and negotiating relationships with the divorced parent and with grandparents who may be hostile to the lesbian or gay relationship. Many professionals believe that gay or lesbian parents should come out to their children as soon as possible (Germain, 1991, p. 117–118).
- Be cognizant of the special stressors that gay and lesbian families are often faced with and work in helping them find healthy coping skills for confronting these.
- Acknowledge the stigma and prejudice that often plague these families and remain aware of your own biases and stereotypical views so as to avoid engaging in any discriminatory behavior yourself.
- Be aware of the prevalent myths about gay and lesbian families and attempt to refrain from perpetuating these in any way.

Communal Living

Communal living in this context differs from communes that were prevalent in the 1960s. Our focus here is on several families living together in a collective as one larger family. We have noted a rise in this type of living situation, brought on by the increased cost of living coupled with financial devastation, for some families who resort to moving in with another to avoid total financial loss or, in extreme cases, homelessness. The complications inherent in this type of family system are clear when you consider the various differences in family lifestyles, values, norms, standards, and so on. Conflicts commonly arise from the simple fact that each family has its own way of functioning that may differ from that of the other families living in the same home. When such differences are extreme or cannot be negotiated, difficulties are amplified.

Treatment Considerations

- Be aware of your biases and values that might affect the way you see communal living or many families living together.
- Explore the most appropriate way to treat such families including seeing all the families living together as a whole, as individual families, in dyads, or as individuals. Regardless of which form is taken, be clear on the purpose and goal of your interventions.
- Support a family's decision to remain or leave the communal living situation as you would any other family.

Extended Life span and Its Impact on Families   Extended life expectancy continues to climb, with advances in medical science and improved living conditions of our society in general. This increase is noted in both men and women and has certainly impacted each gender individually as well as collectively.

Women are having children later in life, and more men and women are finding that they can live to know their grandchildren and even great grandchildren. Some are even becoming parents for a second time either by having children from second marriages late in life or by circumstances that have led to the decision to raise their grandchildren. Both men and women are finding that they may be able to have two different careers. Many are retiring from one and starting a second. Also, many, especially women, are returning to school in their thirties, forties, fifties, and even sixties.

Because people are living longer, they are experiencing problems that were not typical when life expectancy was shorter. For instance, many are living to be 80, 90, and even 100. Some are plagued with ill heath and may need partial help or complete assistance. The financial and emotional demands of these elders often fall on the shoulders of the adult children who may themselves be older and in failing health. The scenario in which 65-year-olds may be responsible for the care of their 85-year-old parents is becoming more commonplace.

Because our society has not been fully prepared to deal with such issues, many find themselves overwhelmed by the stressors of having family members growing older. Whether providing financial assistance and care or having to institutionalize an elderly family member, the emotional demands are great. In many instances, families have chosen to have their loved ones move in with them, thereby changing the nuclear family to an extended one. For some, this transition is smooth and perhaps advantageous. For others, however, this change can create additional problems and perhaps intensify existing interpersonal difficulties within the family. Much will depend on the history of family relationships and the attitudes held by family members about "taking in" their relative. When there has been ongoing strife through the years or when the family feels pressured and burdened, the outcome is generally less than satisfactory. On the other hand, when a strong, positive, and accepting relationship exists, this transition is less traumatic. Moreover, the change can be further facilitated by open communication and realistic expectations by everyone involved.

Treatment Considerations

- Be aware of your own values in relation to older adults and how this could affect the helping relationship.
- Recognize the impact that extended life span may have on individuals and their families.
- Be supportive of the many changes that come with extended living and encourage clients to focus on the strengths versus the weaknesses.
- Similarly, realize that the stressors may at times be too great for certain families to cope with alone; in this case, be prepared to provide additional assistance to families who are more at risk.
- Whenever possible, encouraging open and ongoing communication within the family about life changes can help enhance the adjustment process.

- Consider when you can be most effective and perhaps when you should seek additional referrals to assist certain individuals and families. In other words, acknowledge your limitations and the reality that you may not always be effective with everyone seeking help.

Lack of Economic Stability and the Rise in Violent Crimes   An additional stressor facing many families today is the lack of economic stability as well as the stark rise in violent crimes. It was not uncommon years ago for an individual to work in his or her same job his entire life and then retire. Today, many are finding that their jobs will last only a while before they will have to look for another; the transition from one job to another is a part of life. For many, this reality has created anxiety and a sense of apathy about the future. For the adolescents and young adults who have witnessed their parents' lay-offs and job losses, this may add uncertainty and feelings of disillusionment about their own futures.

The rise in violent crime is real; we are faced with endless evidence of this in daily accounts in newspapers, news headlines, and specialized TV programs such as *Unsolved Mysteries, 48 Hours, 20/20,* and so on.

Additionally, for those who live in lower income and impoverished areas, the incidence of crime is even greater. The social problems engendered by the rise in gangs, drugs, rape, and child abuse are becoming more prevalent social ills we can all attest to. It is not presumptuous to expect that many families are further stressed by these problems and may need additional help as a result.

Cultural aspects are also important in working with families. We cover this issue in depth later in the text; here, just note that cultural issues must be taken into consideration when working with any family regardless of the issues they present.

Treatment Considerations

- Acquire specialized knowledge about the effects of violent crime and/or poverty when working with families more severely affected by these stressors.
- Realize that working with these families may also require additional skills such as social action or community organization.
- Recognize the impact that economic instability and the rise in violent crimes may have on many families you will see in treatment and take these into consideration in your treatment planning and intervention.
- Be aware of your own biased views about poverty and violence and their possible bearing on the treatment interventions you choose and possibly on the helping relationship in general.

WORKING WITH GROUPS

Group work, as with family therapy, is used by a wide variety of helping professionals and is becoming the treatment modality deemed most effective with certain populations. There is much variance in how group practice is conducted; much depends on the specific objectives and the agency setting.

Purpose of Group Work

Generally, there are several purposes that apply to most group work. A group environment helps clients realize that they are not alone with their problems—that others also struggle with similar issues; this knowledge can be very comforting and therapeutic. In this regard, group offers participants the opportunity for satisfying social needs that may not be met as easily through individual therapy. Group treatment also provides group members with the chance to help others by offering support and providing useful feedback and information. Both Yalom (1975) and Lieberman and Borman (1979) suggest that by providing help to others, clients are also helped and are able to observe others achieving their goals as well. Yalom (1975) refers to this process of giving and receiving as an "instillation of hope" that is not often present in individual therapy.

In addition to the general purpose of group, there are many specific reasons that group therapy is the preferred treatment modality. Several of these are discussed below.

First, with children, group work is particularly effective as it can be tied to specific development tasks the children are attempting to master. A group can provide children with the opportunity to learn appropriate social skills; it fosters autonomy, thereby preparing them for extrafamilial groups. Schaefer, Johnson, and Wherry (1982, p. 2) speak to this issue: "The peer group for the child represents a major psychological pathway toward the development of identity, self-esteem, and character formation. A child's character is crystallized in the social arena, and, in a larger number of cases, social deviations can be corrected in group therapy."

Second, group therapy is generally the treatment of choice when clients' main problem centers on difficult relationships with others (Northen, 1982). This is particularly applicable when working with adolescents and involuntary adult clients who consider the clinician an authority figure and will probably be more receptive to the feedback of their peers. Similarly, group treatment may be beneficial for those who are socially isolated, such as elderly clients who have suffered major losses, both children and adults who have been removed from their families for any number of reasons, and those who are hospitalized or institutionalized.

Third, group work allows multiple opportunities for role playing, testing out new skills, and behavior rehearsal in a safe and nurturing environment that encourages reality testing. According to Yalom (1975), it is common for groups to develop into social microcosms that allow members to re-create problems they have experienced in the outside world and that facilitate the working through process within the group environment.

Group work can be effective in many other situations—for instance, in work with incest survivors, with eating disorders, or with men and women's issues. According to Henry (1992, p. 2), groups help people "rehearse behaviors, can assist growth and change, and can be an efficient way to make accurate decisions." In all these cases, however, members should possess some capacity to communicate with others, have common concerns that will allow bonding with other members, and be receptive to the group process. There are people who are reluctant to engage in group treatment for any number of reasons. Some may have an extreme need for privacy or confidentiality and are therefore uncomfortable with any type of group process; some may be so different from others that involvement in group therapy may accentuate their difference further; some are paranoid and may view group therapy in a negative manner. In certain cases clients may need to participate in individual therapy first before they are ready for any type of group involvement; we believe this need should be respected. Regardless of the reasons, we favor engaging in group work with those who are amenable and not with those who do not wish to be a part of a group.

Key Definitions

Thus far, we have used such terms as *group work, group therapy, group process,* and *group treatment* interchangeably. Although we consider all of these relevant to the collective treatment of a number of nonrelated individuals, specific definitions are in order.

Group and Group Practice Defined

Definitions of group have evolved through time. For instance, an early definition of group is provided by sociologist, Eubank (1932, p. 163):

> A group of two or more persons in relationship of psychic interaction, whose relationships with one another may be abstracted and distinguished from their relationships with all others so that they might be thought of as an entity.

Later, in 1959, Olmstead (1959, pp. 21–22), who wrote *The Small Group*, said this:

> A group is a plurality of individuals who are in contact with one another, who take one another into account, and who are aware of some significant commonality. An essential feature of a group is that its members have something in common and that they believe what they have in common makes a difference.

To a certain extent, these two definitions are still valid today. Similarly, numerous definitions for the various types of group practice have been offered. The following definition, proposed by the Association for Specialists in Group Work (ASGW), is one we favor for it is both comprehensive and in line with our focus here:

> A broad professional practice that refers to the giving of help or the accomplishment of tasks in a group setting. It involves the application of group theory and process by a capable professional practitioner to assist an interdependent collection of people to reach their mutual goals, which may be personal, interpersonal, or task-related in nature. (ASGW, 1991, p. 9)

Group Classifications

Typical ways to categorize groups are to divide them into two major types, *task groups* or *treatment groups*; these can then be further subdivided into more specific categories. Task groups are said

> to exist to achieve a specific set of objectives or tasks. Concerted attention is paid to the goals and attainment of the desired ends assumes great importance. The goals determine how the group operates and the roles played by members. (Kirst-Ashman & Hull, 1993, p. 82)

Task groups can serve organizational needs; examples are committees, administrative groups, and delegate councils. They can also serve the needs of clients. They could appear as teams, a treatment conference, or a social action group.

Because this chapter is concerned with therapeutic matters, we have limited our discussion of groups to the second category: *treatment groups*. There are five types of treatment groups that are often highlighted: growth, remedial, educational, socializational, and mutual aid (Barker, 1991; Kirst-Ashman & Hull, 1993; Toseland & Rivas, 1984). The premise for each of these types is that individual members will benefit directly from the group experience.

Both the ASGW (1991) and Conyne, Wilson, Kline, Morran, and Ward (1993) have described specific group specializations as tasks/work groups, guidance/psychoeducational groups, counseling/interpersonal problem-solving groups, and psychotherapy/personality reconstruction groups. These categories reflect the complexity of group practice and the broad use of group work.

Schaefer et al. (1982, p. 2) consider a treatment group to be a *therapeutic group* in that it is "a small, face-to-face group designed to produce behavior change in its members. A therapy group is clearly different from an educational, social or self-improvement group."

In many ways, earlier views regarding therapy groups persist today. For instance, Loeser in 1957 noted that for a therapy group to function, five properties had to be present: (1) dynamic interaction, (2) common goal, (3) size and function, (4) volition and consent, (5) and capacity for self-direction. We discuss each of these as we progress through the chapter.

Corey and Corey (1997) discuss therapeutic groups in terms of increasing self-awareness and knowledge of others, helping members clarify the changes they wish to make in their lives, and providing them with the necessary tools to make these changes. The authors further classify therapeutic groups into (1) counseling/interpersonal problem-solving groups and (2) guidance/psychoeducational groups.

The classic work of Yalom (1975) reinforces the advantages of group therapy. He suggests that therapeutic change is a very complex process that requires 11 primary factors he has termed *therapeutic factors;* these include (1) instillation of hope, (2) universality, (3) imparting of information, (4) altruism, (5) the corrective recapitulation of the primary family group, (6) development of socializing techniques, (7) imitative behavior, (8) interpersonal learning, (9) group cohesiveness, (10) catharsis, and (11) existential factors (pp. 3–4). Yalom considers these factors to represent "different parts of the change process" and to be "interdependent." Further, he contends that although the same therapeutic factors exist in every group, their interplay and importance will differ greatly from group to group.

Differences between *group therapy* and *group counseling* should also be noted. The primary reasons for seeking group therapy are to alleviate specific symptoms or psychological problems, such as depression and anxiety. The purpose of therapy groups is to treat a specific emotional or behavioral disorder believed to cause some type of dysfunction. Because of their curative nature, they are considered remedial and often long term in duration. Moreover, in group therapy, efforts are made to understand unconscious factors and to work on the "reconstruction of the major aspects of the personality" (Corey & Corey, 1997, p. 11). Group counseling, in contrast, has both preventive and educational purposes in addition to being remedial. Group counseling typically focuses on a specific type of problem—personal, educational, social, or vocational. According to Corey and Corey (1997), the main difference between group therapy and group counseling is the level of severity of problems. In the first, problems are more serious; in the latter, they are less so. Group therapy is remedial in nature; group counseling is preventive, educational, and remedial. Corey and Corey (1997, p. 12) state that group counseling focuses on "conscious problems, is not aimed at major personality changes, is generally oriented toward the resolution of specific and short-term issues, and is not concerned with treatment of the more severe psychological and behavioral problems."

Despite the variance found in group work, there are three main goals common to all groups within this domain:

- Helping people develop more positive attitudes and better interpersonal skills
- Using the group process as a way of facilitating behavior change
- Helping members transfer newly acquired skills and behavior learned in the group to everyday life

Due to the current economic constraints within the helping professions, group counseling that is usually of shorter duration is preferred over the more intense and long-term practice of group therapy. Our concentration centers on both these types of group practice because we view them equally as important and instrumental in the change process. Primary differences between these two groups is in terms of goals set, techniques used, length of treatment, the role of the leader, training requirements, and the nature of group membership (Corey & Corey, 1977, p. 9); therefore, we encourage you to select the type of group practice that best meets your specific needs in working with your clients.

Historical Developments of Group Practice

Professional group work of some form has been in existence for over a century. The development of group practice and our current positive view of group work today can be traced back to two main areas: (1) social science researchers who studied groups by experimenting with them in laboratories, and (2) group work practitioners who examined the functioning of groups in practice settings. We emphasize the second area given our emphasis on group practice.

Historically, practitioners from various professions (e.g., social work, counseling, psychology, education, and recreation) within the human services field have contributed greatly to our current knowledge base regarding group work. This precedent continues today, when group practice methods remain a treatment modality used by many professional disciplines.

Group work followed early casework in the late 19th century. According to Toseland and Rivas (1984), group work emerged in England and the United States in settlement houses as an outcome of casework in charity organizations. As noted by such early writers as Brackett (1895) and Boyd (1935), even though group work was also being used for therapeutic purposes in state mental institutions during this time, those who were most instrumental in its evolution were leaders of socialization groups, adult education groups, and recreation groups in settlement houses and youth services agencies.

> Group work, and the settlement houses where it was practiced, offered citizens the opportunity for education, recreation, socialization, and community involvement...settlement houses offered groups as an opportunity for citizens to exercise the power derived from their association for social change. (Toseland & Rivas, 1984, pp. 39–40)

During the early part of the 20th century, various other workers (e.g., adult educators, recreation leaders, and community workers) began to recognize the potential of group work to help individuals participate in their communities, enrich their lives, obtain support, and learn needed social skills and problem-solving strategies. During this era, group work also was beginning to demonstrate positive results in the area of juvenile delinquency and rehabilitation for those who suffered from a mental illness. In the 1940s and 1950s, the trend of using groups to work in a remedial capacity in mental health settings grew. For instance, in 1942, Fritz Redl started a group for the emotionally disturbed that provided specialized diagnostic services unavailable at that time in child guidance centers (Reid, 1981). These newer groups were more insight oriented in scope and focused on diagnosis and treatment of members' problems. This approach differed from the one used with earlier groups in settlement houses, which relied more on program activities, recreation, and education.

The increase in remedial groups in mental health probably resulted from two principal events. The first is the influence of Freudian psychoanalysis and ego psychology, which was flourishing at the time. The second was to respond to the shortage during World War II of personnel trained to provide individual therapy to disabled war veterans. This continued interest in remedial group work was further noted in the 1950s when group work was expanded in psychiatric settings.

Along with the rise in groups within child guidance, mental health, and psychiatric settings, the emphasis of group work in recreational, educational, and community arenas also continued. Groups were prevalent in Jewish community centers, in youth organizations such as the Girl Scouts and the YWCA, and in the area of community development and social action. Also, this era proved to be the golden age of the study of groups (Hare, 1976) given the increase in the examination of small groups as a social phenomenon.

The 1960s saw a gradual decline in the popularity of group services that can be accounted for in several ways: First, there was a rise in work training programs and educational opportunities, and these began to be viewed as more significant than remedial group work. The value of community organizational groups, however, continued to be valued. Second, during this period,

> the push toward a generic view of practice and the movement away from specializations in casework, group work, and community organization tended to weaken group work specializations in professional schools and to reduce the number of professionals who were trained in group work as their primary mode of practice. Taken together, these factors contributed to the decline of group work during the decade of the 1960s. (Toseland & Rivas, 1984, p. 42)

The ebb and flow of group work is noted in the following two decades; use of groups continued to subside in the 1970s and then saw a gradual rise in the 1980s when a principal method of group practice became the remedial approach of helping individual group members improve their functioning. In the 1990s, group work of all types is very much the preferred treatment modality in most arenas within the human services field. Models of practice for remedial groups are based on problem identification, assessment, and treatment and are most appropriate for some populations and in some settings, such as outpatient community mental health clinics and mental hospitals. Emphasis on mutual aid and shared, reciprocal responsibility are more appropriate in community groups and in settings like group homes designed for helping members to live together, to support each other, and to cope with distressing life events. In either case, there is room for the various group types depending on the purposes, practice situations, and tasks facing the groups. Reid (1981) concludes in his comprehensive review of the history of groups that there has always been more than one model of group work in operation to meet the many purposes and goals of group work. As we have seen, group work developed as a response to diverse needs for educational, recreational, mental health, and social services and thus has an eclectic base that allows for the continuing diversity within group practice.

Personhood of the Practitioner

Group Leader as a Person and Professional

Whether a therapist is working with individuals, families, or groups, his or her personhood is integral to the overall success and outcome. According to Corey and Corey (1997, p. 61), "the professional practice of leading groups is bound up with who the counselor is as a person." A helper's ability to establish solid relationships with group members is one of the most important tools of group leadership. Furthermore, the helper's personal qualities, values, and life experiences are brought to every group.

> To promote growth in members' lives, you need to be committed to reflection and growth in your own life. If you hope to inspire others to get out of deadening ruts, it is imperative that you keep yourself vital. You need to be willing to invent yourself if you hope to challenge the members of your group to create meaning in their lives.... The person you are acts as a catalyst for bringing about change in the groups you lead. (Corey & Corey, 1997, p. 61)

Role of Self-Awareness

Integral to understanding oneself is the role of *self-awareness*, which we examined earlier. When working in a group setting, this is especially important because there are more individuals to trigger countertransference reactions. Also, issues of competency are likely to arise out of the dynamic nature of groups themselves; it is not easy to hide or ignore one's feelings of inadequacy in a group. For instance, it is easy to become overwhelmed and question your own abilities when struggling to manage group members or feeling lost about which direction to take. In any of these situations, it is important to be aware of your feelings, thoughts, and actions and to examine ways you can improve. Not only is this important for any helper's emotional well-being and professional growth, but it is also an ethical responsibility and needs to be considered on behalf of the client.

The Role of Values

For practitioners, engaging in this introspective process sometimes reveals certain values they may not have been aware of or that they may not welcome. Certainly, the focus on group work practice is influenced by a system of values. These values impact helpers' intervention styles, the skills they use, and the reactions clients have to them. In this respect, values influence the methods used to pursue group and individual goals. Even when practitioners are completely permissive and nondirective, their actions reveal much about who they are.

According to Morales and Sheafor (1995), a group leader's actions in the group are affected by contextual values, client value systems, and the worker's personal value system. Contextual sources of values include values of society, values of the agency sponsoring the group, and values of the helping professions. Client value systems are also brought into every group. Because each individual has a unique value system, it would be impossible to compose a group whose members all had identical values. Finally, the helper's personal value system also is brought into the practice arena, either consciously or unconsciously. This is evident when practitioners are uncomfortable discussing certain value-laden topics that they skillfully avoid, when they unknowingly impose their own values on the group, or when they debate with those whose values are different from theirs. These situations could be potentially harmful to the group process and to individual members and underscore the importance to practioners of being aware of their values and understanding their own position in working with a group. Supervision and consultation are highly encouraged when clinicians are struggling with some facet of practice that is value based.

Personal Characteristics for Effective Group Leaders

In addition to understanding yourself as person, being aware of your own issues, and realizing the influence of your values in your work, various personal characteristics have been connected to being an effective group leader. The most comprehensive list we found of these traits is offered by Corey and Corey (1997) in their text, *Group Process and Practice*. They identified the following as important: courage, willingness to model, presence, goodwill and caring, belief in group process, openness, becoming aware of your own culture, nondefensive in coping with attacks, personal power, stamina, willingness to seek new experiences, self-awareness, sense of humor, inventiveness, personal dedication, and commitment. Some we have already discussed; others may be new to you. In some cases, as with stamina, sense of humor, or inventiveness, you may never have considered these to be instrumental to group leading. We ask that you reflect on the traits you already possess as well as those you may need to cultivate. Remember that this process is an ongoing one requiring continual self-examination and reflection. For each group you are involved with, you will experience growth in many of these areas, further contributing to the development of these characteristics.

Appreciating Group Dynamics and Group Process

Possession of the personal characteristics listed above will not alone ensure group effectiveness. Understanding group dynamics is also necessary to practice effectively with all types of groups. Clarity about such terms as *group process* and *group dynamics* is in order.

Group Dynamics and Process

A group is composed of a number of people in interaction with one another. These interactions within a group system have often been referred to as *group process*. Group process produces unique forces that influence both group members and the group as a whole, and these are *group dynamics*. According to Schaefer et al. (1982, p. 4), "the term *group dynamics* refers to these interactive forces that operate in a group and influence the behavior of the group's members. This group process itself is often more helpful than the therapist's efforts in getting people to change."

Many authors have written about the therapeutic factors inherent in a group. We have already identified the curative factors proposed by Yalom in 1975. Earlier works by Corsini and Rosenberg (1955) provide a historical glimpse of how groups were viewed and illustrate the continued emphasis on the value of group treatment today. Corsini and Rosenberg divide these therapeutic factors into three categories:

Emotional	*Cognitive*	*Actional*
• Acceptance of, empathy with, and involvement with others	• Spectator therapy	• Reality testing and interpersonal learning
• Altruism or social interest	• Universalization	• Ventilation
• Transference	• Intellectualization	• Interaction

Later, in 1975, the change mechanisms found previously by Corsini, Rosenberg, and Yalom were combined into a single list ranked by degree of consensus among theorists:

Mechanisms with Clear Consensus as Being Most Important	*Mechanisms Considered to Be of Secondary Importance*
• Ventilation	• Universalization
• Acceptance	• Reality testing
• Spectator therapy	• Altruism
• Intellectualization	• Socialization

Despite this division, we believe everyone views group therapy differently and may agree or disagree with which mechanisms are most important. We encourage you to develop your own view based on your personal experiences as to what seems to be the most important factors in the success of your groups.

Four areas of group dynamics were identified by Toseland and Rivas (1984, p. 57) as being particularly significant for practitioners in understanding and working effectively with groups: (1) the communication and interaction patterns occurring in groups, (2) the attraction of groups for their members, (3) the social controls that are exerted in groups, and (4) the culture that develops in groups. Although in a somewhat different format, we review several of these basic tenets of group dynamics below.

The Interaction Process: Active Listening and the Role of Verbal and Nonverbal Communication

Communication has been defined as "the process by which people use symbols to convey meanings to each other" (Toseland & Rivas, 1982, p. 57). Communication in group settings is very important. "On the basis of what is communicated and how communication occurs, a set of people will interact and take on meaning for each other, consent together to attain some goal, and become a group." (Henry, 1992, p. 13).

Communication is a process that involves three main steps: (1) the coding of a person's perceptions, thoughts, and feelings into language and other symbols; (2) the transmission of these symbols of language; and (3) the decoding of the transmission by another person. When this process takes place in a group setting, members begin to communicate with one another and a reciprocal pattern of interaction surfaces. This communication may be suitable to the situation at hand or it can be inappropriate and require the group leader to intervene to clarify meanings and improve future communications.

Substantial research has been done on communication patterns in groups. Many researchers believe that communication influences the emergence of leadership and determines who initiates communication and who will receive it (Henry, 1992). A person's location and involvement in the communication process can also affect the outcome. For instance, if a person is removed from the group and communicates very little, he or she may be viewed as an isolate and thus not considered a leader. Conversely, if a person sits at the center and is actively involved in communicating with other members, he or she is more likely to emerge as a key person in the group and perhaps even as the leader. There are also times when the leader is mostly silent, but when he or she does speak, the experience is very powerful and influences other group members to view the speaker as a leader.

Group decision making is also closely tied to communication. Communication is exchanged among members when they discuss elements connected to group decisions. For instance, members may discuss their ideas, beliefs, and points of view in order to reach a consensus on the group's common goals.

Active listening, which involves much more than just talking through verbal channels, is a special communication skill that is crucial for group leaders. Active listening involves "absorbing the content, noting gestures and subtle changes in voice and expression, sensing underlying messages, and, perhaps most important of all, intuiting what the person is not directly saying" (Corey & Corey, 1997, p. 69).

Furthermore, according to Zastrow (1985, p. 127), every message contains two elements: content and feelings. "Content refers to the topic or idea, while feelings are expressed via smiles, gestures, body posture, and tone of voice" (p. 127). Recognizing the congruence between what is said verbally and what is communicated through body posture, mannerisms, inflection or tone, and gestures can be very useful for group leaders who must continually observe the communication process within the group.

Additionally, by engaging in active listening, the worker can uncover feelings and ideas that may not be expressed through direct verbal means. Active listening is a skill that also allows the development of other important skills such as empathizing, confronting, supporting, clarifying, reflecting, and summarizing; these are based on the helper's ability to understand the clients and the messages they are sending. Because many types of communication are occurring within a group at any one time, the helper is specially challenged to employ active listening.

Group Cohesion (Group Attraction)   Cohesion is an essential ingredient for effective group practice. Yalom (1975) hypothesizes that cohesiveness in group therapy is the analogue of "relationship" in individual therapy. It has been called the glue that holds a group together (Coyle, 1930; Henry, 1992). In this respect, cohesion is considered a "bond that permits members to feel close enough to each other to allow their individuality to be expressed" (Henry, 1992, p. 13). Zastrow (1985, p. 21) defines cohesion "as the sum of all the variables influencing members to stay in a group. . . . It occurs when the positive attractions of a group outweigh any negative implications a member might encounter."

According to the early work of Festinger (1953), cohesion may appear when group members band together in response to a real or perceived threat of being extinguished. People have many reasons for deciding to remain in a group. Cartwright (1968) has identified four interacting sets of variables said to determine a member's attraction to a group:

- Needs for affiliation, recognition, and security
- Incentives and resources of the group such as the prestige of its members, the group's goals, its program activities, and its style of operation
- the subjective expectations of members about the beneficial or detrimental consequences of the group
- a comparison of the group to other group experiences

Henry (1992) cites similar reasons that cohesion develops: attraction of members to each other, to the activities the group members do together, or to the goals the group is working on; cohesion also may develop as a result of external pressures to remain together. Measurement of members' attraction to and involvement in a particular group is suggested in terms of their perceptions of the payoffs and costs of memberships. Below are several of the possible payoffs and costs (Zastrow, 1985):

Payoffs	*Costs*
• Companionship	• Being with people one dislikes
• Attaining personal goals	• Expending time and effort
• Prestige	• Criticism
• Enjoyment	• Distasteful tasks
• Emotional support	• Boring meetings

We believe it is important for the group leader to assess which of these factors may be present in his or her particular group. If the costs are too great, perhaps efforts can be made to alter the group process in such a way that payoffs become greater than costs.

Corey and Corey (1997) believe that true cohesion is unlikely to develop early in the group process. "Genuine cohesion typically comes after groups have struggled with conflict, have shared pain, and have committed themselves to taking significant risks" (p. 153). However, that cohesion comes late does not negate the importance of the initial stages of the group; this is when the foundation of cohesion begins to take shape.

General indicators that suggest that cohesion is developing are: regular attendance and punctuality; members' perception that the environment is safe, shown by members sharing their feelings openly and experiencing a sense of belonging; supportive and caring interactions with one another; and the willingness to risk exposing feelings and perceptions about the group at the very moment they are experienced (Corey & Corey, 1997; Zastrow, 1985).

Cohesion develops internally within the group and is ever changing, as events occurring inside or outside the group may alter how members feel about the group process. Attainment of group and individual goals is more likely when a strong cohesive base exists; therefore, group practitioners must regularly assess the level of cohesion present in their groups. Knowing this will allow them to determine more accurately the changes that might be needed to promote cohesion.

Social Control/Conformity   Social control, or conformity as it is often called, is also important to group practice. Toseland and Rivas (1984, p. 67) describe social control as a "process by which the group as a whole gains sufficient compliance and conformity from its members to enable it to function in an orderly manner." Various factors contribute to such compliance—namely, norms and values, roles, and status. We briefly discuss these below.

Norms and Values   Norms are considered the group's shared expectations and views about the proper and appropriate ways to act in a group setting; they emerge as the group begins interacting. Essentially, they are standards that govern group behavior because they stabilize and regulate behavior in groups; thus, being able to recognize them is extremely important. They may be explicit and thus overtly stated and enforced, as when absences or consistent lateness is confronted. In contrast, norms may also be conveyed in more subtle and implicit ways and thus can be more covert in nature, as when silence occurs when a member arrives late. In either case, norms are significant because they "provide clear guidelines for what is acceptable and appropriate behavior, norms increase predictability and security for members. Norms aid groups in establishing procedures for coordinated action to reach goals. Over all, they are essential for a group's functioning" (Toseland & Rivas, 1984, p. 68).

Roles   The various roles group members play also influence the group. Roles, as with norms, consist of shared expectations of the group. With norms, the expectations are about overall group behavior; with roles, the expectations are more specific and relate to the functions of individuals in the group. In both cases, they are bound to change through time. Examples of specific therapy roles as presented by Schaefer et al. (1982, p. 9) include "the monopolizer, the isolate, the help-rejecting complainer, the self-righteous moralizer, the therapist's assistant, helpful Hannah, and the professional patient."

Roles are important to group process in that they foster a division of labor and the appropriate use of power. Often, the roles played in a group are similar to those performed by group members in their lives outside the group. For this reason alone, they can offer insight into individual members as well as group process.

Status   A member's status in the group, often defined by the person's role, is also pertinent to determining how well conformity will develop. Specifically, status typically refers to "an evaluation and ranking of each member's position in the group relative to all other members" (Toseland & Rivas, 1984, p. 68). Ways in which status is determined will vary among groups; it is defined by group members themselves, although outside factors may also come into play. Conformity to group norms can be determined in part by the status of the individual; low status individuals are least likely to conform, whereas those with medium and high status are more apt to conform. There are times, however, when group members with a higher level of group status will deviate from accepted norms because their relative position within the group makes them feel more secure and free. Observing where each member fits in terms of status is useful to leaders as they develop appropriate group interventions and attempt to understand group dynamics.

Group Culture   Norms, roles, and status in group help solidify a group culture that is unique to the specific group in question. Olmstead (1959) considered group culture to be composed of the values, beliefs, customs, and traditions the groups holds in common. Hartford (1972) saw group culture as the distinctive quality that sets it apart from every other group. Group culture develops gradually as the group develops; it emerges from the members' norm-conforming behavior, although outside influences may also affect the specific culture that unfolds. Understanding group culture is essential in working with the group. For instance, discussions that take place within the group may be unique to the group's culture and therefore not easily understood when related outside the group context.

Group Control and Influence   Another important group dynamic centers on the premise that group control and influence can be great. Henry (1992, p. 15) finds that "each member of a group is influenced by the other members, and in turn influences them in the direction of her or his goals and the goals of the group." Group control and influence may be either negative or positive. Generally, groups that are cohesive and viewed as effective by the group members themselves are likely to be the most influential.

Stages of Group Development   A last essential component of group dynamics is the common stages of group development, the core tasks that group members must manage. The literature is full of material on group development; there are many models, each presenting its own version of how a group develops. Although different in number and specific types and stages, most models propose that all groups pass through similar stages of development. Most see development as a progression from beginning to middle and finally, to end. Because this discussion is limited, we will not elaborate further regarding group development. We do, however, recommend that you seek additional information in this area. Corey and Corey's *Group Process and Practice* is an excellent resource.

Group Leadership

We have seen the complexity of group dynamics that group leaders must understand and become comfortable with. Professional group leadership is the process of guiding the development of the group and its members. The practitioner who is the *formal* designated leader of the group must pilot it so as to ensure that the group as a whole and each of its members achieve their specified goals. Other *informal* leaders may emerge or influence the group process; in fact, many times these informal leaders are extremely instrumental in the group's development. Our focus here, however, is on the professional group leader. We believe that mastering basic leadership skills is essential to becoming an effective leader.

Issues of Leadership and Power   The leadership role brings with it a certain amount of power that must be treated seriously by the leader himself or herself to avoid any misuse. Both attributed power and actual power exist, and they are equally important in facilitating the development of the group and its members. Seven self-explanatory power types have been identified: (1) connection power, (2) expert power, (3) information power, (4) legitimate power, (5) reference power, (6) reward power, and (7) coercive power. Determining which of these you are most comfortable with can help you understand your role.

Most important, however, is how well you as the leader use the power bestowed on you. The needs and well-being of the group must be the priority in the leadership role. When this is no longer true, it is time for any group leader to examine the group process as well as his or her role.

Styles and Approaches to Group Leadership   Leadership styles are certain to impact the overall functioning of the group. Such traits as being open, accepting, responsive, and confident have been found to enhance group cohesion and encourage member commitment. This is less likely when the leader is "neutral, professional, and distant" (Galigor, 1977; Schaefer et al., 1982). In research conducted by Liberman, Yalom, and Miles (1973) results showed that the most favorable outcomes and fewest dropouts were linked to a leadership style of providing considerable cognitive structuring and using only a judicious amount of authority, creating a trusting, open, and fair group environment. Similarly, the leader should not be too dominant or too passive as both can produce negative outcomes (Bole, 1971).

Many approaches to group leadership are well grounded in group leadership theory. We have outlined here four major approaches presented in the literature:

- *The trait approach* assumes that leaders have inherent personal characteristics that distinguish them from others.
- *The position approach* defines leadership in terms of the authority related to a particular position.
- *The leadership style approach* focuses on three styles of leadership: authoritarian, democratic, and laissez-faire.
- *The distributed-functions approach* holds that every member of the group will be a leader at times, taking actions that serve group functions.

Three major models of group work identified by Papell and Rothman (1980) include the social goals model, the remedial model, and the reciprocal model. In each, the role of the leader is different given the specifics of the model.

Leadership Roles   In addition to specific leadership approaches, leaders also assume certain roles. Generally, there are two major types of leadership functions under which most all other roles fall:

- *Task specialist*, which includes the roles needed to accomplish specific goals set by the group
- *Social/emotional specialist*, said to strengthen social/emotional bonds within the group and contribute to group maintenance

In addition to the many roles that fall under either of these two categories, there are often other roles such as executive, policy maker, planner, and ideologist, to name a few that must also be performed, depending on the particular group and circumstances.

In determining factors influencing group leadership, realize that notwithstanding specified models and approaches, group leadership is situational and affected by a variety of interacting factors. A number of models can be used appropriately, depending on the situation facing the group, the capacity of its members, and the skill of the leader. The personality and leadership styles of the helper are important, but equally important are the group participants.

The many strategies for effective leadership are often logical and based in common sense. We present those suggested by Zastrow (1985) to illustrate how basic, yet how important they are to group process. They include using ice-breaker exercises to start; focusing on helping members identify personal and group goals early in the process; not dominating a group; performing task and group maintenance functions and encouraging others to do these as well; matching decision-making procedures to the issues facing the group; creating a cooperative group atmosphere in which communication is open and honest; confronting hostile or disruptive group members; using, and encouraging appropriate self-disclosure; providing relevant information to the group; and bringing closure to each group by briefly summarizing key points of the session.

Skills of Group Leadership   Most of these strategies require certain skills that leaders must possess to be effective. Toseland and Rivas (1984) present an extensive list of leadership skills that they have further subdivided into three major categories shown below:

1. *Skills to Facilitate Group Process*
   Attending Skills
   Expressive Skills
   Responding Skills
   Focusing Skills
   Guiding Group Interactions
   Involving Group Members

2. *Data-Gathering Assessment Skills*
   Identifying and Describing Skills
   Requesting Information, Questioning, and Probing
   Summarizing and Partializing
   Analyzing Skills

3. *Action Skills*
   Directing
   Synthesizing
   Supporting Group Members
   Reframing and Redefining
   Resolving Conflicts
   Advice, Suggestions, Instructions
   Confrontation Skills
   Providing Resources
   Modeling, Role Playing, Rehearsing, and Coaching

This comprehensive list is important to consider when you attempt to identify the skills you still must master. A more recent view of leadership skills is offered by Corey and Corey (1997), who list many of the skills presented above as well as others they feel are necessary for a competent group leader. Included in their survey of leadership skills are active listening, reflecting, clarifying, summarizing, facilitating, empathizing, interpreting, questioning, linking, confronting, supporting, blocking, diagnosing, modeling, suggesting, initiating, evaluating, and terminating. There is considerable overlap in the skills identified by different authors, and this overlap further reinforces the importance of skills that appear in multiple lists. Remember that even though these skills might appear simplistic, they may take time to develop and are not always easy to apply because they may be dependent on conditions such as the leader's emotional state, group conditions, and agency demands or restrictions. We encourage you to develop these skills slowly through ongoing experience in group work. Corey and Corey (1997, p. 68) agree: "Learning how and when to use these skills is a function of supervised experience, practice, feedback, and confidence."

Co-leadership   Co-leadership has stimulated much debate and varying opinions. At one end of the spectrum are those who do not value co-leadership and even find it to be damaging; at the other end are those who favor its use. Practitioners who favor the use of a co-leader cite such benefits as providing the leader with a source of support, feedback, and the opportunity for professional development. It also increases the leader's objectivity by providing alternative frames of references, is an excellent way of training an inexperienced leader; provides group members with models for appropriate communication, interaction, and resolution of disputes; provides leaders with assistance during therapeutic interventions such as role plays, simulations, and program activities; and helps to limit settings and structure the group experience (Corey & Corey, 1997; Toseland & Rivas, 1984).

Co-leadership also has limitations: it is expensive and time-consuming; communication between sessions may be a problem for co-leaders; leaders may not work well together or may have vast differences of opinion that can negatively impact the group; and the group may favor one leader over another, thereby creating conflict and division. To avoid these potential drawbacks, leaders who work together should have complementary styles and personalities. We have found working with co-leaders to be beneficial; however, we have also experienced positive experiences by conducting groups alone. We encourage you to try both to determine your preference.

Overview of Group Planning

Understanding group dynamics and learning the skills necessary to be an effective leader alone are not enough; planning for treatment groups is also essential. Many models illustrate different ways to plan for groups. Although they differ on specifics, they share a belief that the most important element in planning is establishing a group's purpose. Other areas of consensus are the value of the group's sponsoring agency and environment, and the assessment process of members. Of all the models examined, we found a "Planning Model for Group Work" developed by Toseland and Rivas (1984) to be the most comprehensive and applicable to both group therapy and group counseling. Below, we briefly present the basics of their model and expand on some of the more pertinent issues. It has seven major steps.

1. Establishing a Group Purpose and Determining Objectives

Workers must carefully consider why they are forming the group and what general objectives best fit the identified purpose of the group. A statement of the group purpose needs to precede the development of a group and "should be broad enough to encompass different individual goals, yet specific enough to define the common nature of the group's enterprise" (Toseland & Rivas, 1984, p. 118). The statement should contain specifics about the reasons the group is forming, the group process, and the range of individual goals the group might contain. Such a statement gives potential members sufficient information to decide whether the group might serve their interests.

Ideas for groups can come from three sources: group workers, agency staff, and group members.

2. Determining Potential Sponsorship and Membership

Because the agency and its clients are intrinsically linked, it is important to consider both the agency where the group is to meet and the potential types of members available. These may seem very rudimentary considerations, but they can contribute greatly to the success of a group or—even more important—to its demise.

Assessing Potential Sponsorship   In developing a proposal for a group, an important consideration is the degree of support the sponsoring agency will provide. This is essential as "agency sponsorship determines the level of support and resources available to the group.... Treatment groups rely on agency administrators and staff for sanctions, financial support, member referrals, and physical facilities for the group (Toseland & Rivas, 1984, p. 119). Furthermore, it is important to determine whether there is a fit between the agency policies and goals and the purpose of the proposed group. When there is conflict in this area, there are likely to be problems as well.

Assessing Potential Membership   It is equally important to assess for potential membership early as this will alert the leader to the possible viability of the group. Assessment at this stage is simplistic and not as extensive as the screening interview that is often conducted later during group planning. In determining whether a sufficient number of members are available, it is also important to decide whether potential members share in the established group purpose and are amenable to the group plan. There are times, for instance, when a practitioner has a great idea for a group, but it does not match the agency's clientele.

3. Recruiting Members/Attracting Members

Recruitment of members is a very important step in ensuring that a proposed group can actually become a reality. For this process to be effective there must be a sufficient number of potential members for the group. Sources for group referrals both from within the agency and from the outside are explored. Corey and Corey (1997) emphasize finding a practical way of announcing one's group to prospective members; they believe that the way a group is announced influences both the way it is received by potential participants and the type of individuals it will attract. Announcements about the group are best when they are in writing and provide the reader with sufficient information about the group purpose and process. Such announcements may be in the form of fliers to the community, invitations to other community agencies, or memos to colleagues in one's agency requesting appropriate referrals.

Direct contact with potential clients is one of the most effective methods of recruiting, but this may not always be feasible. At times telephone contacts will have to suffice.

Selection Process: Screening Group Members   In selecting group members, you should choose those members who are most likely to benefit from participating in the group. Specific guidelines for selection of members is clearly spelled out by the ASGW (1991): "Insofar as possible, the counselor selects group members whose needs and goals are compatible with the goals of the group, who will not impede the group process, and whose well-being will not be jeopardized by the group experience" (p. 2).

A key point is that the members be appropriate for the group and the group for the members; this is particularly important as a group composed of ambivalent members or those not amenable or emotionally ready for a group may experience conflicts and an ineffective outcome. Because of this, workers may need to establish inclusion and exclusion criteria.

In-person screening interviews are the most beneficial as they allow workers to observe potential members and collect impressions and information about them that can be essential to the selection process. It is very important not to compose a group hastily in response to agency pressures. You must have time initially for proper screening to take place. Also, we recommend that practitioners pay attention to their "gut level" feelings and trust their intuition about members they feel may not be appropriate. As the selecting worker will be leading the group with these very members, a match needs to exist.

4. Group Composition: Forming the Group

In forming a group, you must consider group size, frequency and duration of meetings, time limits, whether the group is open or closed, the demographic characteristics of members, and the space issues and location of the group.

Ideally, group membership is derived according to a set of established criteria that are decided on prior to the actual composition. For many group theorists, the following are important:

- A homogeneity of member purposes and certain personal characteristics
- A heterogeneity of member coping skills, life experiences, and expertise
- An overall structure that includes a range of member qualities, skills, and expertise

Whether you accept all these issues or not, you need to decide on the question of homogeneity versus heterogeneity in the composition equation. In essence, group leaders must assess the potential compatibility of members' "needs and behaviors, similarity of problems, range of tolerance for deviance from behavioral norms, cultural and other characteristics and skills related to the purpose of the group. The aim is to assemble a configuration of persons with the potential to coalesce and function as an entity" (Henry, 1992, p. 5). Also important in the composition phase is the need for group structure that can meet individual client needs and allow the group to accomplish its purpose.

Group size will vary from group to group and is determined in several ways. The age of the clients, the experience of the leader, the type of group, and the problems to be explored are cited by Corey and Corey (1997) as the most essential factors to consider. According to Zastrow (1985, p. 15), "the size of a group affects members' satisfactions, interactions, and the amount of output per member." Some clients and group leaders prefer small groups as they feel more can be accomplished in an intimate environment. Others prefer larger groups with more members for each member to relate to. Although there is no optimal size for a group, more intense therapeutic groups generally are better with small groups whereas psychoeducational groups may do well with larger numbers.

Determining the *frequency and duration* of a group is also a decision that requires careful attention, yet is one often predetermined by the sponsoring agency or by financial constraints. The purpose and objectives of the group as well as the level of intensity expected may help to determine how long each session should last and how often the group should meet. Our rule of thumb is that a more intense group like an incest survivor's group needs to meet weekly for at least 90 minutes to address adequately the many issues likely to surface.

Similarly, the *length of group* and the decision regarding whether the group should be *open or closed* will depend on the agency, the clinician's preference, and the purpose of the group. In some cases, groups are specifically designed to be short term so they remain focused. Also, with the advent of managed care, short-term groups are becoming the norm in most settings today.

Such demographic characteristics as age, sexual makeup of the group, and sociocultural factors are often considered when composing a group. Additionally, we encourage you to seek members who are in similar developmental stages.

Last, the physical setting where the group meets can have a profound effect on the behavior of group members and the group performance. Toseland and Rivas (1984, pp. 134–135) stress that the

> size, space, seating arrangements, furnishings, and atmosphere should all be considered when preparing for a place for the group to meet. Difficulties encountered in early meetings, inappropriate behavior by members, and unanticipated problems in the development of the group can sometimes result from inadequate attention to these aspects of a group's physical environment.

5. Orienting Members to the Group

Once the screening process is complete and the group is composed, members must be properly oriented to the group. This is often done during the initial group session or it can take place during the screening interview. In either case, it is important to explain the purpose of the group and determine group goals, and to familiarize members with the group by discussing procedures and ground rules that often contain limits of the group as well. Also, during this phase, the leader should continue to screen members for appropriateness.

6. Contracting

Whether formalized or informal, contracting typically begins prior to the initial group meetings. In this context, a contract is defined as a verbal or written agreement between two or more members of a group—most often between the worker and the client. The contract not only helps to clarify the purpose of the group but also helps foster commitment and a sense of responsibility in both the members and the leader. Often, contracting occurs at two levels: contracting for group procedures and contracting for individual member goals.

7. Preparing the Group's Environment

Preparing a group's environment typically involves consideration of three factors: (1) preparing the physical environment, (2) securing financial support, and (3) securing special arrangements. Many times the worker may not be able to control factors that exist within the environment; nonetheless, whenever possible, these should be regarded as important to the overall outcome of the group.

Ethical and Legal Issues

Ethical and legal issues as they relate to group practice are similar to those experienced in individual work with clients; at times they are further magnified by the group process when more than one member is present. Grounding in ethical issues is essential in the practice of group therapy or counseling. Corey and Corey (1997) have identified the areas listed here as important for consideration:

- Ethical issues in group membership that must be considered are involuntary membership, informed consent, freedom to withdraw from a group, and the psychological risks of members.
- Confidentiality issues are most important in group work. The leader must not only keep the confidences of members but must also ensure that members maintain confidentiality between one another. Mistrust and betrayal are likely repercussions when confidentiality is violated.
- Uses and abuses of group techniques can occur when the leader is not careful to plan ahead and assess which techniques are most appropriate to the particular group at hand.
- The role of the leader's values is important in a group. When helpers are unaware of their values or push them onto their clients, the result may be problems or unethical behavior.
- Leaders must have knowledge of formal ethical and legal standards.
- Finally, multicultural awareness is necessary in group practice. We discuss this further below.

Diversity Issues in Group Work

"Effective delivery of group-counseling services must take into account the impact of the client's culture" (Corey & Corey, 1997, p. 14). Culture, as it is defined here, includes much more than a person's ethnicity or racial heritage. A person's age, gender, sexual orientation, or socioeconomic status is also included. Culture, defined in this broad way, is bound to impact the behavior of both the client and the helper. Therefore, leaders must have awareness, knowledge, and skills concerning diverse groups for effective practice. They need to be aware of their own biases and prejudices, as stereotyping others will likely interfere with the group process. Many of the problems encountered by minority clients in a group are the result of oppression, prejudice, and lack of understanding; leaders must not perpetuate the problem but instead become part of the solution. Practitioners must remain alert to the many factors, such as a client's race, gender, and sexual orientation, that can impinge on their practice.

As practitioners, we need to understand the cultural values of our clients so that we select interventions that are congruent with their worldviews. Culture influences values, decisions, and actions of all group members; in many cases, it determines what types of problems are brought into group therapy. Each person and cultural group bring their own experiences and expectations into the agency. Attempting to understand the particulars of each group is often necessary to ensure that appropriate treatment will result. All these considerations accentuate the need for helpers to become cross-culturally sensitive or, as we prefer to call it, diversity sensitive.

Varied Group Treatment Modalities

A main theoretical base in which most group practice is grounded is systems theory, which we covered in depth earlier in the chapter. In review, systems theory attempts to explain the group as a system of elements that are interacting. It is one of the most widely and broadly applied theories of group functioning. Other influential theories in the development of group practice are these:

- *Psychoanalytic theory:* group members act out unresolved conflicts from early life experiences in group and in this way reenact family situations within the group environment.
- *Learning theory:* The behavior of group members can be explained by one of three methods of learning: classical, operant, and social learning (Bandura, 1977). Learning theory principles have been useful in helping members to make desired changes.
- *Field theory:* Groups are constantly changing to cope with their social situation although there are times when a "quasi-stationary equilibrium" exists for all groups (Lewin, 1947).
- *Social exchange theory:* Focusing on the behavior of individual group members, this theory suggests that when people interact in groups, each will attempt to behave in a way to maximize rewards and minimize punishments.

To develop and refine their practice skills, group practitioners should examine further any of these theories that seem to fit their orientation more closely.

Current Practice Trends

Group work has become a preferred treatment modality in many human service agencies. This emerging trend reflects the current emphasis in the human services field of providing needed therapeutic services to clients in the most cost-effective and efficient manner possible. In an era when available funding is shrinking and the threat of budget cuts is ever present, many organizations are faced with the dilemma of providing quality and timely treatment while simultaneously limiting the high cost of care. The challenges of these pressures are great for both agency administrators and human service providers, although each is affected in different ways. Human service practitioners must develop sound clinical and specialized group skills as well as learn how to navigate their way successfully within a variety of agency settings if they hope to be effective.

Successful group practice cannot be realized without an understanding of the agency systems in which they exist. Knowledge of group skills alone is often insufficient; the many facets of agency life must also be taken into consideration. Understanding certain agency dimensions—such as the agency structure, policies and history, the current staff makeup, possible political issues unfolding, the formal and informal lines of command, client demographics, and the existing needs of both the clients and the agency itself—are all equally important for a clinician to examine.

Common Types of Groups

Numerous types of groups have been identified in the literature. Knowing the basic premise behind those most commonly used is helpful for practitioners entering group practice. Following is a partial list of the types of groups found in human service agencies, illustrating the diversity of groups available.

- Social conversation groups
- Recreation/skill-building groups
- Educational groups
- Problem-solving/decision-making groups
- Self-help groups
- Socialization groups
- Sensitivity and encounter groups
- Therapeutic groups

Our focus has been primarily on therapeutic groups. To discuss the features of all these groups is beyond the scope of the chapter, so we encourage you to investigate independently those that interest you most.

Developing Your Own Approach

As you continue to practice group work and become more seasoned and confident in your abilities, you will likely find yourself developing your own approach. People develop unique styles reflective of their personalities, theoretical orientations, and individual preferences. Other contributing factors include the agency dictates of where you work and your inclination to engage in short-term or long-term therapy. Ultimately, you will undoubtedly develop an integrative perspective on group practice.

Profiles of Specialized Groups

To provide you with some actual illustrations of group work within agency settings, we present four brief group profiles we have developed and implemented in various human service agencies. They represent different age groups, a number of concerns, and variation in group approach. All are appropriate in today's practice context as they are brief or short term in nature. They include a children's group on developmental and life skills issues, an adolescent anger management group, an adult support group for women incest survivors, and an adult men's issues group.

Children's Group: Development and Life Skills Group

Overall Objective: This group will address general developmental issues, including physical, psychological, and social aspects. A primary focus will be to assist group members gain a better understanding of the overall process of development. Age-appropriate developmental concerns will be discussed with emphasis on acquisition of important life skills. A problem-solving model will be presented, with an individualized plan for each member.

Week 1

- Introduce group and participants/members.
- Define purpose of group.
- Identify group rules, goals, and expectations.
- Contract for commitment.
- Begin development of trust, rapport, and bonding among members. (Introductory process to orient group members to group purpose and process and to encourage commitment and rapport to build on.)

Activity: Present basics about developmental stages via use of a colorful chart. Assist group members to identify the developmental level where they are presently (e.g., latency age, preteen, adolescent, etc.). Ask that they share something they are presently experiencing related to their development (e.g., preteen entering puberty or younger child beginning school).

Week 2

- Outline the developmental process from birth to death.
- Address the biopsychosocial aspects and their importance in a person's overall development.
- Discuss the concepts related to developmental tasks (give ladder example).

Activity: Divide members into pairs and provide them with a picture of an individual with a little synopsis of the person's life (personal, family, social); have each pair identify the biopsychosocial components.

Week 3

- Focus on physical aspects of development and emphasize the particular stage each pertains to.
- Emphasize different growth rates, body shapes, and overall development variance.
- Address the impact of illness or disability on normal development.
- Help members identify their physical aspects and begin to discuss areas where they feel positive and negative.

Activity: Have members individually draw a picture of themselves or share a picture they bring from home. Have them describe what they see in the picture. Are they on target for age? Delayed or advanced? What implication might this have on them overall?

Week 4

- Focus on psychological/emotional aspects of development.
- Outline different emotions and the values for members in being able to express themselves appropriately.
- Discuss individual personality types and acceptance of oneself and of others.

Activity: Using "emotions/feelings" chart, have each member identify a feeling and give a personal example. Also have group discuss what might happen if feelings are not talked about openly. Example of water overflow exercise may be helpful to stress the importance of open expression of emotions.

Week 5

- Focus on social aspects of development including interpersonal relationships and social contacts.
- Define issues related to engaging in relationships with others.
- Address developmental tasks relevant to members' age group.

Activity: Demonstrate how to draw an eco-map and encourage members to draw their own. Have them identify all their roles and types of relationships. Positive (+), neutral (n), or negative (−) signs will be placed after each relationship or interaction.

Week 6

- Discuss integration of three spheres of a person's life.
- Address the concept of overlap as each area is sure to affect the others (systems theory).

Activity: Use systems perspective to show that change in one aspect of a system affects the other aspects of the system.

Week 7

- Address additional issues that affect development:

Ethnicity
Acculturation

Assimilation
Immigration status (legal, illegal, migrant)
Refugee status
Socioeconomic status (SES)
Support systems/positive role models

Activity: Have each member identify unique qualities of themselves by sharing aspects of their culture (e.g., food, clothing, pictures, customs, songs, music, dance, etc.).

Week 8

- Focus on existing concerns, problems, and barriers to fulfilling successfully the developmental stage where members are.
- Discuss the value of being able to identify problems one may have as this is the first step to problem resolution.

Activity: In pairs, have each member choose one area in which he or she feels the need of some help (problem identification).

Week 9

- Focus on realistic goal setting that will lead to goal attainment.
- Identify any barriers or obstacles that may impede goal fulfillment.

Activity: Following problem identification, have each member identify a desired goal that is realistic and attainable. (Use imaging if the group experiences difficulties in identifying a realistic goal.) Emphasis will be on narrowing goals to those that are simple and attainable.

Week 10

- Encourage group to continue working on goal attainment by identifying specific steps each member will take.
- Help group explore possible obstacles that may hinder their success.

Activity: Group will sit in a circle and members will share their goals and steps they have taken to make this goal a reality. The group will brainstorm about possible obstacles and ways to prevent or cope with these.

Week 11

- Encourage group to continue working on goal attainment issues.
- Begin to discuss upcoming termination and any issues this may present.

Activity: This is to be a sharing group with members telling of their progress regarding goal fulfillment. Provide guidance and support when needed as well as positive reinforcement.

Week 12

- (Last group session) Review and prepare for termination of group.
- Review with members the purpose of the group and discuss how they have been personally helped. Emphasize the importance of problem-solving skills.

Activity: Members will be asked to share their experiences; each will be awarded a certificate of merit to honor the group's hard work and commitment. Group will end with a celebration.

An Adolescent Group on Anger Management

Overall Objectives: This group will help members create a definition of anger that is useful. Members will be able to identify popular beliefs and examples of how society has demonstrated ways of handling anger and ways this has influenced them personally. They will become more aware of their own anger by recognizing common and not so common cues signaling anger in both themselves and others. Further, they will learn to associate feelings that may be underlying their anger and to recognize cues that are likely to make them lose control. Members will increase the amount and types of coping skills they use to manage feelings and resolve issues. They will be helped to identify specific problem areas and old habits of coping and be able to practice alternative or refined skills. Last, members will learn to express themselves without attacking and to understand and accept different point of views.

Week 1

- Introduce group and participants.
- Develop group rules, goals, and expectations.

Activity: Use "buddy system:" pair members off and have them interview each other and present to group.

Week 2

- Discuss what anger is. What is anger a response to? What is anger telling you? What have we been told about anger? Is it good or bad?

Activity: Have members create a definition of anger that will be used during the remainder of the group.

Week 3

- Discuss what people do, in general, with their anger. How do they respond and handle it?
- Use books, newspapers, news reports, television, music, and movies to stimulate group discussion.

Activity: Ask group members to think of a newsworthy story related to anger and draw a picture about it that might appear on the cover of a magazine; include a headline.

Week 4

- Discuss manifestations of anger. What are physiological changes that occur in the body when one is angry? What are some automatic thoughts and/or behaviors?

Activity: Encourage members to role play an angry encounter with another member and then document the physiological changes they noted as they became angry. They are also to note the automatic thoughts and/or behaviors and discuss them with their partner.

Week 5

- Assist members to identify feelings behind anger. Anger may only be a smokescreen hiding other things that may be going on (a secondary emotion).

Activity: Ask members to describe five things that make them angry and then identify any other feeling that is evoked.

Week 6

- Ask members to identify what triggers their anger. Ask them to reflect on the things people do, or say, or the things the member does, or does not do, that make each of them angry. What is each one's level of tolerance?

Activity: Ask members to identify their own triggers as well as the ones they know how to use in significant others.

Week 7

- Provide members with specific anger management skills such as "Stop and think techniques," conflict resolution, mapping alternatives, relaxation, and others.

Activity: Give a cartoon handout describing inappropriate ways of handling a situation and ask members to provide an alternative behavior using the skills they have learned.

Week 8

- Role play exercises of actual problem situations and how members would normally handle them.

Activity: Give members a problem sheet and have each one randomly select a problem and demonstrate how the situation would play out.

Week 9

- Continue to work on exploring alternative methods of expressing and coping with anger. Draw on some situations from the previous week.

Activity: Have the group generate multiple solutions to the problems presented and the benefits of using the techniques they have learned rather than their old habits.

Week 10

- Discuss the difference between an argument and a debate.

Activity: Choose a topic to debate, have members ask questions, and elect a moderator to help manage conflict.

Week 11

- Explore and review constructive methods of channeling anger.
- Begin the termination process.

Activity: Group members are encouraged to come up with a chart of constructive ways to deal with their own anger.

Week 12

- Review the group process.
- Share feelings of being in the group and discuss what has been learned.
- Bring closure.

Activity: Closing group activity.

An Adult Group: Women's Incest Survivors Support Group

Overall Objectives: The group therapy, the treatment of choice for most victims of child sexual abuse/incest, will offer members identification with others, which often helps to decrease feelings of being "different" or isolated; immediate relief at the realization that they are not alone; help in breaking the cycle of keeping the secret; safety to explore their feelings related to the incestuous experience; and much needed support, validation, and feedback as well as challenges about their distorted beliefs regarding the sexual abuse.

Week 1

- Introduce members.
- Discuss the purpose of the group.
- Discuss ground rules, including confidentiality.
- Make a contract for commitment.
- Begin development of trust and rapport.
- Allow catharsis/verbalization of pent-up emotions.

Week 2

- Focus on the responsibility of the sexually inappropriate behavior: it was the adult's responsibility, not the victim's. "It wasn't your fault" is to be emphasized, with efforts to begin the empowerment process.
- Use guideline questions regarding the incestuous experience to elicit a beginning discussion about each member's individual experience.
- Begin identification with one another and the similarities members share as victims of childhood sexual abuse.

Week 3

- Begin the process of educating members about the inappropriate touching they may have experienced. (Redirect any inappropriate behavior in session.)
- Discuss appropriate versus inappropriate touching as well as inappropriate sexual acts they may have been exposed to and/or forced to participate in.

Week 4

- Begin to point out how being molested could lead to a variety of feelings, including guilt, anger, sadness, shame, confusion, betrayal, and others.

Week 5

- Focus on feelings about the perpetrator(s). Members will be encouraged to share any feelings, positive or negative, and receive validation. In many instances, they may experience contrasting feelings such as anger and love, which are to be validated and normalized.
- Ask that they discuss how they feel the perpetrators should be treated (reprimanded, punished, disciplined) for the abuse they inflicted. At this point, it may be beneficial to discuss the court system and to stress that a victim's view is not always the same as the position of the court. Stress that this difference does not negate the reality that they sustained the abuse and that it was indeed wrong and inappropriate.

Week 6

- Assist members to talk about the molestation and its impact in their daily lives (How has being molested affected you? What has changed since you were molested? etc.).
- Explore various areas in which being molested has affected them, such as at school, home, with peers, in relation to their self-esteem, trust issues, and so on.

Week 7

- Continue to focus on how abuse can impact various areas of a person's life.
- Encourage members to explore the power they have over changing certain things in their lives. (e.g., doing better at home and work, being less aggressive or passive, asserting themselves more, not blaming themselves by recognizing that they are not to blame for the abuse).

Week 8

- Help members explore how they perceived their caretaker's responses to the disclosure of the abuse.
- Do not focus on blaming the caretaker but on how the "child within" would have wanted the response to be. This helps members to learn how to ask (in appropriate ways) for their needs to be met by significant others who are able to provide emotional support and be nonjudgmental.

Week 9

- Discuss bodily reactions members may have had during sexual acts. The actual bodily feelings may have been pleasurable and therefore confusing. This needs to be normalized. Some emotional features may have been pleasurable as well (e.g., the child may have felt special). This needs to be validated and normalized as well.

Week 10

- Work with women on exploring who they can tell about their abuse and receive healthy reactions from. Explain that people who reacted in ways that didn't make them feel good aren't bad people, but they may not know how to respond due to their own uncomfortable feelings about the subject.
- In an age-appropriate way, educate members about the stigma attached to incest and encourage them to explore ways to reach out that are most likely to be positively reciprocated.
- Remind members that this is the 10th session with only two more before group ends.

Week 11

- Work with members on prevention issues. This will help to empower them and help them gain a sense of control.
- Assist members to explore individual ways to feel safe and prevent further traumatization.
- Continue to work on the termination process by beginning to review the group sessions and to encourage expression of feelings.
- Have group identify ways to empower themselves and to attempt to prevent any further exposure to abuse. Write these on poster board. Ask each member to make her own list on a 3 × 5 card she can take with her as a reminder.

Week 12

- (Last session) Work on anticipatory planning and review of group.
- Encourage expression of feelings regarding termination.
- Encourage maintaining healthy support systems.
- Engage in a graduation celebration during which members complete a "Personal Poster" to keep as a reminder of their group experience. Each member writes her name in the center of a blank note card you provide. The cards are passed around and each member writes or draws a supportive comment/message. Each member

then has a completed card that she will place in an envelope and take with her as a symbolic token of the supportive group process all have experienced.

An Adult Group: A Men's Issues Group

Overall Objectives: The purpose of the group is to provide men an opportunity to work together on common issues such as depression, anxiety, stress, relationships, and unresolved childhood issues. Additional areas of focus include marital and relationship difficulties, parenting concerns, work-related issues, loneliness, isolation, and other issues specific to their male role in society. The underlying belief is that men can profit by talking about their life struggles in a safe and supportive group environment.

Week 1: The Initial Session

- General introductions are made.
- Basic ground rules covering confidentiality, attendance, and basic agency policies are discussed.

Week 2: What It Means to Be a Man

- Gender-related issues are explored, such as what it means to be a man and the messages that men received growing up.
- Men are encouraged to explore how these messages affect them today.

Week 3: What It Means to Be a Man (continued)

- A number of norms pertaining to being a man are discussed.
- Men are asked to explore how these norms influence their daily behavior and how they are related to their current problems.

Week 4: What It Means to Be a Man (continued)

- Men are invited to share their reactions to breaking these norms.
- Individual goals are set that focus on increasing self-esteem, encouraging recognition and expression of feelings, and changing maladaptive thoughts.

Week 5: Relationships with Parents

- Men examine relationships with their parents, particularly with their fathers, as this seems to be a central influence in the lives of most men. Many of the men are disappointed by their father's absence or are angry about the excessive force and aggression that their fathers used to discipline them.
- Men are encouraged to write letters to their fathers expressing their thoughts and feelings.

Week 6: Relationships with Parents (continued)

- Men are asked to read letters they have written to their fathers. This session can be cathartic when unresolved emotions are shared.
- Men explore how they can best apply what they have gained from this letter-writing experience.

Week 7: Relationships with Parents (continued)

- In this session, relationships with mothers are examined.
- Special emphasis is paid to how men dealt with their feelings in their relationship with their mother, and how they separated, both physically and emotionally.

Week 8: Developing Self-Care Skills

- Men explore the importance of self-affirmation and emotional supports.
- The importance of sound nutrition and regular exercise is discussed.
- The need and appropriate use of medication is addressed.

Week 9: Relationships with Significant Others

- Men explore relationships with significant others.
- Some men may be going through a divorce or having marital difficulties, which they openly discuss.
- Others may be gay and may benefit from sharing their relationship struggles or discussing their sexual orientation. (It is important to use language that is free of bias with respect to sexual orientation. Regardless of one's sexual orientation, the men in our groups typically discover they have a great deal in common.)

Week 10: Relationships with Significant Others (continued)

- Because men often experience difficulty in simply recognizing, let alone expressing, their feelings, they are assisted in learning to verbalize feelings in appropriate ways.
- Effective communication skills and assertiveness are discussed. The differences between passive, aggressive, and assertive behaviors are explored.
- Men are encouraged to explore the consequences of these behaviors and to engage in appropriate role plays.

Week 11: Developing and Maintaining Friendships

- Men are encouraged to discuss friendships with other men and women. Because men often receive conflicting messages about appropriate gender roles, it is helpful for them to discuss their struggles in the group.
- The group experience provides men with a rich source of support. Many in the group have difficulty with initiating friendships and maintaining intimacy and are sometimes isolated from other men.
- Men are encouraged to get together outside the group and to develop other sources of support, especially with other men. However, they are cautioned about the drawbacks of subgrouping in a way that could detract from group cohesion.

Week 12: Relationships with Children

- Relationships with children are examined; many of the men enter the group with concerns about being an effective parent.
- The importance of men's relationships with their children is explored.
- Men are taught assertiveness skills to assist them in setting limits and following through on consequences with their children.
- Men often write letters to their children, which they read in group, to help them express and explore unrecognized feelings.

Week 13: Relationship to Work

- Men discuss what role work plays in their lives.
- Members who have lost a job often struggle with feeling devastated over not working and are encouraged to share these feelings.
- Members talk about how work affects their lives at home, especially with their partners and children.
- Men are invited to share their work-related struggles. Men often experience difficulties in working too much, poor limit setting, and frustration on the job. Some report feeling overly responsible and stressed by assuming the role of provider.
- Men are assisted to recognize the importance of a balance between work and play.

Week 14: Sexuality

- Most men have a host of concerns regarding sexuality, so this session is devoted to allowing them the opportunity to explore and deal with these feelings in depth.
- Because men may be reluctant to discuss such concerns openly, small handouts are used to spur the men's thinking on particular topics, such as sexual performance, feelings of attraction to other people, impotence, differences in sexual appetites, and aging.
- Due to the great commonalities among the members, men can explore fears regarding their sexuality and sexual practices in ways that allow them to gain a deeper understanding of their sexuality.

Week 15: Preparing for Termination

- Men are encouraged to begin discussing termination issues.
- Review of group process begins and men are invited to share their feelings about the group experience.
- Issues of loss and separation are broached.

Week 16: Closing of the Group

- The group ends with good-byes and reminiscing among members and leaders. This is a time when valuable feedback is given and the members evaluate their experience. Gains that the members have made are reinforced; tangible items may be given to take with them to remind them of their participation in the group. For example, certificates of attendance with the statement "It is okay to be a man" may be presented to them at closing.
- Men are encouraged to continue in their new behaviors. Potential pitfalls in applying what they learned in the group to everyday living are also addressed.
- A pizza party wraps things up.

With only a select number of groups profiled, we hope you were able to get a glimpse of how effective and useful group therapy can be. We encourage you to consider incorporating group treatment into your repertoire of skills, if you have not; we are pretty certain you will find it both challenging and rewarding in your practice. Below we examine some specific considerations for conducting group work in your agency that we hope can guide you in this direction.

Specific Considerations for Conducting Group Work in Agency Settings

Being able to provide group treatment services adequately is becoming a necessity for human service practitioners. In addition to developing clinical skills and engaging in specialized training for group process and practice, human service workers are most successful if they learn to work within various agency systems. Success is more likely if they understand the agency structure, policies, history, formal and informal organizational lines, client demographics, and existing needs. By being able to maximize the use of available resources strategically and to the fullest, practitioners can realize the optimal goal of providing effective, quality group treatment. Below we present some ideas in question form that might help you determine which groups are the best to pursue in your agency.

Role of Agency Setting

What type of agency? How might the type of agency where the group will be conducted impact group development, implementation, and treatment success? The following are three main types of agencies commonly found within the human services arena: (1) non profit (voluntary); (2) public (governmental); and (3) for-profit (commercial /private).

Determining the Need for Specialized Groups

Why group treatment? It is essential to determine whether the decision to conduct a group is (1) agency directed, (2) staff/clinician directed, (3) client directed, or (4) a combination of any of these. Why would this make a difference? What implications might this have on the development and delivery of group services?

What Type of Group? Groups are extremely diverse and are often developed for very different reasons. Determining what type of group is being considered can be extremely helpful in planning and implementation. How might treatment interventions and outcomes differ depending on the type of group?

The Importance of Planning and Enlisting Agency Support

Are there available resources and support? What are the available resources? How is the establishment of this type of group supported within the agency involved? Below are several areas to consider that will prepare you for the potential problems often encountered in agency settings:

1. *History of group treatment in agency.* Have groups existed before? If so, what was the outcome? Were they considered effective? What was the reputation of both the group and group leaders within the agency, among the clients, and in the community at large? If no group treatment has ever been provided within the agency, what are the prevailing views? Are there fears or stereotypes? Are there any conflicts?

2. Is there any agency pressure to develop or continue with a certain group? Are there economic concerns and/or access issues? Any personal issues?

3. Does staff "resistance" exist? If so, is the resistance from line staff or from administrative personnel? Various reasons for this resistance may exist; some of these include:

- Fear of being more vulnerable than when engaging in one-to-one work
- Fear of incompetence, as designated leaders may not feel adequately trained or prepared to lead groups
- Discomfort with group modality, preference for other types of treatment interventions
- Lack of motivation or support to do group work
- Lack of faith in group process or in the appropriateness for use of group modality to address client needs
- Adjustment difficulties as there is a need to adapt to the different qualities of group process and the changing relationship with clients
- Anger at system for imposing group on staff and/or clients

We hope that together with knowledge, skill, and a desire to engage in group treatment, an awareness of how to work within agency settings will help you engage successfully in group work.

CONCLUDING EXERCISES

1. Draw an eco-map of your own family of origin at some critical point of your development as a child. Use the common symbols to represent relationships and people within your family. As an addition to this eco-map, write one "family rule" that applied to every relationship between that individual and you within the family. After completing this

eco-map, write out any reactions in your feelings and thinking as you made the diagram. What did you learn from this exercise that could be applied in working with others?

_____

_____

_____

_____

_____

_____

_____

2. Now do a second eco-map of your current family. Include any and all significant relationships in your eco-map. Use the common symbols again to delineate each relationship. Now write one "family rule" that you see applying in this current situation. Do you see major differences or more similarities between the rules in the two eco-maps?

_____

_____

_____

_____

_____

_____

_____

3. Using the differentiation-of-self scale, assign numbers to yourself and those in your immediate family based on how you see yourself and them today. Do you find that the members of your family are similar or widely different on these scales? Do you see how these differences could, or possibly actually do, create conflict in the family system? What conflicts do you experience in your own relationships that you could conceptualize as having to do with issues of individuation and/or separation?

_____

_____

_____

_____

_____

_____

_____

5

Macro Level Practice with Organizations and Communities

AIM OF CHAPTER

In this chapter, we discuss working with organizations and communities, we call this the *macro* level of practice. We consider what is necessary to effect change throughout the broad societal spectrum. Theories and models of practice for working with organizations, communities, and social policy are discussed along with current practice trends. Knowledge and skills for effective macro practice are presented as well as an overview of macro-practice roles. Diversity and ethical issues are also explored.

Typically, social workers are at the forefront of macro-practice endeavors; nonetheless, others in the human services field may also be active and instrumental at this level. We believe all human service practitioners can benefit from learning macro-practice skills that will allow them to work not only directly with clients but also indirectly with larger systems that are likely to impact their clients. In our presentation, we consider macro practitioners to be all professionals within the helping professions.

WORKING WITH ORGANIZATIONS AND COMMUNITIES

Working at both the *micro* level with individuals and the *mezzo* level with small groups is fundamental to human services. However, these may often require work at a third level, known as *macro* practice, that involves intervention at organizational, community, and social policy levels. Macro practice goes beyond working with individual clients or small treatment groups and emphasizes efforts on behalf of whole groups or populations of clients. It may also entail examining and possibly challenging major social issues and global organizational policies, thus pushing for change at various levels: organizational, community, and societal/political.

Macro Practice Defined

Numerous definitions of macro practice can be found in the literature. We offer several viewpoints that illustrate the variability among practitioners in their perceptions of macro practice. We begin with Brueggemann (1996). In his view, macro practitioners

> make a difference in people's lives where oppression, intolerance, and insensitivity exist. They try to correct social conditions that cause human suffering and misery. They struggle to get to the

root of the problem by calling attention to injustice, finding out where human needs exist, and working to improve communities and organizational systems. They work to develop new programs and changes in policies. (p. 3)

This view is especially optimistic in encouraging work with vulnerable and at-risk populations. It speaks to the importance of macro practice for those who are most susceptible to the forces of prejudice, discrimination, oppression, and exploitation. Because these are often the clients we see most in human services, there is a need to help them beyond the micro and mezzo levels.

Meenaghan (1987, p. 83), in the *Encyclopedia of Social Work*, identifies four major areas of macro practice: planning, administration, evaluation, and community organizing. Barker's (1987, p. 92) definition describes macro practice as "social work practice geared toward bringing about improvements and changes in general society." Kirst-Ashman and Hull (1993) add four major dimensions they believe must be considered in any definition of macro practice.

The first dimension targets three tasks:

- Changing or improving policies or procedures that regulate distribution of resources to clients
- Developing new resources when clients' needs cannot be met with available ones
- Helping clients obtain their due rights. Changes need to be made in the system so clients' rights can be exercised

The second dimension targets "the system" to determine where and how changes need to be made. *The system* refers to many types of systems, such as social service systems or the legal and political systems. A third dimension centers on advocacy, which involves intervening on behalf of clients to assure that their needs are properly met. This final dimension stresses the importance of macro practice from an organization perspective, with focus on the agencies and organizations where a major part of macro practice actually takes place (p. 117).

Netting, Kettner, and McMurtry (1993, p. 3) highlight the intervention aspect of macro practice. They write that macro practice is a

professionally directed intervention designed to bring about planned change in organizations and communities. Macro practice...is built on theoretical foundations, proceeds within the framework of a practice model, and operates within the boundaries of professional values and ethics. Macro-level activities engage the practitioner in organizational community, and policy arenas.

These definitions make clear the basic tenet of macro practice: Social problems are solved by enacting social change. The various viewpoints also demonstrate the broad spectrum of macro practice.

The Scope of Macro Practice

Macro practice occurs in three main arenas: organizations, communities, and social policy. To help you visualize the depth and extent of working in each of these spheres, we have created Table 5.1 to illustrate specific activities of each of these categories.

This table illustrates some of the many possibilities for enacting change through established macro-practice channels. Rothman, Erlich, and Tropman (1995, p. 9) write that "a major contribution of the social sciences to the practice of community intervention has been in understanding the context in which change agents work—often called arenas of community practice. These are the social settings and systems in which practice takes place."

They add interpersonal relations and neighborhoods to the list of important areas for macro practice. Rothman, et al. also emphasize traditionally recognized organizational and community practice arenas and highlight the instrumental role of small groups. Because

TABLE 5.1   Classification of Macro Practice Activities

Organizational Activities	Community Activities	Societal (Policy-Related) Activities
• supervising professional staff • working with communities • participating in budgeting • writing proposals • writing grants • developing programs	• negotiating and bargaining with diverse groups • encouraging consumer participation in decision making • establishing and carrying out interagency agreements • conducting needs assessments • advocating for client needs	• coalition building • lobbying • testifying • tracking legislative developments that directly affect clients • carrying out other efforts designed to affect legal or regulatory frameworks

SOURCE: Based on Brueggeman, 1996, pp 4–5.

everyone involved in community practice works with one form of task group or another, it is natural to consider these groups viable areas to work in when employing problem solving or engaging in the change process. Examples of such task groups are grass-roots block clubs, agency boards, municipal communities and councils, and staff groups.

Another perspective from which to envision macro practice is the systems impact model (SIM). As described by Zastrow and Kirst-Ashman (1994), this model proposes that, in addition to working with individual clients to solve their problems, the practitioner also works with many other major systems that the client interacts with. SIM stresses the need for practitioners "to work within the institutional and organizational structure on their client's behalf. Often the target must be to change or improve how services are delivered and resources are distributed. Targets of change may also involve improving conditions and services within a community" (p. 25).

A visual representation of the systems impact model shows how the different systems (client, practitioner, organization, community, and institution) interact with one another to achieve their objectives (see Figure 5.1).

We encourage you, whether working in a clinical capacity with individuals, families, or groups, or at the community level with larger systems, to consider an ecological perspective in your work. By routinely examining the many systems that affect clients, you can gain insight into their particular situations and determine more accurately the viability of macro-practice interventions. Such an assessment will help you decide whether to continue pursuing work with these various systems yourself or to ask for help from someone who specializes in this area of practice.

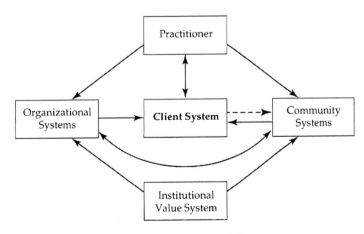

FIGURE 5.1   The Systems Impact Model

A Historical Focus

A Response to Social Problems

In general, macro practice emerged as a way to help alleviate the social problems that plague many in our society. Brueggemann (1996, p. 3) defines these problems as "those conditions of society that create personal troubles and are often embedded in the institutions on which our society is based. Among these social problems are racism, sexism, urban decay, economic injustice, and dysfunctional political systems."

The history of social problems can be traced back to America's early settlers and their ethnic intolerance. English settlers in Jamestown, Virginia, as early as 1716 had slaves who were victimized and blamed for any negative conditions present at the time. During the 1600s, 1700s, and 1800s, Native Americans were also abused and exploited. Not only was their land taken, but many also were forced to move to less desirable locations, some were even exterminated. The "Trail of Tears" symbolizes this eradication process. This is the name given to the 300-mile forced march of 15,000 Cherokees from their ancestral home to Arkansas and Oklahoma—a trek that claimed the lives of over 4,000 of these Native Americans.

From the late 1700s to the mid-1800s, other immigrants became targets of discrimination and oppression. French, German, and Irish-Americans all were victims of "pathological xenophobia" that led to the passage of the Alien, Sedition, and Naturalization Acts. Similarly, as early as the 1500s and especially after the Civil War in the mid-1800s, Mexican-Americans were also exploited; the Treaty of Guadalupe Hidalgo that guaranteed them citizenship was rescinded and settlers from the East took over their lands in the West.

Laws allowed mistreatment of African-Americans to continue well into the 1900s. During this time they were denied participation in the mainstream of American life and opportunity and had to suffer segregation and racial discrimination. The Chinese, who were recruited largely to build American railroads in the 1800s, were also mistreated—victims of violence spurred by prejudice and unfounded fears and suspicions about them. The Japanese suffered a similar plight, coming to this country to work and being exploited and condemned for being different. Their struggles were especially difficult during World War II after the bombing of Pearl Harbor, when the U.S. government forced them into relocation camps.

Many others have suffered exploitation and oppression. These include Jewish-Americans, other European immigrants, immigrants from Latin and South America, and most recently, newcomers from Vietnam, Cambodia, the Middle East, and other countries around the world. Although, we are nearing the 21st century, ethnic intolerance continues to permeate our society. No matter how far we have come, there are always groups who are made scapegoats. Macro-practice efforts specially designed to assist these groups are imperative.

History of Community Organization

Offering another perspective, Garwin and Cox (1987) have traced the history of community organization in the United States since the Civil War. They found that between the end of the Civil War and the beginning of World War I, a number of social issues emerged that greatly impacted welfare policies. In response to these social problems, ideologies were developed and solutions were sought.

Garwin and Cox (1987, p. 67) identified some of these social conditions as "the rapid industrialization of the country, the urbanization of its population, problems growing out of immigration, and changes in oppressed populations after the Civil War." In response to some of the problems of industrialization, poverty, urbanization, race relations, and cultural conflicts that stemmed primarily from immigration, several key ideological conditions developed: social Darwinism, radical ideology, pragmatism, and liberalism. Also, community organizations appeared to help improve the social climate of the times.

Two categories of community organization resulted that can be distinguished as having (1) direct involvement and (2) indirect involvement. The first category addresses efforts carried on by individuals or institutions that relate to present-day social welfare activities. Examples are charity societies, settlement houses, and urban leagues. The second category includes activities conducted by organizations that are not directly affiliated with established community organizations, yet can still be valuable to community practitioners. Political, racial, and other action groups are prime examples.

Through time, the profession of social work itself developed as a response to the social problems of the poor, of racial minorities, of women, and of those suffering some type of disability. In the beginning—in the late 1800s and early 1900s—formal efforts were instituted with the purpose of creating a more humane society.

The period after World War I marked an important era in the development of community organization. This was a time governed by the principles of free enterprise. The country had a profound sense of optimism toward capitalism. Because the societal focus had changed, the social work profession changed also, altering its focus from curing social ills to working with the individual in that era of psychoanalysis, behaviorism, and humanistic psychology. Businessmen were instrumental in organizing efficiency-oriented community chests that let the cost of charitable work be more evenly distributed. There was an increase in the number of welfare professionals who were active in promoting agency development in the area of human service.

The Great Depression brought a change in philosophy and direction for community organization activists. The federal government became involved in welfare planning, and grass-roots activities increased. This shift is also reflected in the strong labor movement of the times. Other important developments in community organization occurred during the period surrounding World War II, as noted in the government's response to the demands of black servicemen and for equality. Training programs for practitioners in community organization evolved, leading to the beginning of professional literature in this area.

The period following saw great advances—first in the civil rights movement and second in the student activism and discontent with the Vietnam war. These events generated a great sense of anger and alienation and an increased professional interest in grass-roots organizing and planning with local citizens.

The next years brought changes in government's views about community organizing that was, in part, politically motivated. As a result, the government reversed its commitment to help certain disadvantaged populations, such as the poor and disabled. Nonetheless, groups such as ethnic minorities and women continued to organize at the neighborhood level.

Today, societal problems still prevail; government, especially at the federal level, has a major role in determining who, when, and how the vulnerable and at-risk populations will be helped. Because of the recent rise in prejudiced and discriminatory behaviors toward minorities, gays and lesbians, those with AIDS, and the poor on welfare, human service practitioners more than ever need to be active in macro practice endeavors. Without our assistance, those who need us most will not be able to overcome these difficult times alone.

Macro Practice from a Community Organization Perspective

Macro practice as we know it today has its roots in community organization. Appreciating the traditional methods of community organization is therefore a prerequisite to understanding the significance of community practice today. Rothman and Tropman (1987) identify social action, social planning, and locality development as the three important approaches to community change. These strategies have been widely recognized and accepted, given their broad applicability across various professional fields and academic disciplines. We briefly review each of these core modes of community intervention here and elaborate on them later in the chapter.

Social Action   Fundamental to social action is the need for advocacy for those most vulnerable in our society. A primary purpose of this approach is to make fundamental community changes that include the redistribution of power and resources and increased access to decision making for marginal groups. A main assumption guiding social action is "the existence of an aggrieved or disadvantaged segment of the population that needs to be organized in order to make demands on the larger community for increased resources or equal treatment" (Rothman, 1995, p. 32).

Areas where social action has been effective include working with AIDS victims, women's rights, gay and lesbian groups, consumer and environmental protection organizations, civil rights and black power groups, and victims' rights.

Social Planning   Social planning employs experts or consultants to work on behalf of disadvantaged clients and thus is data driven and analytic in nature; therefore, it is less likely to involve community participation or input in the problem-solving process. In this regard, it involves "a technical process of problem solving with regard to substantive social problems, such as delinquency, housing, and mental health" (Rothman & Tropman, 1987, p. 4). Also, because social planning is complex and intricate, it is not feasible for use by the general practitioner who lacks specialized training in this intervention mode.

Locality Development   In contrast to the social planning, locality development emphasizes the participation of the community in setting goals and deciding on appropriate interventions. Again, according to Rothman and Tropman (1987, p. 4) locality development accentuates "community change...pursued optimally through broad participation of a wide spectrum of people at the local community level."

Locality development is commonly labeled "community development" and defined as "a process designed to create conditions of economic and social progress for the whole community with its active participation and the fullest possible reliance on the community's initiative" (United Nations, 1955).

In essence, locality development stresses the "moral commonwealth" of the community as described by Selznick (1992). A special quality of this approach is that it seeks to foster unity and active participation of all those within the community. In addition, community development can involve institutional and policy means to strengthen communities, may recommend industrial expansion through economic development, and can have a national or international frame rather than an explicitly local one (Rothman, 1995, p. 29).

Current Focus

Today, the future path of macro practice is marked with conflict and confusion. The traditional methods of community organization discussed above no longer adequately address the pressing needs of present-day society. Local, national, and world politics have changed and resources have continued to decline, creating an even greater need for new approaches to community and organizational practice. As in years past, improvement and change are needed within systems and at the policy level. Also, those who remain oppressed continue to need advocates. "The basic concept of community is no less important now than it was years ago. It remains just as critical to focus on the benefits of large groups of people, their overall well-being, their dignity, and their right to choose" (Zastrow & Kirst-Ashman, 1994, p. 39).

Professional practice at the macro level is multifaceted and complex. To be effective, workers must be both knowledgeable and skilled in understanding and being able to work with organizations and communities. We discuss each of these major components next.

Knowledge Base

Just as a sound theoretical foundation is necessary for practitioners in individual, family, and group therapy, those working at the macro-practice level must have a solid knowledge base regarding organizations and communities and the elements necessary to affect social change. Although macro practitioners rarely focus on just one of these areas, we separate them here for clarity. We begin by providing information about organizations and models for organizational change; this will help you build your knowledge base.

Working at the Organizational Level

By the very nature of our work with people, we are and will always be involved in some capacity with organizations of one sort or another—perhaps directly by working in an agency setting or indirectly through our advocacy and involvement with other organizations on the client's behalf. Understanding the complexities of organizations generally is necessary to be able to work with them specifically. The more we know about the inner workings of the many organizations we will come into contact with, the better equipped we will be to work with them in a positive manner.

Organizations Defined

Even the definition of an organization can be complex. According to Brueggemann (1996, p. 214), "Organizations are artificial, intentionally contrived social tools for accomplishing the goals of their owners and getting work performed." Lauffer, in turn, defines organizations as "purposeful social units, that is, they are deliberately constructed to achieve certain goals or to perform tasks and conduct programs that might not be as effectively or efficiently performed by individuals or informal groups" (1984, p. 14).

More specific to human services, Brager and Holloway (1978) describe human service organizations as "the vast array of formal organizations that have as their stated purpose enhancement of the social, emotional, physical, and/or intellectual well being for some component of the population" (p. 2).

Because our focus is on organizations that provide social services, we must be clear about the definition of social services itself. We prefer the definition by Barker (1991, p. 221), which is comprehensive. He states that social services include "the activities of social workers and other professionals in promoting the health and well-being of people and in helping people to become more self-sufficient: preventing dependency; strengthening family relationships, and restoring individuals, family groups, or communities to successful social functioning."

Overall, organizations are composed of many different individuals who perform specified roles in an effort to provide needed services to certain populations in the community. Typically, the functioning of the agency is organized through the establishment of positions such as manager, line worker, support staff, and others. Ideally, agency staff work together and integrate services in a way that is supportive to all.

To function effectively, or as Woodside and McClam (1994, p. 172) put it, "to be a good bureaucrat," an agency worker should follow these two recommendations. First, practitioners should devote time and energy to studying the structure of the system. They should be aware of barriers, effective procedures, and regulations. They can learn by reading available information about the organization, studying organization charts, and identifying the people in the system who are resourceful and open to sharing their successes. Second, helpers need to be creative and learn to work within the system rather than being paralyzed by it. If you feel a sense of belonging and maintain a belief in your ability to influence the system, you can remain active and make a positive impact. These suggestions advocate your taking a positive approach to understanding and working with organizations. We hope you can focus on the beneficial, constructive, and functional aspects of a bureaucratic organization.

Characteristics and Functions of Human Service Organizations

Understanding the functions of an organization can help you determine which referrals to make for clients who need a particular service and to determine the scope of services available in your organization and others. This information can be useful in assessing whether the agency is fulfilling its mission and can also help you develop a deeper appreciation for the complex nature of organizational systems.

Miles (1975) has identified the following as functions of an organization:

- Direction and leadership
- Organizational structure and job design
- Selection, training, appraisal, and development
- Communication and control
- Motivation and reward systems

Schmolling, Youkeles, and Burger (1993) identify the following as characteristics of a human service organization: a separate identity; its own set of advocates; established purpose and goals; subsystems that often have different purposes; growth or decline; and crises. Recognizing these traits can help you understand the intricacies of organizational functioning.

Organizational Classifications

Society today needs organizations that can address unmet needs or prevalent societal problems. In response, many human service organizations have evolved. Although not perfect or even responsive to all the issues, they do make valuable contributions.

Hansenfeld (1983) views human service organizations as bureaucracies that differ from others in that their raw material consists of people. He classifies agencies in two ways: by the types of clients they serve, and by the procedures and techniques they use to bring about changes in their clients (transformational technologies). He distinguishes among three types of organizations by their use of people-processing, people-sustaining, and people-changing technologies.

Netting, Kettner, and McMurtry (1993) provide a different perspective for categorizing human service organizations, distinguishing among nonprofit, public, and for-profit. These represent the way we perceive the social agencies in which we have worked. Most states devote more than half of their service dollars to private, nonprofit, and for-profit organizations. Examples are managed care organizations, health maintenance organizations, private corporations, and service agencies.

Theoretical Base

Knowledge of the theoretical base from which organizations originated is also important in understanding their current functioning and in developing strategies for working with them successfully. According to Kirst-Ashman and Hull, (1993, p. 121), organizational theory "is concerned with how organizations function, what improves or impairs the ability of an organization to accomplish its mission, and what motivates people to work toward organizational goals." Just as the field of human services has evolved through time, so have organizations. Brueggemann (1996, p. 216) accents this point: "The organization has been shaped by historical conditions and ideologies and has gone through the process of evolution. Not only has our understanding of organizations changed, but also organizations have shaped our views of ourselves and our social world."

When we examine the various theories that have emerged, the many transitions made in organizational theory through the years become evident. There have been many organizational theories proposed to date. Some are classic, such as Weber's work on bureaucracies and the human relations management theory proposed by Mayo. Others that developed with the changing times include the decision making and contingency theories by Bruns, Stalker, Morse, Lorsch, and Thompson. Contemporary issues that are more relevant to present-day organizational focus include: power and politics, organizational culture, and managing diversity.

Some of these approaches have focused on management or leadership style; others have dealt with structural issues as organizational hierarchy, planning, staffing patterns, budgeting, policies, and procedures. Some theorists, such as Netting et al. (1993), make a distinction between *descriptive* and *prescriptive* schools of thought. For instance, a descriptive theory is one that is "intended to provide a means of analyzing organizations in terms of certain characteristics or procedure" (p. 123). Theories such as this often reflect a sociological approach to organizations. Prescriptive theories, on the other hand, are "designed specifically as 'how to' guides, and their goal is to build better organizations" (p. 124). Theories on management and leadership fit this category.

Another feature often considered in the classification of organizational theories is whether the system is open or closed. According to Netting et al. (1993, pp. 124–125), "Open-systems perspectives are concerned with how organizations are influenced by interactions with their environments, while closed-system approaches are more concerned with internal structures and processes." Regardless of how you perceive organizations or which theoretical base you support, it is important to credit all existing views and theories with having a role in molding our current-day perceptions about organizations. In this respect, they all deserve merit. Rather than examining all these theories individually, however, we highlight the four schools of organizational theory and give a brief synopsis of a current-day model.

Brueggeman (1996) see these four schools of organizational theory as (1) the classical bureaucratic school, (2) the human relations school, (3) the decision-making school, and (4) the contingency school. The divergence in theoretical viewpoints is well represented in these four major schools of thought and reflects their evolution. The contemporary model we review is managing diversity, given its relevance to today's diverse world.

The Classical Bureaucratic School   Development of the classical bureaucratic school of thought can be traced to the mid-1850s; it was influenced greatly by such precursors of modern-day technologies as the steam engine, the telegraph, manufacturing factories, business corporations, the organization of the economy, and the bureaucratization of government and political life (Brueggemann, 1996). These advances demanded a more structured approach to understanding and working within organizations. As organizations increased in size and became more diversified, the challenges of management also increased. Questions about how to administer a large organization that continued to be efficient and effective while ensuring that its integrity remained intact inspired the beginnings of organizational theory.

"Organizational structure refers to the way relationships are constituted among persons within an organization;...the coordination of organizational members' activities" when structured properly, enhances that of the others (Netting et al., 1993, p. 125). The main premise here is that when individuals in an organization work together in an orderly fashion they can accomplish more than the same number of individuals working independently. According to Holland and Petchers (1987) and Zastrow and Kirst-Ashman (1994), organizations that are specifically designed, are formal in structure, and have a consistent, rigid organizational network of employees are more likely to run well and achieve their goals than those that are not.

Specifically, a structured approach relates to task specialization, matching a person and position, and leadership. The manager or leader of an organization plays a major role in determining how well an organization functions; a sense of unity, cohesion, and morale all are key to the management philosophy and approach. In 1911, Frederick Wilson Taylor, who is considered one of the great organization theorists of our time, published his book

Principles of Scientific Management in which he described his ideas of making management "scientific."

The classic work of the German sociologist Max Weber is also an important contribution to organizational structure. He introduced the term *bureaucracy* as a theoretical tool to assist in explaining organizational structure; he then applied the label to a particular form of organization. Netting et al. (1993, p. 125) consider the term bureaucracy to represent

> an ideal type, meaning that it is a pure conceptual construct, and it is unlikely that an organization fits perfectly with all the characteristics of a bureaucracy. The bureaucracy typifies descriptive organization theories in that it provides a model against which organizations can be compared, after which they can be described in terms of the extent to which they fit this model.

From this perspective, the concepts inherent in classical scientific management theories are exemplified in traditional bureaucracies. In the field of organizational theory, these two developments marked a new direction that has continued to evolve to its current place. Unfortunately, present-day views of bureaucracies are often negative and fail to recognize the value of this way of conceptualizing organizations. For many, the term *bureaucracy* often conjures up the image of a big, cold, impersonal, uncaring, rule-driven institution. This is lamentable. We have found that bureaucracies in general and human service agencies in particular, possess both strengths and limitations. In the literature, there is agreement that bureaucracies are much misunderstood and unfairly vilified. According to Fairchild (1964, p. 20), bureaucracies are typically described in a negative light and criticized for their devotion to routine, inflexible rules, red tape, procrastination, and reluctance to assume responsibility or to experiment. Typical problems faced by beginning professionals working in bureaucracies include tension between the workers and the administration, the impersonality of the bureaucratic structure, and the bureaucracy's resistance to change (Neugeboren, 1985; Woodside and McClam, 1994). In response to the criticism surrounding the notion of a bureaucratic system, many (Hansenfeld, 1983; Johnson, 1986; Neugeboren, 1985; Woodside & McClam, 1994; Schmolling, Youkeles, & Burger, 1993) have written about the positive aspects of these organizations. To engender a more positive, holistic outlook on bureaucracies and the services they provide, we must be ever vigilant of our own negativity and instead strive to appreciate bureaucracies for their many strengths. The following statement by Hansenfeld (1983, p. 1) illustrates the importance of organizations in our lives:

> The hallmark of modern society, particularly of its advanced states, is the pervasiveness of bureaucratic organizations explicitly designed to manage and promote the personal welfare of its citizens. Our entire life cycle, from birth to death, is mediated by formal organizations that define, shape, and alter our personal status and behavior.

The Human Relations School   Historically, events leading to the evolution of the *human relations school* can be traced back to the stock market crash and resultant Great Depression. After these events, the government, under the leadership of President Franklin D. Roosevelt and Harry Hopkins, successive head of many New Deal agencies, attempted to restore hope in government by developing a wide spectrum of agencies, bureaus, and social administration. During this time, the process of unionization also grew as a viable way for workers to feel protected from the wrath of big business. Most important, however, were the Hawthorne studies conducted by Mayo in the late 1920s. The earlier works of Taylor and Weber of the bureaucratic school had began to suffer criticism for their "rational, structured and rigid" approach to understanding organizations. In response to these, Elton Mayo and his assistant, F. J. Roethlisberger, conducted a series of experiments that considered the role of such environmental factors as intensity of lighting on workers' productivity. Surprisingly, results proved that social forces, not environmental forces, were most influential. "Workers appeared not to respond to the lighting but instead to the fact that they were members of a group to which they wanted to contribute their best effort, and it was this sense of social responsibility that prompted improved performance" (Netting et al., 1993, p. 130).

According to Brueggemann (1996, p. 219), it was this "secondary social system" that complemented the more formal, functional components of organizations. Specifically,

> this secondary system was the informal, natural social system formed out of the needs, interests, and feelings of the members of the organization. So powerful was this human element that it could disrupt or interfere with the effectiveness and efficiency of the organization. The result was the beginning of a healthy respect for the role that group cohesion, control, and conformity can play interfering with productivity.

From this realization, the human relations school emerged. In contrast to the classical bureaucratic school, it viewed social forces as paramount to an organization's functioning. Human relations proponents also considered organizations to be open systems that are adaptable to the world and capable of ongoing growth and evolution. They perceived organizations as "natural" rather than "artificial," and saw change and conflict as a part of any system. This led to the view that organizational politics were a reality and thus must be dealt with and not separated from administration, as had been done previously.

Sarri (1987, p. 31) introduced important concepts of the human relations school including: "employee morale and productivity... satisfaction, motivation, and leadership; and ...the dynamics of small-group behavior." This school of thought emphasized a cooperative work environment and group participation in decision making. Etzioni (1964) summarizes the basic tenets of the human relations approach as follows:

- The level of production is set by social norms, not by physiological capacities. (p. 34)
- Noneconomic rewards and sanctions significantly affect the behavior of the workers and largely limit the effect of economic incentive plans. (p. 34)
- Workers do not act or react as individuals but as members of groups. (p. 35)
- The role of leadership is important in understanding social forces in organizations, and this leadership may be either formal or informal.

Ultimately, researchers began to believe that "satisfied, happy employees will be the most productive" (Zastrow & Kirst-Ashman, 1994, p. 31). This focus on the human element was noteworthy but also open to criticism. Despite their efforts to "humanize the oppressive and impersonal organizational mechanism, they failed to recognize that organizations, at least in their formal sense, are not 'human' at all" (Brueggemann, 1996, p. 220). The notion that organizations possess such human attributes as behavior, rationality, decision making ability, and so on was deemed a theoretical error called *reification*. In effect, "Reification strips people of their humanity, destroys self-determination, and places systems beyond human control" (Brueggemann, 1996, p. 220). These factors eventually led to other theories that focused on the empowerment of employees. In time, the main principles of the human relations school gradually died. Nonetheless, as stated by Netting et al., (1993, p. 132), "the lessons of the Hawthorne studies—the importance of teamwork, cooperation, leadership and positive attention from management—should not be lost on executives and staff of human service organizations today." The emergence of a new school of thought with its structuralist models resulted—namely, the decision-making school.

The Decision-Making School   With the advent of World War II, the basic tenets of the *human-relations* school that emphasized "group process, motivation, inducing the worker to produce, cultivating communication, participative management, democracy in the workplace" were insufficient to cope with the extensive needs of the wartime economy and the efforts of winning the war. Ironically, the principles initially proposed by Weber in the *bureaucratic* model seemed more compatible with the needs of the time, be it "a hierarchy of command, a tight structure, or organizational control" (Brueggemann, 1996, p. 221). These circumstances led to the development of a new school of organizational theory called the decision-making model.

This new *decision-making model*, a neoclassical or neo-Weberian theory, was influenced primarily by theorist Herbert A. Simon, who earned a Nobel Prize for economics. His study of organizations was described in his text, *Administrative Behavior* (1957), that greatly changed how organizations were to be perceived. Simon questioned both the bureaucratic and human relations views of organizations; the classic bureaucratic views were too "rational," yet, those proposed by the human relations school were too "affect" oriented. He viewed his model to be a balanced one that placed organizations in a context of rational action and decision making that one should expect to see in real life (Simon, 1957).

Simon (1976) focused on the individuals of the organization; specifically, he was interested in individual decisions about organizational matters. The analogy is that an individual is much like a computer chip and the organization is like the computer—the individual chips are necessary to make the computer work. Influenced by behaviorist views, Simon "believed organizations can be conceptualized as aggregations of individuals' decisions within the organization, and organizational decision making can be viewed as a behavior that occurs in response to certain stimuli" (Netting et al., 1993, p. 137). Zastrow and Kirst-Ashman (1994, p. 31) emphasize the decision-making model's focus on "both the rational structure of an organization and the more irrational, imperfect behavior of the people involved in that structure." In this regard, the importance of careful decision making is pivotal in ensuring optimal worker productivity. According to Simon (1976, p. xxviii), "administrative theory is...the theory of intended and *bounded rationality*—of the behavior of human beings who *satisfice* because they have not the wits to *maximize*." *Bounded rationality* is a term devised by March and Simon in 1958; they contended that the main premise by which to understand organizational decisions is understanding the constraints that limit decision making. As described by Netting et al. (1993), this decision-making process "may therefore be thought of as a sort of risk-management process...the decision maker seeks to reduce the uncertainty as much as possible in order to make a decision that provides a reasonable likelihood of achieving an acceptable outcome," which is what *satisficing* is all about. In essence, as proposed by March and Simon, "understanding how satisfactory outcomes are sought though decisions made in a context of bounded rationality is the key to understanding organizations" (Netting et al., 1993, p. 138).

In sum, the objective of an organization from the decision-making school perspective is the "total integration and inclusion of individuals into its premises.... [The] organization no longer is a tool serving human ends; it is itself the end, and humans are the means by which the organization functions" (Brueggemann, 1996, p. 223).

The decision-making school, with all its merits, was also not immune to criticism. Continued work in this area led to one of the more recent schools of organizational theory—the *contingency* school.

The Contingency School   In the 1960s, with recovery from World War II well under way, the United States entered an era of expansion, growth, and development, especially in computer and space-age technology. Central to this progressive period were organizations that were being viewed quite differently from organizations of earlier times. Organizational theorists began to consider a more systems or ecological approach in understanding organizational phenomena. For Holand and Petchers (1987, p. 207), a systems approach "construes the organization as a social system with interrelated parts, or subsystems, functioning in interaction and equilibrium with one another. It thinks of the organization as an adaptive whole rather than a structure that is solely rational-legal." This systems approach offered a more flexible and comprehensive way to view organizations. Because the organization is considered a living system that is adaptable to its environment, any dysfunction in one part of the organization will affect the organization as a whole. Because there is much variance in organizations, each can be said to be contingent on a number of factors; thereby, the name *contingency*. In this sense, different organizational styles may be completely appropriate because of the specific circumstances each organization faces. In essence, each organization "must fit and adapt to the conditions in which it finds itself and find a niche in the wider organizational environment" (Brueggemann, 1996, p. 224). Three basic principles of this contingency theory have been proposed by Galbraith (1973) and Scott (1981). The first two principles cited by Galbraith are these: "(1) There is no one best way to organize, and (2) Any way of organizing is not equally effective" (p. 2). In 1981, Scott added a third principle that embodied the open-systems perspective: "(3) The best way to organize depends on the nature of the environment to which the organization must relate" (p. 114). In all three, the central theme reflects the view that "the nature of the organization and its management scheme are contingent upon a variety of factors that are unique to that organization" (Netting et al., 1993, p. 141). Many have proposed different components or factors of an organization that should be considered. Given all these, the number of variations is potentially infinite. In sum, then, to understand the design of a particular organization, one will first have to understand how each component varies in relation to other components. Although admitting that this is a monumental and complex task, organizational theorists appear excited by its potential. As stated by Brueggemann (1996, p. 224), "Contingency-oriented organizational theorists using social ecological metaphors are developing the building blocks for 'anatomy' and 'philosophy' of organizational design."

The Contemporary Theory: Managing Diversity *Managing diversity* is a relatively new organizational model that approaches management of the workforce in the 1990s somewhat differently from other models. Developed by Roosevelt Thomas, Jr. (1991), diversity is identified as the key variable in affecting productivity. Thomas believes management of diverse populations is a critical skill for those working at the organizational level and that it is especially relevant for today's emphasis on a multicultural perspective. Thomas (1991) sees three trends mandating that diversity be dealt with in the workforce:

- American corporations are now involved in business at the global market level, which has become increasingly competitive.
- Diversity is becoming more common and greatly impacting the make up of the U.S. workforce, which is now dramatically different from the workforce of only several decades ago.
- The pluralistic emphasis present today that stresses individual differences and encourages people to take pride in who they are is replacing the outdated "melting pot" concept.

The need to become more diversity sensitive in the workplace is clear. Thomas encourages using the creative resources of the Asian-Americans, African-Americans, Latinos, and other cultural and racial groups that represent today's workforce. To this end, he proposed an analytical framework for understanding organizational cultures that separates diversity into three phases: (1) using affirmative action, (2) valuing differences, and (3) managing diversity.

The first, affirmative action, according to Netting, et al. (1993, p. 150)

> "refers to programs and efforts designed to bring ethnic minorities and women into the organization. The focus is on recruitment efforts, and success is measured by the numbers and percentages of minorities recruited and retained in the organization." (Netting, Kettner, & McMurtry, 1993, p. 150)

When Thomas (1991) developed his approach, he considered affirmative action as a temporary step in moving the organization toward managing diversity. Today, in jurisdictions where affirmative action is no longer legal, other measures will have to be taken to achieve similar results.

The second phase, valuing differences, emphasizes enhancing interpersonal relationships among individual workers. A main objective is to foster acceptance and mutual respect along racial and gender lines; this is achieved primarily through staff training and personnel development. Employees at all levels are encouraged to understand differences, understand their own feelings and attitudes toward differences, and to enhance working relationships among those who are different.

The last phase, managing diversity, is a loftier aim as it encourages a complete evaluation of the organization as a system to learn whether it is sufficiently effective and productive to survive the increasingly competitive environment. For systems that are not, Thomas (1991) recommends that they undertake a long-term strategy to modify their core culture. This venture is not simple as it requires a "full understanding of the existing culture and a planned transition to a new culture where an environment is created that supports full utilization of a diverse work force" (Netting, et al., 1993, p. 150). Moreover, it demands that those involved in the process keep an open mind and be receptive to efforts that enhance, not diminish, acceptance and sensitivity to diversity. Thomas's approach is relevant to today's world and is well worth considering.

Macro-Practice Activities at an Organizational Level

Macro-practice activities at the organizational level are numerous. Specific functions likely for a macro practitioner working in organizations may entail supervision of professional staff, working with communities, participating in budgeting, writing proposals, and developing programs (Netting et al., 1993). Four major areas of practice through which these functions are performed include social planning, program developing, administration, and organization development. We discuss each of these briefly.

Social Planning   Social planning has been described as the "development, expansion, and coordination of social services and social policies" using rational problem-solving methods that are carried out at both the local and societal levels (Lauffer, 1981). "Social planning is a means by which macro [practitioners] carry out their responsibility to provide for the welfare of society—to ensure that social programs, policies, and services meet people's needs to the greatest extent possible" (Brueggemann, 1996, p. 241).

Two types of social planners have been identified: The first is the *staff specialist*, who works in direct service agencies and engages in such practice areas as case management, group work, and clinical services. The function of those specialists is to work directly with the agency director, to examine needs, evaluate services, and engage in grant writing and research as needed. To ensure that clients' needs are being met and that the agency's resources are properly utilized, social planners typically make recommendations based on their assessment. They are also instrumental in assisting agencies to adapt their services to the changing populations that are so common today.

The second type is the *social planning specialist*, who works for an agency solely devoted to social planning. Such agencies are involved in needs assessment, service regulation based on the service jurisdiction, governmental grant reviews and recommendations, the development of new services, and quality control maintenance of mandated areas (Brueggemann, 1996). Social planners of this type must interact with a wide spectrum of agencies ranging from client organizations, other services providers, and ancillary service systems such as universities, governmental agencies, businesses, or community groups. The primary goal of these interactions is to work together to develop comprehensive plans for the joint service areas. These specialists must be able to work well with others in a coordinated fashion.

Social planners, then, must not only be open to examining social problems but must go a step further by finding solutions to these social ills. Because these solutions must be both rational and feasible, the role of a social planner is complex and involves gathering empirical facts, understanding organizational politics, and learning the art of compromise and negotiation.

Planning principles basic to social planning include these:

- Citizen participation, which is considered essential to any successful outcome. Social planning is one of the main ways in which people in the community can feel empowered to assume responsibility for the direction of their lives.

- Commitment to the democratic process—also crucial to the social planner who must ensure that allocation of available resources and services is equitable and fair.
- Mobilization of values, as the values behind macro practice generally and social planning specifically, are equality, justice, fairness, and creating a positive image of society.
- Last, a future-oriented perspective developed by creating a positive image of society that will enhance future efforts for helping people.

Equally as important as these planning principles is the planning process itself. Similar to the problem-solving process of direct helping, social planning is also composed of various steps that build on each other. An adapted version of the steps presented by Simons and Aigner (1985) is offered by Brueggeman (1996), outlined below:

Steps of the Social Planning Process

Step 1: Defining and conceptualizing the problem in the interest of the community at large

Step 2: Building a structure or network of relationships

Step 3: Mobilizing values in particular directions by formulating policy and laying out alternative strategies

Step 4: Deciding on an arena of action

Step 5: Implementing plans

Step 6: Monitoring feedback and evaluation

For best results, social planners need to be flexible and to tailor the plan to the particular situation. Additionally, they need a number of skills to be successful. These skills have been categorized into three areas: (1) technical skills, (2) people skills, and (3) political process skills.

Various approaches to social planning have also been identified. Gharajedaghi and Ackoff (1986, p. 27) listed the following:

- Reactive planning: to restore the past
- Inactive planning: to preserve the past
- Proactive planning: to accelerate the future
- Interactive planning: to create a just society for the future

In sum, social planners need many special skills and a sound commitment to helping those less fortunate through macro-practice methods. As advocates, they strive to equalize the distribution of services, resources, and opportunities. By improving the services that exist as well as developing new programs to fill existing gaps or to prevent problems, social planners foster hope and promise for the future.

Program Development   Program development may be part of social planning or it can be the primary function of program developer. As with social planning, program development is an important part of macro practice.

> The most critical element in any community organization activity is the emergence of some idea and design for a program.... The implementation of such a program, which in almost all instances requires the development of some organizational framework, is in the last analysis the true test of successful community organization, since the program provides in very concrete terms the outputs or services desired and needed by the community. (Hansenfeld, 1995, p. 427)

Certainly, Hansenfeld's view is shared by many who work in the macro-practice arena. As with social planning, program development is also a complex process that requires specialized skills and a strong will and determination to develop an idea and see it through. Furthermore, it may require a prolonged process of negotiation and planning. Not everyone who will be affected in some way by the new program may view it in a positive light. Some may feel they were excluded from the planning process where others may see the new program as a threat to their domain. In both instances, the program developer must be careful to not alienate actual or potential participants and learn to stay optimistic and grounded in the original reasons for the program. "While the planner-organizer may find it necessary to disagree with certain groups who oppose the program, he or she must have enough support and sufficient resources to withstand countervailing pressures" (Hansenfeld, 1995, p. 427).

Several areas of program development need to be taken into consideration. First, it is important to identify the need for a particular program; this is often accomplished through a needs assessment. Needs assessments can be approached in many ways, five of which are listed by Rubin and Babbie (1989, p. 501) as follows: (1) social indicators approach, (2) rates under treatment approach, (3) focused interview on key informant approach, (4) focus groups or community forum approach, and (5) the survey approach. In addition to these, needs can also be identified by communicating with key informants who are either authorities in the field or are knowledgeable about the needs of a particular problem area.

Along with a needs assessment, "every new program requires resources—in particular, money and manpower" (Hansenfeld, 1995, p. 428). There may be a need to work with an advisory group or an action board who could be key to developing a successful program.

Other logistics are also critical to program development:

- *Establishing the board of directors*, a step that comes after you determine what type of agency will be used—for profit or not-for-profit—and whether the organization or agency will be incorporated, meaning that it is officially listed and legally recognized as a corporation.
- *Establishing the agency* by which services are to be delivered. This requires defining the agency's mission, developing the agency structure, staffing the agency, recruiting clients, and obtaining funding. Potential funding sources are numerous: solicitations, membership dues, benefits, fee for service, private foundation grants, government grants, and contracting with other agencies or organizations.
- *Evaluating the program* cannot occur until the program has been successfully established. However, it is essential in ensuring that the program is maintained and that other programs like it can be developed. Types of program evaluation include needs assessment, process analysis, and outcome or impact analysis that examine whether specified goals were met and why or why not. Effectiveness criteria, measurements, and evaluation design are all necessary elements of program evaluation.

Although we have not even scratched the surface in terms of examining the intricacies of program development, we have demonstrated how complex the process can be. Consideration of many inter- and intra-organizational factors is required, and this helps to explain why many ideas for program development never leave the drawing board or fizzle out shortly after their conception. Program development is not an easy task; it is a task that we as human service providers must value and appreciate.

Macro Practice Administration

The success of both social planning and program development is largely dependent on competent administrators. Much of the success of any human service program depends on the quality of its administrative leadership. Administrators are defined here as those who are in charge of helping organizations function; they may be line supervisors, division or departmental managers, all the way up to the executive director. According to Skidmore (1990, p. 163), "administration is the process that has to do with running an agency, and involves goals, policies, staff, management, services and evaluation." Often, the community practitioner is also the administrator; "it is the role, not the person, that shifts" (Rothman et al., 1995, p. 383). Simons (1995, p. 163) observes that "social service administration differs in significant ways from general administration and hence requires a particular type of educational preparation."

The role of a social service administrator is clearly a challenging one that requires general and specialized skills in both administration and human services. Patti (1985) notes an important difference in that administrators of human service administrators assume a commitment to client welfare or service effectiveness rather than to profit or some other outcome. However, Rapp and Poertner (1985) maintain that the social administrator must also be concerned with efficiency, productivity, resource acquisition, and staff morale of the agency he or she is responsible for administering. It is indeed a complex role, which we explore later in the chapter.

Essential to the development of an effective administrator is knowledge and skill in the problem-solving process. Administrators must be adept in such problem-solving techniques as problem identification, goal setting, and plan implementation as well as successful closure (termination) and evaluation.

Tasks of an administrator are numerous. Specific functions of a social service administrator are identified in the preeminent work of one of the earlier management theorists, Luther Gulick (1937, pp. 3–37). Although dated, the particulars presented by Gulick are still applicable today. The functions he described are planning, organizing, staffing, directing, coordinating, reporting, and budgeting. Of these, planning, budgeting, and staffing are especially important in human services.

Planning is often an unnoticed function administrators perform, yet it is considered one of their major tasks. There are many facets to planning. For instance, planning helps provide structure to the agency by determining the activities that will be essential in the agency's future. Written plans, such as the agency's mission statement, are often helpful to employees and clients alike as they provide a sense of direction and purpose. Hersey and Blanchard (1988) highlight the planning function evident in setting goals and objectives for the agency and developing "work maps" that show how stated goals and objectives can be met. Planning also involves writing, altering, or changing internal policies, procedures, programs, and budgets.

Budgeting is a fundamental aspect of administration. Per Netting, et al. (1975) "Budgeting and budget management is often an activity left to upper administration...Fiscal soundness and budgeting practices affect programs and services in a very profound way" (p. 3) and therefore must be taken seriously by administrators. In mastering the budget process, administrators must learn the ways budgeting occurs as well as the different kinds of budgets. They must be able to plan for the moment as well as the future. Often, budgeting decisions create undue resentment and conflict; thus, good administrators will also be able to endure the difficult times resulting from the budget decisions they make.

Staffing or personnel issues are also extremely important to the overall functioning of an agency. Administrators who are adept at working with agency staff are not only better supported but are also more likely to be successful managers. More commonly known today as human resource managers, administrators of social service agencies must be knowledgeable of such principles as the merit concept, the political reward concept, the need concept, and the preference concept.

A major role of any administrator is the decision-making process. Decisions are made every day by administrators, so their skills in this area need to be well developed. Of the variety of techniques used to make decisions in organizations, two categories are often used: quantitative methods and those that are commonly used in meetings. Common quantitative methods involve decision analysis, queuing, decision trees, and benefit-cost analysis. Decision techniques typically used in meetings are brainstorming, reverse brainstorming, and nominal group technique (NGT).

Simons (1995, pp. 165–169) addresses another facet of administration that is extremely vital to the success of any administrator: principles for effective persuasion. These include emphasizing the advantages and rewards, being comprehensible, showing compatibility of values, citing proven results, allowing for trailability, linking messages to influence others, avoiding high-pressure tactics, and minimizing threats to security, status, or esteem. Some of these may seem logical; others may require additional attention.

In all, the role of administrator is rarely easy. As summed up by Simons (1995, p. 163), to be an administrator "requires the execution of tasks such as obtaining funds and clients, supervising and motivating personnel, juggling the conflictual demands of multiple constituents, [and] managing information on program performance." Administrators, although especially susceptible to criticism, must be recognized and credited with the complex role they play. To work effectively with them, we must appreciate how difficult and vital their role truly is.

Organization Development   Organizations are in a constant state of flux; they are subject to changes in the environment and at risk of becoming dysfunctional if not properly managed. According to Harvey and Brown (1992, p. ix, 5), organization development is "an emerging behavioral science discipline that provides a set of methodologies for systematically bringing about organizational change, improving the effectiveness of the organization and its members." A primary role for an organization developer is working with dysfunctional organization systems; both internal and external. An internal organization developer is an employee of an agency that provides direct services; he or she works exclusively with the employees, supervisors, and administrators within the agency. External organization developers, in contrast, are typically private consultants or members of an organization development firm that provides specialized consultation services on such issues as management, training, and problem solving.

There are pros and cons to each of these types of developers. For instance, internal developers understand the agency history and culture better, yet they must be careful not to alienate staff by the recommendations they make for change. Also, because they must continue to work within the agency, they are more subject to scrutiny from peers and superiors. External developers, on the other hand, are less apt to understand the subtleties and intricacies of the agency system they are working with. Nonetheless, they can be more objective and are less liable to be directly affected by the decisions they make on behalf of organizational change.

Understanding the many facets of change is a main function of an organization developer. "There are a number of factors that impel organizations toward change and factors that cause organizations to resist changes" (Brueggemann, 1996, p. 314). Effective developers must be cognizant of both types of forces if they hope to have a positive effect on the system targeted for improvement. Robbins (1992, p. 270) cites six forces within organizations or in their surrounding environment that are likely to spawn the need for change:

- The changing nature of the work force
- Technology
- Economic shocks
- Changing social trends
- The "new" world politics
- The changing nature of competition

Together with forces that engender change in organizations, there are forces that impede it. In some cases, the force may be a natural occurrence, as not all organizations are intended to change or will benefit from it. In other cases, however, resistance toward change can be potentially damaging. Factors that contribute to various forms of resistance include the psychology of the owner, the system's state, its size, its structure, the chain of command, the subordinate position of workers, and the organization culture. (Brueggemann, 1996; Harvey-Brown, 1992).

Similar to working with a client's resistance toward change, macro practitioners must first understand the underlying reasons for the resistance. Once these have been identified, practitioners can develop appropriate strategies to reduce the resistance. Two main strategies are (1) allowing change to occur naturally, or (2) attempting to control change in some manner.

As with social planning and program planning, organization development is also a process, one that must be carefully planned and applied at all levels within the organization (Beckhard, 1969, p. 9). Knowledge of behavioral sciences and organizational theory can help this process, and change must be managed from the administrative level with the primary goal of increasing organization effectiveness and health. Beckhard (1969) suggests also that change be channeled through planned interventions in the organization's processes in order to be effective.

There are many approaches to organizational development. Brueggemann (1996) has identified several that we outline briefly here.

- A *systems* approach: The change strategy could occur in any part of the system, thereby creating changes in other parts of the system.
- A *social ecology* approach: Organizations are considered open systems that are in constant interaction with and adapting to their environment. In this case, the organization developer assists those within the agency to predict likely changes and to be prepared to adjust to them when they do occur.
- An *individual levels* approach: The organization developer views the change process as existing at various levels. However, rather than focusing on all these, he or she chooses to concentrate on one particular level to engage in the change process.
- The *whole systems* approach: The organization is viewed as the sum of its component parts. The organization developer here focuses on one or more of these parts to effect change.
- The *contingency theory* approach: Here the developer perceives an organizational system as having numerous components that must work in a synchronized fashion for the system to function effectively. The organization developer's role is twofold: first, to determine where a breakdown exists within the organization; second, to explore ways to adjust the system so it can return to its previous working order.
- The *therapeutic* approach: The organization developer employs organization, human behavior, and group dynamic theories to diagnose organizational problems. Similar to a clinician, the developer identifies the problems and determines their severity. Then he or she develops an intervention plan to restore the system to its previous level of functioning or perhaps even a higher level. As in clinical practice, the evaluation process, both ongoing and on termination, is necessary to ensure treatment success.

In addition to these approaches, some organization developers have created their own integrated model that consolidates many of these strategies. It is ultimately up to the individual practitioner to determine which procedures best fit the agency's needs and his or her particular style of working with organizations. Some developers focus on the group level; others may decide to concentrate their attentions at the intergroup level.

Working at the group level requires specific techniques that include T-groups, integrated work teams, project management, quality circles, team development, and group conferences. Working at the intergroup level, on the other hand, requires skills in problem solving, boundary spanning, knowledge of the conflict cycle, conflict-resolution strategies, and mutual problem-solving techniques. In some cases, developers may choose to combine these two styles and work with the organization in its entirety. One such method is *management by objectives* (MBO), a process in which each unit of an organization develops long-range goals and short-term objectives (Robbins, 1992).

The role of an organization developer, as seen, requires special training, a sound knowledge base regarding organizations, and skills in the organizational change process. There is often little recognition or awareness of the difficulties of this role.

Working at the Community Level

Macro practice cannot be effective without efforts directed at the community level. Homan (1994, p. 23) considers a community a client to the community practitioner much as an individual is a client to the clinician. He states,

> Just like an individual or a family, a community has resources and limitations. Communities have established coping mechanisms to deal with problems. To promote change in a community, the community must believe in its own ability to change and must take responsibility for its actions and inactions.

Homan (1994), along with many other community practitioners (Brueggemann, 1996; Netting, Kettner, & McMurtry, 1993; Rothman, Erlick, & Tropman, 1995), recognizes many of today's problems as community problems such as gang violence, AIDS, and inadequate health care, to name just a few. It is unlikely that these or any other social problems will be diminished if efforts do not extend beyond the micro and mezzo levels of practice. Community problems must be addressed by the community that feels the problem, this requires the expertise of macro practitioners who are specially trained to work at the community level. We address some of the core issues of community work next.

Communities Defined

The definition of a community varies. Barker's (1987, p. 28) definition is very basic: "A community is a group of individuals having common interests or living in the same locality." A similar definition contends that communities are created when a group of people form a social unit based on common location, interest, identification, culture, and/or activities. Warren (1972) adds to both these views, seeing the community as "the combination of social units and social systems which perform major social functions having locality relevance" (p. 135).

Based on these definitions, Fellin (1995, p. 114) has classified communities into two categories: "those distinguished by common locality or place, and by interest and identification." Zastrow and Kirst-Ashman (1994, p. 36) identify two other variables. One relates to the possibility of interaction: "Due to the common locality, functions, or interests, people in the community interact on some level or at least have the potential to do so." The second variable accents the value of a community working together: "A community can be organized so that its citizens can work together and solve their mutual problems or improve their overall quality of life."

Bellah, Madsen, Sullivan, Swidler, and Tipton (1985) define community as a "A group of people who are socially interdependent, who participate together in discussion and decision making, and who share certain practices that both define the community and are nurtured by it" (p. 110). Similarly, Brueggemann (1996, p. 110) underscores the importance of kinship, shared experiences, and voluntary participation in his view of a community: "Communities, therefore, are natural human associations based on ties of kinship, relationship, and/or shared experiences in which individuals voluntarily attempt to provide meaning in their lives, meet needs, and accomplish personal goals."

No matter which of these definitions you adopt, you will agree that communities can be very powerful and instrumental in problem solving and resolution. As macro practitioners, the likelihood of your being effective at the community level is maximized when you have a thorough understanding of the theoretical perspectives on communities.

Theoretical Base

Several theoretical models have emerged to help explain communities. According to Rothman (1987, p. 309), essentially all communities can be examined from five major perspectives: "structural, social-psychological, people and territory [ecological approach], functional and action process, and social system." Others (Brueggemann, 1996; Fellin, 1995) tend to highlight the ecological and social systems models. Our review features these two models, which seem to incorporate many of the elements described in the other models.

Ecological Systems Perspective   An ecological perspective has been applied to human services at both the micro and mezzo levels and has also proven useful at the macro level. Community, from an ecological perspective, focuses on the relationship of populations to their environment. Germain (1985, p. 7) states that "ecology is the study of organisms: environmental relations." From the ecological position, the community is an organism that has boundaries and that interacts with its environment. As noted in general systems theory, such transactions between the community and its environment ideally result in a reciprocal exchange wherein homeostasis is maintained. In this sense, "a healthy social organism [community] is one in which all of the different components fit together with one another and that has adapted to its environment" (Brueggemann, 1996, p. 110). When this fit is not exact and/or the community has not adapted to its environment, disequilibrium occurs. Stress is typically the culprit, as it is considered one of the main factors affecting this adaptation process. Examples of stress at the community level include pollution, depletion of scarce resources, overpopulation, poverty, unemployment, loss, and devaluation of one group by others as demonstrated by discrimination, oppression, and exploitation of those most vulnerable.

Ecological theory highlights the importance of *habitat* in the social environment. Some communities are overcrowded whereas others may be socially isolated due to their habitat. Furthermore, communities are prone to develop niches in their social habitats. According to Germain (1985, p. 45), a niche refers "metaphorically to the status occupied in the social structure by a particular group or individual and is related to issues of power and oppression." In our society, it is evident that various types of communities exist; those that are fortunate enough to be able to develop their own niches, and those that are not so fortunate. Many are "forced to occupy niches that do not support human needs and goals—often because of sex, age, color, ethnicity, social class, life-style, or some other personal or cultural characteristics devalued by society." Clearly, because of this inequity, interventions at the community level must consider the existence of the oppressive niches—such as slum conditions, poverty, and dejected status—that segments of the population are coerced into.

Fellin (1987) has identified other aspects of the community that need to be considered in addition to habitat and niche. Through his work on communities and ecological systems, he has found these to be important: "population characteristics of a community (size, density, heterogeneity), the physical environment (land use), the social organization or structure of the community, and the technological forces of the community" (p. 3). Equipped with this information, the macro practitioner can better determine "changes in the use of space, the distribution of people, and the movements of people over time" that are helpful in assessing how well the community is adapting to its environment. If there is an imbalance, macro practitioners can zero in on those areas where change and intervention are likely to be needed. Other concepts that human ecologists focus on include "competition, centralization, concentration, segregation, integration, and succession...which not only denote a process of change in communities, but are also used to describe end states, such as integrated or segregated neighborhoods" (Fellin, 1995, pp. 115–116).

A last facet of the ecological perspective we cite includes the concept of "social pollution of the environment" that Germain (1985, p. 37) also addresses in her work. She identifies these pollutants as "poor housing, inferior schools, inadequate health care, poor systems of income security, and inadequate juvenile and criminal justice systems." Also targeted is the "abuse of power, and the oppression of population segments based on race, ethnicity, gender, sexual orientation, disablement, and age." Macro practitioners, especially those working at the community level, must explore ways to rid the social environment of these pollutants or social ills.

Social Systems Model   A social systems framework from which to understand communities is similar to the ecological model. In both, communities are viewed as systems borrowing from general systems theory. From this standpoint, the larger system is the community, and social groups and social organizations make up its subsystems. Based on this conceptualization, the macro practitioner analyzes the way the social subsystems within the community interact with each other, paying close attention to how well both the subsystems and the system as a whole are functioning in meeting their members' needs. Fellin (1995, p. 121) identified three tiers that must be examined: (1) the community subsystems, (2) the social units within these subsystems, and (3) informal groups that also impact the subsystems. These tiers are shown in Figure 5.2.

Integral to social systems theory is the specification of boundaries between the system and its environment. As prescribed by general systems theory, a key function of the community system is boundary maintenance. As noted by Fellin (1995, p. 121), "a community engages in activities that will assure its continuance as a separate entity or social organization. Boundary maintenance is exemplified by physical boundaries as well as legal, political boundaries." In addition to interacting with its subsystems, the community also interacts with outside systems that may include other communities or society at large. Chess and Norlin (1988) have designated this outside system as a "suprasystem" that "provides inputs into a community system and receives outputs" (Fellin, 1995, p. 121). In this context, inputs are forms of resources such as culture, money, material goods, and information that help ensure the community's functioning. Outputs are described as "the result of the interactions within a system, such as the goals of a community and/or its subsystems." Chess and Norlin (1988) see these goals as connected to employment, health, safety and security, social welfare, education, housing, and other indicators of quality of life. A community must interact with both its subsystems and suprasystems and relate to the changes at all its levels, all the while attempting to maintain a state of equilibrium. A community constantly seeks a level of stability or equilibrium. Thus when the various subsystems or the community change, there is an impact on the total community. When the tasks or maintenance functions of the subsystems are not carried out successfully, this leads to a lack of goal attainment and may lead to community disorganization (Fellin, 1995, p. 121).

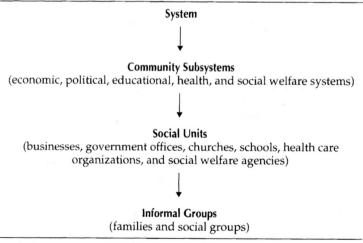

FIGURE 5.2   Fellin's social systems model

In adhering to this social systems framework, macro practitioners must be skilled at various levels of practice. Also, they must be well aware of how sophisticated community systems can be.

Models for Promoting Community Change

Rothman (1995) identifies three core models of community intervention alluded to earlier: (1) locality development, (2) social planning/policy, and (3) social action. Although not exhaustive, these three models offer a broad framework from which to consider community change.

Macro Practice Activities in Community Work

Community work encompasses many functions and professional roles. Homan (1994) stresses the use of traditional approaches to practice in community intervention. He specifies two levels of practice: (1) working as a change agent through direct work with individuals, families, and groups; and (2) working toward community change by engaging in such areas as community organization and development, research and education, and administration.

Netting et al. (1993) see community work as involving a number of components:

- Negotiating and bargaining with diverse groups
- Encouraging consumer participation in decision making
- Establishing and carrying out interagency agreements
- Conducting needs assessments
- Being an advocate for clients in a variety of community systems

For clarity, we have arranged these functions into three macro-practice areas that we address next: research, community development, and social action.

Macro Practice Research

Research is a major area of study and a necessary component of the knowledge base for macro practice. Research at the macro-practice level is applied research as it is centered on gathering information needed to fulfill the goal of bettering society in some way.

Reid (1987) delineates three reasons that research is important. At the basic level, research can help practitioners improve their practice skills. For instance, by structuring treatment interventions so that they can be evaluated through research procedures, practitioners can obtain useful information about the treatment process. They can then learn which techniques were most useful and alter their interventions accordingly. This relates to the second way research is significant as accumulated research helps build a foundation for planning effective interventions.

> Knowledge about what has worked best in the past is a guideline for approaches and techniques to be used in the present and in the future. Research is the basis for the development of whole programs and policies which affect large numbers of people. Such knowledge can also be used to generate new theories and ideas to further enhance the effectiveness of social work [macro practice]. (Kirst-Ashman & Hull, 1993, pp. 15–16)

The third justification for research is also related to learning from what has been effective in the past; determining what has worked in the past can provide current practitioners with clues on how to proceed in the present. Reid's (1987, p. 474) statement underscores the inherent value of research in all areas of practice; "[research] serves the practical function of providing situation-specific data to inform such actions as practice decisions, program operations, or efforts at social change."

The origins of macro-practice research can be traced back to the early work of Dorothea Dix, well known for her efforts in the 19th century to improve the conditions of those suffering from mental illness. Through campaigning and legislative action she succeeded in making major changes in this area. In fact, she is credited for "being personally responsible for the founding of state hospitals of mental patients in nine states . . . and quite a few others around the world" (Trattner, 1989, p. 62). Prior to any campaign Dix would engage in careful research:

> Before starting out, she read widely on the subject and then, wherever she went, studied antiquated and impractical commitment laws, inspected buildings, talked with and observed the patients and their overseers [and once] armed with numerous case studies and statistics depicting the extent and nature of the problem, she made her appeal. (p. 62)

In this way, Dix paved the way for continued research regarding social ills. Her techniques are analogous to such current-day methods as use of existing data, ethnographic research, focused interviews, and participant observation.

Another pioneer in macro-practice research is Paul Kellogg, a social worker and assistant editor of *Charities and the Commons*. He is recognized as among the first to use systematic social surveys to study social problems. He is credited for studying the entire life of a single community, an achievement that proved instrumental in social reform efforts. (Axisn & Levin, 1982, p. 146).

Reid (1987) has identified four major categories of research content specific to macro practice:

- Study of individual clients and their interactions with others in their environment
- Research on service provision that focuses on how services are provided to clients, what type of services these are, and how successful they are in accomplishing their goals
- Study of practitioners, their attitudes and educational backgrounds, and the human services profession in general
- Study of the three basic components of macro practice—organizations, communities, and social policy that examines the effect of the larger social environment on the behavior and conditions of individual members.

Research skills can range from very basic to very involved and complex. The specific research skills you will need will depend largely on the type of work you will be doing. However, for all human service practitioners, fundamental knowledge about research is vital to effective and ethical practice. At the graduate level, human service practitioners, specifically social workers, master three basic research skills: (1) evaluation of practice interventions at both the micro and mezzo levels to determine their effectiveness, (2) evaluation of macro-system effectiveness that measures the quality of service provision and whether social needs are being met; and (3) understanding, analyzing, and critically evaluating social services literature and research. Because a wide array of treatment interventions and approaches is available, practitioners must comprehend basic research questions and methodologies that may help them make informed choices about which treatment approaches to adopt in their practice.

More advanced research skills require specialized knowledge about research design and methodology. For example, research questions alone vary; there are descriptive questions, predictive questions, and prescriptive questions. Action research is a special form of applied research described by Rubin and Rubin (1992, pp. 156–157) as the "systematic gathering or information by people who are both affected by a problem and who want to solve the problem." Elements of this type of research include involvement in what has been called the turbulent action setting, community engagement, use of one's intuition function as well as thinking and sensing functions, attempts to separate values from research, and remaining self-conscious and self-critical throughout the process. The action research process itself is complex and includes various steps: entering the setting, identifying the problem, analyzing the presenting data, and writing the report. The research report must be detailed and comprehensive. Brueggemann (1996, p. 148) offers the following format for writing a research report:

Title Page and Table of Contents
Introduction
Review of Literature
Research Design
Data Collection
Data Analysis
Recommendations
Conclusions

Clearly, research is not only an important aspect of macro practice but also a very intricate science itself. "Research informs and supports intervention approaches. It identifies theories and programs which are likely to be effective. Finally, it helps workers ensure that the client is being helped, rather than hurt, by what workers do" (Kirst-Ashman & Hull, 1993, p. 16).

Community Development

Community development helps struggling communities survive and succeed. Brueggemann (1996, p. 156) defines community development as "a method by which macro social workers apply techniques, develop resources, and promote networks that enable the community to become a source of social, economic, political, and cultural support to its people." Pantoja and Perry (1992) have coined the term *community restoration* for helping people within communities who are both economically dependent and politically disenfranchised. They highlight the following as key objectives in this process:

- To understand the underlying dynamics that have led to the current state of poverty and dependence
- "To mobilize and organize their internal strength, as represented in political awareness, a plan of action based on information, knowledge, skills, and financial resources" (p. 240)
- To dispel myths held by both individual community members and the group culture that influences them to remain participants in their own dependency and powerlessness
- "To act in restoring or developing new functions that a community performs for the well-being of its members—starting with the economizing function" (p. 240).

Based on these views, community developers clearly are at the hub of the wheel; they are actively involved in the community's efforts to restore such ideals as peace, equality, and economic prosperity. By considering the entire community their client, they attempt to make changes that will benefit all its members.

Community development is thus a specialized area of macro practice. Rothman and Tropman (1987, p. 5) consider community development to involve "democratic procedures, voluntary cooperation, self-help, development of indigenous leadership, and educational opportunities." The specific community in question can be chosen in various ways. Often, community developers work for a specific agency, community organization, or community center that may dictate which community requires professional intervention. At other times, professional community developers themselves will choose to work with a community they feel especially committed to helping. Regardless of how the community is selected, the worker must be dedicated to helping its members and devoted to resolving the particular problems they are facing. Because the role of a community developer involves active participation, these practitioners must remain open to working alongside community members and be sensitive to the specific community's history, values, symbols, language, culture, and traditions. For these workers, the inside perspective is preferred over the outside observer view.

Understanding the community is mandatory. Homan's (1994, p. 80) work, *Promoting Community Change*, accents this point.

> The community is a contributor of resources and allies and a provider of pitfalls and opponents. You will want to know where these are—where to go to get what you need and who or what to avoid. The community is, after all, where the need for change, the effort to make that change, and the resistance to change coexist.

Community developers should learn as much as they can about the community they will be working in. A knowledgeable worker will have a better sense of the community setting itself, will be able to put into perspective the work that lies ahead, and will have time to prepare for anticipated problems. There are various ways to learn about the community. Simply being a part of it will give you a sense of the areas that need improvement. Also, asking community members themselves about how they feel and where they envision change to be needed can be helpful. More formal ways of sizing up the community include gathering census data, crime statistics, and the incidence of poverty.

Many community problems are mutually agreed-on by all—the community members, the local government, and the community developer. Examples are the proliferation of gangs, homelessness, the plight of those suffering from mental illness or AIDS, problems of the homebound elderly, crime and violence, drug-infested neighborhoods, racism and discrimination, the high rate of unemployment, and poverty. In some instances, however, there is not a consensus about which problems require immediate attention. In these cases, the community developer will have to discover which problems are most serious and have the greatest impact on the community. Also, he or she will have to determine which problems are most amenable to community intervention. Although not everyone may agree on the specific direction taken, all are likely to welcome some form of intervention.

Equally important to community development is the self-awareness component. Community developers must examine their own concerns, ideological beliefs, and personal strengths.

> If you can't get excited about or invested in an issue, stay away from it and choose something that "grabs you." The community problem should pull at you so that you can engage your emotions and generate compassion for the people involved.... Your own values must [also] be "in sync" with the values of your community and their culture. (Brueggemann, 1996, p. 161)

As we have seen, this same principle is relevant at other levels of practice also. With both individuals and small groups, the practitioner must be committed to the specific population he or she is working with and feel positive about his or her ability to be helpful. When practitioners are not committed or are negative, the chance for change is greatly limited. Helpers may feel that their values are incongruent with their clients' values, or they may believe they are incapable of helping their clients. Practitioners need to be aware of their own values as well as their abilities and strengths. It's true that to be an effective community developer you will need to cultivate many skills, but a good place to start is by recognizing your existing skills and abilities and realizing that additional learning will occur along the way.

Macro practitioners must be adept at developing relationships at various levels. "One of the hallmarks of a community developer is an ability to engage people" (Brueggemann, 1996, p. 162). This involves getting to know people in all areas of the community, identifying those who are key to any proposed change efforts, and recognizing those who might be adversaries. People skills are essential for community developers who must be able to bring people together, to negotiate, and ultimately to motivate people to become organizers themselves. Group meetings are often a vehicle for meeting people and for beginning to develop a cohesive unit of people who will become effective in organizing.

A last important function of a community organizer relates to program development. Once workers have succeeded in organizing around a central issue, they must follow up by developing a plan the community can work toward implementing. Examples include "initiating a community project of volunteer efforts, obtaining a grant for developing a special program, coordinating efforts with other community action projects to form a coalition, and engaging in social action" (Brueggemann, 1996, p. 165).

Social Action

Social action is an approach for those who are deeply committed to helping solve social problems and alleviate some of the ills in society. Social activists are "committed to a process of social change and an end to violence and human misery" (Brueggemann, 1996, p. 179). Saul Alinsky (1974, p. 3), a well-known social activist, captures the essence of this approach in his statement about the challenge of social action:

> to create mass organization to seize power and give it to the people; to realize the democratic dream of equality, justice, peace, cooperation, equal and full opportunities for education, full and useful employment, health and the creation of those circumstances in which man can have the chance to live by values that give meaning to life.

This approach assumes that there is a disadvantaged and oppressed segment of society that needs help in organizing to voice their demands collectively and in such a way that they can be heard. By organizing, these population groups are better able to "pressure the power structure for increased resources or for treatment more in accordance with democracy or social justice" (Zastrow, 1995, p. 315). Types of changes typically pursued through social action measures include redistribution of power, the reallocation of resources, or changes in community decision making (Homan, 1994, p. 29). Social action approaches may seek basic changes in major institutions or in basic policies of formal organizations. Alinsky (1972, p. 130) described the typical process involved: "Pick the target, freeze it, personalize it, and polarize it."

Social activists may take one of three stances in their efforts to effect change: as social prophet, as visionary, and as confronter. The *social prophet* pursues radical change by questioning the status quo and envisioning new alternatives to present-day societal views, attitudes, and policies. A major part of this role is to sensitize society to the plight of the most vulnerable such as the homeless, those with AIDS, those suffering mental illness, and others.

The second, the *social visionary*, envisions a world where certain social problems are nonexistent—"they picture the future as it might be...[and] build the future out of a sense of justice and truth, an end to violence and oppression" (Brueggemann, 1996, p. 180). An important point, however, is that they do not just wish for things to improve; they make the future real by learning from the past. Examples might be evaluating the success of certain programs and then making necessary changes that will enhance their effectiveness or revamping entire programs by starting from scratch.

The third, the *social confronter*, pursues change through direct means. These practitioners not only challenge the status quo, but also directly confront those they hold responsible for the injustice and social inequality that is present by influencing the power structure through persuasion, political action, or power tactics.

If you look carefully through history, it is not difficult to pinpoint events where social action played a major role. Our history is rich with examples of social activism. Chronologically, the most significant examples include the *abolition movement*, which challenged racism and the atrocities of slavery in the Civil War period; the *temperance movement* in the early 1800s, which fought to reduce heavy whisky drinking at the time and began to understand the dangers of alcoholism; the *women's rights movement*, which took hold in the mid 1800's with the first Women's Rights convention that demanded women's suffrage and led to the adoption of a Declaration of Independence; the *settlement house movement* in the early 1900s, spearheaded by Jane Addams, that led to an increase in settlement houses, as well as reform measures to ensure their protection; the *workers' rights movement* that resulted in the legalization of labor unions and protection of workers' rights in the 1930s; the *civil rights movement*, resulting in the passage of the Civil Rights Act of 1964 that prohibited racial, sexual, or ethnic discrimination in employment and in public accommodations; the *protest against the war in Vietnam* in the mid-1960s that led to the American withdrawal from Vietnam; and the more recent protests against the conditions of migrant workers and the adoption of the California Coastline Initiative forcing government to protect the ecology of that state's coast.

Without doubt, the social activists behind these movements have helped shape our society and have provided hope and inspiration for continued efforts to combat the injustices that are ever present. Although the type of social problem changes with the times, these problems always seem present at some level. Today, the plight of the homeless and those living in poverty; those living with AIDS; the horrors of drugs and violence; the rise in racism, discrimination and oppression; the need for adequate housing and health care; the nuclear arms race; and the protection of fragile wilderness areas seem to be current-day issues dictating the need for social action.

The value of social action is evident. Committed and trained macro practitioners, however, are needed to continue chiseling away at difficult societal problems. The process of social action is similar to the problem-solving process used in work with individuals, groups, and organizations. The step-by-step process is as follows:

1. *Unfreezing the system* by defining the problem, engaging the community, empowering the forces of change, and building an action organization.

2. *Moving to action* that involves understanding the forces creating resistance toward change, working toward reducing the resistance toward change, engaging in action strategies and tactics that often result in conflict and require confrontation.

3. *Solidifying change by refreezing it* involves employing political pressure tactics that are intended to change laws and enforce regulations that preserve efforts toward positive change. This stage requires work at the political and policy levels and may involve establishing an agency to operate programs, propose legislation on behalf of the populations organizing for change, and establish methods for monitoring and enforcing set policies and procedures.

4. *Terminating the activist's role* is a move that is determined by various factors: Is the community group sufficiently organized with committed members and ample leadership? Have they been successful in accomplishing set goals? Are they sufficiently involved in the political arena and is the power structure responsive to the organization and to its political strength?

In summary, social action involves three different but related activities:

- Advocacy around specific populations and issues
- Working in local and national elections to elect sympathetic representatives and to support new programs designed to combat large scale problems
- Networking with other groups pursuing a similar agenda (Burghardt, 1987, p. 292)

Advocacy is a major aspect of social action. As defined by Barker (1991, p. 7), it is "the act of directly representing or defending others" by intervening to help clients have their needs met. Two types of advocacy are cause advocacy and case advocacy. *Cause advocacy*, or *class advocacy* as it is sometimes called, relates to efforts by practitioners in pushing for an issue of overriding importance to a group of clients. *Case advocacy*, in turn, relates to activity on behalf of a single case, be it an individual client, family, or small group. Both types are equally important and useful at all levels of practice. Cause advocacy, however, is more widely used in the macro-practice arena.

Working at the Social Policy and Political Level

We have seen the need for social change and the efforts of macro practitioners working at both the organization and the community level to effect such change. A vital part of their role involves working at the policy level, which is the third major area of macro practice. At this level, focus is on the statutes and broader social policies that underlie existing policies. Practitioners working in this area may develop and seek adoption of new statutes or policies and propose elimination of ineffective or inappropriate ones.

Policies can be viewed simply as rules that govern our actions; in practice, they guide our work and our decisions. Categorizing policies as social policies and agency policies is helpful in clarifying the level at which policies exist and where to target our efforts.

Social policy includes "the laws and regulations that govern which social programs exist, what categories are served, and who qualifies for a given program. It also sets standards regarding the type of services provided and the qualifications of the service provider" (Kirst-Ashman & Hull, 1993, p. 17). Integral to working at the social policy level is understanding the intricacies of fiscal planning and appropriations at all levels of government. Morris (1987, p. 644) sees social policy as involving "decisions of various levels of government, especially the federal government, as expressed in budgetary expenditures, congressional appropriations, and approved programs." Because social policies may have a major impact on the allocation and distribution of funds, macro practitioners working to help people obtain needed resources must be well versed in this realm of practice. Defining clients' need is only the first step; helpers must follow by determining what and where the available resources are and actively pursuing ways to help clients secure these resources.

A further distinction can be made between public and social policies: public policies are the "operating principles by which governmental systems carry out their goals," whereas social policies are "operating premises developed to provide direction in the solution of social problems" (Brueggemann, 1996, p. 434).

Agency policies are more specific forms of policy as they include standards adopted by individual organizations, agencies, and programs engaged in service provision. These standards can apply to various aspects of agency functioning; they may define the structure of an agency, the qualifications of its employees, or the rules that govern what a worker is allowed to do within the confines of the agency system.

Regardless of the policy level—public, social, or agency—each is important and plays a vital role in human service practice. It is our moral and ethical responsibility to work toward changing or developing policies that are more amenable to meeting social needs. Likewise, we can have a significant part in ensuring the maintenance of existing policies that are sound.

Theories of Policy Making

Theoretical models are at the foundation of both organizational and community levels of macro practice. This is also the case with social policy, in which numerous theories and models of policy-making exist. Essentially these models characterize the way systems, groups, and people effect policy decisions at the local, state, and federal levels. A brief account of some of the most noted theories or models is presented to give you a better sense of the broad range of views regarding social policies.

Elite theory maintains that policy is in the hands of those who are most powerful in government, business, and other powerful systems. Because elitists consider the population in general to be "largely passive, and apathetic, and ill-informed," they view their approach to policy making to be the best. Therefore, from this perspective, public or social policy does "not reflect demands of 'the people' so much as it does the interests and values of the elites" (Dye, 1975, p. 25).

The *institutional model* assumes that policy making must be understood from the perspective of formal institutions of government where policies are developed and enacted.

As it is the legislatures at both the state and federal levels that officially create and execute laws, policy makers must understand the internal workings of governmental institutions that are, in effect, responsible for policies and laws becoming a reality.

Dye maintains three reasons that the institutional model is a valid one: (1) only governmental institutions can officially create public and social policies and "lend legitimacy to them" (Dye, 1975, p. 18); (2) only government includes everyone within its jurisdiction; and (3) because of the publicly sanctioned power of government, it has the direct authority to compel people to comply with set policies.

Interest-group theory contends that in a pluralistic society, such as in America, policy making is greatly influenced by special interest groups that together with the government determine the direction of public policy. In this respect, many believe that a more democratic system of policy making can exist because government helps mediate and manage conflicts between the various interest groups representing certain causes.

There are as many advocates of this theory as there are opponents. Advocates, for instance, believe that

> because of the fundamentally pluralistic nature of the American political environment, policy making occurs as various interest groups press for policy outcomes on both the institutional as well as agency structures of government. Policy making is a reflection of the success or failure of these groups having their particular preference schedules enacted into legislation. (Brueggemann, 1996, p. 348)

Opponents of this view are critical because they believe it does not truly represent a democratic approach, as the interest groups that often succeed in having their policies enacted into law are those with the most power. Thus, when "elites appear as leaders of groups and compete with other elite-led groups for policy rewards" those who are less powerful are less likely to be heard (Morrow, 1975, p. 51).

The *bargaining and negotiation approach* is a simplistic one that seems more neutrally accepted. Basically, this view stresses the necessity for policy makers to learn that they will have to compromise and modify their demands to ensure that some aspects of their policies will be accepted. Although persuasive skills integral to compromising, bargaining, and negotiating are important for practitioners to master at all levels of practice, they are especially valuable for policy makers at the macro-practice level.

Systems theory, which we have alluded to before, is helpful here in viewing policy making. It is perhaps the most comprehensive approach as it incorporates many facets of the other social policy theories. When a systems approach is applied to the policy process, recognizing all the players and their influence in determining policy becomes much easier. To be exact, "the policy process is seen as a total system in which the variety of political actors such as legislatures, administrative agencies, political parties, interest groups, elected executives, and private citizens all operate in various ways to determine policy" (Brueggemann, 1996, p. 350).

Understanding systems theory is useful for those working not only with individuals and families but also in communities and at the policy level. We highly recommend that all practitioners learn the theoretical underpinnings of the systems approach, as it is likely to serve them well throughout their human service careers, whichever path they take.

Along with understanding the theoretical views and models underlying social policy, practitioners must also be skillful in *policy analysis*. This requires "a critical assessment of policies in terms of their effectiveness in achieving the welfare of the people they are intended to benefit" (Brueggemann, 1996, p. 335). Policy analysis therefore is concerned with discovering what is wrong with current policy, exploring ways policies can be changed or improved, and ultimately deciding among competing policy preferences.

The specific steps of the policy analysis process are analogous to the problem-solving approach. First, the policy problem is defined; next information and facts are gathered to allow a redefining of the problem and a relabeling. Once these steps are complete, practitioners determine solutions, often needing to choose among alternatives by considering their political and economic feasibility. Finally, alternatives are ranked and then policy implementation can take place.

Once a policy proposal is written, it will not necessarily be implemented automatically. Policy planners must often engage in extensive negotiation to assure implementation of their proposed policy. This may require involvement in case conferences and in additional fact gathering, position taking, community work, petitions, and at times, even media campaigns. Political pressure tactics may have to be used. "To make lasting changes, organization members need to think about ways to change laws, regulations, and levels of enforcement.... Political pressure tactics are intended to change laws and enforce regulations that preserve the short-term victories" (Rubin & Rubin, 1992, p. 266).

Often, direct work within the political system is also necessary. This may entail lobbying, letter writing, telephoning (or faxing in today's technology), and testifying. We explore some of these further below.

Macro-Practice Activities at the Policy and Political Level

Two major activities in this area of practice are coalition building and lobbying.

Coalition Building   Coalition building is a form of collaboration. Netting et al. (1993, p. 113) describe a coalition as "a loosely developed association of constituent groups and organizations, each of whose primary identification is outside the coalition." Mizrahi and Rosenthal (1993, p. 35) write that "coalitions are complex organizations, born out of a desire or urgent need for intergroup action and cooperation." In some cases, "agencies created to serve the needs of a special population collaborate to assess needs, to examine the fit between needs and services, and to present a united front and a stronger voice in pursuing funding for programs" (p. 35). Clearly, coalition building is an important function of anyone working at the social policy level. According to Mizrahi and Rosenthal (1993), "The coalition is a vital and increasingly utilized mechanism for collective organizing and policy formation. Through coalitions, separate groups can develop a common language and ideology with which to shape a collective vision for progressive social change" (p. 12).

Even though a practitioner's desire for social change may be strong, engaging in coalition building is a complex matter that requires knowledge, specialized skills and a greater commitment toward internal collaboration and planning. Organizing and maintaining social change coalitions "requires both an understanding of the theory underpinning such interorganizational entities and the application of the practice principles derived from collective experience" (Mizhari & Rosenthal, 1993, p. 35).

Lobbying   Lobbying is a specialized form of persuasion that is central to the pluralistic nature of American government. As a method of persuasion, lobbyists seek access to lawmakers to influence policy making and decision making. They try to persuade elected officials to consider and to press for policy considerations that the lobbyist feels are important. Netting et al. (1993, p. 254) state that lobbying is "a form of persuasion that addresses policy change under the domain of the controlling system. The action system will have to determine if new legislation is in order to achieve their goal." The main purpose behind lobbying is to provide needed services for people who may not be able to engage in political bargaining themselves. Lobbyists come in many shapes and forms; they may be "large business corporations who work in their own self-interest to influence policy in their favor" or they may be individual practitioners, often social workers, who "lobby public governmental officials in the public interest" (Brueggemann, 1996, p. 358).

Haynes and Mickelson (1986, 1991) have written extensively about the lobbying process; they delineate three very basic concepts that are essential for practitioners to consider: (1) be factual and honest, (2) make forthright presentations and support them with data, and (3) address issues of cost and social impact as these are critical concerns of decision makers. In addition, they cite specific guidelines for successful lobbying:

- Lobbyists should know their issues thoroughly.
- They should identify a core group of committed and effective workers to aid in the lobbying.
- They should enlist support from a lawmaker who is sympathetic to the issue at hand.
- Lobbyists must be familiar with the political process—specifically—the formal legislative structure and procedural steps a bill must go through in order to become law.
- Lobbyists must be willing to invest time and energy working at the capitol building where they will be in a better position to answer any legislative questions or intervene in the political process as the need arises.

The value of coalition building and lobbying is without question. Other important functions of a policy maker include testifying, tracking legislative developments that directly affect clients, and carrying out other efforts designed to affect the legal process. Not everyone is suited to work in this area, but for those who do, it can be a very exhilarating experience that is well worth while when a bill one has supported is eventually enacted into law. Without human service practitioners to work at this level of social policy, many of the policies we have today would not exist. Whether we are interested in this aspect of macro practice or not, we must appreciate it as a vital part of the helping process.

Skill Development

Micro, mezzo, and macro skills are all useful to the macro practitioner. To be truly effective in working with organizations, within communities, and in the policy arena, you will need to know how to work well with individuals, families, small groups, and larger systems.

Micro Skills

Key skills to working with individuals—warmth, empathy, genuineness, objectivity, sensitivity, acceptance, patience, a desire to help, and a commitment to ongoing self-awareness—are all important when working with those in organizations, the community, and at the political level. Regardless of the level, direct contact with people will be necessary; this may involve work with clients, co-workers, administrators, community leaders or groups, and legislatures. Knowing how to communicate appropriately, form positive relationships with others, and demonstrate a commitment to the helping process, whatever the level, is central to a helper's ultimate effectiveness and success as a macro practitioner. We concur with Homan (1994, p. 52) that "the fundamental skills you need to work at the individual level are the same skills you use when you work for community change."

Mezzo Skills

Often, working with families and small groups may lead to working at the macro-practice level on their behalf. Ironically, the same skills that are effective with small groups can be useful when working with larger systems. For instance, skills in problem identification, goal setting, coordination, decision making, and conflict resolution are common to both treatment and task groups, and are also valuable for organizational and community change.

We encourage you to recognize the wealth of knowledge you have acquired about working with people and the repertoire of skills you have developed; the challenge is often in knowing how to transfer these skills. You may still need to cultivate special macro-practice skills, but it is important that you not overlook what you already know.

Macro Skills

Throughout this chapter, we have identified macro-practice activities specific to the three main areas of macro practice. For instance, social planners, program developers, and administrators are needed at the organizational level; in community work, researchers, community developers, and social activists are necessary; and work at the social policy and political levels requires expertise in coalition building, lobbying, and testifying. In our discussion of each of these, we have touched on the skills needed to be effective. In review, the major skills important to macro practice as defined by Kirst-Ashman and Hull (1993, p. 126) are "building and maintaining organizations, evaluating results, fund-raising, budgeting, negotiating, mediating, influencing decision makers, needs assessment, planning, political skills, and working with coalitions." Even though this list is not exhaustive, it does provide you with an idea of the breadth of macro-practice skills.

Last, we have alluded to two additional skills at various times—empowerment and collaboration. Because we believe they are central to your work as a macro practitioner, we explore them further.

Empowerment

Empowerment is a goal many helpers strive for with their clients who are in some way oppressed and vulnerable. Essentially, "empowerment practice strives to develop within individuals, families, groups or communities the ability to gain power" (Gutierrez, Parsons, & Cox, 1998, p. 4). Netting et al. (1993, p. 252) believe that empowerment "involves enabling people to become aware of their rights, and teaching them how to exercise those rights so that they become better able to take control over factors that affect their lives." Individuals who develop stronger beliefs in their own personal power and the power of an organized group are more likely than others to succeed in becoming empowered. Gutierrez, Parsons, and Cox (1998, p. 20), who have written a sourcebook on empowerment practice, describe the empowerment process as involving these four basic components: "(1) a critical review of attitudes and beliefs about one's self and one's sociopolitical environment, (2) the validation of one's experience, (3) an increased knowledge and skills for critical thinking and action, and (4) action taken for personal and political change."

On a similar note, Homan (1994, p. 129) believes that "empowerment involves overcoming sets of beliefs, oppressive structures, and stifling routines that keep people and their concerns isolated from one another." He highlights five factors that are necessary to empower those participating in a change effort:

- Personal interest or investment in the project; a feeling of being an important part of things
- A belief in the possibility of a successful outcome
- Development and recognition of individual and group resources
- An opportunity to take action and to make meaningful contributions
- The recognition of common interests and common risk taking

Keeping these basic factors in mind can help practitioners in empowering even the most vulnerable. Further efforts must also be taken to afford individual clients or groups with the opportunity for decision making and performing tasks; provide continual encouragement; and engage in ongoing recognition of individual or group achievements through positive reinforcement measures.

Empowerment practice can occur at many different levels, can be performed by many types of helpers, and can range in complexity. As a skill or practice method, it is essential to practitioners at all levels.

Collaboration

Collaboration, like empowerment, is a key skill of human service practitioners. In simplest terms, it is the procedure in which two or more persons work together for a common purpose. It implies working together in a joint venture. "People and groups work together when they recognize mutual needs and recognize that working together will help them meet those needs better than working alone" (Speeter, 1978, p. 92).

Specific tactics for effective collaborative work include "problem solving, joint action, education, and mild persuasion, each involving a more or less active attempt to influence another party" (Brager & Holloway, 1978, p. 131). According to Warren (1975, p. 138), use of any type of strategy must be "based on the assumption of a common basis of values and interests through which substantial agreement on proposals is readily obtainable." Collaborative strategies should be exhausted before using adversarial approaches because they are "good faith efforts to change the target system based on positive assumptions about the target. Only when they fail should other approaches be used" (Kirst-Ashman & Hull, 1993, p. 137).

As noted throughout this chapter, becoming an effective macro practitioner is not a simple matter. For effective practice at this level, professionals will need to acquire a vast knowledge base about organizations, communities, social policy, and the political process. Also, beyond micro and mezzo skills, they will need to cultivate numerous skills and roles specific to macro practice. Several of the key roles important in effecting community change are initiator, negotiator, advocate, spokesperson, organizer, mediator, and consultant. Although you may not consider yourself an expert in any one of these areas, you may have assumed facets of these roles in some aspect of your work without even realizing it. We encourage you to consider which roles you are capable of assuming as well as those you need to continue working on. Knowledge, skills, and a sound commitment to becoming a competent macro practitioner are the basic ingredients you will need as you venture into the macro-practice arena. At times, this path may seem long and treacherous, but as in any area of practice, professional development begins with a practitioner having faith in the process and being receptive to learning as he or she travels on the journey.

Ethical Issues

Questions regarding values and ethics permeate every aspect of human service practice at the macro level. In fact, "half of professional decision making requires ethical rather than scientific judgement" (Hokenstad, 1987, p. 4). However, like most practitioners, we often do not recognize the significance of our values or ethical perspectives until we are directly confronted by them. Because we must routinely address practical matters of values and ethics in our daily practice activities, we must make an effort to be more aware of them.

Values tell us what is important, the "shoulds" and "oughts;" "ethics are the operationalization of values or behaviors that occurs in carrying out those values" (Netting et al., 1993, p. 57). This is true, whether working at the micro, mezzo, or macro level of practice. In Chapter 7, we consider in great detail the nature of ethical and legal issues of direct practice. Here, we only highlight the importance of adhering to a firm code of ethics to remain ethical in any practice endeavors you pursue. This requires being cognizant of ethical dilemmas as they arise and taking responsibility for engaging in sound ethical decision-making practices whenever necessary. Many times, practitioners will face situations requiring them to choose between equally important values; to confront these, they will need to have developed well-thought-out strategies that will allow them to make informed decisions with which they can feel comfortable.

Kapp (1987) presents three ethical principles that are useful to professionals in analyzing macro-practice decision making. These are autonomy, beneficence, and justice. According to Netting et al. (1993, p. 60),

> Balancing autonomy, beneficence and justice demands an analytical approach to decision making and intervention. Inevitably, the macro practitioner will face ethical dilemmas that go beyond the bounds of the codes of ethics [of his profession]. This requires that he or she have a strong professional identity.

Homan (1994, p. 349) emphasizes the need for practitioners to root themselves in a fundamental code of ethics. He suggests that helpers review their commitment to ethics and keep it alive. "Your steadfast commitment to integral ethical principles will guide you to act differently as you face different challenges to your beliefs."

Diversity Issues

In working in human services, practitioners at all levels of practice will be faced with diverse populations who are in need of some type of professional assistance. To be effective, practitioners must be aware of diversity issues and have a respect for those who are unlike them. In the following chapter, we discuss at length the many facets of diversity and the importance of engaging in diversity-sensitive practice that requires being aware of your own issues and biases, having a sound knowledge base of the specific populations you will be working with, and acquiring the necessary skills to work effectively with these populations.

Working with Diverse Groups

Macro practice with diverse groups entails working with various communities. We consider communities of specific ethnic background, those who live in poverty, those struggling with AIDS or some other physical or mental disability, gay and lesbian groups, and other oppressed groups such as women, children, and the elderly.

In working with diverse communities of color, for instance, practitioners need to know specific facts, according to Rivera and Erlick (1992, pp. 14–22). Homan (1994, p. 22) summarizes these as follows:

- Familiarity with customs, traditions, social networks, and values
- An intimate knowledge of language and subgroup slang
- Leadership styles and development
- An analytical framework for political and economic analysis, including an appraisal of who has authority as well as who has power and the sources of mediating influence between the ethnic communities and the wider communities
- Knowledge of past organizing strategies and their strengths and limitations
- Skills in conscientization (developing a critical consciousness) and empowerment; (this process implies that both the organization and the community learn from each other the problems at hand and the strategies and tactics employed)
- Skills in assessing community psychology
- An awareness of self and personal strengths and limitations

In working with all diverse groups, a process similar to this would take place. The key factor, regardless of the specific community, is the commitment to helping the individual group coupled with an understanding of the issues that are specific to their plight.

Following are brief synopses of articles related to working with varied population groups; they offer insights into the broad range of interventions and activities macro practitioners can engage in when working with diverse groups.

Macro Practice and AIDS

- Hooyman, Frederiksen, and Perlmutter (1995), *"Shanti: An Alternative Response to the Aids Crisis"*

Shanti is a Seattle-based alternative agency that focuses on meeting the emotional needs of people with life-threatening illness and those grieving the loss of someone who has passed away. In addition to providing emotional support to individuals and their families, Shanti also serves the larger community through offering public education, information and referral, and providing training and technical assistance to community organizations.

- O'Hare, Williams, and Ezoviski (1996), *"Fear of AIDS and Homophobia: Implications for Direct Practice and Advocacy"*

The authors discuss the studies of health care trainees and professionals and their response to patients infected with HIV infection and AIDS. The survey they conducted examined the relationship between homophobia and the fear of AIDS within the context of several categories; the results showed that "those who more liberally endorsed the rights for gay men and lesbian women were less homophobic and had less fear of AIDS." The authors offer suggestions for sensitizing human service professionals to the "inhibitory effects of homophobic fear of AIDS when working with HIV-infected clients or people in the gay community."

Macro Practice with Children

- Freeman (1996), *"Welfare Reforms and Services for Children and Families: Setting a New Practice, Research, and Policy Agenda"*

Freeman addresses the barriers to family and community self-sufficiency. He asserts that many of these barriers are a result of policy makers' "biased definitions of self-sufficiency and related social program values." He further contends that macro practitioners "need to understand the worldviews and values that underlie political definitions of self-sufficiency to have a greater influence on social policy." Most important, the author proposes new research and policy agendas for the human services [social work] profession that can enhance social justice and integrate policy development into clinical and community practice.

- Kamerman (1996), *"The New Politics of Child and Family Policies"*

In this article, Kamerman talks about the "slowdown in the legislative process and the absence of any closure on the proposed congressional initiatives on social safety net programs." She discusses the unclear social policy developments that exist as well as the United States' entrance into a "period of social policy devolution, federal funding cuts for social benefits and services, and a major philosophical shift in the premises underlying social policy." She emphasizes the need for human service practitioners to become informed about these developments to remain effective in the practice, education, and advocacy arenas.

Macro Practice with Ethnic Groups

- Kaul (1995), *"Serving Oppressed Communities: The Self-Help Approach"*

Kaul discusses the importance of organizing people around common concerns in geographical areas where they live. He encourages identifying attainable problem-solving programs that can be initiated and accomplished on a self-help basis.

- Gutierrez, Alvarez, Nemon, and Lewis (1996), *"Multicultural Community Organizing: A Strategy for Change"*

This article focuses on demographic projections and reduced government spending and their impact on minority groups. The ethnic, cultural, and racial diversity of U.S. society poses challenges for policy makers, social service administrators, and human service practitioners. The authors state that "social workers [human service practitioners] must possess skills for working in a multicultural environment if communities are to be well represented in the planning process."

Macro Practice with People with Disabilities

• Mackelprang and Salsgiver (1996), *"People with Disabilities and Social Work: Historical and Contemporary Issues"*

The authors address recent developments in the disability movement, including independent living. They compare these new approaches to traditional ones and make a plea for practitioners and those in the disability movement to "combine their efforts to enhance the lives of people with disabilities."

• McEntee (1995), *"Deaf and Hard-of-Hearing Clients: Some Legal Implications"*

McEntee highlights the passage of the Americans with Disabilities Act that increases the need for social service agencies to become more aware of the needs of their deaf and hard-of-hearing clients. She reviews federal legislation and statutes and their impact on service and legal programs. She states, "As service providers and advocates, social workers [human service workers] need to ensure that their own services adequately meet legal and ethical obligations and that the profession advocates for other agencies to do the same."

Macro Practice with the Elderly

• Barusch (1995), *"Programming for Family Care of Elderly Dependents: Mandates, Incentives, and Service Rationing"*

Barusch surveys and critiques three approaches: "increasing family care of elderly people: filial support legislation, incentives for family caregivers, and service rationing provisions." Her conclusions indicate that policies requiring or encouraging relatives to provide care may have adverse consequences for elderly people and their families. She contends that human services advocates "should support policies and interventions that provide universal access to a continuum of care alternatives that facilitate rational health care decision making by families." Also, she encourages empowerment and sustenance of family members who choose to care for elderly relatives.

• Ozawa (1995), *"The Economic Status of Vulnerable Older Women"*

The author focuses on the needs of economically deprived older women. Black women and Hispanic women, particularly if unmarried, are most at risk as their economic status remains extremely low. Efforts to help them at both the practice and social policy levels are suggested.

Macro Practice with Gays and Lesbians

• Proctor and Groze (1994), *"Risk Factors for Suicide among Gay, Lesbian, and Bisexual Youths"*

Proctor and Groze share results from their study, which was a convenience sample of 221 self-identified gay, lesbian, and bisexual youths who attended youth groups across the United States and Canada. Participants were given the Adolescent Health Questionnaire, which assesses family issues, the social environment, and self-perceptions. The results showed that the youths' scores were significantly associated with suicidal ideation and attempts, suggesting that immediate involvement by humans service practitioners is warranted.

• Ball (1994), *"A Group Model for Gay and Lesbian Clients with Chronic Mental Illness"*

The gay affirmative group model suggested by the author focuses on "the unique social, developmental, and psychoeducational needs of lesbian and gay clients who attend psychiatric day treatment." Ball suggests modifications of traditional group interventions that are gay sensitive and enhance the mental health of this population.

Macro Practice with Health and Mental Health Issues

• Davitt and Kaye (1996), *"Supporting Patient Autonomy: Decision Making in Home Health Care"*

Davitt and Kaye share their study, which "examined the policies and procedures that home health care agencies have developed to handle the incapacitated patient and life-sustaining treatment decisions." Results indicate the need for policies that are consistent among social workers, nurses, and other staff to handle such difficult ethical dilemmas. The authors also review specific policies and recommend future policy changes and development.

- Edlefsen and Baird (1994), *"Making It Work: Preventive Mental Health Care for Disadvantaged Preschoolers"*

The authors believe that by combining educational and mental health services to disadvantaged preschool-age children, significant benefits for both the children and their families can result. The authors contend that "such a partnership recognizes the use of each profession's expertise to create a more balanced, efficient, and realistic approach to serving children," especially needed in the present time of fiscal constraints.

Macro Practice and Poverty

- Coulton (1996), *"Poverty, Work, and Community: A Research Agenda for an Era of Diminishing Federal Responsibility"*

Coulton discusses research issues as they pertain to the increased poverty of many of today's families. She stresses the need for "a research agenda that fosters employment and builds community resources" that can provide preconditions and preparation for employment along with access to employment and ongoing support for working families.

- First, Rife, and Toomey (1994), *"Homelessness in Rural Areas: Causes, Patterns and Trends"*

In this article, the authors examine homelessness in rural areas. They report their findings from a 1990 statewide study of rural homelessness funded by the National Institute of Mental Health. Results from the survey suggest important implications for short-term emergency assistance and longer-term policy making with this homeless population.

Macro Practice with Women

- Gutierrez (1995), *"Working with Women of Color: An Empowerment Perspective"*

The article highlights the need for empowerment of minority women (e.g. African-American, Latina, Asian-American, and Native American) who compose at least 20 percent of the total female population of the United States. Gutierrez suggests that empowerment practice with these women of color be on the "personal, interpersonal, or political level." She also advocates the need for human service workers to move beyond working with individual women and to engage them "in group efforts toward both individual and community change."

- Lazzari, Ford, and Haughey (1996), *"Making a Difference: Women of Action in the Community"*

The authors describe a qualitative study that "documented the contributions of 21 Hispanic women who were identified as being active in their community." The findings suggested the need for added research in the area and the importance of being active in one's community.

These are just a few of the journal articles we came across in our literature review illustrating how important macro practice is for both direct and indirect practice efforts. We hope this glimpse of macro practice in action can inspire you in your future practice endeavors.

CONCLUDING EXERCISES

1. What do you really know about the city/community you are practicing in? Do you know what social service agencies are available for clients in your community? What resources

can you turn to if you have to make a referral? Are there listings of social service agencies for your use? Consider visiting a social service agency in your community as a way of getting to know your community better. Another possibility is to invite an agency representative to visit your class to speak about the agency and the population it serves.

_____

_____

_____

_____

_____

2. If you could change the world in some significant way, what would you most hope for in terms of human change? Let's say that you have been given the power to pick one human suffering to be able to change. What would you use your power to change? Discuss the area of change you would focus on and ways to effect such change most successfully.

_____

_____

_____

_____

_____

Consider sharing in class your views, purpose, and direction. Also consider what agencies/efforts are already underway to work on your particular dream. If there is none, how could you begin to formulate a plan so that you could make some small difference in the area you wish to change?

_____

_____

_____

_____

_____

3. Teamwork, cooperation, leadership, and positive attention from management have importance that should not be lost on executives and staff of human service organizations today (Netting et al., 1993). Discuss what you believe this means and if you see this being implemented in the agencies you are practicing in today. If not, what impact do you think the current methods of management have on the staff of the agency? What do you see as the challenges of developing teamwork and cohesion within an organization?

_____

_____

_____

_____

_____

6

Practice with Diverse Populations

AIM OF CHAPTER

Learning to work with diversity is a prerequisite for anyone embarking on a career in human services. For the practitioner working with different populations and a myriad of issues, the value of becoming diversity sensitive, knowledgeable, and skilled is crucial. Practice is less likely to be effective unless the helper considers the multidimensions of people's lives. This chapter examines the many areas of diversity practitioners are apt to encounter in their daily work with individuals, families, and groups.

Diversity conjures up many thoughts and feelings, some more positive and accepting than others. For many, the notion of becoming diversity sensitive is tied specifically to ethnicity and embodies a multicultural perspective. We prefer to consider diversity in a much broader perspective in which ethnic sensitivity is only one important facet. Other areas of diversity that are equally important are those related to age, gender, socioeconomic status, and sexual orientation. We discuss each of these areas individually, presenting the salient issues, illustrating these with case scenarios, and providing suggestions for effective interventions.

PRACTICE ISSUES RELATED TO RACE, CULTURE, AND ETHNICITY

The most notable areas of diversity involve race, culture, and ethnicity. To become diversity sensitive, human service practitioners must understand and respect those who are different from themselves. The importance of integrating awareness, knowledge, and skill is emphasized by various experts in the field of multiculturalism. We concur with this view and highlight the need for practitioners to (1) become aware of their own biases, prejudices, and possible resistance toward becoming diversity sensitive; (2) be willing to gain knowledge about the specific cultural and ethnic populations they will be working with; and (3) be open to developing the skills necessary for effective practice with specific cultural and ethnic groups.

To provide you with a beginning framework for working with different racial, cultural, and ethnic populations, we present a brief history of various ethnic groups that highlights their plight and continued struggles in becoming part of mainstream America. Our intent is to help sensitize you to the many issues inherent in working with diverse groups and to encourage you to develop a more accepting, respectful, and flexible approach in your work with others different from yourself.

Historical Perspectives

Learning about the history of the particular ethnic group you will be working with is crucial to understanding the issues specifically tied to their ethnic development and cultural norms and values. Failing to consider a person's historical roots is much like conducting a clinical assessment with a client without asking about the person's social and childhood history. Many facets of one's past are instrumental in exploring present-day issues, so much would be lost if little attention were given to one's history. This is true both individually and collectively.

Successful work with clients from different racial and ethnic groups—immigrants, first-born, or fifth-generation Americans—requires a willingness to understand their experiences from a historical perspective. Taylor (1994) questions the actual existence of the "American family" as a "monolithic unit, that is, as a fixed and uniform, structure" (p. 1). Instead, he suggests that the "American family" be considered as "experienced differently by people in different social classes, different racial and ethnic backgrounds, and even by gender" (Baca Zinn & Eitzen 1990, p. xiii). Taylor sees

> the United States [as] a mosaic of family patterns created by a history of incorporating diverse racial and ethnic groups that came to this country at different periods and under vastly different social and economic conditions. These groups differ on a number of important attributes, including cultural histories, socioeconomic characteristic, kinship structures, and intergenerational relationships, and patterns of residence. (Taylor, 1994, p. 1)

Every person must therefore be understood on various levels: individually; within the family; within the larger social context in which they live; and also in regard to their racial, cultural, and ethnic group status. Our focus here is on the last level that takes into specific account a person's ethnic group.

Understanding Key Concepts

To understand fully the complexities involved in working with different ethnic groups, you need to be aware of certain terms. In the next sections we offer various definitions and examples that will be helpful.

Ethnicity and Ethnic Groups   Ethnicity has been defined in many different ways, ranging from a broad emphasis on traits that make someone different from ourselves (Gilman & Koverola, 1995) to a more specific view of ethnicity in which it is the "characteristics of people that distinguish them from others on the basis of religion, race, or national or cultural group" (Brill, 1995, p. 249). According to Gordon (1964), an ethnic group is

> any group which is defined or set off by race, religion, or national origin, or some combination of these categories.... All of these categories have a common social-psychological referent, in that all of them serve to create, through historical circumstances, a sense of peoplehood."(p. 27)

This definition recognizes an ethnic group as having a sense of common identity or "peoplehood" that is seen in feelings of togetherness and of being a special group.

Although many who belong to the same ethnic group are also of the same racial background, there is a difference as the two are not necessarily the same. An ethnic group shares some common ground that can be their religious beliefs and practices, language, historical continuity, and common ancestry or place of origin; a racial minority differs from the majority in its physical characteristics. Not all ethnic groups are minorities. It is not a group's ethnic or racial status that makes it a minority per se; rather, it is "a group's restricted access to sources of economic and political power and marginal location in the social order" that designate it a minority (Taylor, 1994, p. 2). An example might be Roman Catholics; they make up a specific ethnic group given their common sense of identity based on their religious views, yet they may be of different racial groups.

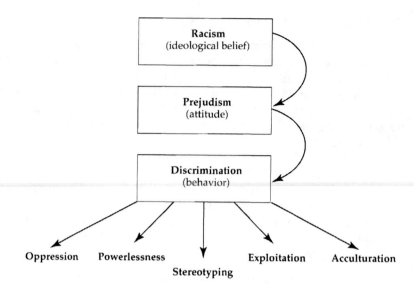

FIGURE 6.1   Institutionalized Racism

Race   Those who have a common set of physical characteristics are believed to be of the same race. However, the members of a racial group may or may not share the same sense of togetherness or identity that holds an ethnic group together. For instance, Latinos from various parts of Latin and South America may be of the same race yet they may not be of the same ethnic group.

Ethnocentrism   A common outcome of strong ethnic ties is the feeling of ethnocentrism that is defined as "the tendency to view the norms and values of one's own culture as absolute and to use them as a standard against which to judge and measure all other cultures" (*Encyclopedia of Sociology*, 1974, p. 101). This belief—that certain ethnic groups feel superior to others—is not a new notion; it can be systematically traced in the history of civilization. Nonetheless, such an elitist attitude could be potentially negative as it often gives rise to prejudicial views and discriminatory behaviors. As human service workers, we must guard against developing or perpetuating ethnocentric beliefs and instead work toward dispelling myths and stereotypes about culture, ethnic, or racial groups. In particular, we must be aware of any prejudiced views we may hold, as they may affect our work with certain clients.

Racism   Similar to ethnocentrism, racism has been defined as "the belief that race determines human traits and capabilities and that particular races are superior to others" (Brill, 1990, p. 249). When a person holds such beliefs, he or she will logically expect the domination of one racial or ethnic group by another. Lum (1996, p. 57) stipulates that racism is used as an "ideological system to justify institutional discrimination of certain racial groups against others." The visual representation in Figure 6.1 may help to explain the process whereby *racism* is considered an ideological belief that leads to an attitude of *prejudice*, which in turn results in the behavior of *discrimination*. Expression of this discrimination results in *oppression, exploitation,* and *stereotyping,* by the oppressors, and *powerlessness* or *acculturation* for the oppressed.

Prejudice, Discrimination, Stereotyping, and Oppression   *Prejudice* has been defined in many ways. Simply, "prejudice is an unfavorable attitude toward people because they are members of a particular racial or ethnic group" (McLemore, 1994, p. 121).

Prejudice involves the act of prejudging in which one makes a judgment in advance of examination. Racial or ethnic prejudgments usually involve thinking negatively of others based solely on their race or ethnicity with no real justification. The judgment is unfounded and likely to be accompanied by scorn, dislike, fear, and aversion (Zastrow, 1996, p. 376). A person's skin color and other physical characteristics often give rise to the belief that these traits determine their particular behaviors, values, intellectual functioning, and attitudes. Such prejudiced views are based not only on erroneous and unjustified assumptions but can also be extremely detrimental, both psychologically and socially, to the well-being of the particular group being targeted.

Theories of prejudice center on three direct causes: (1) cultural transmission, (2) personality traits, and (3) group identity. The first viewpoint contends that "the building blocks of prejudice are contained within the society's traditions or culture and are transmitted to children in a natural way as they are exposed to those traditions in the home and community" (McLemore, 1994, p. 123). Scenes in the movie *Mississippi Burning* exemplify this concept well; during a Ku Klux Klan rally, a father is holding his toddler in his arms and is encouraging him to cheer, which the child gladly does.

Personality theories

> reflect the idea that some kind of conscious or unconscious personality need or problem lies behind racial and ethnic prejudice.... Prejudice performs some important functions for the personality of the prejudiced person; it serves in some way to help the person cope with his or her inner conflicts and tensions. (McLemore, 1994, p. 129)

Group identity theories of prejudice consider that the person's group membership is a major molding force in how each member views himself or herself, those of the person's own group, and all others. Group identity is tied to ethnocentrism which we discussed above. In essence, this theory reflects the belief that, "members of an in-group ordinarily have feelings of loyalty and pride toward their own group and feelings of superiority and contempt toward members of out-groups. The in-group is seen as possessing the right, the natural, the human ways of living and thinking" (McLemore, 1994, p. 133). Because of the powerful nature of this type of thinking, we must be cautious in our own behaviors and avoid perpetuating these views.

A common consequence of prejudice is *discriminatory* behavior whereby members of a particular minority group are denied access to opportunities and rights solely because of their racial or ethnic group status. When individuals are judged because of the social group they belong to, rather than on their own merit, discrimination is at play. McLemore (1994, p. 121) defines discrimination as "an unfavorable action toward people because they are members of a particular racial or ethnic group."

When you work with others different from you prejudicial attitudes and discriminatory behavior can result. Because these can be subtle, being self-aware and cognizant of one's own views and behaviors in working with others is extremely important. Discrimination is rampant in our society. Examples are noted in minority group members not being chosen for a particular job, denied access to certain privileges, and excluded altogether from consideration due to their race or ethnicity.

Stereotyping is also interwoven in prejudicial attitudes and discriminatory behaviors. McLemore (1994, p. 125) defines a stereotype as "a shared, but not scientifically validated, belief concerning the characteristics of the members of different racial or ethnic groups as compared to some reference group." General stereotyping is a common behavior we all engage in at some time or another. It can be a helpful generalization. However, when we engage in negative stereotyping about particular groups of people, we can cause harm and inadvertently perpetuate existing stereotypes even further.

Oppression is an ultimate outcome of discriminatory behaviors. Zastrow (1996) defines oppression as "the unjust or cruel exercise of authority or power. Members of minority groups in our society are frequently victimized by oppression from segments of the white power structure" (p. 377). Other terms for oppression include *persecution, subjugation, domination, tyranny,* and *cruelty,* all suggesting the unfavorable and objectionable quality of this behavior. Despite our objections, it is fairly easy to note oppressive groups in our society today.

Discrimination and Oppression in Historical Context

Throughout history, societal acceptance of racial and ethnic minority populations has differed. Every individual as well as every ethnic group that has entered the United States has experienced their entry in a unique fashion molded, in part, by the various sociopolitical factors present during their arrival. According to Lieberson (1980), Steinberg (1981), and Wagley and Harris (1956), the acceptance of different racial and ethnic groups has varied along a continuum in that those groups who were closest to the national ideal in language (English), religion (Protestant), descent (Northern European), and physical appearance (light Caucasian) were preferred over those groups who were the furthest from the norm. These preferences, so to speak, were further reinforced by the enactment of various laws and policies through the years such as the Naturalization Law of 1790, Civil Rights Act of 1866, and the Immigration Act of 1965. Although clear progress has been noted in these legislative actions, discrimination and oppression still exist today for those who remain furthest from the norm. In human services practice, you will likely work with such disenfranchised ethnic groups and therefore must be cognizant of their specific group's experiences in regard to their acceptance or lack thereof on entering the United States.

Many reasons for racial discrimination and oppression have been promoted, including projection, frustration-aggression, insecurity and inferiority, authoritarianism, competition and exploitation, socialized patterns, and history. *Projection* is noted when a minority group may serve as a projection of a prejudiced person's fears. *Frustration-aggression* occurs when some who are frustrated displace their anger and aggression onto a scapegoat. *Insecurity and inferiority* are seen in those who discriminate to counter these feelings within themselves. *Authoritarianism* is connected to a distinct type of personality associated with prejudice and intolerance. *Competition and exploitation* result when certain groups justify their discrimination by developing an ideology that their group is superior and that it is right and proper for them to have more rights, goods, and so on. *Socialized patterns* are noted in the learned behavior phenomenon that is transmitted from generation to generation. Last, acts of discrimination and oppression are well documented in *history.* A look at the past of many minority groups illustrates how those viewed as "second-class citizens" have been those most conquered, enslaved, or admitted into our society on a subordinate basis.

Impact of Migration, Immigration, or Refugee Status

We can gain a better understanding of those from different races, cultures, and ethnic groups if we look at their particular place in history and their experience with discrimination and oppression. Knowing their specific immigrant status is also important in understanding their specific plight. According to Drachman (1995), "Familiarity with the issues surrounding the different immigrant statuses and an understanding that entitlements vary from status to status are necessary for effective service delivery." She encourages the development of programs aimed at gathering information about the entire migratory process that will help workers assist and connect individuals to appropriate resources and services. Drackman also challenges us, as human service workers, to become actively involved in advocacy efforts that support the immigrants who are most in need of services.

To help you better understand the immigrant populations, we offer an overview of the various immigratory statuses that exist today.

A *refugee*, as defined by the Refugee Act of 1980, is a person who is outside of and unable or unwilling to avail himself or herself of the protection of their home country because of persecution or fear of persecution on account of race, religion, nationality, membership in a particular social group, or political opinion. These individuals often have had to flee or been forced to leave their country suddenly, leaving with few possessions and without proper good-byes and closure to life in their native land. Because of their sudden and, at times, unforeseen departure from their homeland, refugees often are in need of special types of assistance not necessary for other immigrants. These services, included in standard resettlement services for which refugees are typically eligible, consist of relocation assistance, financial aid, medical care, English language training, employment and vocational counseling, and job placement. These services, extremely necessary for survival, do not address the emotional needs of these individuals. This is where human service workers can be instrumental in helping refugees successfully adapt and assimilate. Also, the specific reasons for their refugee status may have a significant bearing on how they are accepted and assisted on their arrival into the United States. For instance, refugees from communist countries may experience their entrance into the United States differently from those from a war-torn country such as El Salvador.

A *parolee* is an individual who is admitted into the United States under emergency conditions or is admitted if the person's entry is considered in the public interest. Typically a parolee is not entitled to the same benefits as a refugee, although they may be from the same country. This discrepancy has at times led to a parolee feeling rage, depression, confusion, struggle to find work, and an overall sense of impotence in the resettlement experience (Drachman & Halberstadt, 1992).

An *immigrant* is an individual lawfully granted permission to reside permanently in the United States. There are different ways to gain such permission. Access and entitlement to services may also vary in accordance with the way this status is achieved. Most immigrants are required to have a *sponsor* to be eligible for citizenship. Also, services to these individuals are often restricted because of their immigratory status. Understanding the reasons for their decision to become immigrants in the first place is all-important in understanding their particular situation. Those who did not choose to immigrate but felt forced to, or who came expecting something different are likely to be more unhappy, frustrated, angry, or depressed than those who came on their own accord and were realistic in their expectations upon arrival.

Temporary residents have legal permission to live and work temporarily in this country. Many are here on student or work visas; others are undocumented and have gained legal status through the government-established amnesty program. Typically, the stay for these individuals is time limited, yet they are eligible to become permanent residents after a period and usually must do so to maintain their legal status. For some, the prospect of having to return to their country is a given. This may be experienced with enthusiasm as they may feel home sick and after a taste of American life, prefer to return to their homeland. For others, however, the transition back to their country is difficult and painful. This is seen when the individual is forced to return, or is not ready to do so but is bound by time restrictions.

Undocumented persons are those who have no current authorization to be in the United States. This group comprises two main segments: (1) those who enter illegally, and (2) those who enter legally under a nonimmigrant status for a temporary period (e.g., students and tourists) and who remain in this country after their visas expire. The plight of the undocumented is perhaps the most difficult of all as they are subject to deportation and are ineligible for many of the services refugees and legal immigrants are afforded. They often live in fear of being caught, which fosters a tendency to avoid needed services such as health care; police protection in cases of assault, rape, and other crimes; and educational services for themselves and their children. As a result of their undocumented status, many are also subject to exploitation and oppression such as being paid less than minimum wage for working in substandard conditions, being charged high prices for rent while living in inhumane conditions, and being subject to scams related to becoming legalized. Perhaps most troubling for these immigrants is the extent of the discrimination they must face. Many are taken advantage of physically, emotionally, and economically and used as scapegoats for existing societal problems. Most are considered inferior and thus more easily mistreated and subjected to the perils of prejudicial attitudes, racist views, and discriminatory practices. Negative stereotypes prevail about those ethnic groups most commonly associated with undocumented residents, and these individuals often have feelings of powerlessness and shame. As human service workers, we must be humane in our treatment of this population and not condone societal persecution and maltreatment. We must see them as human beings first, then as undocumented individuals. The societal climate noted in passage of California Proposition 186 in 1996 clearly illustrates how negatively the undocumented are viewed. Not only are they blamed for certain problems in our society but they are denied the same services afforded those who are documented and citizens.

Stress Factors   For all immigrants, regardless of their immigratory status, stress is a universal phenomenon. The loss issues involved in relocation can be great as many immigrants leave family, friends, community, and homeland. On arrival in the United States they must face such stressors as adjusting to a new country, a new environment, and a new language while simultaneously having to find housing and secure gainful employment. For some, the migration process itself was traumatic, as they had to leave their homelands abruptly and with little warning. The loss of lives of family members or friends while en route to the United States is often experienced by the undocumented who enter this country illegally and thus must take heavier risks that involve the threat of deportation, violence, and in some cases, death. Stepick (1984) has found that the adjustment is more difficult when the cultures of the countries of origin and the policy of the receiving country are harsh on the newcomer, thus complicating adjustment.

Hulewat, in her article "Resettlement: A Cultural and Psychological Crisis," speaks directly to the issue of resettlement that most racial and ethnic groups entering the United States must face. She notes that resettlement is a crisis for both the individual and his or her family. This crisis is natural for individuals who must leave behind all that is familiar to them and start a new life in a new country with a different language and culture. "Resettlement is a crisis that can offer enormous opportunities as well as many risks. The resettlement worker's task is to maximize the opportunities and minimize the risks by helping families with the many concrete tasks of immigration while remaining attentive to their many emotional needs" (Hulewat, 1996, p. 129). She further identifies various areas that she believes must be attended to by helpers to be effective in their work with people who are resettling. This help includes the three central concepts below:

1. Understanding the five stages of resettlement that illustrate the different emotional and concrete tasks necessary for the completion of each stage. The goal for the helper is to assist the immigrants "to integrate their past cultural experience into their new life and culture" (p. 129). Briefly, these stages are as follows:

• *Stage 1—Preimmigration or Preparatory Stage* during which "splitting," considered an adaptive tool to maneuver defensively to protect against intolerable feelings, is often used. Although considered an adaptive mechanism, splitting can also have pathological connotations; this is true especially when used by individuals with borderline personality traits and when it is used long term. Ideally, splitting occurs as a temporary defense against a transitory crisis—in this case, the resettlement process.

• *Stage 2—Actual Immigration*, which will vary from one particular group to another and from one individual to another.

• *Stage 3—Arrival in the New Home Stage*, which can be both exhilarating and exhausting. Often it is during this stage that people are unaware of the cumulative effects of the stress they have been under. Efficiency is encouraged and the dominant mood is anxiety for most, with optimism in some. Sometimes anxious depression ensues when the tasks of looking for a job and learning a new language are fully realized. This is often the stage when workers are likely to be the most involved and apt to be of most help.

• *Stage 4—Decompensation Stage* occurs when the family begins to realize and experience the loss it has suffered. This is a time to make peace, as the ability to balance the past and the present is crucial. Ambivalent feelings must be acknowledged and efforts made toward adaptation and acceptance.

• *Stage 5—Transgenerational Stage*, in which unresolved conflicts from the immigration experience are passed on to succeeding generations. (Hulewat, 1996, pp. 130–131)

2. Understanding the cultural styles and psychological dynamics of the population being resettled. "The degree to which those dynamics create internal dissonance or consonance in the immigrant as he or she attempts to adjust to life in the United States is significant in determining how the resettlement process will proceed" (p. 129).

3. Understanding the individual dynamics of the family. "Each family brings its own cultural understanding as it has been processed through the unique life experiences of its members. These individual dynamics affect the family's ability to tolerate cultural dissonance and to manage the tasks necessary to proceed through the stages of resettlement" (p. 129).

Adaptation, Assimilation, and Acculturation Issues

Resettlement can indeed be a crisis time for many families in which they face adaptation, assimilation, and acculturation. For instance, adaptation, which refers to the capacity to adjust to surrounding environmental conditions, is certainly a process that most immigrant families face on their arrival in the United States. Adaptation involves changing to "fit in" and survive in the surrounding environment. The degree to which such adaptation is accomplished will vary from family to family and culture to culture. Much will depend on the immigrant status of the particular family; their similarities to Americans in appearance, family values, customs, and language; their desire to enter the United States; the realistic nature of their expectations; and the specific political, economic, and racial climate of the surrounding environment where they plan to migrate. Accordingly, factors that influence both migration and adaptation include the language, race, country of origin, and place of residence. McGoldrich, Pearce, and Giordano (1982) elaborate on these, suggesting that they are also factors that affect ethnicity; others include socioeconomic status, educational achievement, and upward mobility of family members and the political and religious ties to the ethnic group.

Examples of both a difficult and a smooth adaptation to migrating to the United States reveal how individual this process can be. In the first instance, we have an undocumented immigrant family from Latin America who do not wish to enter the United States but do so because they are fleeing violence and warfare within their country. They really did not want to leave their homeland, and on their arrival notice in America how different they are in terms of racial and physical appearance, values and customs, language, and so on. They feel unwelcome and discriminated against almost from the instant they arrive.

The second example illustrates a smoother adaptation. Here we have an immigrant family from Canada who wish to enter the United States legally seeking increased educational and occupational opportunities for their family. They speak English, look like Americans, and have many of the same cultural traditions and values. They find adaptation relatively easy.

Piaget, who writes extensively about the adaptation process as it relates to cognitive development in children, would say that adaptation is composed of two processes: assimilation and accommodation. We highlight the assimilation and accommodation processes as they relate to ethnic issues.

Assimilation in its most basic form refers to taking in new information and integrating it into the schema or structure of thought. This occurs when a person is exposed to a new situation, event, or piece of information. Not only is information received and thought about at a conscious level, but it is also integrated unconsciously into a new way of thinking. Regarding ethnicity, assimilation relates to how well immigrants and racially defined minorities have been able to incorporate into society—in essence, how well they have been able to absorb, integrate, and blend into the American way of life. The notion that all immigrant and racially different families were subject to a universal process of incorporation into society was a key assumption of the now famous theory proposed by Park in the 1950s on racial and ethnic assimilation known as the "race-relations cycle" (Baca Zinn & Eitzen, 1990). In his view, when a dominant group (Americans) comes into contact with certain minority groups, the two groups enter a series of interactions characterized by competition, accommodation, and eventual assimilation. Park viewed this cycle as progressive and universal: "Customs regulations, immigration restrictions, and racial barriers may slacken the tempo of the movement; may perhaps halt it altogether for a time; but cannot change its direction, cannot, at any pace, reverse it" (Park, 1950, p. 150). Park's views contained a strong "assimilationist bias" in which he insisted that assimilation was a necessary and desirable outcome of dominant-minority group interactions. Although this theoretical perspective dominated the literature for over half a century, the reality of this so-called assimilation process was questioned by the very policies enacted by the United States that promoted the integration of only certain ethnic minorities within the American mainstream. The existence of such laws as the Naturalization Law of 1790, the Civil Rights Act of 1866, and the Immigration Act of 1965 clearly illustrate how truly difficult it is to undergo the assimilation process proposed by Park. For decades, the assimilation process of free "white" immigrants has been easier than for those who are either "not free" or "not white." Many would venture to say that this subtle yet pervasive policy continues to prevail today and will persist as long as discrimination and oppression coexist.

Accommodation, according to Piaget's schema, refers to the process by which children change their perceptions and actions to think using higher, more abstract levels of cognition. Children take in new information and eventually accommodate it; they build on the schema they already have and use new, more complex ways of thinking. Regarding ethnic and racial issues, many immigrant and minority families begin the assimilation process by learning to accommodate to the American way of life: learning the language, abiding by American laws and standards, incorporating certain American traditions into their own family traditions, and so on. Many factors will influence how well certain families and individuals learn to accommodate: Some may be unwilling or unable to engage in this process; others may find it relatively simple.

Acculturation applies to how well cultural assimilation has taken place. This is directly tied to how well immigrants have been able to incorporate the new culture versus retaining the culture of their country. Acculturation occurs at different levels for different minority groups and even within family systems. Retention of the language and cultural patterns of the country of origin often impede acculturation. Also, in many geographic areas where a certain immigrant or minority group is strong in number, institutionally complete enclaves have developed where immigrants are insulated from the dominant culture. Examples include Miami's "Little Havana" and San Francisco's "Chinatown." Also, the perception of the immigrant population as either "reluctant" (i.e., Cubans) or "unwanted" (i.e., undocumented workers from Mexico and other parts of Latin and South America) has a bearing on how willing or motivated they will be to assimilate and become acculturated into this society.

A common outcome of the acculturation process is intergenerational tensions within the family where the parents/elders are more likely to hold onto strong cultural values that may be different from the American views their children have begun to internalize. According to Bernal (1982, p. 197), an example of the conflicting value orientation with respect to dependence and independence is clearly noted in many Cuban families, where the culture fosters the continued dependence of children and teens on their parents. In these families, the children are more likely to have internalized the norms of independence commonly found in U.S. society, thereby creating undue stress within the family system.

As an example of the difficulties of acculturation, a Latino female moves out of her home at 22 to attend graduate school. Although a positive step by most Western standards, this was not an acceptable behavior within her Latin culture. She had acculturated, but her parents had not, thus creating much conflict and resistance.

The option of biculturality has been recommended as a way of adaptation that reduces intergenerational tension. From this framework, each generation adjusts to the other generation's cultural preference or "parents learn how to remain loyal to their ethnic background while becoming skilled in interacting with their youngsters' Americanized values and behaviors, and vice versa" (Szapocznik & Hernandez, 1988, p. 168).

Similarly, Spencer and Dornbusch (1990) consider biculturalism an option to avoid complete assimilation, in that each person learns to negotiate both his or her ethnic values and the values of the dominant culture. We tend to favor this view as it allows the value structures of both groups to be respected, retained, and available to the individual. This lets the person negotiate which standards to use depending on the situation and his or her preference. The approach is not only more realistic but also seems more in line with a self-determination stance and with the "cultural pluralism" perspective we discussed earlier. Furthermore, for many minority groups who view as a priority the retention and maintenance of their culture, language, and traditions, this bicultural option lets them remain true to the culture of their homeland while allowing them to learn American culture and values. As human service providers, we must respect others and assist them in the ways they wish to be helped.

Cultural Variations of Family Systems and Social Supports

Specific Functions of the Family: Family Values, Structure, Rituals, and Socialization

In working with individuals and families from different cultures, we must understand the specific functions of the family such as family values, rituals, and aspects of socialization.

Family Values   Most families possess a set of family values that dictate how certain events and experiences are to be interpreted, believed, and performed. Environmental, educational, and experiential factors all play a part in the development of many of these family values. Typically, as pointed out by Trotzer (1981), parents are the "carriers, monitors, developers, reinforcers, and interpreters of their children's values" whereas the "children merge their experiences with the expectation of their parents and significant others" (Tsing & Hsu, 1991, p. 93). The maintenance and transmission of values within the family are important family functions and deserve careful attention from practitioners who will bring their own values into their work. Family and gender roles, power and authority issues, and family activities all reflect family values.

Family Structure   Family structure will differ from family to family; however, commonality seems to exist among individual ethnic groups. Lum (1996, p. 10) considers the ethnic family structure to be "based on such cultural values as the relationship of humanity to nature and the environment, harmony, past-time orientation, collective relations with people, being-in-becoming, and a positive view of the nature of humanity." Some ways to learn about traditional family structure are to observe how children are cared for, what the parent-child roles are like, and how sibling relationships are structured. In life transitions or times of crisis, how does the family respond? Do they turn to extended family for support? Do they seek help from religious healers or leaders? Do they ask for professional help?

Family Rituals   Family celebrations and family rituals will vary even within the same ethnic group. Family rituals as described by Tseng and Hsu (1991, p. 96) are "repetitious, highly valued, symbolic occasions or activities observed or undertaken by a family." According to Bossard and Bell (1950), family rituals are a part of family culture through which enduring values, attitudes, and goals are transmitted. Carter and McGoldrick (1989, p. 83) believe "it is extremely important to reinforce a family's use of rituals so as to reinforce their sense of identity." Clearly, rituals are very important in promoting a sense of family identity, pride, cohesiveness, solidarity, and continuity. Tseng and Hsu (1991) have further categorized family rituals into (1) culture-related family celebrations, (2) family-observed ritual activities, and (3) patterned interactions practiced by the family.

Each ethnic group, and for that matter, each family engages in their own rituals and chooses to celebrate certain events in their own unique way. Generalizations can be made, however. For instance, according to McGoldrick, Pearce, and Giordano (1982, p. 10),

> Cultural groups vary also in the emphasis they place on different transitions. The Irish have always placed most emphasis on the wake, viewing death as the most important life cycle transition. Italians, in contrast, emphasize the wedding, while Jews often give particular attention, to the Bar Mitzvah, a transition most groups hardly mark at all. Families' ways of celebrating these events differ also with [the] Irish [by] drinking, the Poles by dancing, the Italians by eating, etc.

Holidays are celebrated differently by each group as well. Some cultures emphasize Christmas, others may highlight Easter, and others may honor the Jewish or Chinese New Years as the most important. "These customs evoke deep feelings in people that relate to the continuity of the rituals over generations and centuries" (McGoldrick, Pearce, & Giordano, 1982, p. 10). Because of their special meaning, these holidays should be respected by human service practitioners, who should be careful to not criticize them unduly. There may be times when helpers will not understand the purpose behind a ritual and thus find it irrational or unnecessary. In such situations, they should examine the cultural rituals more carefully and perhaps research their significance before passing judgment on their relevance.

Similar to family rituals are "rites of passage" that are often influenced by ethnicity. Generally, regardless of cultural background, the universality of rites of passage suggests their importance to the human species.

Family Socialization   Understanding the way family members socialize with one another as well as with those outside the family can provide valuable information about the family system and how it functions. Assessing family socialization patterns can provide insights about family and friendship bonds and the typical division of labor and problem-solving styles used within the family. For practitioners, it is especially important to understand the inner workings of a family, as this will provide a more comprehensive view and allow a tailored treatment approach.

The many studies about family socialization are too numerous to cite here, so we offer only a few examples. In British West Indies families, socialization emphasizes the obligations children have to their families, especially when the family has sacrificed for the child to get ahead (McGoldrick, Pearce, & Giordano, 1982). In Japanese families, the division of labor between husband and wife is very clear, placing them in separate relational worlds. This division of labor is not quite as clear in middle-class West German families, where the husband is much more involved in family life (Tseng & Hsu, 1991).

Cohler and Lieberman (1980) concluded from their studies that when efforts are made to maintain strong family ties, individual family members are better adjusted personally and can overcome stressful life events better due to the support and concern they are apt to receive during times of crisis. Even so, in some instances, emphasis on maintaining such close family ties can actually be a primary source of family strain (Tseng & Hsu, 1986). Overall, the presence of an extensive social network has been found to be equally important to an individual's adjustment and life satisfaction (Pakel, Prusoff, & Uhlenhuth, 1971).

Sociocultural Issues and the Family

Immigrant and minority families experience numerous types of unique sociocultural experiences that have a significant impact on their lives, both individually and within the entire family. Three commonly noted examples are sociocultural change, migration, and the meaning of being a minority. We have already touched on issues related to migration as well as those related to discrimination and oppression faced by many minorities. We have not, however, dealt in any length with the ramifications of sociocultural change.

Sociocultural Change   Sociocultural change includes changes in family structure, living conditions, and values systems. For instance, the American family has undergone many changes since the 18th century. "Prior to the 19th century, the American family was a more rural family unit that emphasized cooperation and self-sufficiency as well as the economic arrangement aspect of marriage, and viewed familism and authoritarianism as two basic principles to govern internal organizations" (Tseng & Hsu, 1991, p. 100).

Family life changed greatly after World War II, and even since 1960 there have been many changes in America that have greatly influenced family life. Many have resulted from the sharp decline in fertility rates, an equally sharp rise in the divorce rate, a large increase in the labor force participation of married women, and the growth of nonmarital cohabiting relationships (Cherline, 1983). Because of the emphasis on individualism and democracy,

> the contemporary American family must face transitional conflicting values from authoritarian familism to democratic individualism, changing values that challenge marital stability and the need for a meaningful marital relationship, the pressure of the romantic love concept, the economic independence of women, and parental confusion and conflict about child rearing. This indicates that a society's family system will change in association with its historical path. (p. 100)

Consider the impact of change on a family from another country who, in addition to the above, must also have to change many other facets of their lives. The physical, psychological, and societal stressors are vast for anyone in this situation.

At times, change occurs at such a rapid pace and is so great that it is likely to affect most families within that society. Often change is more easily accepted and accommodated by the younger generation who are more able to absorb it swiftly and reframe their views. In contrast, it is commonly members of the older generation who feel especially challenged and perhaps even intimidated by changes in social structure, hierarchy, lifestyles, common beliefs, and value orientations that threaten their authority and traditional value system. It has also been noted that young males are more sensitive and vulnerable to change when it is rapid and results in disorganization and family confusion. Although both genders respond better to structure, clear direction, and goals, and the benefit of parental guidance and support, males more than females tend to manifest open conflictual and rebellious behavior toward their parents when change occurs rapidly within the family structure or there is change in the family's value system.

Stressors and Coping Styles

Family Stress   Family stress occurs in most families, yet because of the added stressors related to migration and being a minority, immigrant families and racial minorities are often at greater risk for suffering from stress-related conditions. Family stress involves any strains, burdens, problems, or conflicts that seem to cause considerable discomfort, tension, or frustration for family members. Some stressors are clearly culturally related; for instance, issues tied to resettlement and minority status are cultural. For some families, other cultural issues may cause stress as well. Examples include the pressure for some ethnic groups to have large families, given their religious views on procreation; other issues can arise from strong cultural views of insisting that extended family live in the home, or that women should stay home and care for their children whereas the male is to be the primary breadwinner. None of these are necessarily negative or stressful in and of themselves; yet when placed in the context of contemporary American life that minimizes these views, they are likely to cause stress. In an effort to be ethnic sensitive, human service practitioners have to be in tune with the common stressors faced by minority families and be able to assess existing stressors when working with a particular family system.

Family Coping   Coping, which relates to the manner in which one attempts to reduce or manage a demand, is often thought of as an individual matter; but family methods of coping also exist and can be valuable to assess and work with. Tseng and Hsu (1991) consider family coping to be evidenced in the "collective family-group action to eliminate or manage demands, either through coordinated family-group behavior or complementary efforts of family members" (p. 106).

According to Patterson (1988), there are different ways coping can be accomplished on a familial level:

1. direct action to reduce the number and/or intensity of demands;
2. direct action to acquire additional resources not already available to the family;
3. maintaining existing resources so they can be allocated and reallocated to meet changing demands;
4. managing the tension associated with continuing strains; and cognitive appraisal to change the meaning of a situation to make it more manageable. (pp. 202–237)

Tseng and Hsu (1991) add a fifth component

5. promoting changes in family structure and relations that give the family more strength and skills for dealing with strain and demands effectively. (p. 107)

In all these coping styles, cultural issues may be present. In some cultures, ties with extended families are thought to be a source of emotional and instrumental strength (e.g., upwardly mobile black families) versus a serious strain. Similarly, some cultures seek spiritual support in times of crisis (e.g., Hawaiians) whereas others (Chinese Americans) prefer to mobilize the resources that exist in the family. These few examples suggest that in many situations there are culture-preferred patterns or culturally sanctioned ways for families to cope with stressful life situations.

Special Issues   Special issues that must also be taken into consideration when helpers work with different ethnic groups include intercultural marriages, cultural differences between generations, and problems of culturally uprooted families. Practitioners must learn about the specifics surrounding these issues and develop the necessary skills to help clients work through them.

Culturally Relevant Treatment Considerations

Considering Culture in Assessment

When working with those who are different from us, we must extend our orientation and understanding of individual and family dynamics to accommodate the racial, cultural, or ethnic differences they present. According to Lum (1996, p. 216), "Culture consists of the ideas, customs, skills, and arts of a group of people that have been cultivated and passed on from generation to generation. Cultural beliefs and practices have enabled people of color to persevere in the face of racism, prejudice, and discrimination." Therefore, to be culturally sensitive when working with different cultures, practitioners need to focus on cultural assets from a strengths perspective and the natural supports systems common to many ethnic groups.

Typically when working with families, practitioners often focus on family development, family subsystems, and the family group itself. With immigrant or minority families, however, focusing on these issues alone is not enough. Additional attention must be paid to their marriage forms, kinship relations, marital residential choice, family authority, and family values. Tseng (1986) cautions that without focus on the cultural aspects involved, the danger of "cultural myopia" is heightened. Cultural myopia occurs when helpers judge the health and functioning of their clients by the helpers' own standards, thereby failing to recognize the clients' cultural and family variations.

Four major components of cultural formulation that must be considered in any psychosocial assessment include the cultural identity of the individual, cultural explanations of the individual's illness, cultural factors related to the person's psychosocial environment and levels of functioning, and cultural elements of the relationship between the individual and the clinician (American Psychiatric Association, 1994; Lum, 1996). Incorporating these four aspects of cultural formulation into your practice is highly recommended.

Congress (1994) proposed using a culturagram that consists of 10 categories of cultural information for a specific family:

1. Reasons for immigration
2. Length of time in the community
3. Legal or undocumented status
4. Age of family members at time of immigration
5. Spoken language
6. Contact with cultural institutions
7. Health beliefs
8. Holidays and special events
9. Crisis events or stressors
10. Family, education, and work values. Whether used as a brief assessment tool or in lieu of a more traditional psychosocial assessment, this culturagram can provide the practitioner with relevant information about acculturation issues and functioning level of each family member.

As important as considering the relevant cultural issues is the need to clarify *normality* and *dysfunction* as they pertain to the specific culture being considered. For instance, it would not be uncommon for helpers raised in the Western culture to value independence and thus have a tendency to consider enmeshment and dependency within a family as pathological. In many cases, this assessment would be erroneous as these behaviors may be considered to fall within the norm of expected behaviors in a specific culture.

It is also important to recognize that cultural variations may exist within families. As helpers, we must assess each family individually as well as considering the possible influence of its cultural background. Another important aspect of working with those who are culturally different is that as helpers, we carry our own culture-influenced perceptional bias. Because of the inherent danger that we might act solely on our biases, those of us working with ethnically different populations must be self-aware and remain open to adjusting and correcting our biases, regardless of how insignificant they may seem. We agree with Tseng and Hsu that "the perceptions of a family's interaction pattern is subject to and influenced by the evaluator's own cultural background, and is in need of cautious adjustment in the actual transcultural practice of family assessment" (1991, p. 174).

Tseng and Hsu (1991) also address the issue of a cultural blind spot. They believe that no matter how hard helpers try to be culturally sensitive, they may still have a cultural blind spot that limits their ability to understand completely the cultural meaning of certain behaviors in their clients. Avoiding this completely is impossible, so it becomes even more imperative for helpers to remain alert to their own as well as their clients' reactions that may signal a misunderstanding or misinterpretation. When helpers are willing to inquire into and explore meanings within the cultural context, they can minimize misunderstandings and further enhance the trust between helper and client.

Process of Culture-Adjusted Family Assessment

When practitioners work with culturally different individuals and their families, they must be able to adjust their assessment methods to fit the particular needs of the client's culture. Tseng and Hsu (1991) highlight several issues that need to be considered with families of different cultural backgrounds. Although some may seem obvious, others may not. The issues are listed here:

- *Cultural engagement* whereby the helper engages the family in accordance with the background of the family members (e.g., personal style, educational level, and social class) and culturally sanctioned social greetings and etiquette.
- *Gearing to existing authority hierarchy* involves being able to recognize hierarchical patterns quickly at the outset and attempt to use them in working with the family system.
- *Culture appropriate communication* must be recognized, observed, and respected. This may involve the channels of communication, the language of expression, and the way meanings are transmitted.
- *Culture-relevance to exposure to private matters* relates to being sensitive to certain subjects that may be private for certain cultures; these rules must be observed and respected.
- *Establishment of an appropriate role for the therapist*, depending on how the cultural system views treatment.
- *Clarification of approach, procedure, and goal* is essential in orienting clients from different cultures to the concept of treatment, its purpose, and the importance of goals.

We end our discussion on assessment by presenting an outline of a cultural family assessment form presented by Tseng and Hsu (1991). We believe it consolidates all the major factors necessary to engaging in a thorough and comprehensive assessment of clients from different cultures.

Cultural Family Assessment Outline

1. Basic Information
2. Presenting Situation
3. Ethnic/Cultural Background
4. Current Family Functioning
5. Family Interactional Functioning (Family Mental Status)
6. Family Stressors/Problems
7. Problem-Solving Patterns
8. Family Strengths
9. Family Satisfaction
10. Family Diagnosis and Dynamic Formulation of Problem
11. Suggested Interventions

In sum, to conduct a culturally relevant and clinically comprehensive assessment that will aid in appropriate treatment endeavors, helpers must consider existing cultural factors along with those normally considered in a typical family assessment. These include ethnic/cultural background of the client, the family coping patterns, and family strengths. Cross-cultural issues must also be considered when results are interpreted because many standardized assessment forms and testing measurements do not accurately reflect the particulars of the client's cultural background.

Considering Culture in the Development of Treatment Interventions

In working with different races, cultures, and ethnicities, helpers must take care to consider the most appropriate ways to engage in professional helping. Helpers need to explore the most meaningful and effective ways of helping clients that take into consideration and respect their cultural background. This sometimes requires alternative and creative ways of helping clients who are not receptive to traditional helping modalities. Whether working with individuals or families, practitioners must consider each family member's views of professional helping. Sometimes family members are opposed to seeking help from the outside and may not welcome your assistance. There may be times when individuals with this negative view can undermine the entire therapeutic process.

The two major factors that are important at this juncture are the therapeutic relationship and the selection of treatment modalities. The therapeutic relationship is a key ingredient for effective treatment outcomes. This is true in any helping relationship but is perhaps even more critical with clients of different cultures and ethnicities. Proctor and Davis (1994, p. 314) assert that

> race continues to have significant—sometimes volatile—effects on personal relationships, both in society at large and in professional practice.... Often, the racially dissimilar social worker [practitioner] and client approach each other with little understanding of each other's social realities and with unfounded assumptions.

Some theorists (Banks, 1972; Jones, 1979; Siegel, 1974) have addressed this same issue, indicating that "client-practitioner rapport may be difficult to establish because of racial in-group-out-group mistrust" (Davis & Gelsomino, 1994, p. 321). Nonetheless, no one has proven that racial dissimilarity necessarily impairs treatment outcome; in contrast, evidence suggests that practitioners who are experienced, sensitive, and skilled can work effectively with clients from whom they are racially dissimilar (Davis & Proctor, 1989). For this to occur, however, helpers must consider several factors:

- Practitioners must be responsible for developing a personal comfort with racially dissimilar clients.
- They should appreciate the stressors associated with racial dissimilarity and be willing to examine critically the range of their own exposure to minorities and cultures. When a practitioner's experiences are limited, he or she is encouraged to engage in interethnic exposure.
- Practitioners must acknowledge the racial differences with the client and convey their own goodwill and understanding.
- Practitioners must take responsibility for acquiring the knowledge and skills necessary to work effectively with different clients. This behavior communicates professional competence and helps foster trust in the helping relationship.

Tseng and Hsu (1991) have also discussed the importance of the client-helper relationship and suggest attention to three areas:

- Be alert to cultural transference of the therapist-client relationship on the therapeutic process.
- Respect and utilize the culturally defined and sanctioned family hierarchy and relations already existing within the family.
- Be aware of the effect of the therapist's cultural background and value system on the therapeutic relationship.

Selection of treatment modalities is a complex process that requires consideration of numerous factors. Practitioners must be comfortable and well versed in treatment strategies, and the client system must be willing to accept the specifics of the treatment process. Practitioners must match the treatment modality with the specific individual or family and deal with family matters according to cultural priorities.

Overview of Frameworks for Ethnic-Sensitive Practice

There are many different views about how to engage in ethnic-sensitive practice. A most renowned framework was proposed by Devore and Schlesinger (1981), who wrote one of the first books on ethnic social work practice. Their views rest on two basic propositions: (1) that ethnicity and social class shape life's problems and influence problem resolution, and (2) the problem-solving process must focus simultaneously on micro and macro problems. They emphasize the need for helpers to be well versed in two main areas: general knowledge of human behavior and specific knowledge of ethnic group and social class.

Green (1982) is also to be credited for his model of *help-seeking behavior*. He recognizes the diverse perceptions of ethnic groups in today's society and encourages practitioners to be sensitive to this. Many ethnic groups are reluctant or resistant to seeking professional guidance of any type so helpers need not only to acknowledge this but also to attempt to understand the underlying reasons for this behavior. Green's model comprises the following main components:

- The client's definition and understanding of an experience as a problem
- The client's use of language to label and categorize a problem
- The availability and use of indigenous community resources and the decision making involved in problem-intervention strategies
- The client's cultural criteria for determining problem resolution

This model highlights use of ethnographic information in treatment planning, intervention, and evaluation. Ethnographic researchers "seek to learn from those they are studying and to understand the perspectives, customs, behaviors, and meaning within the context of their culture" (Thornton & Garrett, 1995, p. 68). By observing clients in their environment, helpers can obtain important information about a person's or group's cultural values, family structure, child-rearing practices, religious beliefs, experiences with discrimination and oppression, and level of acculturation. By making an effort to understand your clients' point of view, you convey the message that they are the experts and you're the learner; perhaps most important, you indicate that you are truly interested in understanding their situation and respect their worldview.

Another pioneer in the field of ethnic-sensitive practice, Ho (1987) conceptualized a multicultural family therapy framework for working with culturally diverse families. His framework takes the following six basic factors into consideration: (1) ethnic-minority reality; (2) impact of external system on minority culture; (3) biculturalism; (4) ethnic differences in minority status; (5) ethnicity and language; and (6) ethnicity and social class. Many of these are reminiscent of the culturagram proposed by Congress, discussed earlier.

Westwood and Ishiyama (1990) offer practical suggestions for working with culturally different clients.

- Encourage clients to speak their own language (words and phrases) to best illustrate how they are feeling at the moment.
- Because nonverbal communication is often culturally based, check for accuracy of your interpretations when in doubt about clients' nonverbal communication.
- Be open to using alternate modes of communication, other than a solely verbal exchange, to increase client's comfort and involvement.
- Attempt to understand and learn culturally meaningful expressions used by clients.
- Encourage clients to bring in material, such as books, photographs, and other articles of significance, that might help you understand their perspective.

The last framework we consider is an integrated approach that we have developed as a comprehensive view of ethnic-sensitive practice. In this approach, we integrate significant insights from various authors with our own suggestions. In contrast to the other frameworks, we do not elaborate on specific treatment considerations, but present an overview of the most important factors of working with diverse populations. Our intent is to capture the essence of Sue and Sue's recommendations of being aware, knowledgeable, and skilled.

Know yourself: Human service practitioners must be willing to engage in self-definition and introspection. Gilman and Koverola (1995) list the following ways to work on knowing who you are.

- Having a clear sense of your own ethnic and cultural background—getting in touch with your roots
- Being clear about your religious and cultural belief systems
- Examining your own value system
- Remaining open to examining prejudices and negative stereotypes that you may associate with certain cultural groups
- Accepting yourself, with your strengths and limitations; being open to examining your underlying motivations for working with others

Use empathy versus sympathy: It is important to be empathetic, not sympathetic, with all clients, especially those from an oppressed population. "Sympathetic [helping] is simply another form of oppression" (Gilman & Koverola, 1995, p. 381). Being empathetic involves being sensitive to clients' circumstances but does not include feeling pity or claiming you understand how they feel.

Being yourself: Allow yourself your own mannerisms, style of dress (keeping it appropriate), and sense of humor. You won't work better with certain clients by changing who you are. You will do best to be yourself and model the importance of being authentic to your clients.

Avoid stereotyping: Refrain from making assumptions that may be incorrect or even prejudiced.

- By engaging in continual self-awareness, you can remain open to examining your views; this will help you avoid stereotyping clients.
- Know the policies of the agency you represent.

Know your clients: Be willing to learn about your clients so you can be understanding and sensitive to their special needs.

- Recognize that we all possess the capacity for thought, feelings, and behavior.
- Develop an appreciation for different cultures. According to Brill (1995), it is essential "for workers to be equipped with knowledge of the past history of the particular ethnic group" and to "be knowledgeable about and sensitive to the meaning of the present status and situation of the ethnic group and individuals with whom they are working" (p. 251).
- Be knowledgeable about different cultures. Avail yourself of any opportunity to learn more about people who are ethnically different from yourself.
- Be sensitive to the different communication styles of different cultures.
- Learn from your experiences. Allow your clients to teach you about their culture.

Make appropriate assessment of your clients' needs: Individualize each case by learning to take into consideration the clients' personal circumstances, their culture, and their value systems. Be aware that ethnic groups are not monolithic (Brill, 1995).

Cases in Point

To illustrate working with various ethnic groups, we present a brief overview next, highlighting four major ethnic groups: African Americans, Native Americans, Asian Americans, and Hispanics/Latinos. The overview of each group illustrates its experience with racism, discrimination, and oppression. We also offer a list of suggested treatment interventions for working with each population group.

Ethnic-racial minorities number well over 40 million in the United States today and are represented in all the special population groups. They are subject to many of the same stressors others experience such as sexism, AIDS, homelessness, ageism, and poverty.

African Americans   African Americans, this nation's largest minority, have suffered great discrimination since their arrival as slaves almost 400 years ago. Morales and Sheafor (1995, p. 276) consider them one of the "most brutalized minority groups in the United States." Taylor (1994, p. 12) notes that their history of servitude, followed by a prolonged period of legal discrimination, enforced segregation, and exclusion, has profoundly affected the institutional properties, integrity, and functioning of African American families and continues to influence their collective fate in the United States."

Today, African Americans continue to suffer some form of discrimination. Barringer (1992, p. A1) notes that economically there is a difference between what working African Americans earn versus what their white counterparts earn. This disparity is reflected in the 1990 figures showing that the median income of African American families was $24,089 whereas the median for white families was $35,811—a difference of over $11,000. This illustrates that African American families earned 63 cents for every dollar earned by white families. Reasons for this difference are unclear. However, members of a given minority—in this case, African Americans—"usually have received lower economic rewards than they would have in the absence of discrimination" (McLemore, 1994, p. 145).

Treatment Considerations

• Some African Americans who enter treatment have been sent by someone else who has defined their problem for them; therefore, practitioners should attempt to reframe the problem from the client's perspective. A satisfactory working relationship with African American clients requires the helper to consider the multiple problems they face as a group.

• Some African Americans may not view therapy as a solution to emotional problems, so the helper will have to explain the benefits of engaging in treatment.

• Helpers should recognize the basis on which clients will permit them to influence their lives. This relates to developing a positive therapeutic relationship. Different methods for promoting a positive relationship have to be explored. Helpers must be willing to enter the life space of the African American client and to learn the intricacies of that person's unique style of expression. Helpers must personalize their approach to meet the specific needs of the individual client.

• Helpers must not pass judgment without understanding deeper issues. They need a keen sensitivity to in-group diversity and must not quickly label all African Americans the same.

• Because African Americans have experienced years of servitude and white superiority, they may respond better if the therapist takes on a peer-collaborator role where there is mutual respect and mutual sharing of information. Helpers need to challenge the invisible hegemony of privilege and help identify and build on the strengths of African Americans.

• Practitioners must be committed toward empowering African American clients and their families; a primary goal of treatment might therefore be to help clients achieve a sense of control over their lives.

• In accordance with empowerment practice, practitioners should avoid focusing on problems of the client that are not specifically relevant to the presenting problem.

• A study of African immigrants in the United States (Kamya, 1997, p. 154) explored the relationship among stress, hardiness, self-esteem, spiritual well-being, and coping resources. The results showed that "spiritual well-being was significantly related to lower stress levels and greater hardiness and self-esteem." Based on these results, the author recommends that practitioners need to "pay attention to this population by examining the interactive processes of their experiences."

• The role of religion, spirituality, music, and African heritage are also very important to many African Americans and thus must be understood from their frame of reference and considered integral in the treatment process when applicable.

Native Americans   Native Americans differ from African Americans in that they are *not* immigrants. Their plight began when their land was conquered. They have been subjected to both overt and subtle forms of discrimination through the years. In California, Native Americans numbered about 250,000 in the late 18th century, yet by 1900, only about 10,000 were left. Their decline can be traced to diseases transmitted from European settlers, dehumanizing and decimating conditions, warfare, slavelike treatment, and genocide by the whites (Daniels & Kitano, 1970; Taylor, 1994, p. 14). "Racism toward Indians has been expressed in the lack of comprehensive, progressive federal policy over a period of many years, a lack that has severely hampered Indians' ability to move forward in the United States" (Morales & Sheafor, 1995, p. 272).

Berry (1965) has identified five major governmental policies toward Native Americans as (1) extermination, (2) expulsion, (3) exclusion (reservations), (4) assimilation, and (5) self-determination. The first three policies reflect the majority views about this minority group and illustrate how severely they were treated; the last two provide some hope.

Taylor (1994, p. 14) notes the ultimate outcome of years of maltreatment toward Native Americans:

> Generations of exploitation and neglect have had a deleterious effect on the cultures, lifestyles, and well-being of Native Americans. Expelled from their native lands and resettled on prison-like reservations where they were prohibited from practicing native religions and other displays of Indian culture, Native Americans have experienced great difficulty in sustaining and preserving their ethnic identities.

Treatment Considerations

• Practice with Native Americans "must be harmonious with the client's environment and degree of acculturation" (Williams & Ellison, 1996, p. 147). Treatment interventions should include the clients in planning and implementation that will help them feel more in control of their lives and thus less oppressed. Consider involving the Native American community in the planning and implementation of programs, and include peers, family members, and community representatives in these interventions when appropriate.

• Show your desire to learn about Native American traditions.

• Because intervention strategies will likely be influenced by numerous cultural beliefs, customs, and values, make an effort to understand them. Determine what importance the culture may assume in ongoing care. Design interventions that are sensitive to the value constraints of both traditional and Western cultures.

• Because there is much variation among tribes, be aware of these and avoid making generalizations.

• Confidentiality is important and well-defined boundaries in terms of sharing information must be taken seriously. Because of the importance of the tribal influence, it is easy to breach confidentiality without realizing it.

• Sensitivity on the part of the helper is necessary in recognizing the many obstacles Native Americans face on a daily basis that may interfere with their ability to follow through with treatment (e.g., economic problems, transportation).

• Be sensitive to and respect the various spiritual beliefs and traditional medicine practices important to clients. Include ceremony and ritual in intervention when appropriate and consult with traditional healers.

• At times, helpers may need to work hand in hand with medicine men or shamans. At the very least, learn about their practices and respect them as important to the clients. Avoid criticism or ridicule as this will certainly impede a trusting relationship from developing.

• In working with Native Americans, focus on their belief in restoring spiritual balance to their lives. Perhaps this balance was altered by personal, family, or extended family problems. Aim interventions at restoring a balance between physical well-being and spiritual harmony. Understand the clients' definition of illness and fit the interventions with that definition.

• Be aware in your assessment that many Native Americans are more present oriented than past or future oriented; therefore, they may not be as receptive to discussing the past as the helper might want them to be. Similarly, during treatment planning, clients may not be very interested in preparing for the future, an attitude that must also be respected. Emphasize the present and existing problem-solving skills.

• Recent efforts to work with Native Americans on grief issues have proven effective. Because of the many losses they have sustained, "grief groups are being viewed as therapeutic resources for Indian people who are finding support and understanding in releasing the accumulated pain and sorrow that has so profoundly impacted their emotional and physical well-being" (Morales & Sheafor, 1995, p. 509).

Asian Americans   Asian Americans is a term that commonly refers to Chinese, Korean, Japanese, Vietnamese, Thais, and Pacific Island persons (Hawaiians, Guamanians, Filipinos, and Samoans). These groups are very homogeneous in their experiences in this country, even though each group has its own unique language, history, culture, religion, and appearance. They share many of the same stressors of other minority groups. For instance, they have had to endure similar immigration and assimilation struggles, they have suffered many forms of exploitation, and they have also had to face the consequences of racism. According to Murase (1977, p. 953), Asian Americans have been the victims of "humiliating, repressive, and vicious acts of racism" for over 100 years. Every group, including the more recent influx of Vietnamese, have had to endure some form of discrimination and oppression.

Chinese Americans, who were the first of the Asian minority groups to arrive in the United States, were exploited for their labor and subjected to many restrictions. Chinese immigrants were commonly depicted as "backward, immoral, filthy, rat-eating, opium-crazed heathens" (Dower, 1986; Miller, 1969; Saxton, 1971; Taylor, 1994). At the time, they were considered a "yellow peril," "an unassimilable horde whose willingness to work long hours for low pay threatened the livelihood of white working men" (Taylor, 1994, p. 115).

Japanese Americans were a second group of Asians who immigrated to the United States at the turn of the 20th century. Numerous laws were passed to limit their numbers and to prevent them from participating in certain occupations and agricultural pursuits. "Picture brides" (selected from pictures in a publication) were common and reflect the exploitation of Japanese women. Through time, Japanese Americans became known as the "model minority," meaning that they were "a nonwhite group that has overcome all obstacles through hard work and determination." Many debate this notion, and question the extent, cost, and meaning of Japanese American success (McLemore, 1994, p. 185). Economic discrimination is noted in the discrepancy between what a Japanese American makes compared to a white American (Kitano, 1981, p. 130).

Treatment Considerations

• The success of Western approaches to working with Asian Americans is connected to the degree of the clients' acculturation, so it is important for practitioners to be aware of this level and to tailor treatment based on this. The more acculturated clients are, the more receptive they will be to Western psychotherapeutic techniques (Mokuau & Matsuoka, 1992).

- It is important to understand the difference between Pacific and Asian worldviews and those of the Western world. Yamashiro and Matsuoka (1997, p. 176) contend that "in the West, it is presumed that people react emotionally only after considerable cognitive operations have occurred." In contrast, however, it is presumed that those in the East "respond emotionally to visceral cues or sense memory." Based on this, use of relaxation or meditation techniques may be more useful when working with Asian Americans than the application of cognitive strategies common to Westerners.

- In a similar fashion, Western views often promote independence and view dependence in a negative light. For many Asian and Pacific Americans, interdependency (Jung, 1998) is very valued; thus helpers should learn what this means and include this view in the treatment process.

- Any attempt to develop culturally appropriate interventions when working with Asian Americans must start with a survey of Eastern approaches to child rearing and human development, experiences with oppression and trauma, reactions to stress, views on family pathology, and typical coping styles (Yamashiro & Matsuoka, 1997, p. 176).

- With Asian Americans, a more holistic approach to treatment may have to be adopted. The Eastern orientation seems to value a more intricate connection between systems, so any therapeutic approach must involve the various systems that are important to the clients being served. This is especially important with Asian and Pacific clients, who believe in the practices of prayer and meditation and ritual or offerings that are considered "means by which individuals can extricate themselves from underdesirable life situations" (Yamashiro & Matsuoka, 1997, p. 176).

- Another prerequisite to working with Asian and Pacific Americans is understanding the fundamental norms and values that govern their behaviors. Respect their brevity of speech, their different views about eye contact, and limited discussion of private matters. Helpers must learn to appreciate these as part of the clients' cultural background and worldview.

- Issues of confidentiality are extremely important to Asian and Pacific Americans. "The issue of shame and losing family face must be addressed in all measures that ensure confidentiality during intake and service" (Yamashiro & Matsuoka, 1997, p. 176).

- Whenever feasible, use bilingual and bicultural practitioners to provide clients with a climate of familiarity and help to ensure that the intricacies and nuances of their behaviors and systems will be understood. This does not preclude non-Asians as practitioners, but it highlights the need for cultural sensitivity when working with this population.

- Recognize that many Asian and Pacific American communities believe that dysfunction and mental illness occur when individuals separate themselves from their culture that is so important in providing psychological and spiritual support. Treatment efforts must address this issue whenever it is relevant.

- Last, as noted with both African Americans and Native Americans, it is important to recognize the diversity within the Asian and Pacific American communities. Cultural values, norms, and rituals will vary among the different ethnic subgroups that compose the larger group of Asian Americans.

Hispanics/Latinos   The last minority group we consider is the Hispanic or Latino group. Taylor (1994) cites Americans of Hispanic or Spanish origins as one of the fastest growing minority populations in the United States, second in number only to African Americans. As with other minority groups, Latinos are very diverse and heterogeneous. Included in this major ethnic category are Mexican Americans, Puerto Ricans, Cubans, and others from Central and South America. Each of these groups is unique; even though they are all considered Latinos, they can be very different from one another. Similar to the Native Americans, some Mexican Americans were natives and therefore are not truly immigrants. "It was largely through military conquest and the annexation of their homelands in what is now the American Southwest that Chicanos [Mexican Americans] were involuntarily incorporated into the United States" (Taylor, 1994, p. 13). Other Latinos did immigrate, however. Some came voluntarily, and others fled their homelands because of war and famine; still others came as political refugees. Regardless of how they arrived, Latinos have historically suffered racism, discrimination, and exploitation. Furthermore, they tend to suffer the effects of poverty much more than the general population. In addition to racism and discrimination, they typically are affected by low income, unemployment, underemployment, undereducation, prejudice, poor housing, and cultural-linguistic barriers. The unfortunate results of living under these conditions have been the rise in alcoholism, substance abuse, juvenile delinquency and gangs, and a high percentage of imprisonment in the adult population; these conditions tend to perpetuate racist views and discriminatory behaviors (Morales & Sheafor, 1995, p. 277).

The plight of Mexican Americans specifically has been well documented.

> As a social category Mexicans experience distinctive treatment in virtually all areas of social life.... [They] fall well behind non-Hispanic whites on most indicators of status and well-being. They face obstacles to entering the economic mainstream of society. They have lower levels of educational completion, lower incomes, lower standards of living, and lower life expectancy. These inequalities reflect ongoing patterns of institutionalized discrimination that create serious problems for families." (Taylor, 1994, p. 66)

Specific to immigration is the issue of undocumented or illegal immigrants. There is much controversy about how to reduce the immigrant population and how to deter the undocumented from illegally migrating to the United States. This is especially noted in border states such as California, Arizona, and New Mexico where there is a much greater influx of Latin Americans. Some see these immigrants, documented or not, as convenient scapegoats for our country when economic times are tough (Morales & Sheafor, 1995, p. 277). Once again, this demonstrates that discrimination and exploitation remain prevalent in our society today.

Treatment Considerations

• Working with Hispanic/Latino clients can be a very challenging experience for any practitioners, regardless of whether they themselves are Hispanic/Latino. A primary reason is the diversity of this group, which has as many socially important differences as similarities. These differences impact provision of services and may also affect service effectiveness and utilization. For instance, work with an upper-middle-class, university-educated documented immigrant from Latin America may require a different approach from the one that will be effective with an uneducated, impoverished, undocumented immigrant or refugee from another part of Latin America. Although both deserve equal respect and equal treatment, therapeutic approaches will have to be tailored to meet their individual needs and circumstances.

• The tendency to assume that people who are defined as Hispanic have a common Spanish culture is offensive to some in this group. Examples might include clients who call themselves by another name, such as Latino, or who feel their unique heritage is lost when they are generalized as belonging to a large group of Hispanics. Hispanic/Latino practitioners, even from the same country of origin as their clients, must be careful not to make generalizations or assumptions that they already know what their clients are experiencing. A thorough assessment of their individual issues remains necessary.

• In a similar manner, when all Hispanics or Latinos are placed in one category, it is often assumed that they also share the same cultural traits. In many cases this is not true at all. An alternative approach that practitioners must consider "emphasizes the need not only to value but also to expect diversity in a group. This view conditions exceptions and encourages development of strategies that avoid stereotypes when addressing the needs of the client" (Castex, 1994, p. 290).

• To develop culturally appropriate and sensitive treatment interventions when working with Hispanic/Latino clients, practitioners must consider various ethnically significant features that include national origin, language, family names, religion, racial ascription, and immigration or citizenship status. We discuss each of these briefly below.

- National origin is very important to consider because Hispanics/Latinos do not all come from the same country of origin; some may even view those from certain Latin or South American countries in a negative light. The different historical experiences of each country can affect the ethnic self-identification of clients. This historical experience of immigrants may be especially significant to their current status and views.

- Because of the variation in national origin among Hispanics/Latinos, practitioners must ask clients where they are from and what their nationality is to learn whether they consider themselves members of an ethnic group within that nationality. The helpers must become familiar with the group history and its immigration experiences and be able to identify formal or informal service providers for this national group.

- Do not assume that all Hispanics/Latinos are fluent and literate in Spanish. Some may speak languages other than Spanish or may speak a regional Spanish dialect. Even when all members are Spanish speaking in a group, they may not speak the same exact language; different slang and nonverbal expressions may be used that create a very heterogeneous group. Be sensitive to this and do not generalize a group because they are all "Spanish speaking." To be responsive to language differences, helpers need to find out what language the client communicates best in, be sensitive to the possibility that people in crisis may be able to express themselves only in their native language, and be open to enlisting the help of trained interpreters when necessary.

- Family name may mean a variety of things for many Hispanic/Latinos; for some, it may be their surname; for others, their married name; and for still others, a new name given to or taken on their arrival to the United States. We have also found that many Latin cultures used affectionate nicknames that may seem inappropriate to others not familiar with the culture or family dynamics. On the basis of these variations, we suggest that practitioners "make no firm assumptions about language use, ethnic status, or recent heritage based on a name" (Castex, 1997, p. 292).

- Racial ascription must also be considered when practitioners work with Hispanic/Latino clients. Because racial ascription can have a bearing on how a client is viewed and views him or herself, practitioners should consult with their clients on how they regard themselves in terms of racial status. Sometimes, clients who were accepted and felt comfortable with their racial status in their homeland may no longer feel the same after entering the United States. Instead, they may be faced with unexpected racism and discrimination feelings of inadequacies, conflictual relationship with others, and perhaps even limited opportunities.

- Religion is another aspect of the Hispanic/Latino culture about which many assumptions are made. For instance, many believe the majority in this ethnic group are Roman Catholic; this may not be the case. Practitioners should make no assumptions about clients' religious affiliations but should ask them directly about their religious preferences during the assessment process. This may be a valuable piece of information that can help guide treatment and practice.

- Ascription by others needs consideration. Because the term *Hispanic* can have so many meanings, it is important that the helper "let people identify themselves, to remember that ascriptions may vary by social context, and to remember than individuals may not see themselves as members of the group they have been placed in" (Castex, 1994, p. 294).

- Immigration or citizenship status can be very important on many levels. Documented versus undocumented status may impact clients' access to public services, mobility, employment opportunities, and position in society. Also, the immigration experience may be very important to their present status and the issues they bring to treatment; thus, it must be carefully examined once the client is securely involved in treatment. Because of the feasibility that immigration or citizenship status may be an important issue to Hispanic/Latino clients, practitioners need to learn which resources are available to help clients of a particular status.

• Be sensitive to the realistic fears clients may have on revealing their status and help foster trust by attempting to maintain confidentiality within legal constraints. When working with an immigrant population, become familiar with current immigration regulations as well as sociopolitical issues tied to their country of origin.

• Be aware of the many struggles this minority group may be facing economically, socially, and emotionally. Poverty, low income, single-parent families, and a very young demographic profile characterizes the life of many Hispanic/Latinos. Assessing which of these apply to one's clients is important in both assessment and treatment endeavors.

AGE

As with race, culture, and ethnicity, age-related issues affect us all. In relation to our age, we are all at a specific developmental stage that is ever changing. In working with people, whether individually, in families, or in groups, we must recognize the unique aspects of each major stage of an individual's development. This knowledge will help us understand them and develop appropriate treatment interventions.

Developmental Perspective: A Biopsychosocial Slant

We take a developmental approach in discussing the various stages of life that we all experience. Development is the "systematic changes and continuities in the individual that occur between conception and death" (Sigleman & Shaffer, 1995, p. 3). A stage of development is "a period of life that is characterized by a specific underlying organization" (Newman & Newman, 1995, p. 41). According to stage theorists, there is a specific direction for development whereby an individual progresses through set stages. Ideally, at each new stage a person incorporates the gains made during earlier stages. For instance, at each stage, individuals are confronted with certain tasks that require the integration of their personal needs and skills with the social demands of their culture, the end result being acquisition of new capabilities, enhanced skills, and a new way of interacting with others.

The various areas of human development fall into three broad realms: physical, psychological, and social. Although understanding all three is essential, we focus primarily on the psychosocial sphere. Theorists see human development as a product of the interaction between individual (psycho) needs and societal (social) expectations and demands (Newman & Newman, 1995, pp. 39–40). We will also note the most salient biophysical issues as they relate to psychosocial development.

Erikson (1950), one of the most famous of the stage theorists, postulated the "Eight Ages of Man;" here he proposed eight stages of development with corresponding ego conflicts at each stage that if successfully mastered are said to produce new ego skills. His stages are these:

Stage 1—Basic Trust versus Mistrust
Stage 2—Autonomy versus Shame and Doubt
Stage 3—Initiative versus Guilt
Stage 4—Industry versus Inferiority
Stage 5—Identity versus Role Confusion
Stage 6—Intimacy versus Isolation
Stage 7—Generativity versus Stagnation
Stage 8—Ego integrity versus Despair

Currently, many theorists believe that physical and cultural evolution has created a need for the expansion of these stages. Some have added new stages. For instance, many have included the prenatal period from conception to birth in their schema. Newman and Newman (1995), who propose an 11-stage system of development, not only include the prenatal stage but have also divided adolescence into early and later periods; additionally, they have a stage for the "very old" to accommodate those who are 75 and older. We find Newman and Newman's postulation to be the most comprehensive, yet because it is so extensive we have modified it in our discussion to include children, adolescents, young adulthood, middle age, and later life. Below, we address these stages with regard to the typical difficulties faced by individuals at each of these life phases, paying particular attention to the issues most relevant to human service practice.

Children

Children are often the most vulnerable and at risk for emotional, physical, and sexual abuse or exploitation of some sort. Because children have not fully developed the skills and defenses needed to deal with the stressors of life, they are more likely to need our attention, guidance, and support. According to Schmolling, Youkeles, and Burger (1993), "Children are endangered not only by poverty but by illness, rejection, lack of understanding, the inability of parents to socialize them properly, and many other factors" (p. 57). Because they are a high-risk group for developing both physical and emotional problems, we must be available to them and cognizant of the special issues they may be facing.

Childhood is broad and encompasses vast change and growth, given the physical, cognitive, and psychological maturation that occurs between birth and adolescence. It is important to be aware of each child's particular stage of development. In working with children, practitioners need to assess the developmental milestones that have been mastered and the ones yet to be conquered. Likewise, they must determine which factors in the child's life may promote development and which may serve to impede it. Numerous texts have been written specifically about the normal developmental tasks of childhood; therefore, we will not elaborate on these. Instead, we will address a few of the most significant issues affecting children today that require professional intervention.

Biological Aspects   Prenatal care can greatly impact children and have a significant bearing on their overall development. It is not difficult to understand how substance abuse and poor nutritional and environmental circumstances for the mother can impact the child. Some are born with noticeable birth defects and ailments that can be directly tied to the prenatal conditions. For others, however, the ill effects of prenatal care are more subtle and difficult to determine. An example is the many children who exhibit learning disorders on entering school. Often when carefully examined, their difficulties can be traced back to substance abuse by the parents.

Environmental conditions can also affect the child's health and development. This is most vividly evidenced in children who live in poverty and suffer from malnourishment. Many health and emotional problems can be directly connected to the impoverished conditions in which the child lives and grows.

Psychological Considerations   The psychological ramifications of poverty and impoverished environmental conditions have been extensively written about, suggesting the continued need for work in this area. Children with physical, developmental, and emotional disabilities may also experience additional psychological difficulties that are directly related to their condition. For instance, those who suffer some developmental delay or a developmental disorder must face the prospect of ridicule and stigmatism for being different. They may also experience frustration and low self-esteem because of their inability to function at the level they aspire to. Problems are often intensified when a child suffers from a specific type of psychological disorder such as attention deficit hyperactivity disorder. These children may encounter additional stress in daily living resulting from their impairment in functioning. They may have difficulty learning and thus have academic problems. They are likely to experience a wide range of behavioral and social problems that alienate others and frustrate parents and other caretakers. Ultimately, because of their overall difficulties in functioning, they may develop strong feelings of inadequacy and/or low self-esteem. In many such cases, specialized treatment including individual and family counseling, parent education, school interventions, and medication management are all suggested.

Social Dimensions   Earlier we have alluded to children who live in poverty or in single-parent families. Clearly, these conditions can severely affect children's development. Physically, they may be prone to illness and malnutrition that are likely to impact their cognitive and psychological growth. The additional stressors they must face can contribute to many problems, ranging from medical concerns to emotional and mental dysfunction. You can help these children by emphasizing their strengths and providing them with healthy options, outlets, and support. Most important, however, is helping to instill in them a sense of hope so they can come to feel capable and optimistic about their future.

Children who are victims of child abuse or violent crimes must confront additional challenges to overcome the potentially negative effects of being abused or witnessing abuse. Some children are able to go forward successfully despite the abusive living conditions they have had to endure. Others may never be able to erase the physical and emotional scars of such abuse or violence. Some may continue to live feeling insecure, inadequate, and unworthy—thus, more subject to depression and other emotional or mental illness. Others may repeat in their own interpersonal relationships the abuse they sustained or witnessed. Still others may engage in criminal behavior, demonstrating little regard for others and displaying antisocial traits.

Many children who are victims of severe child abuse may be placed in foster homes or given up for adoption. The emotional repercussions of these events can be great. Feelings of rejection and abandonment may dominate their lives and ultimately affect their interpersonal relationships and social functioning.

In earlier discussions of divorce and blended marriage (Chapter 4), children affected by divorce, custody battles, and the remarriage of their parents are subject to additional stressors and challenges. Often, the ability to overcome these life events is dependent on the specific circumstances surrounding them and especially on the existence of supportive and caring significant others.

Working with Children

Certain characteristics are especially desirable for those who work with children:

- A willingness to know oneself.
- An ability to be imaginative and accept the role of the child, regardless of how bizarre the child may appear initially. This requires an empathetic, accepting, and understanding approach.
- An optimism that allows the adult to instill hope in the child and provide a positive role model.
- An awareness by helpers of their own social situation and of the ways their position influences their perspective. This includes being aware of their own prejudicial views and a commitment to remain as objective as possible in work with others, especially if they are different from the helpers.
- A disinclination to take a negative stand toward the patient's parents and siblings. Respect the child's love for his or her parents and be careful to not undermine their relationship.
- A willingness to work enthusiastically with a tentative hypothesis prevents the stultification of treatment. Take care to not make premature assumptions or judgments about a case as this will surely influence the course and outcome of treatment. Acknowledge when one is not sure or fully understands what is going on. (Adams, 1982, pp. 45–49)

In addition to these traits, Adams (1982) highlights specific factors favoring a good outcome of treatment in working with children. We have incorporated these into the following three major areas:

1. Acknowledgment and recognition of the child's parents and siblings. Rather than viewing them as the enemy, consider them allies who can enhance the treatment outcome with the child. If this is impossible, at least do not alienate them. Be willing to recognize their strengths and don't focus only on their limitations.

2. An attempt to understand the child, to recognize his or her strengths, and to work toward emphasizing these. A good assessment will typically allow the clinician to consider ways in which he or she can treat the child optimally and most successfully enter the child's world. The helper may have to develop a repertoire of skills to do this, especially in play therapy and the behavioral and cognitive therapies found most effective in work with children.

3. Recognizing the importance of proper referrals and the transition that may be necessary to actualize these. Practitioners who work with children need to be knowledgeable of other sources of help when it is necessary to refer the child for additional treatment or when parents or other members of the family need their own help. Being aware of those in the community who can provide such services and whom you respect can aid in this referral process.

Treatment Considerations

• A thorough, comprehensive assessment must include information on pregnancy and early developmental milestones. Information on childhood and family history is crucial in better understanding the overall world of the child.

• It is essential to see childhood "not as a zone for adults to colonize but as a way of life in its own right" (Adams, 1982, p. 1). From this perspective, to serve children most completely, helpers must be willing to enter their world. This requires being respectful, spontaneous, allowing fantasy and play, and acknowledging the here and now.

• Although children have much in common, note their differences and recognize them as unique individuals with their own set of values, temperaments, and issues. Diversity is just as evident in children as it is in adults. As helpers, we must recognize them as separate, individual beings whose treatment must also be individualized.

• Self-awareness is crucial in work with children as it may trigger many of our own family of origin issues and perhaps create countertransference reactions. Also, we must be aware of our particular style in relating to children and decide honestly whether our approach is helping. Adams (1982) cites several ways of relating to children that may not be therapeutic in the long run including "Truth-Faith Healer," "Me Adult/You Child," "Good-Enough Parent," "Big Sibling," "Babysitter," "Teacher," and "Cop."

SCENARIO

WORKING WITH CHILDREN

A young girl, age 6, is brought to the mental health center by her new foster parents. She was given to them by the children's service worker about 2 months ago after she was found abandoned in a park in another state. They report that it has been a difficult adjustment as she seems to be quite regressed and was constantly hungry when she was first brought to them—that she was famished all the time. Her verbal skills were limited, although they are improving. At first she did not talk at all; she was quite withdrawn from others, though now she has warmed considerably. She had alluded to some possible sexual abuse at some points saying that someone "touched her there" (pointing to her genitals) while she had been taking a bath with her foster sister. Despite her obvious difficulties, she appeared in your office as an active, cute, 6-year-old girl who seemed, at initial appearance, to be perfectly happy. Based on the foster parents' description, though, she did seem to be lagging developmentally and at the recommendation of the kindergarten teacher, they had decided to bring her in for an evaluation because of the difficulties she had been having. The foster parents also tell you they were given almost no information about her when she came to live with them, and that the children's service worker apparently knew very little about her history. How would you work with this client? Paying attention to the dynamics and issues in this case, what do you think would be the challenges in treating a child like this? Would you be inclined to take a negative or positive view of the natural parents, given the above information?

Adolescents

Adolescence is often a turbulent and stressful time, yet it can also be a joyous period of experimentation and adventure. As with children, the diversity among adolescents is great, with some successfully traveling through the adolescent years with little difficulties and others facing adversity and struggling at every turn. For most, though, peer approval seems to be essential and thus must be acknowledged.

Adolescence can be viewed as a journey on the road to find oneself, given the search for identity common at this point in a person's life. Erikson labels this stage as "Identity versus Identity Confusion," indicating the focus on identity development. Along with this task, adolescents must also begin individuation from their family, which may not always be easy. Carlson and Lewis (1988) have eloquently addressed the plight of adolescence:

> Adolescents are unquestionably at a vulnerable developmental stage as they attempt to navigate the difficult transition from childhood to adulthood. Any life transition holds the potential for danger or for growth, but adolescence represents probably the most crucial transition, combining a general life adventure (bidding farewell to childhood and joining the adult world) with a series of specific changes (leaving one school for another, entering the work world, learning to think differently, meeting new expectations, seeking independence from parents—even adapting to a new body. (p. 1)

Because of this tumultuous state, many adolescents engage in risky behaviors and are therefore in danger of experiencing additional difficulties. Before exploring these in more detail, we have summarized the main struggles faced by most adolescents. They fall into the following three categories that make up the biopsychosocial characteristics.

Biological Aspects   Teens typically experience major growth and physical maturation during their adolescence; the growth spurt is often rapid resulting in both height and weight gains for most, as well as the hormonal changes generated by the pubertal process. The growth and hormonal changes that occur during this stage can be overwhelming and emotionally difficult to cope with.

Risks: The growing adolescent has a great nutritional need, and the reaction to this may be problematic. First, parents may feel angry and perplexed by the large amount of food their teenager is eating. Second, some teenagers develop poor eating habits that may last a lifetime (e.g., developing a craving for junk food) and can predispose some to an ongoing battle with obesity. Last, despite their need for additional calories, some adolescents have an obsession with their body image and are intent on not gaining weight; they can develop eating disorders of some type (e.g., anorexia or bulimia).

Some female teens (33%) may be plagued by dysmenorrhea—painful menstruation that can affect them physically, emotionally, and socially. Many teens (80%) also have acne. For some, this is simply a nuisance; for others with severe acne, the stigma and ridicule can be great, leading to embarrassment, social withdrawal, and low self-esteem.

Psychological Considerations   Along with physical growth and maturation, adolescents also experience cognitive development as they learn to use complex and abstract thinking;

this allows them to acquire such skills as assimilation and accommodation that lead to formal operational thinking. They learn enhanced communication skills, such as the use of metaphors and satire. Some teens practice expressing themselves in intense ways that may often seem quite dramatic and eccentric as in their speech, dress, and choice of music. Peer approval is very strong for most early teens; later, dating and the awareness of their own sexuality may take precedence, and ultimately the task of developing their own identity and making their own decisions surfaces. As a result, attitudes and emotions may undergo rapid change. These young people will have mood swings and emotional turmoil—and in the context of the many changes they are experiencing, these may indeed be quite appropriate. Self-esteem issues are often at the core of most problems they face.

Moral development also typically occurs during this stage, with egocentrism often quite high. The desirable outcome is for them to become conscious and self-aware.

Risks: Adolescents are prime candidates for experiencing emotional issues, given the many changes they must undergo in their transition from children to adults. For many, the stressors they face along their journey is great. Some tend to externalize their frustrations, hurt, and anger; they act out by engaging in delinquent behaviors. Others, also affected by their life situation, may internalize their pain and become depressed; these teens may possibly resort to suicide or drug use as a way to cope. In between these extremes are adolescents who suffer from loneliness, sadness, nervousness, tension, sleep problems, and interpersonal difficulties. If not attended to early, these can lead to more severe problems.

Suicide prevention is an especially important area to focus on if you plan to work with adolescents. Understanding the problem is a first step, and it requires you to learn the truth concerning certain myths about adolescent suicide. Causes of adolescent suicide have been divided into three areas: (1) adolescent struggle to develop and integrate a unique identity, (2) familial factors, and (3) environmental factors (Carlson & Lewis, 1988, p. 43). Carlson and Lewis cite the following as general areas that may contribute to the decision to attempt suicide: self-esteem issues; poor communication with parents, other adults, friends, and loved ones; achievement orientation; poor problem-solving skills; narrow commitments, and high stress. Family factors include poor communication, loss, dual-career families, single parenting, blended families, mid-life transition of parents, and abuse. Environmental factors that have been noted are pressure to achieve, mobility, uncertainty about the future, graduation from high school, the fear of nuclear war, and drug abuse.

Successful intervention must incorporate various components: support from family and friends needs to be sought along with involvement of others such as school counselors, therapists, physicians, and church personnel. Efforts must be made to notice changes in teens' behavior and personality traits as well as picking up verbal cues and themes or preoccupation in thinking. Finally, crisis intervention and management skills must be employed; these include being a good listener and staying nonjudgmental, providing support, conducting a thorough risk assessment, considering the situation a crisis whether the suicide attempt was just a gesture or acting-out behavior, and being able to make decisions relative to needed interventions (e.g., hospitalization or no-suicide contract). Because the stress involved in such situations is often quite high and the decisions never easy or clear-cut, helpers should also seek their own support, guidance, supervision, and/or consultation. Being flexible enough to work with colleagues or crisis center staff or to refer the teenager to a more appropriate helper are all essential.

Social Dimensions   The role of the family is quite influential in the life of the adolescent. Because the adolescents' tasks are tied to separation and individuation, developing autonomy from the family is essential. Also, the resolution of parent-adolescent conflicts and attachments so commonly seen at this stage are necessary for the adolescent to progress successfully into adulthood. Unfortunately, this point is often where many problems result as either the family, the adolescent, or both are unable to proceed without additional conflict and turmoil.

Social supports for the adolescent are found primarily in their peers at first; then, they may center on their love relationships and perhaps in relationships with such mentors as teachers, counselors, or church clergy. Because peer pressure and conformity are so strong, it is important to help adolescents define their own identity and learn to make rational decisions about their behavior and choice in friendships.

Risks: Substance abuse is considered by many to be the number one social issue facing young people today. Both drinking and the use of drugs is high during this developmental stage. Some engage in substance abuse as a form of experimentation and find that peer pressure and the need to conform are great pressures. For some, alcohol and/or drug use is only transitory; for others, it leads to dependence and a serious problem they will take with them into adulthood.

Changes in adolescents' bodies as well as the onset of puberty can heighten their awareness of their sexuality. Some sexual experimentation can be healthy, but undue pressure from peers can lead to problematic sexual behaviors. For some, it poses no major problems and can be a positive part of their maturational process. For others, however, there are often additional problems resulting from sexual involvement. Some teens feel guilty and perhaps ashamed about their sexual behavior. For those who are survivors of childhood sexual abuse, sexual encounters typical during adolescence may trigger repressed memories of their prior abuse which can be very traumatic and overwhelming. Sometimes, teens faced with unwanted feelings and flashbacks may combat these by increased promiscuity and sexual experimentation, or on the other extreme, complete abstinence and social withdrawal.

Adolescent pregnancy, abortion, sexually transmitted diseases, and AIDS are possible results of unprotected sex that seem to intensify during the adolescent period. Each of these circumstances can be extremely difficult to deal with, especially in the case of contracting a serious disease, the extreme case being AIDS. The importance of helping teens learn about the consequences of irresponsible sexual behavior and the need to be more responsible with their sexuality cannot be understated.

Some teens react to their life difficulties through delinquent behavior. Types of this behavior include running away, school suspension, dropping out, gang involvement, and engagement in violent activities. The resulting legal issues when teens are arrested for criminal activity add to their already difficult situations.

Treatment Considerations

One way to consider treatment is to decide on the most appropriate focus: individual, family, or school. In some cases, an integration of all three will be best. Below are some considerations that can help practitioners working with this at-risk age group.

- Be aware that there is great diversity among adolescents. For some, this time period can be turbulent and stressful; for others, it can be a positive period in their lives.
- Many adolescents will be working on finding their identity, and many of the problems they present may be tied to this developmental task.
- In addition to identity issues, many adolescents are also struggling to individuate themselves from their families and may sometimes engage in rebellious behaviors as a way to separate.
- Because of the high likelihood for adolescents to engage in risky behaviors, helpers must be prepared to deal with the additional difficulties these behaviors pose.
- Be cognizant of the biological aspects, psychological considerations, and social dimensions that are common to this age group along with the accompanying risks in each of these categories.
- In regard to biological risks, be knowledgeable about teens' nutritional needs, tendency to engage in poor eating habits, risk of developing an eating disorder, the problem of dysmenorrhea for some female teens, and the plight of having severe acne.

- Psychological risks to consider include the emotional issues tied to maturation. Some teens externalize these frustrations by acting out or engaging in delinquent behaviors. Others may internalize these and resort to suicide or drug and alcohol use to cope. Some, for a variety of reasons, may feel that life is overwhelming and not worth living, and these may attempt suicide.

- Social risks include substance abuse and sexual acting out; both can lead to increased problems in a variety of areas such as drug addiction, adolescent pregnancy, abortion, sexually transmitted diseases, and AIDS.

- In working with teens, it is especially important to enter their world and make an attempt to understand them from their perspective. Because they will test and challenge you, it is even more necessary that you make an effort to gain their trust. This is often possible by respecting them and refraining from becoming an authority figure.

SCENARIO

WORKING WITH ADOLESCENTS

A teenage girl, age 17, is brought in for evaluation due to excessive anxiety and fears that she has recently expressed to her parents. The mother of the client brings her in and states that recently her daughter has expressed fears about someone lurking outside her window at night. Each time she has expressed this fear, her mother has found it to be unfounded. She is confused and uncertain as to what is happening with her daughter. The daughter was seen by a psychiatrist before the referral to you, the mother states that the psychiatrist told her that her daughter may be hallucinating and may need medication. The daughter stated that she did not want to take medication and wanted to talk with someone else. The mother then requested that another counselor evaluate her daughter. The daughter always has done well in school and has several friends that she is close to. She has no family or prior history of depression or psychosis. The daughter talks to you about her fear and states that she has had that fear recently and it is always a man that is lurking outside her room. When you ask for her history, she tells you about both her parents, who are divorced. She lives with her mother who has custody and visits her father infrequently due to her perception that he is "demanding and bossy." She also reveals reluctantly that her father touched her once on her breasts when she was younger and that she never said anything about it. She says as well that she often feels "uncomfortable" around him due to his demanding, intrusive behaviors with her. She often perceives that he is attempting to get her to do things his way with no consideration for her feelings. She does not want to take medication but wants to continue seeing you to talk further about how she feels. What would you do if you were expected to treat this adolescent? How would you approach the mother about what you believe is happening with her daughter? What special considerations might you consider due to her status as an adolescent?

Young Adults

Young adulthood is also a frequently difficult phase as it is a transition from focus on oneself to focus on others. Although not always emotionally or financially prepared, many young adults are expected to become economically independent from their parents as well as make other decisions independently. Intimacy issues are often strong at this stage and may lead to marriage and the decision to have children. Work and career issues also take on a more important role.

Biological Aspects   It has been said that "young adulthood represents the peak of physical development across the lifespan" (Ashford, Lecroy, & Lortie, 1997, p. 407). Health-conscious behaviors may replace those less-than-sound nutritional habits of adolescence.

The reproductive system is at its prime and this is often a time when pregnancy and childbirth take place. The impact of hormones on women may be especially strong during this period.

Risks: Cancer is sometimes a health hazard of the reproductive system for both men and women, seen in the increased incidence of testicular and breast cancer during this stage of life. Also, for some adults the presence of a mental illness (e.g., schizophrenia, depression, bipolar disorders) is first diagnosed or suspected in these years.

Psychological Considerations   Cognitively most young adults are expected to be able to engage in formal operational thinking, with an emphasis on problem solving, although this is not always the case. Communication issues that surface often center on the differences in how men and women communicate and the miscommunication that can result.

According to Erikson's schematic view, young adults are struggling with issues of "intimacy versus isolation." The search for intimate relationships and perhaps even love is strong at this stage. The realization that ideal love may not truly exist can be especially painful for some. Others may not find a partner and may be left with feelings of isolation and loneliness.

Risks: Because making major life decisions and taking on major life responsibilities is an integral part of this stage in life, many stress-related illnesses or conditions are likely to surface. For some, coming to terms with a serious mental illness can be especially stressful and emotionally difficult. Early intervention is often the key to successful treatment and a favorable outcome.

Social Dimensions   Often at this stage, young people separate from their family of origin, although it is not uncommon today for many young adults to remain at home due to economic factors. For many, becoming independent from family is a strong force. For others, the additional focus on selecting a partner, marrying, and forming a family of their own comes into play. These changes can result in various adjustment difficulties that may require professional intervention. Marital, parenting, and family counseling are often necessary to prevent further difficulties. In some cases, the issues related to infertility or voluntary childlessness may also put additional stress on the young adult.

Many may not be ready for marriage but still want companionship. Issues tied to being single or developing gay and lesbian relationships may surface at this juncture.

Last, young adults who are said to be at the prime of their career development may find this period especially difficult. For some, the desire to continue their education may not be possible because of economic pressures. For others, work-related problems may take on a larger role, and those uncertain of their future may have difficulty knowing which direction to take.

Risks: Additional social as well as emotional stressors are seen when a family has to face the illness (physical or mental) of a significant other, whether a spouse or a child. Similarly, when divorce occurs, there are social and emotional factors that must be dealt with.

Other issues young adults may have to face are incidents of sexual harassment, the acquisition of sexually transmitted diseases, rape and sexual assault, and spousal abuse or battery. All can be difficult to cope with and depending on the situation may require professional intervention and assistance.

Treatment Considerations

- Recognize that young adults are as diverse as adolescents, with some very mature at age 18 and others are still like adolescents in their behaviors at age 25.
- Acknowledge the pressures that many young adults face and the impact this may have on them physically, psychologically, and socially.
- Realize that some young adults may have to juggle many responsibilities that they may not be ready for, such as work, marriage, or parenthood.

- Be aware of the drug and alcohol problems that may be present at this stage when freedom from parental control opens the door to increased experimentation.
- Be prepared to help young adults deal with intimacy issues and relationship struggles they may never have experienced before (e.g., in a marriage or in work).
- Recognize signs of mental illness that may first surface during this stage.

SCENARIO

WORKING WITH A YOUNG ADULT

Karen is a 22-year-old, African-American female who enters a local mental health center for counseling after her recent breakup with her high school sweetheart whom she had dated for 5 years. She is very upset about this loss and questions her ability ever to find someone else to be close to. She feels that no one is supportive of her as both family and friends felt Todd (her boyfriend) "was a loser anyway." Although Karen agrees that Todd was probably not the best influence on her, she still misses him and feels great loss. She is even more anxious because she has not had a menstrual cycle in over a month and fears she may be pregnant. Furthermore, she had heard rumors that Todd had been "sleeping around" prior to their breakup and worries that he has not used protection and may have infected her with something.

Complicating Karen's sadness is news that her parents are contemplating a move out of state because of a job transfer her father has been offered. They are selling their home. Karen has been living with them and now they would like her to move out as they plan to rent a small condominium until the out-of-state move. Although she is gainfully employed, Karen does not feel financially or emotionally ready to move out on her own. She is hurt and angry at her parents for their sudden decision to move and not consider her plight in their plans. She also feels abandoned by several of her close friends who grew tired of her on-and-off relationship with Todd. She feels all alone and seeks help to know how to cope. How would you proceed if you were to start working with Karen? What are some of the most important young adult issues she is dealing with? What treatment approach might you consider?

Middle Age

The definition of middle age has changed with the increase in life expectancy. Today middle-age adults would have considered themselves old just a few decades ago. Unlike childhood, adolescence, or even young adulthood where certain stressors are characteristic of the specific developmental stage, middle age is a much more diverse period in life that is affected primarily by each person's experiences and life choices. Typically, however, those at this stage are involved in launching their children, career development, and developing a new perspective about who they are. Reviewing their past and looking to the future can precipitate a crisis for some who may be saddened by the past or hopeless about the future. As seen by Erikson, issues relevant to "generativity versus stagnation" are at the forefront.

Biological Aspects   During middle age, physical growth and maturation have already occurred. At this stage, many must face the aftermath of previous behaviors (e.g., poor eating habits, excess tanning, excessive substance use, sexual acting out) experienced as cancer and heart disease. Stress can be great at this stage and may also account for the medical and psychological problems faced by many middle-age individuals. For women, menopause poses additional physical and emotional stressors, and both men and women must deal with changes in their own sexual response.

Risks: During middle age, various life-threatening or chronic illnesses become even more of a concern for certain individuals. The incidence of breast cancer is greater for women and prostate cancer for men. Also, heart problems and those related to menopause intensify.

Psychological Considerations   For some, entering middle age may create a state of crisis, better known as the midlife crisis. The body is changing, children are leaving home, parents are growing older and perhaps more frail, and individuals are more realistic about their life goals and their own mortality. Facing these issues often requires individuals to review their past to renew and reintegrate their identity. Even those who experience no major difficulties at this stage often have a period of reflection that can be emotionally stressful. Many find that they are more serious about their social and moral responsibilities, resulting in increased involvement in socially responsible activities such as coaching, serving as representatives on boards and task forces in the community, participating in fund-raisers and volunteer activities, or working for a particular cause or charity.

Risks: For many, alcohol and drug addiction is at its peak, leading to additional physical, psychological, and social problems. For some, this is the time when they can no longer ignore or deny the consequences of their substance use; also, for many, denial no longer works and they must accept their addiction. Loss of health, jobs, relationships, and freedom are often the most potent incentives for individuals with an addiction finally to seek help and acknowledge the seriousness of their condition. Also, co-dependency issues must be confronted by significant others who have previously placed all the blame on the alcohol or drug user.

Social Dimensions   Many experience divorce and remarriage during their middle years. As discussed in previous sections on single parenting, divorce, and blended families, the stress felt can be great. For some involved in or victims of extramarital affairs, spousal abuse, and rape, the pressure may be even greater. Additionally, there are the potential conflicts that may result from relationships between midlife parents and their adolescent teens, adult children, and aging parents. For many, the additional issues of being grandparents may also come into play.

Risks: In addition to the family breakup, many individuals also face job loss through layoffs, downsizing, and personal crises. In extreme cases, these individuals may not be able to recover and may end up homeless.

For middle-age adults who had children early, their children are now adults and may be having their own problems with divorce, substance use, or criminal activity. Not only does this put additional stress on the parents but it can also lead to their taking on the added responsibility of raising their grandchildren until the time when their adult children can resume this task. In some cases, these young grandparents must take on the role of parents permanently for the second time.

Another issue becoming common in middle age is the need to care for one's aged parents who are frail and alone. The strain of this added responsibility can be quite difficult, especially when the parents have Alzheimer's disease or serious medical or mental illnesses.

Treatment Considerations

- Middle age must be recognized as a diverse period of life and one that does not have the specific stressors seen in adolescence and young adulthood.
- Often during this stage, middle-age adults are more intensely involved in launching their children, in pursuing their own career development, and in discovering a new sense of identity.
- Crisis is common at this stage, with high-risk behaviors involving substance abuse, job loss, and marital conflicts often resulting in extramarital affairs, divorce, and remarriage.

- Because of the extended life span, middle-aged adults often have to care for both their teenage or adult children and their aging parents; the combination of these tasks requires additional skill and much needed support and validation.

SCENARIO

WORKING WITH MIDDLE-AGE ADULTS

Jim, who is 45, comes referred by his medical doctor after suffering from severe migraine headaches for several months. He has been married for 25 years and is having marital problems and increasing conflicts with his teenage and young adult children. It seems that his 16-year-old is involved with the "wrong crowd" and has been caught drinking and smoking marijuana. His 23-year-old has not decided what he wants to do with his life and is currently just working part time and spending all his extra time with his girlfriend. His 21-year-old daughter moved out of the house last year to live with her boyfriend but is now asking to return home to escape his abusive behavior and unfulfilled promises to get a job. She is unsure whether she is pregnant with his child. Complicating these stressors with his children is Jim's sense that his wife is increasingly distant from him and the family as a whole. Since she returned to college, she is more involved with study groups than the marriage or the family. Whenever Jim brings up his concerns about her lack of involvement, she becomes defensive and accuses him of not wanting her to grow and of being fearful of her becoming independent. Although he is a bit frightened by her newfound freedom, he does not believe he is as threatened by her success as she feels he is. He basically feels alone in the marriage and in tackling the family problems that seem to be at an all-time high. He's also a little jealous of her future as he is stagnant at work, with no room to grow professionally but increasing pressure to produce. He would also like to return to school, but then who would be the breadwinner? He feels caught and unsure of who he is any more. With all the stressors at home, he often finds reasons to work late or to go out drinking with some of his co-workers, and this infuriates his wife. Also, he has become increasingly fond of a new female co-worker whom he has been assigned to orient to the job. He is fearful of his impulses and yet suddenly feels excitement about going to work again. He seeks your guidance to help him sort out his struggles and develop more appropriate coping skills. He is aware that if he's not careful, his drinking could get out of hand and his temptation to have an affair could become a reality. How would you approach working with this middle-age adult? What age-specific issues can you see that may need to be addressed in treatment?

Later Life

Later life now spans a large time frame as people are living longer than ever before. Some have divided this stage to include the young old, old, and very old. Because this age range is so broad, there is great diversity among people in their later years. According to Erikson, the goal in this stage is "ego integrity versus despair," reflecting the importance of the life review process.

Biological Aspects   Biophysical changes are well noted during this stage; as the individual ages, functions of the various biological systems of the body decline. Chronic health problems may develop or worsen and physiological changes may impact sexual activity, although "older adults do remain interested in and capable of sexual activity" (Ashford, Lecroy, & Lortie, 1996, p. 498).

Risks: The results of age together with chronic health problems such as arthritis and hypertension are common concerns during later life. The presence of psychopharmacology is often an added hazard resulting from the desire to treat these health problems.

Psychological Considerations   Growing old may be met with much emotional grief and turmoil. For some, it may be a relatively gratifying period when demands are lighter, leaving more time to enjoy life. Such positive attitudes and experiences in later life are of course closely related to good health and strong support systems. Those who are healthy and involved in positive relationships with their mate, adult children, and friends seem to do much better than others in accepting their age and enjoying some of the inherent benefits of retirement and grandparenthood.

Mental decline, sensory deficits, and memory problems are variable and may affect some much more than others. When medical problems or social deprivation and excessive stress exist, these problems are certain to be greater. For most elderly, reminiscing is a part of their communication and is said to be instrumental in helping them feel positive about their achievements and to bring closure to their life.

The common theme of loss is felt by most older adults, whether the loss is physical, emotional, or social in nature. Physically, the elderly lose some of their vigor, vitality, and ability to function as before and may be fearful of the ultimate loss of life—death. Emotionally, many may lose spouses, contemporaries, and sometimes even adult children. Socially, many feel the loss of their role as a productive member of society and thus may experience loneliness and regret on retirement. Regardless of the type of loss experienced, grief reactions are common and must be attended to to prevent further psychological impairment.

Risks: Depression secondary to various forms of loss is a major psychological problem of many older adults; it can result in suicide, which is quite prevalent among the elderly. Suicide is often disguised, the elderly are less apt to reach out for help, or they may be especially isolated; therefore, suicide intervention does not occur as readily with older people as with other age groups.

The prospect of living with Alzheimer's disease or having to be the caregiver of a loved one diagnosed with this condition can be quite overwhelming and psychologically difficult to cope with. Support, guidance, and education are crucial for these elderly.

Social Dimensions   Socially, some elderly may find themselves surrounded by loved ones and praised for their accomplishments; others may feel isolated, consider themselves failures, and remain homebound. Family and other support systems are pivotal at this stage when older adults may need increased assistance in various aspects of daily functioning. Also, because the transition from working to retirement may be difficult for some, this difficulty must be acknowledged and recognized.

Risks: Some elderly who are too frail or ill to care for themselves may become increasingly vulnerable to elder abuse, a form of maltreatment in which older persons may be subjected to physical, sexual, emotional, and fiduciary abuse as well as neglect. Laws similar to child abuse exist in most states to help reduce elder abuse.

An added concern faced by many elderly is the prejudicial attitudes and discriminatory practices they experience because of their age and vulnerable status. Ageism is common in our society and can affect the elderly directly or more subtly. In either case, the damage from ageism can be great.

Last, we look at the financial limitations of many elderly who must live on a fixed income. A number of older people live in poverty as a result and thus cannot obtain the benefits of medical interventions and nutritional supplements available to their more financially stable counterparts. Also, the elderly living in impoverished neighborhoods are prone to be the victims of theft, mugging, and sexual assault.

Treatment Considerations

- When interviewing an elderly client, follow these specific interviewing techniques:

  a. Because the elderly person has decreasing energies to cope with his or her everyday problems, helpers may have to invest proportionately more energy into the interview.

b. The helper needs to pace the interview according to the client's fluctuating energy levels and physical conditions.

c. Sensory losses need to be taken into consideration. Communication can be maximized if the helper sits close to the older person and faces him or her directly. Be aware of "recruitment deafness"—when an individual can't hear at a certain level but shouting is too loud.

d. The use of touch can be a meaningful communication bridge.

e. Avoid information overload by speaking slowly, using short sentences, dealing with one thought at a time, and asking for feedback to be certain meaningful communication has taken place. The elderly typically take 15% more time to respond than do younger adults.

- Enhance the older person's self-esteem by encouraging his or her maximum participation and acknowledging the person's role as an authority on aging.
- Initially, when working with the elderly, focus on the problems they can handle with success. This will relieve their feelings of helplessness and disorder, which otherwise might lead them to withdraw from treatment.
- Set realistic goals; aim to restore a degree of functioning that was recently lost or that is now realistically possible, not what you would like to see happen.
- Recognize the importance of choices for the elderly. Acknowledge your confidence in their ability to make choices and to follow through. Allow older individuals to choose their priorities. Because some may have limited energy, let them decide how and where to spend it.
- Be sensitive to the need for sexual expression in late adulthood.
- Be aware of possible changes in mental functioning and the processing of information.
- Understand that many elderly are coping with loss and the management of grief and depression.
- Use of reminiscing and the life review are important techniques for working with the elderly. They may be helpful in linking relevant past events to the present as well as helping to bring closure to an individual's life.
- Some elderly do not have the strength to cope with the confusion of bureaucracies; thus, human service providers need to act as their advocates whenever necessary.
- Crisis intervention with the elderly may be necessary, so helpers should cultivate skills in this area.
- There may be a need to work with families whose older member has Alzheimer's. This will require supportive interventions for both the caregivers and patient.

SCENARIO

WORKING WITH THE ELDERLY

A 70-year-old man named Bill applies for service at the local Community Mental Health Center after experiencing some depression and loneliness he could not seem to shake on his own. Bill tells you that he is separated from his former wife, who lives in the home they worked for together. He has had some prior experiences with counseling as a part of the Scientology religion but has had no mental health counseling in the past. He is mostly concerned about his loneliness and isolation, living alone in an apartment and having little contact with others. He does state that he goes to a dance once in a while sponsored by the local senior center but states that he would never go to regular activities there as he is "not that old yet and can't identify with the other elderly people there." He states that he has few hobbies other than writing and says that he often writes to local newspapers with his opinion about what is happening in the world today. He hopes that someday he will become well known for his writing abilities, but when you look at what he presents as his writing, it appears to be quite disorganized and disjointed. He asks for your assistance in helping him know what to do to get his writing published and to be less lonely. If this were your client, how would you approach him using the skills and intervention techniques suggested in the text? What would be some realistic goals for this client considering what he says he wants to accomplish?

GENDER ISSUES

Gender issues may play a significant role in our work with clients. Although both men and women must be seen first as individuals with their own unique issues and personal make-ups, it is nonetheless important to acknowledge their gender and its possible impact on their current life situation. Despite present-day efforts that emphasize male-female similarities, there are notable differences biologically, psychologically, and socially.

Gender differences are clearly noted in physical characteristics that separate males from females. Much has been written on the psychological differences in male-female emotional development and expression as well as the formation of a specific gender-based identity. Socially, communication styles, career choices, demographic patterns, and socialization itself seem to vary between men and women. To some extent, these differences can lead to commonly noted generalizations about either sex and may also give rise to gender-role stereotypes. It appears that many of the gender roles and stereotypes existent today are rooted in early childhood experiences during which boys and girls are taught about their differences based on gender.

Traditional views of men and women continue to prevail, yet changes in gender-role identity are continually evolving, never in a fixed state. "As these changes become more prevalent, women and men are more likely to take on new personality styles. In fact, some research seems to indicate that individuals who take on multiple roles, including nontraditional sex-roles, may be more satisfied in life than those who do not" (Neukrug, 1994, p. 196).

In light of both these differences in gender and changes in gender roles, we believe there are two main goals that must be met: (1) We must become knowledgeable about the differences between men and women along with developing an appreciation for and understanding of the historical aspects underlying these gender differences, and (2) we must remain open and willing to acknowledge the changing demographics and understand their impact on each gender as well as on society in general.

Understanding and appreciating the consequences of gender roles and beliefs on individuals and their development is essential in assessing human behavior. Because gender issues incorporate all the factors that structure the relationships between women and men, we must recognize the extent of their influence on the particular clients we hope to help. Gender differences "can be used to distort many aspects of social life." This is best exemplified in studies about the differences between male and female aggression or sensitivity (Ashford, Lecroy, & Lortie, 1997, p. 126). Clearly, in this as in other gender-based research, our biases can influence how we interpret results. The information on averages resulting from many such research studies can mislead us into discounting the many realities of men and women and ultimately influence us to continue holding onto traditional biases that exist about each sex.

To minimize the preponderance of such biases and to help you understand the nature and impact of gender issues in your work with clients, we present below an overview of various key points related to gender.

Gender Identity and Gender-Role Stereotypes

A not-uncommon first question asked of a new parent is "Is it a boy or a girl?" In fact, even before the delivery, there is often much focus on what the sex of the child will be, and although many parents say that they are more concerned that their child be born healthy, the sex of the child is nonetheless an important issue. A central factor in identity development is one's gender. Cyrus (1993) illustrates this point well in her discussion of gender identity:

> As we grow older, we identify ourselves and each other as boys or girls, as women or men. In American culture, gender is the most salient feature of one's identity. It shapes our attitudes, our behavior, our experiences, our beliefs about ourselves and about others. Gender is so central to our perception of social reality that we often are not even conscious of how it shapes our behavior and our social interactions. (p. 59)

Before elaborating any further, let us clarify a few basic terms necessary to understanding the complexities involved in gender identity: *Sex* relates to the biological aspects of an individual's physiological and hormonal characteristics. In essence, sex refers only to the biological status of being male or female. *Sex roles* are tied to differences in reproductive traits between men and women. *Gender* relates to societal categories that ascribe roles, appropriate behaviors, and personality traits to women and men (Cyrus, 1993, p. 59). "It represents a socially constructed concept and not a fact of nature with specific biological imperatives" (Ashford, Lecroy, & Lortie, 1997, p. 126). *Gender roles* thus refer to the masculine and feminine behaviors that are based on social definitions of male and female.

Gender-role stereotyping is said to begin at birth and continue through childhood, adolescence, and adulthood. Basically, gender stereotyping involves expectations about how people should behave based on their gender. Gender identity development is directly connected to such categorizing in that boys and girls often learn what it means to be male and female based on those stereotypes they have witnessed, heard about, or been told to follow. Transmission of these stereotypes occurs throughout one's development and lifetime. Parents, friends, neighbors, teachers, coaches, religious leaders, mentors, the media, and society as a whole are all potential agents in conveying messages regarding these roles and stereotypes; many do so unknowingly or with the best of intentions at heart.

Male and female stereotypes vary from culture to culture and society to society, yet many tend to transcend most cultures and societies. According to Ruth (1990), the following are various predominant male/female stereotypes many of our clients will have been introduced to early on in their lives: Males are to be "powerful, creative, intelligent, rational, independent, self-reliant, strong, courageous, daring, responsible...forceful... authoritative...[and] successful"; in stark contrast, female stereotypes include being "nurturant, supportive, intuitive, emotional...needful, dependent, tender, timid, fragile... childlike...passive...obedient...[and] submissive"(p. 126). You might react with feelings of anger, dismay, or disbelief that such stereotypes are still sanctioned by many, including many within the helping professions. If you take a closer look at your own views, you may find that you too still hold some of these stereotypic beliefs. Becoming conscious of this is the first step to dispelling the myths that result in our own generalizations based on gender.

These stereotypes have several troubling aspects. First, many of them are not even based on the individual and his or her own personality, strengths and weaknesses, or likes and dislikes. Furthermore, the pressure to conform to established sex-role stereotypes can contribute to many psychological problems such as depression, sexual dysfunction, obesity, and other clinical concerns (Rothblum & Franks, 1983). Gender has also been found to influence self-esteem, psychological well-being, and physical health.

Second, such gender-based stereotypes "often limit people's alternatives" as seen in the pressure many feel to conform to gender-based expectations, and this conformity leads to the continued perpetuation of the stereotypes. (Zastrow & Kirst-Ashman, 1994, p. 351). Last, the false or exaggerated labels we place on men and women may color our own views of what is right or wrong, preferable or nonpreferable. When our views are affected, discriminatory practices can follow. Sexism is a type of discrimination that can be directly tied to our adherence to gender stereotypes.

Sexism

Sexism is a complex, long-standing problem. Basically, it is considered to be discriminatory behavior toward women by men. Although sexism may take on many forms, universally it continues to support a system of oppressive gender relations for women (Ashford, Lecroy, & Lortie, 1996, p. 129). Sexism has been considered similar to racism in that it is also "a system of beliefs and behaviors by which a group is oppressed, controlled, and exploited because of presumed gender differences" (Anderson & Collins, 1995, p. 67). The statement by the male authors Blood, Tuttle, and Lackey (1995) clearly illustrates how profound this issue actually is:

> Sexism is much more than a problem with the language we use, our personal attitudes, or individual attitudes, or individual hurtful acts towards women. Sexism in our country is a complex mesh of practice, institutions, and ideas which have the overall effect of giving more power to men than to women. (p. 154)

By virtue of the power men hold, both formally and informally, one can easily see how such a power differential can result in sexist behavior. Our definition of power echoes Blood, Tuttle, and Lackey's views that connect power with the ability to influence important decisions—politically, economically, interpersonally, and, many times, even sexually. The two main areas where male dominance seems to prevail involve the economic arena and the area of violence directed toward women. Women have less earning power than men and still continue to experience discrimination even when they are equally as skilled and capable as men in their respective positions. Also, women continue to face a constant threat of violence and sexual aggression seen in rape, sexual assault and harassment, and domestic violence.

In her article, "Sexism in English: A 1990's Update," Nilsen (1993, p. 159) focuses on language and its connection to sexism. She found that because "the language a culture uses is telltale evidence of the values and beliefs of that culture . . . a careful look at English [can] reveal the attitudes that our ancestors held and that we as a culture are therefore predisposed to hold." Below are the three main points identified in her study of the English language:

1. Women are sexy; men are successful. This is sometimes seen in the American culture where a woman is valued for the attractiveness and sexiness of her body, whereas a man is valued for his physical strength and accomplishments.
2. Women are passive; men are active. Women are often expected to play a passive or weak role while men play an active or strong role. Nilsen highlights the wording of the wedding ceremony itself as an example of how this view continues to prevail, even in the midst of our more liberated ideas.
3. Women are connected with negative connotations, men with positive connotations. Many positive connotations are associated with the concept of *masculine,* whereas either trivial or negative connotations are linked to the corresponding *feminine* concept.

Perhaps the most revealing finding of this study is that even in our language, sexist attitudes prevail. "Language is like an X ray in providing visible evidence of invisible thoughts" (p. 160). Only when people begin to pay attention to their own sexist language can they begin to change their thoughts and expectations, thereby reducing sexism itself.

Overall, in the English language as well as in the political, economic, interpersonal, and sexual arenas, sexism continues to predominate. Male dominance coupled with physical strength, voice, and acculturated ways of dealing with anger, certainly explain how easy it would be for many women to feel intimidated, passive, and fearful of challenging this power differential. Furthermore, because of the subtle nature of some of these factors, it is understandable that some women continue to struggle with issues tied to this pervasive and at times covert system of unequal power. As practitioners, we are in an advantageous position as we can help both men and women recognize their own behaviors and explore ways of becoming more compatible and equal in status and respect.

Our focus has been on women and their struggle with oppression, but men also can fall victim to sexist behaviors and feel the pressure of conforming to the ideal male role stereotype. Thompson (1993), who writes about male role stereotypes, addresses this issue by demonstrating the high cost men must pay for the ways they have been stereotyped and for the roles they play. He cites a male "code of conduct" that outlines six major male role stereotypes:

1. Act tough
2. Hide emotions
3. Earn big bucks
4. Get the "right" kind of job
5. Compete—intensely
6. Win—at almost any cost
   (Thompson, 1993, p. 147)

The cost of this code is further illustrated by what men must give up to continue abiding by these prescribed norms. For instance, men who become highly competitive may lose their perspective and good judgment. Those who hide emotions can suffer both psychological and physical problems. The pressure to earn "big" money can influence a man to pursue a career he truly does not want or to work long hours and perhaps two jobs to provide for a family he rarely sees. Furthermore, many men who wish to work in a less "masculine" type of job may feel pressured to do otherwise, only to feel frustrated and unfulfilled. In all, men who succumb to the pressure of being competitive, winning, earning big bucks in the right kind of job, and avoiding showing their emotions pay a terrible price in terms of their physical health. Studies have shown that men fall victim to serious disease much more often than women and on the average, die 8 years sooner than women (Thompson, 1993, p. 147). Certainly, "loss of life is a high cost to pay for following the code of the male role stereotype" (p. 147). What is the alternative? We all know that when men do challenge these stereotypes, they are apt to face other types of pressure and ridicule that can be just as damaging as abiding by them. In essence, men are caught in a double-bind that society itself has placed them in. As human service providers, we must be sensitive to their plight and work toward helping shape a more appropriate role for men that allows flexibility, sensitivity, and acceptance.

Cultural Implications

Our discussion regarding sexism has centered primarily on the American culture. As helpers, we must also be sensitive to sexist issues in other cultures because many clients from other countries and cultures will not hold the same views regarding sexism that we do. Some women are perfectly content with their role, even though we may view it negatively. Before assuming they are oppressed and unhappy, we must first explore their views and feelings. In an effort to help these women, we may be creating more confusion and possibly more harm than we would if we learned to view life from their perspective and to accept their preferences.

Male-Female Similarities and Differences

We have just seen some of the issues both men and women face in regard to sexism and have found that both may be subject to a certain amount of pressure to conform to predisposed societal standards—although the pressure is different for the two sexes. In terms of development, there are also similarities and differences. Until recently, developmental theorists have focused primarily on male development (Bernard, 1975). According to Carter and McGoldrick (1989),

> Female development was seen only from an androcentric perspective and involved learning to become an adaptive helpmate to foster male development.... Whereas separation, differentiation, and autonomy have been considered primary factors in male development, the values of caring and attachment, interdependence, relationship and attention to context have been primary in female development. (p. 32)

These differences in development are important to consider when working with either gender; taking the time to explore what it means to be either male or female for a particular client will allow you to understand the person better individually and in terms of gender. Gender differences in personality, ability level, communication, and so on may also impact the way clients behave or view the world. We begin by looking at the actual psychological differences that influence the way men and women are socialized.

Psychological Differences

Research regarding the psychological differences between men and women is abundant and fuels the ongoing controversy about these differences. We have compiled a list of the differences most commonly agreed on in the literature. For clarity, these differences have been grouped into categories: physical, personality, ability level, and communication.

Physical Differences

• The most obvious physical differences between men and women are undeniable as seen in chromosomal, hormonal, genital, and reproductive differences. "Males typically grow to be taller, heavier, and more muscular than females, although females may be the hardier sex in physical disorders" (Sigelman & Shaffer, 1995, p. 300).

• Boys are more developmentally vulnerable to prenatal and perinatal stress, disease, and disorders such as reading disabilities, speech defects, hyperactivity, emotional problems, and retardation (Henker & Whalen, 1989; Jacklin, 1989).

Personality Differences

• Males are more physically and verbally aggressive than females, starting as early as age 2 (Maccoby & Jacklin, 1974; Sapiro, 1990).

• "Some evidence shows that females tend to have less self-esteem than males, and perhaps because of this are also less assertive" (Block, 1976).

• Boys, starting in infancy, appear to have a higher activity level than girls (Eaton & Enns, 1986; Eaton & Yu, 1989).

• Girls are more compliant with the requests of parents, teachers, and other authority figures (though not with the demands of peers).

• Girls are more likely to use tactful and cooperative, rather than forceful and demanding tactics when attempting to induce others to comply with them (Cowan & Avants, 1988; Maccoby, 1990).

• Girls rate themselves higher in the traits of nurturance and empathy (though their behavior does not always differ from the boys' behavior) (Fables, Eisenberg, & Miller, 1990).

Ability Level

• Recent studies reflect very little difference, if any, in ability level between the sexes. Clearly, there are no differences between males and females in intellectual ability, yet many continue to argue that girls have higher verbal abilities than boys and boys have more mathematical ability. Current thinking suggests that there are very few actual differences.

The majority of the discrepancy can be accounted for by the way in which boys and girls are socialized, not on their actual ability level.

• Males outperform females on tests of visual/spatial ability. These differences can be detected in early childhood and persist across the life span. (Maccoby & Jacklin, 1974).

Communication Styles

• Communication styles between males and females do seem to differ, as evidenced in verbal and nonverbal communication. However, some of these differences are the opposite of what many of us would have expected:

1. Men talk more than women and tend to interrupt more.

2. Women are more likely to give information or make self-disclosures than men are (Cozby, 1973). "The less people adhere rigidly to traditional gender role stereotypes, the more flexible they are in their level of self-disclosure (Gerdes, 1981).

3. Nonverbal behavior also differs between men and women. Men more frequently touch other people; women are most likely to be the ones who are touched. Also, touch has been associated with status, with those with higher status more likely to touch those of lower status than the other way around.

4. Women tend to be more coy in their behavior. "Coy here is defined as involving gestures of submission including lowering the eyes from another's gaze, falling silent when interrupted or pointed at, and cuddling to the touch. Seen as 'feminine' type behaviors, they tend to convey deference, dependence, and a sense of needing leadership and protection" (Zastrow & Kirst-Ashman, 1994, p. 355).

Additional areas where communication differs between the sexes are noted by Beck (1993):

• Women seem to regard questions as a way to maintain a conversation; men view them as requests for information.
• Women tend to connect "bridges" between what their conversational partner has just said and what it is that they have to say.
• Men tend to ignore the preceding comment by their partner.
• Women seem to interpret aggressiveness by their partner as an attack that disrupts the relationship. Men seem to view aggressiveness simply as a form of conversation.
• Women are more likely to share feelings and secrets. Men like to discuss less intimate topics, such as sports and politics.
• Women tend to discuss problems with one another, share their expectations, and offer reassurances. Men, on the other hand, tend to hear women (as well as other men) who discuss problems with them as making explicit requests for solutions rather than simply looking for a sympathetic ear (Beck, 1993, pp. 84–85).

Demographic Characteristics

• The growth rate of women in the labor force has doubled the rate of men.
• Today, more men take part in child care compared with the past (although women continue to do most of it).
• Men own the vast majority of corporations in the United States.
• Regardless of occupation, women on average, earn less than men.
• Women have lower death rates than men.
• Loss of a spouse by death is much more likely for women than men.
• Although women and men start college in equal numbers, women are less likely to finish.
• Elderly women are less likely to commit suicide than elderly men (Tauber, 1991).

Certainly, if all these differences are factual, they represent notable variations among the sexes. Except for physical characteristics, whether these differences are due to biological/genetic factors or social forces in how boys and girls are perceived and raised remains unclear. It is obvious, however, that the way men and women approach life is different. "Generally, men are more restrictive in their ability to express emotions, are less communicative, have more difficulty with intimacy, and are more competitive than women, whereas women are usually more nurturing, more concerned about how their actions affect others, and have more difficulty expressing anger than men" (Newburg, 1994, p. 196). Regardless of the cause, many of these differences have considerable implications for male-female development. Because of the pressure on both men and women to abide by the traditional gender-role stereotypes, both groups have problems. In the workplace there is sex discrimination, as men earn significantly more than women. There are double standards of conduct for males and females, and power struggles also exist. Overall, oppression is greater for women; but men are also at risk because of the stress of having to fit the traditional male stereotypes. As helpers, we must be in tune with how these gender-based differences influence our clients and how we might intervene to lessen any negative ramifications.

Because men and women do differ on various levels, as we have seen, a separate look at each gender seems indicated. Next, we briefly touch on some of the most important issues faced by both women and men that we as human services professionals must be keenly aware of.

Women's Issues

Societal changes have brought dramatic changes for women and the roles they play within the family, at work, and in society in general. Women continue to make up more than half the population—about 51.4% (Morales & Sheafor, 1995, p. 267). They have become an increasingly significant part of the workforce and are a large segment of students pursuing a college education. Women are becoming more involved in previously male-dominated professions such as business, medicine, law, and law enforcement. They are learning to manage dual careers—working both outside and within the home. Despite gradual changes in gender roles, many women who have a full-time job or career still continue to be the primary child caregiver within the home. These changing roles, in many ways applauded, have also caused additional stressors women did not experience years ago when their primary role was to be a mother and wife. The stress of having to be a "working" mother can be great. Working women often feel guilt at being away from their children; frustration at having to deal with work stress, child care issues, and money concerns; and sheer exhaustion from having to balance two very complex and expansive roles (worker/professional and wife/mother). This stress is significantly higher for single mothers or women with additional pressures of work difficulties, marital conflicts, separation or divorce, or physical or sexual abuse.

Traditionally, women have suffered from economic exploitation, inequality, discrimination, and oppression. Because of their more vulnerable status, they have been subject to both economic and physical abuse, seen in lower wages, domestic violence, sexism, sexual harassment, and rape. Additionally, they may suffer more than men from physical and mental problems: depression, eating disorders, and certain health problems such as infections and respiratory diseases; chronic conditions like arthritis, osteoporosis, and diseases of the urinary system; and death from childbearing, abortion, and breast cancer. Sadly, women are the fastest growing population of people to acquire AIDS, and as women grow older, they constitute the majority of older Americans. "Because of their relatively low incomes and greater likelihood of being widowed, older women have higher rates of poverty than do older men" (Morales & Sheafor, 1995, p. 306).

Women today, despite the women's movement and the sexual revolution, still face some of the same struggles their ancestors experienced. In many instances, the personal and social problems of women are linked directly to gender inequality and institutionalized sexism. As emphasized by Morales and Sheafor (1995), to be effective in our work with women, we, as human service professionals, need (1) to understand the cultural and social context for women's problems, (2) to develop strategies that can meaningfully change women's lives, and (3) to continue to identify the ways in which sexism pervades practice.

To understand the cultural and social context for women's problems, we can look at women's traditional roles of wife, mother, and homemaker. Historically, gender inequality has existed in most life spheres for women. Social inequality, social powerlessness, and social dependency, although less than in years past, still persist. As a result, many women are still submissive, have low self-esteem, and are forced into having primary responsibility for the caretaking and nurturing role. These conditions, coupled with the continued transmission of negative cultural beliefs about women through gender-role socialization, have greatly contributed to their current-day struggles, as their feelings of powerlessness and personal devaluation continue to be reinforced. Survival issues and those of personal growth are clearly impaired by the confining social roles and limited options women still face in both the educational and employment arenas. If change is to occur, a strong focus must be on changing these deeply rooted views about women and the roles they "should" or "ought" to assume. This brings us to the second goal: developing strategies purposefully geared toward helping women change. To this purpose, work at various levels of practice—from micro to macro—could be considered. Below we highlight some of the points to consider in working with women.

Treatment Considerations

• Specialized knowledge about women and women's issues should be incorporated into any treatment plan or change process. Also, the cultural background of each client should be considered as you develop ethnically sensitive treatment interventions.

• Women tend to bring five common problems to treatment:

1. Powerlessness
2. Limited behavioral and educational options
3. Anger that is often turned inward, resulting in depression
4. Inadequate communication skills
5. Failure to nurture self (Collier, 1982, pp. 57–68)

• Women's functioning at home, at work, and in the community needs to be evaluated in terms of existing gender-role norms and discriminatory practices.

• Social, economic, and legal issues that impact women must be closely examined. Be aware and sensitive to possible discriminatory practices and sex- or gender-biased attitudes within the community specific to women. Also, the roles women play within their own families must be explored to determine their restrictive or nonrestrictive nature.

• Helpers need to be open to working with a wide range of behavior and roles. We need to be as nonsexist and nonbiased as possible in our approach. This entails incorporating a nonsexist knowledge base and value system in our practice.

• Empowerment of women needs to be a major emphasis in helping them free themselves from the dependent roles they have typically played. Empowerment can take place when you remain positive, hopeful, appropriately self-disclosing, and focused on women's strengths and potential for growth and change.

• Professionals can help women build their self-confidence and self-esteem. Positives need to be emphasized in treatment.

• Assertiveness training can provide women with guidance on ways to handle difficult or uncomfortable situations.

• Focus on being able to express anger in appropriate ways and in understanding underlying dynamics.

• Helpers can teach women to take care of themselves; this will help them in beginning to believe they are worthy.

- Women can be helped to recognize the various alternatives available to them and to evaluate the pros and cons of each, thereby learning decision-making and problem-solving skills.
- Female practitioners are often positive role models for other women, so they must be aware of how they present themselves. An increased chance for empathy and shared experiences is possible.
- Male practitioners can also be instrumental in helping women by remaining conscious of their gender differences and continually working to ensure that traditional male-female power relationships do not prevail.
- In work with families, the helper must carefully assess the family's views on gender roles and attempt to work within these confines. Helpers who are aware of the many types of family forms that exist are more likely to be open and receptive to working with the entire family in an effective manner. Family history related to male-female roles and cultural issues must also be carefully considered as you develop goals that are realistic and ethnically sensitive.
- In both family and group work, the helper should be aware of gender issues within the group. This applies to all-female groups as well as mixed groups. Take care to prevent sexist behaviors and attitudes from negatively affecting the group and interfering with group cohesiveness and growth.
- Change can be difficult and may present additional problems. Encouraging change without considering all the ramifications, both positive and negative, is not only unethical but is also likely to be ineffective. Helpers must be honest with their clients about the many potential outcomes of change. Sometimes in our excitement to help women become more independent and self-sufficient and to like themselves, we neglect to consider the impact these changes may have on their relationships with spouses, children, extended family, friends, employers, and others. Preparing women for possible reactions, conflicts, or resistance is an important part of the treatment process. Furthermore, if a woman once apprised of these possibilities chooses not to pursue certain growth-fulfilling goals, we need to respect this decision.
- Service provision must be expanded by providing more resources and services, such as provision of affordable housing, child care, educational opportunities, and financial and legal assistance.
- Often, services cannot be provided without changes in existing services and in community, state, and national policies.
- New programs and policies that are geared toward combating victimization and are responsive to women's needs may also be necessary.
- Practitioners working at a macro level need to know the impact of discrimination and oppression and work to fight it. Efforts involving social planning, program development, and policy analysis are usually those most effective in creating this type of change.
- Finally, empowerment needs to be global. Work with both women and men at various levels must be considered. Also, a focus on helping children and adolescents learn to view gender roles in a more egalitarian manner is an important step in preventing continued sexist behaviors and problems.

As reflected in many of these interventions, helpers must be aware continually of how sexism pervades practice, and they need to remain committed to combating this in a professional manner.

SCENARIO

WORKING WITH WOMEN

Grace is a 32-year-old Latina female who applies for services from a Catholic Charities agency. She states that she and her husband are having severe marital difficulties and that she has been quite upset recently. She says that her husband is very domineering and expects her to have the house spotless and dinner cooked when he comes home, even though she also works 20 hours a week outside the home. She says that he yells at her a lot and puts her down when he gets upset with her. They have two children, ages 6 and 10, who are watched by her mother when Grace goes to work. Her husband is very jealous of her and will not allow her to have her own friends or activities. She would like to go out with some of the female friends she has made on her part-time job, but they have stopped asking her; as she always refuses out of fear of her husband's reaction every time they ask. She states that her husband has threatened to leave her in the past and that this scares her because she has "no skills" and believes she would not be able to support herself and the children if he leaves. When she told her mother what was happening in their marriage, her mother said it is the duty of the wife to take care of her husband and that she must stay and try to make the best of it because "that is how God would want it." Grace is in good health but does not have time to exercise with all her responsibilities. She has been feeling more and more sad recently; she is not sleeping well and has begun to think that there must be something wrong with her because she cannot give her husband what he wants. She is often so tired by the time she gets home from her job that she has little energy left to keep up the house and cook as her husband wants her to. If you were a counselor at a Catholic Charities agency, how would you handle this client? What would be some of the gender issues you see entering into this situation? How would you handle your own values in reference to working with the issues this client presents? Would it make a difference if she was seen by a female or a male helper?

Men's Issues

Similar to women, men have also witnessed a change in roles in recent years and this has caused stress and conflict. The recent men's movement seems to have been aimed more at men's internal change than any kind of external change. The two popular icons of the men's movement are Robert Bly who wrote *Iron John* and Sam Keen who wrote *Fire in the Belly*. Bly is primarily concerned about how our culture contributes to the separation of boys from their fathers. His book has generated a movement for men to recover their manhood by denouncing contemporary culture and rediscovering their lost masculine identity. Keen encourages men to break free of cultural boundaries in a similar fashion that will allow them to reassert their masculinity and male role. Many theorists believe that

> in childhood, boys prematurely dissociate from their mothers and enter the aggressive, competitive world of men. The boy desires to turn to his father but finds him emotionally distant and unavailable. The result is that his developing masculinity is defined by turning away from anything feminine. (Ashford, Lecroy, & Lortie, 1997, p. 438)

As with women, men are being challenged to change the stereotypic images that are intricately woven into every role they play. This type of change, although welcomed by many, is bound to be challenged by an equal number who remain true to traditional male-role stereotypes. The presence of conflict is therefore a given. Men, like women, will have to battle negativity and differences in opinion in their pursuit of a new male identity and role. As helpers, we will have to be both supportive and sensitive to the specific struggles inherent in their efforts to change.

Many of the issues men face today are also directly connected to gender-based stereotypes noted in the "Male Code of Conduct." Traditional stereotypes must therefore be challenged and men supported as they explore new roles and identities.

Fanning and McKay (1993) capture the essence of the typical male stereotype in the following:

> The old masculine stereotype describes a tough, lonely man who works hard and suffers in silence. The dark side of this stereotype is the man about whom women complain so much; the

man who is violent, abusive, lustful, and untrustworthy. This typical man is a kind of zombie: uncommunicative, dead inside, apparently incapable of an inner life. He is friendless, unavailable to his children, and estranged from his own father. He is almost completely out of touch with his feelings, unable to express any emotion except anger. He drinks and works too much. (p. 3)

As noted in this view of a stereotypical male, various traditional roles and norms stand out. Levant (1992) cites seven traditional roles and norms based on a review of the literature. These include:

- Avoiding femininity
- Restrictive emotionality
- Seeking achievement and/or status
- Self-reliance
- Aggression
- Homophobia
- Nonrelational attitudes about sexuality

Many of these we have already discussed. Most seem still to prevail in some fashion today; they succinctly describe the messages men receive about what it means to be a man from the time they are young boys. The crisis of masculinity evidenced in the past several years given the changes in male-female roles at a societal level has fueled increasing interest in providing needed services to men as well as women. The question of what it means to be a man, particularly for those men in mid-life, seems to be one of the most persistent unresolved issues that leads men to seek help.

Raised to be like their fathers...good providers, strong and silent, they were discouraged from expressing vulnerable and tender emotions and taught to avoid anything that seems feminine. With women now working outside the home more and more and being able to provide for their own living, men cannot fit as easily into the old "good provider" role. Men are being asked to take on roles and show caring in ways that violate the traditional male codes, such as in nurturing children and showing vulnerability. Although in many instances positive, the result of this change can lead to a loss of self-esteem and uncertainty about what it means to be a man."(p. 381)

Men have traditionally been the breadwinners and authority figures in their families. This is not necessarily the case today when women are also having or choosing to work to help with financial demands. Men are taking on more caretaking and housekeeping roles because women are working outside the home. Also, they may have to relinquish the primary authority role for a shared role with their mates. For some men, this has not been a difficult transition; for others, it is a stressful change that they have met with strong resistance and anger. Men are also frequently given double messages like "Be strong and don't cry" with "Be more sensitive and emotional." As a consequence, many men struggle with knowing how to express themselves. Sometimes, they may communicate their hurt in an aggressive manner for which they are further criticized. Having seldom been encouraged to express feelings other than anger, it is not surprising that some men may not know how to acknowledge and express the range of feelings they possess. To compound this confusion further, men are typically not encouraged to seek mental health treatment as readily as women and are often unjustly criticized or humiliated when they do. Clearly, men are in transition as much as women are; they require the same consideration that women do in terms of understanding, sensitivity, and specialized knowledge and skills for effective practice.

To help men in the most effective way possible, practitioners must develop a positive image of masculinity that can restore men's self-respect. This excludes any approach that reverts to an outmoded model of aggressiveness, dominance, or disconnected manhood. "We must walk a fine line in terms of crediting men for what is valuable in masculinity on one hand...to learn to celebrate these qualities and to be able to identify those aspects which are obsolete and dysfunctional" (Levant et al., 1992, p. 385).

In alignment with the strengths perspective, Levant et al. recommend that the following positive male attributes be celebrated in our work with men:

- A man's willingness to set aside his own needs for the sake of his family
- A man's ability to withstand hardship and pain to protect others
- A man's tendency to take care of people and to solve their problems as if they were his own
- A man's ability to express his love by doing things for others
- A man's loyalty, dedication, and commitment
- A man's stick-to-it-iveness, his ability to solve problems and think logically
- A man's ability to rely on himself
- A man's ability to take risks and stay calm in the face of danger

Countering these positive attributes are such obsolete or dysfunctional traits as these:

- Men's relative inability to experience emotional empathy
- Men's difficulty in being able to identify and express their emotional states
- Men's tendency for anger to flip into rage or violence
- Men's tendency to experience sexuality as separated from relationships
- Men's difficulty with emotional intimacy
- Men's difficulty in becoming full partners with their mates in maintaining a home and raising children

Because many of these traits have been traditionally accepted as part of being a "true man," men may need to challenge them and learn new skills along with engaging in emotional work that may be foreign and uncomfortable to them. In our overzealous desire to help men, we might rush to encourage a focus on emotions that they may not be particularly ready for and which might deter them from pursuing any further treatment. We must use caution in how we pace our interventions and remain alert to any signs of discomfort or frustration in our male clients.

Treatment Considerations

- Specialized knowledge of men's issues should be a part of any approach to working with men. Race, culture, and ethnicity must also be taken into account. Because male and female stereotypes about men directly impact their use of social services, human service professionals must analyze their own perceptions and biases. It is neither easy nor the norm for men to seek professional help. To compound this situation, helpers, female and male alike, may not fully challenge men to engage in emotional work where they can learn to articulate their feelings.
- Female helpers need to be aware of their own biases about a man's not being capable of emotionality and guard against the belief that expressing feelings openly is a sign of dependency or weakness in a man. Clearly, these views will hinder men from learning to feel and express themselves in appropriate ways. Women clinicians also need to be careful about giving male clients double messages: Verbally they may encourage men to express their emotions; however, when men express hurt or anger, the helpers may show nonverbal disapproval or discomfort.
- Male practitioners may also have negative stereotypes about their male clients. This is evidenced in the fact that men are often diagnosed with more serious problems (e.g., affective, impulse control, or personality disorders) than they merit.
- Helpers must be willing to make treatment appealing to men. They may do so possibly by encouraging a man to work toward change through various avenues, without solely emphasizing the need to be emotional. In other words, they should start where the client is instead of pushing too quickly for change that may be unrealistic.
- Seeking help needs to be reframed as a sign of strength, not weakness.
- Family history and cultural issues may be intricately involved in the male role of clients and thus must be carefully examined with particular clients. As with women,

encouraging a man to change may be met with conflict, resistance, and disappointment by those around him. Helpers need to be sensitive to this possibility and take it into consideration during their treatment planning.

• As with women, work with men can be effective at various levels; micro, mezzo, and macro. Many of the same suggestions given for working with women pertain to men as well.

• An all men's group has been found to be a very effective treatment modality. (In Chapter 4, we highlighted a men's support group).

SCENARIO

WORKING WITH MEN

Samuel, a 34-year-old Latino male, enters treatment on the recommendation of his wife, who also entered treatment at the Catholic Charities agency about one month ago (see preceding scenario, "Working with Women"). He says that he has never come into counseling before and doesn't really think he needs it now, but agreed to come because his wife's counselor "thought it would be a good idea." He says his wife is depressed and that he does not understand why, but is willing to help her if he can. He states that his wife has recently begun to express dissatisfaction about their marriage, which he is quite concerned about. He states that they have been married for 12 years and have two children, ages 6 and 10. He had been the sole provider in the relationship for years, until about 4 years ago when his wife decided to go out and get a part-time job against his wishes. He dates the beginning of the problems in their marriage to that time. He thinks she became more unhappy and depressed after that time and wishes she would quit working to stay home with the kids. He states that he is not depressed and is doing well in his job as a baker in his family's business. He says that to make the business successful, he has to be away long hours and that he is under a lot of pressure there constantly. He would like to put more time into his home but really can't right now. His wife has also been telling him that he doesn't talk with her or spend enough time with her, but he wants her to realize that these things are beyond his control due to work and his schedule. He says that to him everything appears fine and he thinks that maybe his wife is just depressed. He is confused about her saying that she is unhappy in their relationship because she has never said she was unhappy before. He does say that he is willing to help his wife in any way he can, even if it means coming to counseling by himself or with her. He says that he cares for his family and loves his wife and wants her to get better. How would you approach this individual, considering what he says he wants out of the therapeutic relationship? Considering specific gender issues unique to men, what would be some of the influences that you could see at work in this situation?

SOCIOECONOMIC STATUS

A very important part of becoming diversity sensitive involves assessing an individual's socioeconomic status and understanding the many ways it may impact the person's overall functioning. Simply put, socioeconomic status (SES) refers to the position people hold in society based on such factors as income, education, occupational status, and prestige of their neighborhoods (Sigelman & Shaffer, 1995, p. 104). Historically, our society has distinguished among upper, middle, and lower socioeconomic groups, the last including both working-class people and families living in poverty. Although we should refrain from discriminatory practices that may result from such categorizing, it is helpful to know where a person lies in the spectrum ranging from poverty to upper-middle class. Insight into an individual's particular SES can alert human service practitioners to the possible stressors the person may be facing and help in determining appropriate interventions, at both a micro and a macro level. Clearly, there will be different stressors at each level and variations among the levels themselves.

Individuals from different socioeconomic levels may also differ in their values, socialization goals, and ultimately their view of the helping process. The helper needs to be aware of these differences and work toward helping clients at the level where they are. Below we take an in-depth look at one of the major SES groups: the poor. This is not to deny that being in the middle and upper class does not have its own stressors; thus we will address these briefly as well.

The Poor

According to the literature (Cooper, 1991; Neukrug, 1994; Zastrow & Kirst-Ashman, 1994), over 15% of the population of the United States is poor. Cooper in 1991 found that 34 million Americans (nearly 14%) were living below the poverty line. The poverty line is the level of income the federal government considers sufficient to meet the basic requirements of food, shelter, and clothing. This definition suggests that there are many more who may fall just above this level and thus must also struggle with limited income and lower living standards. A sad statistic indicates that despite being one of richest countries in world, the United States has experienced a gradual increase in poverty in the past decade (Zastrow & Kirst-Ashman, 1994).

Causes of Poverty

Through the years, many attempts have been made to understand the causes of poverty. There are many variables that contribute to whether one lives in poverty and the many differences among those who are poverty stricken. Some have lived in poverty all their lives; others have experienced it temporarily due to situational and environmental circumstances. Each different condition has its own influence on intervention strategies, so helpers should be aware of these distinctions.

One well-known explanation for poverty, known as the "culture of poverty," was proposed by Oscar Lewis in 1966. He determined that the poor are poor because they have a distinct culture or lifestyle. He concluded that poverty results from extended periods of economic deprivation in a highly stratified capitalistic society. This type of economic deprivation results from high unemployment rates and low wages for those who remain employed. According to Lewis, when such economic deprivation persists, people develop attitudes and values of despair and hopelessness. He states:

> The individual who grows up in this culture has a strong feeling of fatalism, helplessness, dependence and inferiority; a strong-present time orientation with relatively little disposition to defer gratification and plan for the future, and a high tolerance for psychological pathology of all kinds. (Lewis, 1966, p. 23)

Once developed, this culture continues despite the economic status. Some believe that these very attitudes, norms, and expectations limit the poor from seeking opportunities to overcome their SES. As you can imagine, Lewis's postulation sparked much controversy. Many, like Leacock (1971), argue that the distinctive culture of poverty is the result of continuing poverty, not the cause. She further stipulates that the reason the poor do not hold middle-class attitudes and values is because these standards are irrelevant to their situation. In contrast, if they were able to acquire good paying jobs, they would probably be more apt to take on the values of the middle class.

Others, such as Ryan (1976), are even stronger in their criticism of the "culture of poverty" phenomenon, calling it a classic example of blaming the victim. He contends that it is our social system that allows poverty to exist; the poor are poor because of limited resources, not because of their culture.

In exploring the causes of poverty in children, policy analysts have focused on three main factors: changes in demographics, changes in economic conditions and changes in government spending on social welfare programs (Jones, 1995).

Who Are the Poor?

Certain population categories have been identified in attempts to delineate the poor. These include one-parent families, often headed by women with limited skills and resources; children; the elderly; large families; minorities; and the homeless. Other factors also related to poverty are educational level, unemployment, and place of residence.

Many within these categories have already experienced difficulties due to their vulnerable status. The additional stressors of poverty can only weaken their status even more. A telling predicament, for instance, states that "children raised in poor families are likely to remain poor in their adult years." Specific stressors resulting from living a poverty stricken life include these:

- Poor diets and nutritional status. The poor have higher rates of malnutrition and disease.
- Less access to medical care and lower quality care from health care professionals (Medi-cal stigma).
- Substandard housing, possibly with no running water and the infestation of rats and cockroaches. Living in unsanitary conditions.
- Exposure to higher levels of air pollution, water pollution, and crime.
- Living with no heat in the winter and no air conditioning in the summer.
- Limited privacy; the walls of many dwellings are too thin to keep outside noises out.
- Limited belongings and clothing, which is often ragged and old.
- Increased susceptibility to emotional disturbances, alcoholism, and victimization by criminals, and a shorter life expectancy.
- Living in slum housing, unstable marriages, and little opportunities to enjoy things many of us take for granted like traveling, dining out, seeing movies and plays, shopping, and attending sports events.
- Schools that are of low quality and offer few resources. This most often leads to poor academic performance and increased chance for becoming a high school dropout.
- Inability to obtain stable, well-paying jobs because of limited skills.
- The poor are more likely to be arrested, indicted, and imprisoned, and are given longer sentences for the same offence than more affluent perpetrators.
- Emotionally, poverty can lead to feelings of despair and low-self esteem that further impede possibilities for physical, emotional, intellectual, and social growth. At its worst, poverty can instill the feeling that a person is a "second-class" citizen who is neither deserving or able to change his or her life circumstances.

Homelessness in America

One possible consequence of being poor is becoming homeless. Some estimate that as many as 3 million Americans may have no homes (Neukrug, 1994, p. 268). Although there has always been homelessness in America, it has increased in recent years—in part, because of deteriorating economic conditions affecting mostly those who are poor. Also, the stereotypic view of the homeless as being hobos, skid row bums, or alcoholics, no longer applies. According to Neukrug (1994), the homeless of today include these:

- Children who have run away from home
- Intact single-parent families
- Intact families who have no place to live
- Poor single men and women
- Those who have minimum paying jobs but cannot afford shelter

- Those with a substance abuse problem
- The deinstitutionalized mentally ill

Additional studies report that the homeless of today are more likely not to have any shelter, are younger, are less apt to find employment, and are heavily minority (Axelson & Dail, 1988; Rossi, 1990). Many of the same stressors faced by the poor also apply to the homeless but are further compounded by their homeless status. Homeless are at greater risk of developing AIDS, tuberculosis, and other diseases. Children who are homeless are more likely to do poor academically because of retarded language and social skills; they are more prone to suffer abuse and experience delayed motor development. Emotional ramifications can result in feelings of despair, depression, and a sense of hopelessness.

Treatment Considerations

- Helpers need to be aware of their own biases regarding socioeconomic status and to keep these in check when working with those whose status is different from their own.
- Be aware of "clinical cooling out" of poor people in agency settings. This happens when poor people are discouraged from seeking further help. They may be considered unworkable and can be given a subtle message that they are inadequate patients. As helpers, we must prevent this from happening in agencies where we work.
- Be respectful and sensitive to their needs and circumstances. Be careful not to make rash judgments regarding their situation without considering possible underlying issues.
- Be knowledgeable of the many stressors the poor and homeless must face. Work with them at their level by remembering Maslow's hierarchy of needs philosophy.
- Take care not to talk down to the poor; they are equal to us regardless of their financial status.
- As helpers, we must not adopt a "blame the victim" philosophy.
- Because there is just as much variance among the poor and homeless as in the general population, helpers must not overgeneralize or stereotype. Individual assessments remain essential for each case.
- On a macro level, helpers can be advocates for change and engage in program design and development that can help these individuals cope and perhaps even be lifted out of poverty.
- Policy changes are needed. Efforts must be made at a broader level by exploring ways to increase resources for this population, especially the provision of affordable housing.

SCENARIO

WORKING WITH THOSE OF LOWER SOCIOECONOMIC STATUS

Julie, a Caucasian female age 29, is a single mother of five children who applies for service at a county social service agency after she loses her housing. She had been living with her brother for over a year after her separation and divorce in her second marriage. She says that she and her brother "came to blows" after a disagreement over the rent money she owed him. She says that she had been collecting a limited amount of money from Aid to Families with Dependent Children, but that her brother said the amount wasn't enough for her and her children to stay there any longer. She became very angry and a physical fight between them ensued. She denied that she had been hurt physically, but she did have several visible bruises on her arms. Julie states that she does not know what she will do now, as she and the children have nowhere to go. Her five children are all fairly young; the ones from her second marriage are 1, 3, and 4 years old; the children from her first marriage are 7 and 9. The children look tired and haggard; their clothing is quite old and tattered. Julie says that her second husband couldn't be located to help her with child support and that he was physically abusive with her and the children when he lived with them, anyway. When she tells of her own childhood, she describes living in poverty, with her parents always fighting physically. As a child, she would run and hide when the fighting got too bad. She begins to cry when telling you about this and says that she is really overwhelmed now and doesn't know what to do. If you were treating this client and her children, how would you proceed in helping her?

Middle and Upper Class in America

Being middle or upper class can also pose stressors that must not be overlooked in working with either of these populations. Although they do not suffer the same conditions as the poor, higher SES people can feel overwhelmed, frustrated, or depressed. The fallacy that "money buys happiness" illustrates a common problem for the middle and upper class in America. Often the pressure of maintaining a certain standard of living places undue strain on families. The stress of having to strive continually to move up the ladder of success, following the American Dream, often takes its toll on individuals and families. Many may succumb to alcohol or drug use to cope with these pressures. Marital conflicts resulting in divorce are common. Abuse of children in many ways can result. Legal problems may be present, and emotional problems such as depression and anxiety may develop. You will need to determine which stressors your clients may be experiencing as they relate to socioeconomic status.

SEXUAL ORIENTATION

Working with diversity also involves working with gay and lesbian individuals, families and groups. To be effective, nonbiased, and nonjudgmental, helpers must understand issues of sexual orientation and homosexuality.

Historical Views

Historically, the mainstream of our population has defined themselves as heterosexual. Although most of us think of ourselves as purely heterosexual, in reality we are probably neither purely heterosexual or homosexual. It may be simply easier for most of us to place ourselves in a neat category that we can feel comfortable with. In the famous Kinsey studies of the 1940s, Kinsey himself found that "most individuals are not 'purely' homosexual or heterosexual but rather are on a continuum between homosexual and heterosexual" (Neukrug, 1994, p. 196). Only in recent years, however, has it become a little easier to acknowledge being homosexual. Americans are beginning to accept or adjust to the reality that homosexuality is indeed a sexual orientation of many in our society. In the 1989 Gallup poll, 47% of Americans felt that homosexual relations between two consenting adults should be legal, 36% said they should not, and 17% had no opinion. Even more surprising were results indicating that 71% of Americans felt that homosexuals should have equal rights relative to job opportunities. These statistics suggest an increased acceptance of the gay and lesbian lifestyle and sexual orientation in our country. In the medical and psychiatric fields there have also been advances. Until recently, being homosexual or bisexual was considered a mental illness and warranted a psychiatric diagnosis. Today, both the American Psychological Association and the American Psychiatric Association view homosexuality as a sexual orientation, not an illness.

According to Moses and Hawkins (1982), sexual orientation refers to an individual's preference for partners of the same sex, opposite sex, or both sexes for sexual and affectional relations (pp. 43–44). It is important to emphasize that the desire to share affection or become life partners plays a significant role, as does sexual attraction, in the determination of sexual orientation. Another very important distinction must be made between sexual preference and sexual orientation. In the past, individuals were believed to choose to live a homosexual lifestyle. Today, research findings increasingly suggest that "homosexual feelings are a basic part of the individual's psyche rather than something…consciously chosen"—hence, the term *sexual orientation*. "Although it is still unclear how we obtain our sexual orientation, it appears now that sexual orientation is determined very early in life and may even be related to biological and genetic factors" (Neukrug, 1994, p. 197). Biological theories tend to center on genetic and hormonal factors, and psychosocial theories emphasize that homosexual behavior is learned or results from certain patterns of family relationships. Storms (1981) proposes an interactionist theory that focuses on the interaction of biological predisposition and the effects of the environment. At present, there is no consensus as to which, if any, of these theories are correct. However, many experts do agree that homosexuality probably results from some mixture of both biological and psychosocial variables. The exact nature of the mixture is still unknown (Crooks & Baur, 1993; Gooren, Fliers, & Courtney, 1990).

All these changes in how society views homosexuality indicates progress in terms of acceptance. Unfortunately, despite these advances, being homosexual in our society is still very controversial. Heterosexism, defined as the belief that heterosexuality is or should be the only acceptable sexual orientation, still is prevalent. This belief results in discrimination, harassment, and acts of violence toward homosexuals and is encouraged by fear and hatred (Blumenfeld, 1992). The next section is a brief overview of the critical issues facing many gay and lesbian individuals today.

Being Homosexual

Coping with Homophobia, Discrimination, and Oppression

To "come out" and acknowledge that one is gay or lesbian is difficult and can create much stress and anxiety. Many homosexual individuals will be confronted with negative stereotyping, prejudice, discrimination, and even rejection. Incidents of oppression, exploitation, and stereotyping continue to exist despite what the statistics tell us about acceptance. Many who are homosexual are still living a heterosexual lifestyle because of the stigma and shame they fear will be placed on them should they openly acknowledge their sexual orientation. It is not uncommon for many to marry and live a heterosexual lifestyle to avoid internal and/or external pressures resulting from coming out.

The term *homosexual*, which resulted from a medically oriented, heterosexual perspective, is no longer universally accepted. Many prefer the term *gay* that has become the most common and popular term for those who define themselves as homosexual. Although the term can be used to describe both men and women, the term *lesbian* is preferred by many homosexual women. It is important to be aware of how individuals refer to themselves and to respect and be sensitive to their preferences.

Homophobia—the fear and hatred of sex with a same-sex partner—continues to be a prevalent feeling of many toward the homosexual population. According to Ashford, Lecroy, and Lortie (1997),

> In our society, power and privilege are structured in terms of sexual preferences, and our system has institutionalized heterosexual forms of gender identity. In fact, a heterosexual gender identity has been considered the yardstick for determining normality and has contributed to the isolation and oppression of gays, lesbians, and bisexuals. (p. 130)

Homophobia can be viewed as an irrationally negative attitude toward those who are homosexual. As long as homosexuals continue to be labeled pejoratively as *fag, faggot, sicko, effeminate, butch,* or *dyke,* homophobia remains a major obstacle to accepting homosexuality within our society. Homophobia often gives rise to internalized oppression, which creates problems with self-esteem and self-image for homosexuals. Herek and Berrill (1990) cite various factors that appear to be related to homophobia. These include authoritarianism, highly religious views, relating to others with negative views toward homosexuality, limited or no personal contact with gay or lesbian people, and in recent years, the epidemic nature of AIDS, which many erroneously attribute to immoral sexual practices between gay men. It is often the homophobic attitudes and beliefs that gay and lesbian people encounter that are a problem, not their homosexuality itself.

As helpers working with all areas of diversity, we must look at our values and biases and determine whether we have any homophobic tendencies. Honest self-awareness can lead us to corrective action and minimize the risk of discrimination on our part, regardless of how subtle it may be.

Gay and lesbian individuals are confronted with many special issues. Some who risk being open about their sexual orientation are likely to experience family rejection, hostility of peers and friends, and verbal and physical abuse by people who have difficulty accepting their different orientation and lifestyle. For many, the result is to feel confused, angry, and fearful, emotions that further reinforce their self-doubt and perpetuate feelings of low self-esteem. Practitioners must help individuals cope with any of these issues if their practice with them is to be effective.

Human service professionals should also be aware and knowledgeable of the history of oppression homosexuals have endured. Gay and lesbian individuals experience this negativism in many ways—from being disqualified for a job or academic program due to sexual orientation to being violently and psychologically abused because of their homosexuality. Secrecy and fear often prevail in the world of a homosexual, especially during adolescence when issues of identity formation are so crucial. Homosexual adolescents often experience alienation and face developmental problems because they may repress their identity, face peer rejection, and feel the pressure of cultural sanctions against intimate homosexual relations (Ashford, Lecroy, & Lortie, 1997, p. 348).

The prevalence of and risk for depression and suicide among homosexual adolescents is higher than in heterosexual population and is thus a serious area for human service practitioners to be aware of. Additional areas of high risk are "destructive alcohol and drug abuse, being victims of crime, sexual diseases including AIDS, and school dropout. These risks can be compounded if the youth is a minority" (Ashford, Lecroy, & Lortie, 1997, p. 349). According to McManus (1991), gay and lesbian minorities or youth of color often live within three communities: their ethnic community, the gay and lesbian community, and the majority community. For some, maintaining this balancing act can be very stressful, especially because each of these three worlds fails to support significant aspects of a person's life.

As with any vulnerable and oppressed group, the history of the gay and lesbian movement is complicated and multifaceted. The way in which homosexual individuals cope with their sexual orientation differs from individual to individual. Some have chosen to remain invisible to avoid oppression and stigmatization; others have decided to fight actively for gay and lesbian rights. Many fall in between. Most significant is that this minority group is no longer silent and hidden. "The visibility and influence of the lesbian and gay minority will continue to grow" (Morales & Sheafor, 1995, p. 339). We must strive to respect members of this group by recognizing their individual differences and preferences. If their behavior becomes problematic and they are at risk for increased difficulties, we need to help them explore more appropriate channels, not focus on their sexual orientation as the source of these problems.

Gay and Lesbian Relationships, Couples, and Families

Relationships   Relationships among gay and lesbian couples vary as do those within the heterosexual community. There is no one standard relationship style, although many myths and stereotypes exist that would lead you to believe that these relationships are dysfunctional and pathological in nature. Here are a few of the most common stereotypes:

- *The queen and the butch:* This refers to the views that gay males are very effeminate and lesbian women are masculine looking. Although some individuals do fit these stereotypes, many do not. In fact, only a small percentage of gays or lesbians could be identified by appearance.

• *Playing male and female roles:* There is a false belief that gay and lesbian people assume "male" (dominant) and "female" (submissive) roles.

• *The myth of child molesting:* A very dangerous and stigmatizing myth is that homosexuals are inclined to molest children. The reality is that a majority of child molesters are heterosexuals (Newton, 1988).

• *Gays and lesbians must have AIDS:* Because gay men were among the first groups to be afflicted by the HIV virus, they are forever associated with it. In reality, AIDS is now spreading more quickly among heterosexuals than in any other group, with women much more likely to contract the disease from heterosexual men. AIDS is therefore not a "gay disease" (Morales & Sheafor, 1995, pp. 360–361).

In general, gay men are reported to have a greater number of partners than heterosexual men. Lesbians, in turn, seem to have substantially fewer partners than gay men. This difference may be more gender based than due to sexual orientation. A study of homosexual relationships conducted by Bell and Weinberg (1978) resulted in the development of five major categories for gays and lesbians. Although the study has been widely criticized, some of its findings are worth mentioning. These include close couples, open couples, functionals, dysfunctionals, and asexuals. Perhaps the most notable finding of all is that most gay and lesbian relationships do not differ greatly from their heterosexual counterparts. The major difference seems to be in societal acceptance, seen in the fact that marriage of homosexuals is illegal in most states and social obstacles to be together seem greater.

Gay and Lesbian Parents and Families   Despite the impossibility of marriage and the societal stigma of the alliance, many gay and lesbian couples live in monogamous relationships that to them carry the same commitment that marriage does. Many have families that vary greatly. Some had children in a heterosexual relationship before coming out and now live in a homosexual relationship and continue to parent their children. Others create relationships with partners who have children from prior relationships, and still others adopt or consider invitro fertilization or the help of surrogate mothers to have a child of their own. Because of the present-day stigma that continues to affect how society views homosexuality, custody and child visitation issues are complicated. Also, adopting or pursuing other novel ways to have children are difficult endeavors.

Working with gay and lesbian couples and families will require helpers to be professional, accepting, and cognizant of the additional stressors these families are likely to face.

Treatment Considerations

• Helpers need to be aware of their own biases and keep them in check. As with any other area of diversity, if helpers find being non-judgmental too difficult, they should consider a referral. Because of the hidden nature of our biases, there may be times when we are not completely aware of them until they stare us in the face.

• We encourage helpers to seek consultation and supervision when a situation such as this arises. It is important not to refer too quickly nor to hang on too long to a case the helper feels inadequate in treating. It would be ideal if the helper could feel comfortable with the client's homosexuality. This may not always be the case. Even without a feeling of comfort, workers may still be effective in working with gay and lesbian clients if they remain open to exposing their values and allowing their own beliefs to be challenged.

• Understanding the likely homophobic reactions of others and the impact this can have on the individual client will be helpful for practitioners.

• Effective work with gay and lesbian individuals requires a dual focus: The practitioner must be able to see the ways that sexual orientation may affect a client and ways that it is not an issue.

• One of the most common mistakes made by human service professionals is failing to consider that a client may be gay or lesbian. Despite stereotypes, most gays and lesbians are not visibly identifiable and may not identity themselves at first. This is especially true when the problem that brings them to therapy is not related to their sexual orientation. Not knowing poses a challenge for the helper who must delicately assess the client's sexual orientation.

• Clearly, an accepting and open attitude in the helper will ease clients' disclosure of their sexual orientation.

• When the sexual orientation of a client is unclear, helpers should have a positive attitude and use language that conveys acceptance of both heterosexual and homosexual orientations.

• Clients may not wish to share their sexual orientation, especially at first. Explore the situation with an open mind and allow clients to convey only what they are comfortable with.

• Because the developmental pathways to gay and lesbian identity are numerous, helpers should understand and accept the many types of gay and lesbian identities and experiences.

• Knowing the typical developmental issues based on the individual's age is important. Issues of individuation and separation are especially valuable in assessment and treatment planning.

• Helpers must be aware of the many problems and vulnerabilities that both the individual and his or her family are likely to experience as they face the challenges of coming out.

• In working with this population, helpers need to attend to the broad range of issues relevant to being gay or lesbian.

• At all levels, consider the role of social supports and ways to reduce isolation.

• On a macro-practice level, efforts must be made to work toward policy changes that emphasize the values of justice, independence, freedom, community life, self-determination, and change. Building coalitions and supporting legislation are other ways to help.

• Advocacy, social planning, and community change are important in combating the negative stereotypes about gay and lesbian individuals.

• Working with gay or lesbian couples and families may require additional skills in couple and family treatment. Working with these couples will require special knowledge regarding the added stressors they are apt to face due to their homosexual status.

• In working with gay and lesbian parents, practitioners can help these parents identify and emphasize the joys of parenthood rather than focusing solely on gay or lesbian issues.

• Workers can help these parents address the issue of coming out to their children. This should be done soon to avoid additional problems related to hiding and keeping secrets.

• Issues tied to child management and blended families can be addressed by helpers when gay and lesbian clients are experiencing this type of situation.

• Because of the potential for prejudice and discrimination professionals should help parents teach their children situational ethics. This means that parents are open with their sexual orientation, yet they teach their children when and where it may be appropriate to discuss their relationship openly and when it may not. Some agree with this notion; others see it as connoting shame.

SCENARIO

WORKING WITH GAY AND LESBIAN CLIENTS

Yvette is a 28-year-old female who comes to a local community mental health center for services. She states that she is quite upset about the breakup of a relationship with her partner, Susan, whom she has been with for seven years. She says they had been having difficulties recently as her partner had become quite distant and withdrawn and was staying out until very late, dancing at the local club. Although she had made numerous attempts to discuss what was going on in the relationship, she was always met with anger and defensiveness. Then one day last week she came home to find all her partner's things removed from the home with a note stating that Susan was leaving her, that she had found someone else she loved and wanted to be with. Yvette states that she has not been sleeping well since that time and has been sad and depressed much of the time. She has little appetite and has missed several days of work as she felt she could not concentrate well enough to be there. She is an accountant for a major corporation and is worried about missing so much work. She says that her family has not been supportive of her relationship as they did not agree with her "lifestyle." When she told her father of the breakup, he said that he never liked Susan anyway and that he hoped now she would decide to lead a "normal" life. She mostly wants your help with her depression and sadness and would like to discuss how best to take care of herself during this difficult time. If you were the therapist in this situation, how might you handle this client? What are your thoughts and feelings about treating an openly gay or lesbian individual? Would this situation present any value conflicts for you? If so, what might you do to handle the situation?

OTHER AREAS OF DIVERSITY

Other areas of diversity that deserve mention include those related to disabilities (e.g., developmental, learning, mental, and physical), addictions (e.g., substance use, eating disorders), religious and spiritual preference, and geographic considerations (e.g., rural versus urban life). Although we do not elaborate any further here, we encourage you to consider these as important areas of diversity as well.

CONCLUDING EXERCISES

1. In line with becoming more fully diversity sensitive, consider visiting and interviewing a professional from a community agency that serves a population that you are not familiar with or that is different from you. Maybe you are not familiar with the Hispanic/Latino, Asian-American, Native American, or African-American populations in the community where you practice. How could you take one small step to become more familiar with community services while at the same time exposing yourself to people different from you?_____

    _____

    _____

    _____

2. List the different cultures that you consider influential to your development and/or that you identify with. For instance, Randy considers a few of the cultures that influenced him to be Italian, English/Irish, men, American. Consider as many of these as possible and briefly share your identification process. _____

    _____

    _____

    _____

Imagine you moved to a new country and lost your identification with two of your cultures above, perhaps the first two on your list. How do you think you would feel?

_____

_____

_____

3. Using your class as an avenue to explore issues of diversity, consider sharing your lists of cultures that you identify most with, but only after the group attempts to guess what identifications they imagine each class member to have. Note how often the group is correct and how often they were either incorrect or did not get the identification altogether. This exercise could help to highlight the cultural diversity present right there in the classroom setting. It may also help to show that we cannot always know immediately the cultural identifications of the people we come into contact with. _____

_____

_____

_____

_____

7

Ethical and Legal Issues in Advanced Practice

AIM OF CHAPTER

Effective clinical practice requires helping professionals to be well versed in ethical and legal principles and guidelines. In fact, to ensure quality treatment and reduce the potential for harm to clients, each discipline within the human services has established its own formal code of ethics and standards of practice. We have included the codes of human services, counseling, and social work in the appendix. Considering the array of issues helpers are likely to face in their day-to-day work with clients and their families, they must be able to recognize and deal with any ethical or legal issue that may arise. Some of these issues may be clear and easy to recognize; others can be much more complex and difficult to tackle. It is these difficult issues that we focus on in this chapter.

For each major ethical or legal issue, we first present a review of the most pertinent information and discuss relevant treatment considerations when applicable. The particular issue is illustrated by a scenario that challenges you to apply the information you've learned. We begin with a brief overview of the tenets of ethical decision making that should help you to develop an effective treatment approach.

ETHICAL DECISION MAKING

Knowing the process for sound decision making provides the foundation for identifying and working through specific ethical and legal issues (Cournoyer, 1996). The particular steps in this process will vary from practitioner to practitioner and agency to agency. Further, this is an evolutionary process that changes as a helper gains practice experience and make a commitment to remain abreast of current ethical and legal mandates applicable to both the helper's state and specific profession. We encourage you to develop an approach that fits your own style and is congruent with the agency standards where you practice.

Ethical Decision-Making Models

The nature of ethics within human services is broad and complex. Numerous authors have written extensively about the topic and have developed models for ethical decision making. We look briefly at two models described in the literature: those of Van Hoose and Paradise (1979) and Hill, Glaser, & Harden (1995). They overlap a bit as both address ways to make sound decisions in a dilemma of an ethical nature. The main features of each model are shown in Table 7.1. Both models emphasize the importance of being thorough and honest in the decision-making process. Taking one's role seriously in making ethical decisions is essential.

Engaging in a decision-making process requires you to accept your level of responsibility as a helper and recognize the limits of your professional role. An approach we have found useful in our agency work and with the diverse client populations we have treated centers on the need for individual responsibility in the helper. Below is an outline of our approach; it is divided into three major steps that build on each other. Sometimes you may find that only the first or second steps are necessary; each situation will require a distinct approach.

Step 1: Assessment, review, and action

Helper's individual responsibility:

- Identify the problem, dilemma, or conflict
- Delineate the potential ethical or legal issues involved
- Assess the existence of possible moral or value conflicts
- Review the relevant ethical or legal guidelines and specific agency policies and procedures
- If possible, determine an appropriate plan of action

Step 2: Consultation

Helper's individual responsibility:

- Seek guidance from supervisor and/or fieldwork instructor
- Consult with peers or colleagues
- Pursue feedback from agency administrators
- Seek legal advice from legal counsel provided by agency
- If possible, determine and proceed with a plan of action

Step 3: Involvement in counseling or therapy

Helper's individual responsibility:

- Determine possible underlying issues that may be affecting the helper's ability to make certain ethical decisions
- Work through unresolved issues or deal with unfinished business that may be triggered by a particular client situation or interaction
- Understand emerging themes surrounding certain ethical or legal issues
- When and if possible, pursue a determined plan of action

TABLE 7.1   Ethical Decision-Making Models

Van Hoose and Paradise	Hill, Glasser, and Harden
1. Identifying the problem/dilemma 2. Examining existing codes, standards, rules, or guiding principles 3. Generating possible courses of action (getting input from supervisors, colleagues; seeking legal advice) 4. Considering the consequences (who is responsible?); selecting a course of action (justification may need to be given here, either to oneself or others)	1. Recognizing the problem 2. Defining the problem; collaboration with client is imperative here 3. Developing solutions with clients 4. Choosing a solution 5. Reviewing the process with client and rechoosing 6. Implementing and evaluating with client 7. Continuing reflection

How to Handle Unethical Behavior

The issue of unethical behavior in yourself and/or in others should be addressed early in your educational process. Ogloff (1995) indicates that "there are two general ways to prevent illegal or unethical behavior by therapists: ethical guidelines and the law (criminal and civil)" (p. 369). However, such guidelines and legal stipulations alone are insufficient. Most crucial is that the human service worker be well informed regarding these guidelines and possess morally sound principles on which to base competent judgments and decisions.

Unethical Behavior in Oneself

When you encounter a situation in which you find yourself bordering on unethical and illegal actions, you must stop and explore your behavior and possible motives. It is important to engage in serious self-inquiry that can allow you to consider what underlying issues may be present. Second, seek consultation and direction from others (supervisors, legal counsel, ethics board members, etc.). Their viewpoints will help you assess the situation more objectively and reconsider how you might proceed. When helpers fail to explore their own issues and behaviors, they often begin to establish a negative pattern of unethical or illegal behavior. Often, these result in a malpractice suit against the helper, and most unfortunate, harm to the client.

Unethical Behavior in Others

If you are aware that a colleague or administrator is engaging in unethical behavior, the best recourse is to bring up the issue with the individual and encourage him or her to discontinue the behavior immediately. Depending on the nature of the situation and the outcome of your interaction, you may be obligated to report your colleague's misconduct to a professional ethics board (Ogloff, 1995). Either action may seem difficult or unrealistic.

Most cases involving ethical dilemmas are not pleasant experiences. You may feel anxiety, confusion, frustration, and even anger when you encounter unethical behavior in a colleague. When you observe this unethical behavior in yourself, the negative feelings are usually intensified and accompanied with guilt, shame, and embarrassment. With all ethical dilemmas, we highlight the need for on-going self-awareness, supervision, consultation, and a commitment to the values, ethics, and purposes of the human service profession itself.

Informed Consent

A prerequisite to engaging in clinical practice involves understanding the specifics of *informed consent.* Considered a very important client right, informed consent require practitioners to disclose to clients pertinent information about the treatment they are pursuing. Such disclosure can help clients make informed decisions about entering and remaining in treatment. This process ensures clients' right to self-determination and autonomy and promotes respect for clients and their desires.

Facets of Informed Consent

Informed consent has many facets. An important aspect is your role and the particular agency you are working in. Below is an overview of the types of issues clients should be informed about as they are most relevant to clinical practice. We have adapted them from the American Counseling Association Code of Ethics (1995).

Informing Clients about Helper's Role Qualifications and Experience   Clients have a right to know the training and experience of the professional who will be treating them. This includes informing the client when the helper is an intern or volunteer.

Treatment Considerations

- From the onset of your interaction with a client, make an effort to inform the client about your role, qualifications, experience, and orientation to helping.
- Clients will appreciate your openness and respect your candor. This is especially important for beginning helpers or interns who may need to consult with their supervisors about their clients.
- For interns, this is an opportune time to inform the client of your limited stay at the agency. If you do not tell the client initially, doing so can become increasingly more difficult with the passage of time.
- Informing the client initially opens the door to discussing the feelings associated with developing a therapeutic relationship.

SCENARIO

A client is not informed of your inability to prescribe medications and assumes this is something you are qualified to do. One day the client calls in crisis and is further upset to learn that he will have to be referred elsewhere for a medication evaluation. This scenario might have been avoided if you had informed him about your qualifications initially.

SCENARIO

As an intern, you have been seeing a client for a 3-month period and have been instrumental in helping the client overcome some very emotionally difficult times. You are approaching the date when you will be leaving the agency to pursue another internship as required by your educational program. At your next visit with the client, you introduce the issue of terminating or transferring his case as you will be leaving soon. The client is caught off guard and initially reacts with shock and then becomes angry and ultimately thinks he has been betrayed and abandoned by you. If you had informed him initially of your intern status and limited stay at the agency, you might have helped the client to deal with his emotions more effectively. (Note: Clearly, no amount of preparation can completely eliminate the natural emotions clients feel on termination.)

Providing Clients Information about Services to Be Rendered   The specific information to be provided about services will depend on the particular agency setting and professional role of the helper. Most often, it will include educating clients about assessment procedures or intake policies, techniques to be used, testing to be performed, if any, or referrals to be made. It may also involve describing the specific services your agency provides and the scope of these services.

Treatment Considerations

- Make an effort to incorporate information about services into your introduction as this will allow the therapeutic relationship to progress more smoothly.
- We define this process as "structuring the interview" and find it most valuable in developing a positive working relationship with clients.
- Take care to offer such information in ways that clients will comprehend and encourage them to ask questions.
- Often we may believe we have satisfied the disclosure requirements, yet clients may feel hesitant or intimidated about asking for further clarification.

- Be aware that people seeking help may not have the necessary communication skills to be direct or assertive about their needs, especially those who are in crisis or experiencing difficulties at the time of their application for treatment.

SCENARIO

A client seeks treatment hoping to be seen immediately and have his issues resolved quickly. When he enters your office, he is already irritated by the lengthy questionnaire he was asked to complete on arriving. He cannot understand why all this information is necessary. This could be a good time to inform the client about the intake policies and the process that will follow. Taking time to listen to his concerns and explain the intake process in a sensitive manner can ultimately save you much time and energy, although it may not immediately lessen the client's frustration in this situation.

Providing the Client with Information Regarding Goals of Treatment, Limitations, and Potential Risks   Depending on the individual case, the helper needs to ensure that the client understands the implications of treatment and diagnosis, reports to be made, referrals to be given, and so on.

Treatment Considerations

- At times, clients could be seeking a particular type of service and may not be aware that such a service does not fall within your expertise.
- If you were not clear about the services you provide, the client may not have the opportunity to be referred to an appropriate provider.
- By taking time initially to explain treatment goals and limitations as well as problems that might result from treatment, miscommunication can be minimized.

SCENARIO

The client wants someone who will help him with detoxification from substance abuse and you work primarily with the counseling department at your agency setting. If you had not informed the client of your position at the start of treatment, he might have continued seeing you for some time before realizing that he needed a referral to a substance abuse treatment center.

Treatment Considerations

- Often in counseling situations, clients are not aware that they will need to work on changing their behaviors and instead enter treatment with the assumption that you will either "solve" their problems or "cure" them or significant others.
- Educating clients about the counseling process is essential in assisting them to learn what helping entails.
- You may not be able to avoid situations in which clients have unrealistic expectations. However, in anticipating such scenarios, you can become better prepared to respond more ethically and sensitively.
- It is important to recognize the cultural implications at play when considering the development of a treatment plan and objectives. Working with clients from multi-ethnic backgrounds requires your awareness and sensitivity to the underlying cultural issues.

SCENARIO

A client has heard about the new antidepressants and enters your agency with the belief that if he just takes some of these pills he will be "cured" of his depression and will never feel sad again. Education about medication is crucial here.

Treatment Considerations

- From the very onset, it is important to note the differences between you (the helper) and the client. Recognizing these and making an effort to understand the specific cultural, ethnic, or racial differences that exist may enhance the therapeutic/working relationship.
- By commenting on these differences, you will likely create a climate of openness that will encourage clients to share their feelings and viewpoints as well. Moreover, this will facilitate the development of the most effective treatment goals and interventions as they will be tailored to meet the specific needs of the client.

SCENARIO

A Latino couple from Mexico enter counseling with marital disagreements. The wife is now going to school and pursuing a career. As they have been married for four years, the husband believes it is time to start a family and that she should stay home to raise the children. The helper, who encouraged the wife, now feels conflicted about how to proceed.

Sometimes, in our eagerness to help others, we encourage clients to pursue certain goals without informing them of the limitations of such goal attainment. In this case, understanding the cultural issues related to gender expectations may have helped to prepare you for possible outcomes in this situation. Orienting this couple to the potential positive and negative consequences may have prepared the wife for the difficulties she has encountered and have given her the choice to decide if this was the direction she wished to take.

Ensuring That Clients Understand the Fees and Billing Arrangements   In many agencies, practitioners may be asked to be directly involved in collecting the payment fee services as it may be considered part of the therapeutic process. In this event, the helper must be knowledgeable about the fee or billing procedures and be prepared to share this with the client at the outset of treatment.

Treatment Considerations

- As part of your introductory comments, inform the client about the cost of the services and the acceptable method of payment.
- This will allow clients to make an informed choice about their ability to pay for services and/or any need to apply for assistance.

SCENARIO

A client may assume the treatment he is seeking is included in his insurance plan, therefore he does not inquire about the cost. The worker may also make an assumption that the client is aware of the costs involved. This client may become quite upset when he receives a bill for such services which he did not expect. The worker, as well, may be upset by this scenario and may end up not being paid for the services provided.

Providing Clients with Clear and Specific Information about Their Rights to Confidentiality as Well as the Exceptions   Clients are often unaware of their right to confidentiality. Also, they are frequently misinformed about what confidentiality means. It is important to explain confidentiality, including spelling out all the instances in which patient information may be shared with others, such as in a treatment team or within a supervisory relationship.

Treatment Considerations

- It is crucial to inform the client from the start about your role as an intern, as an unlicensed professional seeking supervision hours, or your responsibility to the agency.

- Allowing clients to make their decisions to pursue treatment knowing these aspects of confidentiality is an important step in respecting their rights and feelings.

SCENARIO

An intern in an agency setting finds herself perplexed about an issue a client has raised. She tells the client she will have to consult with her supervisor to get further information about how best to help her. The client becomes upset, for she does not want personal information shared with other staff members.

Discuss the Exceptions to Confidentiality When You Introduce the Topic in General
Including information about the exceptions to confidentiality is just as important as discussing clients' right to confidentiality itself. Some believe this will deter clients from sharing certain potentially "reportable" incidents; nonetheless, it remains an important part of informed consent. Our experience as social workers and psychotherapists as well as the experience of the many students and workers we have supervised suggest that this is not the case. Clients usually share such information regardless of knowing whether it will be reported. We believe this demonstrates the clients' greater desire for help in their lives.

Treatment Considerations

- As with the other areas of informed consent we have mentioned, workers have a responsibility to inform clients fully about the privileges and exceptions to confidentiality.
- Some agencies have written forms to describe these expectations to clients. If this form is presented at the onset of treatment, it demonstrates professionalism as well as respect for the client's rights.

SCENARIO

A client you are quite bonded to and have been working with for a long time reveals to you at a recent visit that she lost control and physically abused her young son leaving visible marks. She feels tremendous guilt and seeks your emotional support and guidance. When you notify the client about your need to report this incident to children's protective services, she becomes irate as you had never before disclosed this type of exception to confidentiality.

Informing Clients Regarding the Need for Their Written Consent before Releasing Client Records/Information   Clients will often be upset when they ask you to release or obtain information about their situation and you cannot until they sign a release form. "This form exists in many versions in as many different settings. Very few agencies and practitioners use it appropriately and abuses of consumer privacy occur almost daily" (Wilson, 1978, p. 57). Wilson further stipulates that general forms are in reality useless and truly do not constitute consent. Consequently, to meet the informed consent criteria, agencies should develop a specific form tailored to the client's unique needs. Your agency may have already developed its own forms for this purpose.

Treatment Considerations

- When you initially broach the issues of confidentiality, you can also include an explanation about the need for a signed release to communicate with others about the client's case.
- Clients often feel pleased to learn that such care is being taken to keep their identity and treatment confidential.

SCENARIO

You are working with a child in a children's center who was brought in by the father. As a part of his explanation about the family circumstances, he tells you that the mother and he are divorced because the maternal grandmother was using drugs and was physically abusing the child. Later in your work with the child, a telephone call comes in from the grandmother inquiring about the child and details of the current situation with the child and the father. As the worker how would you handle this telephone call? Should you talk to her?

Ensuring that Clients Recognize Their Right to Obtain Case Records   Often, helpers themselves are unaware of the client's right to obtain case records and may discourage or deny clients the privilege of viewing their records. This issue of allowing or being required to permit clients access to their records remains controversial. There are as many advocates for open access as there are opponents. Regardless, it is a legal right of clients to receive copies of their records in many instances; it is often an ethical issue as well.

Treatment Considerations

- Although you may believe it is inappropriate for clients to review their records, you need to be prepared for the reality that they may ask to do so.
- If you believe such a release would be harmful to the clients' emotional state, you may want to consult with a supervisor or legal counsel regarding your legal obligations.

SCENARIO

A client who is quite emotionally disturbed and exhibits paranoid ideation insists on seeing your evaluation, progress notes, and treatment plan. You are very uncomfortable with this demand as you feel the client may misinterpret certain information that is documented and become even more irrational and upset. How would you handle this? How would you feel in this situation?

Allowing Clients to Participate in Developing the Treatment Plan   Allowing your clients to have a say in the services you are developing for them encourages a cooperative atmosphere that can enhance the helper-client relationship and assist the helper to do continuous evaluation of the success of treatment strategies or services rendered.

Treatment Considerations

- If from the outset you inform clients of their right to participate, and the importance of their being active participants in treatment, you will minimize potential conflicts later when and if the client is unhappy with the services provided.
- This cooperation also allows you to include an evaluative component in your work with the client that could ultimately be a positive factor toward constructive change.

SCENARIO

You are working with a teen client in a chemical dependency unit. You are new to the unit and are unfamiliar with the procedures. On one of your first cases, you are more empathetic than confrontive with the client only to experience difficulties later when the clients' progress is evaluated and found to be minimal. The client then becomes upset as you were not clear from the onset about her need to be an active participant in her treatment. She had assumed you would be the one to make her "well again."

Informing Clients about Their Right to Refuse Any Recommended Services and Be Advised of the Consequences of Such Refusal   Often clients are not aware that continuing treatment is their choice. They may not want to remain in treatment, yet they do so anyway. This behavior could result in clients' remaining in treatment beyond the point that is helpful to them. Conversely, clients may refuse treatment without being apprised of the possible consequences of this withdrawal. This behavior primarily pertains to agencies where voluntary services are provided.

Treatment Considerations

- Taking care to explain clearly about services offered along with any possible consequences of refusing treatment is also essential at the onset of treatment.
- Clients are likely to appreciate the time and effort you show in informing them about these issues. They may not be aware of certain aspects of treatment and will benefit from a clear explanation.

SCENARIO

You are working in an agency that monitors visitations of parents whose children have been removed from their care because of some form of child abuse. You have learned in your training that as a worker in this setting you must make clear to the parents the need for them to comply with the court stipulations. For instance, parents may be required to attend parenting classes, engage in counseling, attend drug and alcohol diversion programs, and maintain regular monitored visits with the children. By letting them know that if they do not follow through with the mandated services they may jeopardize the reunification process, you are ensuring that these parents fully understand the choices they have.

Despite the apparent drawbacks and ethical dilemmas related to informed consent, there are also benefits. First, informing clients fully about services helps establish trust between the helper and the client. Second, in providing this information to clients, the helper is able to observe their response and reactions. If the response is negative, this may signify that a problem exists and it needs to be explored further.

Understanding Your Agency's Policy on Informed Consent

To respond appropriately to your agency's policy on informed consent, you must first understand the policy and how others in the agency generally implement it. Different agency settings will require different levels of disclosure depending on the types of services they provide. Although there is great variance in how agencies adhere to this ethical code, most agencies do comply in some manner. Take the initiative to learn the policies of your agency and the underlying reasons for them. Equally as important as learning about the specific policy on informed consent is discovering the informal norm for how much information is disclosed within the agency and how it is used. Consulting with your supervisor about this issue when you first begin your internship is perhaps the most sensible way to gather information on your specific agency's policies on informed consent. Asking direct questions of others in the agency can also help you learn how the agency functions in this regard.

Confidentiality

Closely related to informed consent is confidentiality, which pertains to a client's right to privacy. Considered a most important concept within human services, confidentiality is both an ethical and a legal issue. Ethically, helpers can foster a strong and trustworthy relationship with their clients by responsibly protecting their identity and treatment history.

Legally, helpers in most jurisdictions are bound by specific penal code statutes that prevent disclosure of any information (with certain exceptions) about the client without his or her signed authorization (Leslie, 1989; Corey, Corey, & Callanan, 1998). For helpers who work in a clinical capacity, knowledge of the specific aspects of confidentiality is especially important, given the confidential nature of the relationship itself. If a trusting relationship is to develop and clients are to feel confident in sharing their innermost feelings and concerns, they must know their privacy will be respected. In this context, confidentiality can be seen as a fundamental cornerstone of the helping relationship. Because of the many circumstances surrounding this issue, however, adhering to the strict principles of confidentiality is not always possible nor is it always a simple, straightforward matter.

Confidentiality in Agency Settings

In theory, maintaining the confidentiality of our helper-client interactions may seem easy enough; in reality, this process may not be simple at all. There are many matters surrounding confidentiality that are ambiguous in nature and require careful scrutiny and possible consultation with supervisors, colleagues, or legal counsel. It is therefore important to understand the many intricacies involved. Helpers will benefit from learning what constitutes a breach of confidentiality and what the specific exceptions to confidentiality are. Additionally, details regarding release of information procedures, safeguarding records, and the sharing of client information with supervisors and colleagues are important to know. The issues surrounding confidentiality may be handled differently in different agency settings and from state to state, so it is essential that helping professionals become familiar with the specific policies and procedures of their respective agency and setting. All these aspects of confidentiality are important to understand as they are issues you will inevitably face in your clinical work with clients. Acknowledging confidentiality in clinical work is a basic aspect of creating a safe working relationship with your client. Below, we will take a closer look at each of these issues.

Breach of Confidentiality

Breaking confidentiality can have serious repercussions. On a clinical level, clients may feel violated if the helper has breached their right to privacy by releasing information about them, or sharing their case situation openly with others. Clinical implications include the erosion of trust in the relationship that will likely hamper future efforts severely; it can even result in termination of the relationship altogether. Even more serious, inappropriate disclosure, whether done knowingly or inadvertently, can result in possible harm to the client or the client's family. In all cases, a breach of confidence is a critical matter that must be avoided. On a legal level, there may be serious liability issues. In many states, violation of confidentiality laws can lead to criminal prosecution of the practitioner. In fact, breach of confidentiality is one of the most common reasons for malpractice lawsuits by clients against helping professionals (Leslie, 1989).

Treatment Considerations

- Resist temptations to discuss clients in social situations where nonclinical professional people may overhear.
- We have found it better to err on the side of caution whenever releasing information to others, even with signed releases.

SCENARIO
At a social occasion with some friends from school, a student intern begins to talk about a client she saw earlier that day. She begins to give details about this client's sit-

uation, and another student suddenly recognizes the client as a relative. The student who is treating this client becomes extremely anxious as she realizes she has disclosed personal family information that had been kept secret for years.

Release of Information

As evident in our discussion of breach of confidentiality, releasing information about a client, written, verbal or otherwise, is generally prohibited legally unless there is explicit written and signed permission by the client (Baird, 1996). In many instances, clients will want certain information released and will gladly provide authorization. At times, when clients want information released as this will benefit them, they may feel frustrated when asked to give written permission. We encourage you to emphasize to clients their right to privacy stipulated by ethical and legal principles of confidentiality.

Treatment Considerations

- Informing clients of the need for signed releases before information can be sent to another person may be helpful. This won't prevent all instances of clients becoming angry, but it will be effective with some.
- A client's reaction can always be used to observe and discuss how he or she typically manages thoughts, feelings, and actions.

SCENARIO

A client calls and asks you to send a letter to the court system verifying her treatment, and she tells you only a few days before the letter must be received by the court. You tell her you are glad to comply but she will have to come in to the office to sign a release form before you can send the letter. She becomes angry that she has to return to the agency. She believes that giving verbal permission on the phone should be sufficient.

Safeguarding Records

Often, a breach of confidentiality results from improper safeguarding of client information that is most often filed in client records. Assessment and intake information, case notes, testing, documentation of phone contacts and correspondence with the client or on the client's behalf, and any other client records must be kept in a safe place where access is available only to authorized personnel. Efforts to ensure the confidentiality of these records requires helpers to take certain precautions they may have not previously considered necessary. Suggestions include keeping the records in a locked file cabinet, developing an efficient, reliable checkout system, being careful to not leave client files or any other client information within sight of others, labeling client records confidential, and—specifically for students—disguising the identity of a client, his or her situation, and the agency in any process recording or journal assignment (Baird, 1996, p. 33). Although it may be impossible to safeguard records completely in all situations, taking precautions can certainly help minimize any infractions.

Treatment Considerations

- Be cautious about where clients' records are kept; ensure that they are in a secure place.
- Use agency safeguards that may already be in place to protect client records.
- Resist temptation to be loose with client files such as leaving them places where clients may access them.

> **SCENARIO**
> An angry client comes in complaining about the treatment he received from his last worker, who is not in the agency. Your supervisor asks you to see the client, which you agree to do, despite your reluctance. The client goes on and on about how angry he is at the former worker. He demands to see the records that are on your desk. When you get up to consult with your supervisor on the advisability of allowing him to see his records, the client grabs the chart on your desk and runs out of the agency.

Sharing Information with Supervisors and Colleagues

Working in human services most often requires working collaboratively with others in some capacity. This collaboration typically requires sharing information about clients with supervisors, colleagues, or other professionals. The reason for such disclosure is usually for the direct purpose of helping the client in some way. Information may also be shared for your own learning, with more emphasis on what you as the worker, were doing. In most cases such disclosure must be explicitly authorized by the client in both written and verbal form. In real-life agency work, however, this policy is not always followed. For instance, helpers often share information with colleagues about their work with clients in staff meetings, peer review seminars, case conferences, and individual and group supervision. If this is the case, it is important to inform the client of this in a general manner during the person's initial visit. This is part of the informed consent process.

In some cases, contact with others may be necessary to ensure the client's safety from self or others or because of legal mandates, such as subpoenas. Usually these situations involve exceptions in which breach of confidentiality is allowable.

Treatment Considerations

- Telling clients of disclosures to be made can prevent future problems.
- In an agency setting, it becomes even more important to inform clients of the limits of confidentiality and with whom information about their case will be shared.

> **SCENARIO**
> A divorced father brings in his 10-year-old daughter for counseling. After your session, the client informs you that the child's mother has primary custody of her and will be calling to receive information on the treatment offered. He would prefer that you not share any details with her. You inform him that because of the legalities involved you will have to discuss this further with your supervisor to determine the best course of action. He does not seem happy with this but accepts it as a part of your job.

Confidentiality and Agency Policies and Procedures

Working in agency settings can present helpers, especially those who are new or just beginning their fieldwork experience, with complex, unclear, and often ambiguous situations. Many agencies have policies and procedures regarding the most common ethical and legal issues. For instance, a number have developed specific forms for release of information. Others have set rules about safeguarding case records, some of which are more stringently enforced than others. Formal and informal procedures for sharing client information with colleagues are often well established. Nonetheless, it will be up to you to learn your agency's policies on confidentiality through direct questioning of others, including your supervisor; observation; and reading the agency policy and procedure manuals. Most agencies are bound to have had some experiences involving a breach of confidentiality or a test of the limitations of confidentiality. You need to understand the specifics involved as well as learn about the agency history in dealing with these situations (Woodside & McClam, 1994).

Exceptions to Confidentiality

Thus far we have maintained the importance of confidentiality both ethically and legally within human services. This mandate must take precedence in all situations with the exception of only those matters stipulated by the particular state and county statutes where you practice. In most states, these include cases of child, elder, and dependent adult care abuse; potential harm to self and others, such as suicide or homicide; gravely disabled individuals; and court orders and subpoenas requesting client information. Disclosures in these situations are mandated, whereas permitted disclosures include cases of emotional and spousal abuse and those involving the reporting of AIDS.

ADVANCED TREATMENT CONSIDERATIONS

Child Abuse

Most states have specific laws about child abuse. Typically these laws require those in the helping and teaching professions to report their suspicion or knowledge of abuse of anyone under the age of 18 to a designated agency, usually the children's protective services. The agency is then responsible for investigating the matter to determine whether protection of the child is needed. In most instances, the identity of the reporter remains anonymous and laws exist to protect the reporter from civil liability if he or she made the report in good faith and without malice. In contrast, failure to report a suspected or known case of child abuse can result in criminal penalties and civil action against the worker who suspected abuse but did not act (Baird, 1996, p. 35). Professional involvement in child abuse cases has both ethical and legal implications. Given the legal reporting stipulations, it is imperative that helping professionals understand the complex issues encompassing child abuse. Working within human services, helpers must learn to recognize signs of all types of child abuse and know how to make a child abuse report.

Understanding Child Abuse

A thorough understanding of child abuse may take years of direct experience working with children and their families as well as involvement with child welfare services. Because of the limited scope of this text, we can provide only an overall view. We begin with the historical perspectives of child abuse that illustrate the intricacies of this issue.

Historical Perspective

Although child abuse has existed throughout time, prior to the 19th century it was primarily a private matter. Parents and other family members had the right to care for and discipline their children in the manner they saw fit. Little or no outside interference existed to minimize or prevent the maltreatment of children.

Not until 1895 was the first Society for the Prevention of Cruelty to Children established in New York City; it grew out of a concerned citizen's efforts to seek public protection for a young child, Mary Ellen Wilson, who was brutally abused by her caretaker. From this case, a precedent was set and legislation was enacted to protect of all children from neglect and abuse (Filip, Schene, & McDaniel, 1991; Johnson, 1986, p. 83).

The 1960s brought a heightened awareness of child maltreatment. Child welfare and medical professionals are credited with increasing public awareness about child abuse that have since resulted in specific child abuse laws in every state. Each state has its own set of laws that define child maltreatment, reporting responsibilities, and procedures as well as the overall purpose, focus, and organization of children's protective services.

Child Abuse Defined

Child abuse involves neglect or the physical, sexual, or emotional maltreatment of a child by other than accidental means by another person, usually at least 4 years older than the child. Child abuse is considered to exist when there has been "an act of omission or commission that endangers or impairs a child's physical or emotional health and development" (Kamerman & Khan, 1976, p. 46).

Characteristics of Family Dynamics

Some of the more common dynamics found in abusive families can be helpful to workers trying to understand this complex problem. Some of these are listed here:

- The adult abuser's need for nurturance and dependence were unfulfilled, with a history of deprivation or abuse in the adult's own childhood quite likely.
- Many abusing parents had no positive role models; these parents may not know nurturing, healthy, and appropriate child-rearing practices.
- Abusing parents may have a strong fear of relationships, resulting in isolation and obstacles to engagement.
- Often because of isolation and limited resources, abusive families may lack support systems.
- Relationship and marital problems may be severe, placing the child at further risk because of the threat of domestic violence.
- Life crises seem to characterize some abusive families. Because of limited coping skills, little or no support, and poor self-control, these individuals may find themselves in constant crises.
- Often these parents are so overwhelmed with physical and emotional survival themselves that they have little to give their children, who often are left to fend for themselves.

Types of Child Abuse

The various types of child abuse have been written about extensively. Next, we explore each type, with attention to the subtleties sometimes involved.

Physical abuse involves physical injury to a child by another in a way that was not accidental. This maltreatment can range from mild physical abuse to severe corporal punishment. Many cases of physical abuse are caused by well-meaning parents who either do not have adequate parenting skills or who lose control in an effort to discipline their child. These parents may feel remorseful after the fact and become frightened of their potential to cause harm. Also, they may be willing to seek help and may benefit from parenting classes and counseling.

Severe corporal punishment or abuse is seen in more extreme cases when parents resort to unreasonably harsh forms of discipline on the child. This is most likely when the parent is extremely agitated or angry and lashes out at the child in an abusive manner. The presence of alcohol or drugs in the home can increase the risk of this type of abuse. Furthermore, there may be generational patterns of abuse in the family. This type of physical punishment and rage often perpetuates violence into future generations.

Physical abuse can also take the form of physical assault, defined as a more intentional, deliberate assault on the child. This may include burning, biting, cutting, poking, twisting limbs, or otherwise torturing the child. Many parents who engage in this type of severe abuse were likely abused themselves as children and may also experience some emotional or mental problems that affect their parenting abilities (California Department of Justice, 1976; Crime Prevention Center, 1988, p. 2).

To simplify your efforts to detect this type of abuse, we have compiled information on physical and behavioral symptoms along with caretaker indicators in Table 7.2.

Treatment Considerations

- Attempt to remain calm. Discuss confidentiality and exceptions to it honestly.
- Allow the client to ventilate his or her feelings of fear, anger, hurt, confusion, being overwhelmed, and so on.
- Positively reinforce the fact that he sought your assistance and has been open and honest.
- Provide support and validation for his or her struggles, yet uphold clear boundaries with regard to child abuse laws established to protect children.
- Explore with the client ways to cope with the situation at hand. Discuss possible consequences of reporting.

SCENARIO

You are conducting an assessment with a man at a homeless shelter when he tells you that he "lost control" with his daughter who was having a tantrum. He used an extension cord to whip her. Although he feels bad about this, he explains how overwhelmed he is since his wife left him with the children. He was unable to afford child care and subsequently lost his job, and that has now led to his homeless status. He seeks your guidance on how to handle his current life situation.

TABLE 7.2   Physical and Behavioral Indicators of Physical Abuse in Children's Caretaker Behavior

PHYSICAL INDICATORS OF PHYSICAL ABUSE

Physical abuse is suspected when any of the following occur:

Damage to Skin and Surface Tissues

Unexplained Bruises or Welts

- Bruises resulting from abuse are found on multiple surfaces of the body, particularly the buttocks, back, genitals, and face. Sometimes they are also seen on the lips, mouth, torso, or thighs.
- Bruises on several different surface areas of the body may indicate the child has been hit from several different directions.
- Bruises may appear in a characteristic pattern such as the outline of a hand, paired bruises from pinching, or the impression of an item of jewelry (such as a ring).
- Sometimes the bruises may be clustered, forming rectangular patterns that may reflect the shape of the article used to inflict the abuse such as an electrical cord, belt, or buckle.
- Multiple bruises may all be the same color or different colors if the injuries were sustained at different times. Red, red-blue, violaceous, black-purple, green tint, pale green, yellow, or bruises with fading margins reflect various stages of healing. When bruises reflect differences in coloration, it is useful to take color photographs of the bruises for investigation of suspected child abuse.

- Blows from a heavy object such as a baseball bat or fist on soft tissue may result in deep muscular bruises or hemorrhage. These are rarely discolored and, in time, may be seen on x-rays.
- Bruises may appear regularly after absence, weekend, or vacation.

Unexplained Lacerations/Abrasions

- Lacerations are often to mouth, lips, gums, eyes, or to external genitalia.
- As in bruises, the multiplicity and location of wounds should be considered. Previous injuries, repeated, and different age abrasions are often noted.
- Wrap-around injuries (from a flexible object such as a strap, belt, or cord).
- An imprint of an object on skin (belt buckle, hose coupling, hand, spoon).
- Odd shapes and sizes of injuries.

Unexplained Burns

- The location of a burn and its characteristics (shape, depth, margins, etc.) may indicate abuse.
- Children instinctively withdraw from pain, so burns, without some evidence of withdrawal, are highly suspect because a child will usually try to escape being burned—this will often result in splashes, uneven burns, and sometimes burns on the hands.
- Scalding a child with hot liquid is the most common abusive burn. Young children are scalded by

(continued)

TABLE 7.2   (continued)

immersion and older children by having liquids thrown or poured on them.
- Immersion burns include
  - Glove or sock-like appearance (from immersion in hot liquid).
  - Doughnut-shape on buttocks or genitalia (held down in hot liquid).
  - Pointed or deeper in middle (hot liquid poured on).
  - Jackknife position burns are often seen when a child is held in water and only the buttocks or genitalia may be burned.
- Shape of object burns (poker, heater grill, utensil)
  - Patterns like an electric burner, iron, etc.
- Rope burns appear on wrists or ankles when children are tied to beds or other structures. They may also be found on arms, legs, neck, or torso.
- "Zebra burns" also indicate abuse. Such burns result when a child is held by his or her hands and legs under a running hot faucet. The tissue on the child's abdomen and upper legs folds up, preventing burning in the creases. The resulting "zebra strips" from scalding of exposed tissue are clearly evident.
- A child is often burned by a cigar or cigarette, especially on the soles, palms, back, or buttocks. These types of burns are difficult to diagnose, but when inflicted they are often multiple. There is a searing effect, perhaps with charring around the wound.
- Infected burns may indicate that they were not treated.

Bite Marks
- Bite marks may be found on any part of the body and may be described as doughnut-shaped or double horseshoe-shaped.
- In some bites, tooth impressions may be seen and used to identify the perpetrator.
- Time is of the essence in recording bite marks as they become less distinct with time.

Damage to Brain

Head Injuries
- Signs of intracranial injury should be ruled out by a careful examination of the child's eyes and nervous system. For certain groups of suspected victims, a skull x-ray may be an important part of the examination.
- Serious intracranial injury may occur without visible evidence of trauma on the face or scalp. Such injury may cause brain damage or death if undetected and untreated.

- Of great importance, head injuries are the most common causes of child-abuse related deaths and an important cause of chronic neurological disabilities.

Whiplash Shaken Infant Syndrome
- These types of injuries occur when an infant is shaken with excessive force.
- There is an apparent contradiction in the essential elements of this syndrome as intracranial and intraocular hemorrhage occurs in the absence of signs of external injury to the head.
- The only indicators of this symptom may be vomiting, rapidly enlarging head, and subtle neurologic signs causing the injury to go undiagnosed for years.
- More severe deficits such as blindness, deafness, or paralysis may appear sooner, or may be first manifested at school age as minor neurologic deficits or learning problems.

Damage to Other Internal Organs

Internal Injuries
- Blunt blows to the body may cause serious internal injuries to the liver, spleen, pancreas, kidneys, and other vital organs and occasionally may cause shock and result in death.
- Detectable surface evidence of such trauma is present only about half the time. Physical indicators or serious internal injuries may include distension of the abdomen, blood in urine, vomiting, and abdominal pain.
- Internal injuries are the second leading cause of death for victims of child abuse.

Damage to Skeleton

Unexplained Fractures/Dislocations:
- Any unexplained fracture in an infant or toddler is cause for additional inquiry or investigation.
- "Spiral fractures" of the long bones, which result from twisting forces, are almost always caused by abuse when they occur before a child begins walking.
- Other fractures that raise suspicion are "chip" fractures at the end of long bones; multiple rib fractures, especially back rib fractures; and healing or healed fractures without explanation, revealed by x-ray.
- Any fracture to the skull, nose, and facial structure are suspicious.

BEHAVIORAL INDICATORS OF CHILD PHYSICAL ABUSE

The actual physical abuse may not be evident, yet there may be behavioral signs that can lead you to suspect possible physical abuse. They are as follows:

Low self-esteem
- Feeling of child that he or she deserves punishment.

TABLE 7.2 *(continued)*

- Negative self-statements and self-concept.

Extreme Reactions to Adults/Parents

- Avoidance of physical contact with adults (child may shy away from adult who attempts to get close as he or she may fear physical harm). Generally fearful of adults, nonspontaneous, unwilling to speak in front of his or her parents.
- Conversely, a child overly eager to please adults when asked to perform menial tasks.

Fearful Reactions

- Is frightened of parents/afraid to go home.
- Has vacant or frozen stare.
- Lies very still while surveying surroundings (infant).
- Responds to questions in monosyllables.
- Is apprehensive when other children cry.
- Reports injury by parent.

Demonstrates Behavioral Extremes

- Aggressiveness or withdrawal.

Inexplicable Changes In Child's Behavior

- Hyper- and hypoactivity.
- Over- and undercompliance.
- Indiscriminately friendly approach to adults.

Excessive Need for Attention

- Inappropriate or precocious maturity.
- Manipulative behavior to get attention.
- Indiscriminate attention-seeking.

Inability to Trust

- Capable of only superficial relationships.

CARETAKER'S BEHAVIORAL INDICATORS IN PHYSICAL CHILD ABUSE

Adult has history of abuse as a child.

Inability to explain child's injury appropriately

- Parent unable to explain injury.
- Parent offers illogical or unconvincing explanation.
- Parent offers contradictory explanation (there are many discrepancies in the story).
- Parent blames third party.
- Explanation is inconsistent with medical diagnosis (i.e., impossible for injury to have been sustained as parent relates).

Parents' Overreactions

- Parents are extremely nervous.
- Parent may attempt to conceal the child's injury or to protect identity of the person responsible.
- Parents may bring the child to medical facility for unneeded treatment.

Parents Exhibit Behavioral Extremes

- Overly loving and caring.

- Unconcerned about the child; distant, cold.

Unrealistic Expectations or Views of Child

- May significantly misperceive child (e.g., sees the child as bad, evil, a monster).
- May have very high expectations of the child that are developmentally impossible and inappropriate (i.e., the parent believes that a 7-year-old should be the caretaker of siblings and clean house).

Violent Threats or Behavior

- Parents verbally threaten child's life.
- Parents use harsh discipline inappropriate to children's age, transgression, and condition.

Other

- Psychotic or antisocial behavior.
- Misuse of alcohol or other drugs .

Sexual Abuse

Sexual abuse exists when there is any sexual activity between a child 18 years and younger and an adult or person 4 years or more older than the child, including exhibitionism; lewd and threatening talk; fondling; and oral, anal, or vaginal intercourse (Chaidez, 1991; Downs et al., 1996). Thus, sexual abuse encompasses a broad spectrum of behaviors, any of which can have occurred on one occasion or for a longer period of time, as seen in incidents of chronic molestation. The specific type of sexual abuse inflicted or the frequency and length will differ in every case. Likewise, every individual child will respond in his or her own way. Although it is important to understand the degree and frequency of the abuse to develop an appropriate treatment plan, what is most important is recognizing the trauma that is likely for each child regardless of whether they were fondled only on one occasion or severely abused for years. As helpers, we must be careful to not negate the impact of any type of sexual abuse and conversely to be cautious not to predetermine the level of trauma the child ought to be experiencing. Our work with child abuse has taught us that each child will experience the abuse in a unique manner.

Sexual abuse is believed to be the most unreported and disbelieved type of child abuse. It is the least talked about and often the most secretive form of assault on a child. The taboo regarding sexual abuse remains in part because of the difficulty in accepting that sexual abuse can be inflicted on innocent children. This seems especially true when the abuse is done by family members. Disbelief may also be affected by the degree that sexuality is covered over or denied in our culture.

The various forms of sexual abuse include incestuous/intrafamilial abuse, child molestation, child rape, and sexual exploitation/child pornography. Some children may be exposed to one type whereas others are victims of multiple forms. In almost all situations, the resulting trauma to the child is great. Emotional and psychological repercussions are often manifested. It is not uncommon for those who have suffered from child sexual abuse to feel anger, guilt, shame, and disgrace. They may believe that they are not worthy and may suffer from low self-esteem. There are times when the impact of this form of abuse can result in more serious problems with depression, anxiety, substance abuse, eating disorders, suicidal and acting-out behaviors, and dissociative disorders. For this reason alone, human service practitioners must be involved early with children in their recovery process in an attempt to prevent or at least minimize long-term problems. Indicators of sexual abuse in children and abusers appears in Table 7.3.

Treatment Considerations

- Listen carefully and encourage further discussion.
- Conduct a thorough assessment of the client and his or her history.
- Support the client's feelings while challenging statements of blame and responsibility.
- Follow any legal mandate to report if there is suspicion or evidence of abuse.

S C E N A R I O

A latency-age girl was brought in by her mother because the girl was having problems with peers in school. The girl reveals to you that "naked pictures" of her were taken in a day care center where she was left when she was 5 years old. Now that she is a little older, she is aware that this was wrong and feels very ashamed and guilty.

TABLE 7.3   Physical and Behavioral Indicators of Sexual Abuse in Children; Caretaker Behaviors

PHYSICAL INDICATORS OF CHILD SEXUAL ABUSE

Sexual abuse is suspected if one or more of the following is present:

Appearance of Clothing

- The child has torn, stained, or bloody underclothing.
- Semen is found on clothing.

Physical Appearance of the Child

- Vaginal or penile discharge.
- Bruises, bleeding, or lacerations in external genitalia, vagina, or anal areas.
- Difficulty sitting or walking.
- Swollen or red cervix, vulva, or perineum.
- Semen around the mouth or genitalia.

TABLE 7.3   *(continued)*

Physical Conditions

- The child has venereal disease, especially in preteens.
- The child experiences pain or itching in the genital area.
- The child feels pain on urination.
- The child has frequent bladder infections that cannot be explained by any other cause.

- The child has poor sphincter tone.
- The female child is pregnant.

(Caution must be taken not to make assumptions immediately regarding sexual abuse without first conducting a thorough physical and/or psychological evaluation.)

BEHAVIORAL INDICATORS OF CHILD SEXUAL ABUSE

Often, no physical indicators of sexual abuse are present, yet such abuse has occurred. In these instances it is often the child's or caretaker's behavior that may offer possible clues.

Child's Statements

- Child states that he or she has been sexually assaulted by a parent or by some other person. The child may report sexual activities to a friend, classmate, teacher, friend's mother, or other trusted adult. The disclosures may be direct or indirect, such as: "I know someone who...," "What would you do if...?" and/or "I heard something about somebody...," Furthermore, it is not uncommon for the child victim of chronic, severe, or acute sexual abuse to be developmentally delayed.
- You have knowledge of child's history of previous or recurrent injuries/diseases or sexual abuse.

Sexual Behaviors of Children

- The child exhibits precocious sexual knowledge seen in detailed and age-inappropriate understanding of sexual behavior (especially in younger children).
- The child displays inappropriate, unusual, or aggressive sexual behavior with peers or toys (may be reenacting the abuse they sustained).
- Child may engage in excessive, compulsive masturbation which is often indiscrete.
- The child displays excessive and age-inappropriate curiosity about sexual matters or genitalia (self or others). Displays precocious sexual interest.
- The child shows precocious, bizarre, sophisticated, or unusual sexual behavior in being unusually seductive with classmates, teachers, and other adults. The child may express affection in ways inappropriate for age.
- In extreme cases, a child may become excessively promiscuous or be involved in prostitution.

Body-Related Behaviors

- Enuresis, fecal soiling, or fascination with feces or urine.

- Distorted body image and/or self-consciousness of body beyond that expected for age.
- Frequent tiredness often from nightmares or sleepless nights.

Fearful Reactions

- Extremely fearful in general, increased phobias or other compulsive behaviors.
- Fearful of falling asleep due to nightmares. The child often has a sleeping disturbance where he or she is fretful of falling asleep or may sleep long hours.
- Fearful of showers/restrooms.
- Fearful of home life demonstrated in arriving at school early or leaving late, or by spending as much time as possible away from home.
- Suddenly fearful of other things such as going outside, participating in familiar activities.
- Extraordinary fear of males (in cases of a male perpetrators and female victim).
- Excessive concern or fear of homosexuality (especially for boys).
- Fearful of using toilet, afraid of having a bowel movement; child will often avoid using toilet or refuse toilet training.
- Child may express fear of a person or intense reluctance to being left somewhere or with someone.

School Problems

- Significant change in school performance; sudden drop in grades and change in attitude.
- Frequent absences and tardiness.
- Poor concentration, which affects learning.
- Refusal to dress for physical education.
- Nonparticipation in sports or social activities.

Self-destructive Behaviors

- Accident prone
- Self-inflicted injuries such as cutting oneself.
- Suicide gestures or attempts.
- Use of alcohol or drug abuse.
- Unprotected sexual activity.

(continued)

TABLE 7.3   *(continued)*

Acting Out or Delinquent Behaviors

- Oppositional behavior at home, school, with peers.
- Runs away.
- Is cruel to animals.
- Sets fires.
- Shows conduct disorder-type behaviors such as stealing, physically or sexually assaulting a more vulnerable person, etc.

Eating Disturbances

- Overeats to obesity as this may be a way to build a physical wall around self.
- Undereats to anorectic proportions as this may be a way to remain asexual.
- Eating habits fluctuate related to anxiety level.

Behavioral Extremes

- Either pseudo-mature or regressive behaviors such as bedwetting or thumbsucking.
- Overly compliant or extremely hostile and defiant.
- Clinical depression or hypomanic-type behavior where child acts too dramatic.
- Poor hygiene or excessive bathing.
- Extremely sexual or fearful of any type of touch; for teens, reluctance/refusal to date.

Peer Relationship Problems

- Poor peer relations and social skills; inability to make friends.
- Aggressive or delinquent behavior.
- Withdrawn and distant, hence unable or uninterested in engaging with others.

CARETAKER'S BEHAVIORAL INDICATORS IN CHILD SEXUAL ABUSE

Additional clues that sexual abuse exists are often gathered by observing the caretaker who may behave in a variety of ways, including the following:

- The caretaker/parent may be extremely protective or jealous of a child.
- The caretaker may encourage a child to engage in prostitution or sexual acts in the presence of the caretaker.

- The caretaker was a child victim of sexual abuse.
- Caretaker/parent may be experiencing marital difficulties.
- The caretaker misuses alcohol or other drugs.
- The caretaker is frequently absent from the home.

Emotional Abuse

Emotional maltreatment is often the most difficult to detect, prove, or treat. The two main categories are *emotional assault/abuse* and *emotional deprivation*. The first, emotional assault, and is said to exist when there are excessive verbal assaults on a child such as belittling and blaming or using excessive sarcasm; continual negative moods in the caretaker; constant family discord; and double message communication (Chaidez, 1991; Downs et al., 1996).

A child's self-esteem and sense of trust in others can be scarred by this type of emotional assault. Emotional maltreatment can impact a child emotionally, behaviorally, and intellectually, even though the scars are not readily visible. In fact, severe psychological disorders have been traced to excessively distorted parental attitudes and actions. According to a crime prevention report, many children who have emotional or behavioral problems have experienced some form of emotional abuse by their caretakers (California Department of Justice, 1976; Crime Prevention Center, 1988).

Emotional deprivation occurs when children are not provided with normal parenting, and feel unloved, unwanted, insecure, and unworthy (California Department of Justice, 1976; Crime Prevention Center, 1988). When parents tend to ignore their children, whether intentionally or not, the result can be a form of emotional starvation that has been directly linked with delinquent behavior in adolescents and criminal behavior in adults. Even if a child does not exhibit aberrant behavior, the trauma to the child can nonetheless be marked. For children to develop normally, they need emotional involvement from their parents as much as they need proper nutrition.

In clinical work, you will probably witness the aftermath of some form of emotional abuse in a number of your clients. This can be true whether you work with children, adolescents, or adults. You may find that part of your role is to help repair the damaged self that is an outcome of your client's childhood experiences with emotional abuse. Unlike the specific definitions of physical and sexual abuse discussed above, the determination for emotional abuse is quite vague and will differ from parent to parent and family to family. Cultural implications may also be present. Also, because action is rarely taken when this type of abuse is reported, it is more difficult for helpers to treat this type of abuse directly. A more common occurrence is to find emotional abuse along with the other types of abuse. As human service providers, we must intervene whenever possible.

Working with children who have been abused by their parents can engender feelings of upset and anger toward the parents. We as helpers must be careful not to become so negative toward these parents that we lose sight of the positive traits they may also possess. Furthermore, it is more valuable to see the parents as allies in treatment than as the enemy. This will require us to maintain a more balanced view and to acknowledge that they, like us, are imperfect—with both strengths and limitations. Indicators for emotional abuse are shown in Table 7.4.

Treatment Considerations

- Listen carefully and validate the client's emotional struggle.
- Do not minimize the abuse simply because it is not physical or sexual in nature. Also, emotional abuse frequently co-exists with other types of abuse; hence, it is important to determine whether other abuse also exists.
- Provide the client with education about the emotional abuse he or she has sustained while emphasizing that he or she is not responsible or at fault for this.

SCENARIO

A teenage boy you work with at the shelter for teen runaways tells you that he left home because he could no longer handle his home situation. Although he has not been physically abused, he says that he is constantly yelled at, belittled, and verbally attacked by his father who has a drinking problem. He admits that his father is an alcoholic but will not seek help. His mother also is frequently abused verbally and is afraid of standing up to the father.

TABLE 7.4   Physical and Behavioral Indicators of Emotional Abuse in Children; Caretaker Behaviors

PHYSICAL INDICATORS OF EMOTIONAL MALTREATMENT

Speech Disorders

- Little or no verbal or physical communication with others.
- Delayed speech or speech deficits.

Appearance

- Shallow, empty facial appearance.

- Frail appearance as child may refuse to eat adequate amounts of food.
- Delays in physical development; the child may be unable to perform normal learned functions for a given age such as walking, running, or jumping.

Failure to Thrive

(continued)

TABLE 7.4 *(continued)*

BEHAVIORAL INDICATORS OF EMOTIONAL MALTREATMENT

Perhaps more telling are behavioral clues children display that alert human service professionals to possible emotional abuse.

Overly Adaptive Behaviors

- The child is inappropriately adult-like as seen in parenting other children.

Behavioral Extremes

- Child does not change expression.
- Child may be overly compliant and passive.
- Child may be aggressive and demanding.

Conduct Problems

- "Acts out" and is considered a behavioral problem.
- Displays antisocial behavior (aggression, disruption) and obvious delinquent behavior (drug abuse, vandalism).

Mood

- Child is withdrawn, depressed, and apathetic.
- Child may have sleep disorders.
- Child may feel inhibited to play and may seem lethargic.
- Child may resort to attempted suicide if depression and despair are great enough.

Other Psychological Reactions

- Child may show hysteria.
- Child may display obsessive and/or compulsive behaviors.
- Child may be unusually fearful and develop phobic reactions.

Developmental Lags

- Child may be emotionally delayed.
- Child may exhibit intellectual or mental delays.

CARETAKER'S BEHAVIORAL INDICATORS IN EMOTIONAL MALTREATMENT

Appearance

- Caretaker seems cold and rejecting of child.
- Caretaker and child do not appear to have an emotional bond.
- Caretaker seems unconcerned about the child's problems, often more focused on their own.

Treatment of Child

- Caretaker tends to blame or belittle child.
- Caretaker appears to treat siblings unequally.
- Caretaker's expectations are often inappropriate; he or she shows no regard for the child. An example would be to expect an 8-year-old to care for younger siblings although child may have to miss an important school event to do so.

Neglect

Neglect, better described as *physical neglect*, occurs when a parent or caretaker fails to provide a child with adequate food, shelter, clothing, protection, supervision, and medical and/or dental care. It is generally a matter of omission in child care. Legally, neglect is considered the negligent treatment or maltreatment of a child by a parent or guardian under circumstances that present harm or threaten harm to the child's health and welfare (California Department of Justice, 1976; Crime Prevention Center, 1988). The two categories of *severe neglect* and *general neglect* have been identified.

Severe neglect means (1) the negligent failure of a caretaker to protect the child from severe malnutrition or (2) medically diagnosed failure to thrive. Also, severe neglect can result when a parent or caretaker willfully causes or permits the child to be placed in a situation where endangerment is possible.

General neglect is the negligent failure of a parent or caretaker to provide adequate food, clothing, shelter, medical care, or supervision when no physical injury to the child has occurred. This can result when parents or caretakers, either deliberately or due to severe inattentiveness, permit their child to experience avoidable suffering and/or fail to provide one or more of the ingredients generally considered necessary for developing an individual's intellectual and emotional capacities.

Specific types of neglect delineated by Zastrow (1990) are:

- Child abandonment
- Letting a child live in filth, without proper clothing, unattended, unsupervised, and without proper nourishment
- Educational neglect, in which a child is allowed to be excessively absent from school
- Medical neglect, in which no effort is made to secure needed medical care for the child (p. 171).

Often parents/caretakers cannot afford adequate child care and either leave their children in hazardous situations or alone to care for themselves, better known as 'latchkey children.' At other times, caretakers may be unaware of the care needed by the child or too caught up in their own personal struggles to show much concern. It is important to conduct a careful assessment before immediately judging a caretaker to be intentionally neglectful of the children they are responsible for. As human service providers, part of our role is to educate and assist those in need of learning more adaptive coping, problem solving, and parenting skills. Also, it is our responsibility to inform our clients about the law and the possible consequences of neglect.

Some of the conditions described in Table 7.5 may exist in any home environment, yet it is the extreme or persistent presence of these factors that generally indicate some degree of neglect. In extreme conditions, the home may be declared 'unfit' and the environment may be decided to constitute a form of 'severe neglect' that may justify the need for protective custody and dependency proceedings for the children.

Treatment Considerations

- Carefully assess the home situation to determine the existence of actual neglect.
- Explore the whereabouts of the parents and the underlying dynamics for this.
- Assist the child in recognizing that he or she is not at fault for being left unattended.
- Notify the authorities of the neglect situation.

SCENARIO

You work in the school setting to provide counseling to children referred by their teachers. One such child, an 8-year-old boy is referred because of his teacher's concerns regarding his poor attendance and school performance. She notes that he often arrives late, falls asleep in the classroom, appears unkempt, and never turns in homework. She has attempted to contact his parents but has received no response. When you interview the child, he tells you that he never receives his parents' help because they are always gone. When the situation is explored further, he reveals that he is often left alone all night under the care of his 11-year-old brother. When asked about his parents' whereabouts, he answers, "They are out partying, I guess."

TABLE 7.5   Physical and Behavioral Indicators of Physical Neglect in Children; Caretaker's Behavior

PHYSICAL INDICATORS OF PHYSICAL NEGLECT

Physical Appearance
- Underweight, poor growth pattern; e.g., small in stature, wasting away of subcutaneous tissue, abdominal distension, bald patches on the scalp

- Consistent hunger
- Constantly tired, sleepy, or listless
- Poor hygiene; the child is always dirty, may be infected with lice, tooth decay is serious

(continued)

TABLE 7.5 *(continued)*

- Inappropriately or inadequately dressed
- Evidence of unattended medical or dental need

Poor Housing Conditions

- Conditions in the home are unsanitary, such as garbage, or animal or human excretion everywhere.
- The home lacks heating or plumbing.
- There are fire hazards or other unsafe home conditions.
- The sleeping arrangements are cold, dirty, or otherwise inadequate.

Lack of Supervision

- There is evidence of poor supervision seen in repeated falls down stairs, repeated ingestion of harmful substances, and/or a child who is cared for by another child.
- The child is left alone in the home or unsupervised elsewhere—such as left unattended in a car, stores, or a street.
- In extreme cases, there is total abandonment.

Unmet Nutritional Needs

- The nutritional quality of food in home is poor.
- Meals are not prepared, children may either not eat, may cook for themselves, or snack when hungry.
- There is spoiled food in the refrigerator or cupboards.

Exploitation

- Children are overworked.
- The child is kept from attending school to help caretaker work or care for younger siblings.

Psychosocial Failure to Thrive

- Children born with low birth weight and perhaps short lengths and small head circumferences continue to have problems with growth.
- These children are small because their nutritional and/or emotional needs have not been met.
- They may demonstrate delayed development or abnormal behavior.
- Hospitalization may be required to screen for significant medical illness and, more important, to see whether the child responds to adequate nutrition and a nurturing environment with a rapid weight gain and more appropriate behavior.
- Evaluation consists of more than weighing and measuring the baby. The behaviors of the child and the parent should be observed, ideally during feeding, and the child's response to feeding should be noted.

If physical neglect is left untreated, the health or life of the child may be endangered and emotional disorders, school problems, retardation, and other forms of dysfunction may result.

BEHAVIORAL INDICATORS OF PHYSICAL NEGLECT

There are certain behaviors that children may display if they are survivors of physical neglect.

Delinquent Acts

- Stealing food, vandalism.
- Alcohol or drug abuse.
- Prostitution.

School Problems

- Frequent absences.

- Extended stays at school, seen in early arrival and late departure.
- Constant fatigue, listlessness, or falling asleep in class.

Children's Statements

- Child states there are no caretakers.
- Child begs or steals food due to hunger.

CARETAKER'S BEHAVIORAL INDICATORS IN PHYSICAL NEGLECT

Many times, the caretaker will exhibit certain behaviors which together with your observation of the child will clue you into the possibility of some form of neglect.

Inappropriate Behavior

- Misuses alcohol or other drugs.
- Maintains chaotic home life.
- Frequently leaves home to "party" or because of sexual/emotional involvements.

Emotional Problems

- The caretaker may be depressed and show evidence or apathy or futility.
- The caretaker may be mentally ill or of diminished intelligence.
- The caretaker may have been a victim of neglect as a child.

Child Abuse Risk Assessment and the Law

Risk Factors: Engaging in professional assessment and treatment of child abuse will require you to be able to identify risk factors and be knowledgeable about the legalities within the jurisdiction where you work. Determining whether the risk in a case is high, moderate, or low can be important in developing appropriate treatment plans and in knowing how best to proceed in a particular case. Below are some brief guidelines that may help:

- Severity and/or frequency of abuse
- Severity and/or frequency of neglect
- Location of injury
- Child's age; physical and mental ability
- Perpetrator's access to child (opportunity)
- Child's behavior
- Child/caretaker interaction
- Child's interaction with siblings, peers, other adults
- Caretaker's capacity for child care
  (Filip, Schene, & McDaniel, 1991)

Legal Issues: Child abuse laws vary from state to state. You must be well informed of the child abuse reporting requirements of the state in which you practice as well as the policies of your agency. You must also keep up with any changes in the law.

SCENARIO

In California, all mandated reporters are required to telephone a child abuse report immediately to children's protective services. When the situation is potentially dangerous, a call to law enforcement is also made at this time. A written report is to be made within 36 hours of first suspecting or encountering the abuse. Failure of a mandated party to report a known incident or suspicion of abuse is a misdemeanor punishable by up to 6 months in the county jail and/or a $1,000 fine. When reporting as required by law, mandated reporters are immune from civil and criminal liability, even if the report proves to be unfounded.

Factors in Working with Child Abuse Cases

Reporting a suspected or unsubstantiated incident of child abuse does not mean that civil or criminal proceedings will be initiated against the suspected abuser. At times, there will be inaction that is upsetting when the reporter believes action should have been taken. Conversely, there are times when the action taken may seem severe in relation to the abuse. In either case, children's protective services and/or law enforcement officials are the ones in charge of the formal investigation that may lead to court proceedings. The human service professional, depending on his or her role, may not have much control over the ultimate outcome of the report.

Difficulties in Making a Child Abuse Report

The difficulties inherent in making a child abuse report are numerous. First, deciding when to make a report of this nature is at times a complicated matter. Deciding when to make a child abuse report is one of the most difficult responsibilities a human service worker can face. Determining whether the instance is reportable can be complex as the cultural standards of a community may be involved.

When cases are clear-cut, the helper may be confident that a report must be made. However, at times the situation is murky and the worker must rely on his or her judgment about making a report. We encourage helpers to consult with supervisors, colleagues, or children's protective services staff if they are uncertain of how to proceed.

In addition to the logistics of making the report, emotional factors may also come into play. Workers may experience anxiety, fear, and sadness when dealing with a child abuse situation. When the abuse is severe, helpers may experience feelings of shock, disbelief, and anger. Some helpers may find that they have internal barriers to reporting. In the literature, these have been identified as follows:

- *Denial* by the worker of the abuse or its severity
- *Rationalizing* by accepting unrealistic explanations for how an injury occurred; also, minimizing the incident in our own minds, or justifying it
- *Betrayal* felt when contemplating having to report a child abuse case
- *Fear of breaking up a family* if report is made, although this rarely happens

Child abuse is so vast and diverse that it requires treatment at various levels. Depending on the severity of the abuse that has been reported and substantiated, there may be a need for immediate action. The local protective children's services agency will likely intervene to ensure the safety of the child. There may be a need for "crisis nurseries, extended day care centers, and emergency foster homes which will provide short-term counseling to relieve a potentially damaging crisis situation" (Zastrow, 1990, p. 177). Crisis intervention is almost always provided to the child victim. Often parents as well as the family are also treated through a crisis intervention framework.

Children and families are often referred to counseling or mental health centers for therapeutic interventions. Psychological or clinical treatment is provided in a variety of modalities: individually with the child, or the parent/caretaker (perpetrator), conjointly with both parents in marital or couple's treatment, in family therapy where part or the entire family is seen together, and in group settings where either the child is a member of a group for abused children or the parents are in a parenting group often mandated by the courts. Interventions range from crisis intervention and brief therapy to short-term treatment or long-term counseling. It is quite common for parents to be involved in a parenting effectiveness class or behavior modification programs where modeling and role playing are used to change the negative behavior of parents toward their children. Self-help groups such as Parents Anonymous have been developed by parents themselves in an effort to help them gain the needed support to cope with their parenting roles, learn more adaptive coping, and have immediate access to others in times of crisis.

Support, education, and advocacy often go hand in hand in assisting these families to overcome their crisis and develop healthier coping skills. To be effective in working with these families, helpers must possess strong case management skills. They need to be able to identify the service needs of their clients and be advocates in the community for the development, adequacy, and availability of such services (CWLA, 1990, Sections 1.17 and 6.3). These services should include the following:

- in-home services
- day care services
- respite care
- intensive crisis counseling (individual and group)
- medical services
- parent education
- mental health services
- substance abuse services
- family planning
- parent self-help groups
- legal services
- housing services

- educational services
- emergency financial services

Additional areas of concern equally important to address are these:

1. The symbiotic behavior by the parents and in the family system
2. Isolation
3. Poor marital communication
4. Poor impulse control (temper/impatience)
5. Little knowledge of child development and management
6. Job or unemployment stress. (Justice & Justice, 1990)

You may find this list quite applicable to many child abuse cases. Not all cases of abuse involve these, however, nor is this list exhaustive; often, many more factors contribute to the abusive situation. Nonetheless, the list can be a guideline to use in considering treatment options. Last, we feel strongly that your success in working with abused children and their families will be determined partly by your ability to respect and support them as well as demonstrate a genuineness in wanting to help them.

Treatment Considerations

- Be open with parents/caretakers and the child about the reporting requirements.
- Utilize contracts with clients in which treatment goals and expected outcomes are spelled out.
- Set limits on clients' self-destructiveness or self-defeating behaviors. Many times such limit setting helps clients feel protected and cared about.
- Use authority in an appropriate way to gather information about the abuse and in making the report. Although this may entail being firm, you can still be supportive and reassuring, showing interest in wanting to help the family in a professional capacity.
- Be ready to face denial in parents and refrain from forcing the issues. Use your clinical skills in working with clients by also addressing their strengths.
- If you are making a child abuse report, attempt to stay with the clients emotionally if possible. If not, provide them with appropriate referrals.
- Coordinate with others to avoid chaotic treatment and unnecessary disruption of a family that is already in crisis. It is important to work as a team with others involved in the case so that clear communication exists.
- Seek consultation whenever necessary. Helpers should be willing to reach out for support, objective feedback, and guidance. Taking time to take care of yourself will ultimately help your clients.
- Seek continuing education in the child welfare field. Learning about changes in legalities and treatment approaches can be useful and help you prevent burnout.

SCENARIOS

CHILD PHYSICAL ABUSE

Mrs. Gomez entered a local mental health center because she felt overwhelmed and angry about a recent incident with her 10-year-old daughter. She explains that she and her husband, who are both from Mexico, had been feeling very upset with their daughter for her repeated stealing and negative attitude. They had attempted various ways to discipline her and earlier this week, on learning of another theft incident, they felt that a "good spanking" was in order. The next day their daughter told her teacher about being spanked and when the nurse examined her, she found visible marks of the discipline the child had received. Children's protective services were notified and a home visit was conducted. Mrs. Gomez was at work when the child welfare worker came to the home, but she was informed by her mother who provided child care that the worker stipulated that if any further spanking occurred, the parents could be arrested or the children taken away. Mrs. Gomez was still to meet with the worker.

In this initial session with Mrs. Gomez, she appeared stressed by this incident in addition to job and financial pressures. She shares her disappointment and anger at her daughter and seems to blame the entire incident on her without any awareness of the part she and her husband may play in the family conflicts. She holds strongly to the belief that corporal punishment is necessary and she feels it was her daughter who was in the wrong, first by stealing and second by telling the authorities at school. Now she is considering punishing her daughter for this behavior by withdrawing a promise to go on a planned vacation in several weeks. If this were your client how would you proceed? What are the child abuse issues involved? Are there any cultural issues to consider? If so, what are they?

CHILD SEXUAL ABUSE

Beverly is a 14-year-old Caucasian female who enters treatment at a family agency to deal with her upset feelings; these are secondary to her recent disclosure that her step-father had molested her for a period of years. Beverly's parents divorced when she was 3 and her mother remarried several years later, as did her father. Shortly afterward, her mother and stepfather had their own child, whom Beverly was very protective of. Because she had never been very close to her father, she had chosen to live with her mother and only visited her father sporadically. Apparently, Beverly had never told anyone of the molestation by her stepfather that had been occurring for years—since she was 6 till recently when she turned 12. She was fearful of not being believed and of upsetting the family, especially her Mom who seemed content in this marriage. Beverly finally told a good friend at school when she had become worried about the stepfather possibly beginning to molest the younger sister who was approaching age 6. Her friend told her mother, who then notified the school; a school counselor ultimately made a child abuse report to children's protective services. Beverly was upset that this secret had been divulged, but she did acknowledge that the abuse had happened when questioned by the authorities. Within a matter of a few days, she was removed from her home and taken to her maternal grandmother on a temporary basis. Once her father learned about the abuse, he insisted on having Beverly come live with him. Beverly felt she had no choice, but she had a very difficult time adjusting to moving away from both family and friends to go live with her father and his new wife. To make matters worse, Beverly did not get along with her father or stepmother, who were stricter and more protective than her mother had been. Most upsetting, however, Beverly's mother stood by her husband and denied that he had ever abused Beverly; she called Beverly a liar. Beverly felt betrayed and when her father pursued a court trial against the stepfather, things became even more intense. Gradually, Beverly learned that she did not have support from any of her family on her mother's side as they chose to stand by her mother and stepfather and even testified against Beverly in court. She felt very alone as she no longer felt accepted by her mother's family and did not fit in with her father's. When she enters treatment, she feels confused, guilty, angry, sad, and frustrated. How would you help Beverly? What are some of the key issues you would focus on?

CHILD NEGLECT

Jon and Judy were married very young, both trying to escape the abuse of their own families. Even prior to their marriage they had experimented with drugs and alcohol, but once married their habits seemed to grow into an addiction. Furthermore, the circle of friends they developed reinforced their drug and alcohol lifestyle. When Judy became pregnant with her first child, Sue, she did attempt to reduce her use of both drugs and alcohol and succeeded in delivering a healthy baby girl, who although small in size and weight had no other problems. This successful birth proved to both Judy and Jon that a little drug use was not harmful. On the conception of her second child a few years later, Judy did not curb her alcohol or drug use as much as she had done with the first, yet the baby was also born small but without any noticeable problems at birth. It was not until her third child was conceived, about the same time that her first daughter started school, that problems were noticed. It seems that Sue always arrived late to school, was usually dirty, and wore clothes that had not been washed and were not appropriate for her. In class, she would fall asleep often and seemed ravenous when it was time for a snack. The teacher felt concern, which only grew when she learned that Sue was often not picked up at the bus stop and had to walk a long distance alone. The school counselor was notified and chose to make a child abuse report given the neglect that seemed to be present. When the child welfare worker interviewed both Sue and her parents, he confirmed that neglect indeed was evident, and given his concerns for both Sue and her younger sister, they were removed from their parents and placed in a temporary shelter. Jon and Judy were shocked and angry because they did not understand why their children had been removed, this despite the fact that drugs were found in the home and both parents tested positive for drug use. Judy who was now 4 months pregnant with her third child did not believe that using would affect the baby and resented the authorities for trying to tell her how to live her life. If you were working with Jon and Judy how would you proceed? What are the salient issues you would focus on?

CHILD EMOTIONAL ABUSE

Jeff is a college student who seeks counseling at the school counseling center. Initially, he enters reluctantly given his fear of not being understood and being viewed as "weak" for seeking help. He tells the counselor that he is finding it difficult to concentrate on his studies and feels no energy or motivation at all to continue. Furthermore, he doubts his abilities and questions whether college is right for him. He describes feeling depressed and angry at his parents for their lack of support and ridicule of him through the years. When he talks of his childhood, he becomes visibly upset and explains that his parents always told him he would amount to nothing. They never encouraged him, and when they learned he was going to college they laughed at him, stating, "Son, college is not for you, you're a fool, you're only going to fail and make a laughingstock of the family." Because of their criticism and lack of emotional support through the years, Jeff had moved out when he was 17. He had had very little contact with his parents since, but he recently went back home to spend time with his ailing grandmother who had been the only one in his family who had been kind and supportive of him. When he saw his parents after several years, he felt all the old hurt surface and he sadly realized that they would never change. On his return to college, he found himself even more depressed and disillusioned with his life, questioning his abilities to succeed. If you were assigned to work with Jeff what might be some interventions you would consider? What are the most pressing issues to work on? Where would you begin?

Other Forms of Abuse: Elder and Dependent Adult Care

Elder Abuse

Another vulnerable segment of our society is the elderly (age 65 and older). As with children, older adults are often victimized and subjected to abuse within and outside the family. Human service professionals need to be aware of this serious problem. We are both ethically and morally responsible to act as advocates for those most susceptible to victimization. Legally, too, we are mandated reporters, as with child abuse. The main difference in reporting is that elder abuse reports are directed to adult protective services whereas child abuse is investigated by children's protective services.

The continued rise in elder abuse in the United States is supported by statistics indicating that an estimated 2 million elder abuse reports are made each year (American Association of Retired Persons [AARP], 1993). Although alarming, this number is probably much lower than the actual abuse cases. Most statistics are believed to underestimate the prevalence of the problem and by some estimates, five out of six cases go unreported. This inaccuracy may be related to insufficient data, due partly to differing definitions of the term *elder abuse* (AARP, 1993).

As the elderly population increases, cases of elder abuse are likely to mount. The increase of elder abuse is likely to be one of the negative consequences of this population growth. Along with the need to work with the elderly on many other issues, elder abuse must also be targeted and treated.

Elder Abuse Defined   O'Malley (1979, p. 2) defines elder abuse as "the willful infliction of physical pain, injury or debilitating mental anguish, unreasonable confinement or willful deprivation by a caretaker of services which are necessary to maintain mental and physical [health]."

Types of Elder Abuse   The term *elder abuse* is commonly thought to signify acts of physical violence against older persons; in fact, it encompasses many different forms of dangerous behavior. According to Morales and Sheafor (1995), elder abuse includes "physical assault, verbal harassment, malnutrition, theft or financial mismanagement, unreasonable confinement, oversedation, sexual abuse, threats, withholding of medication or required aids, neglect, humiliation, and violation of legal rights" (p. 270).

In a recent AARP "Public Policy Institute FACT SHEET," common forms of elder abuse and neglect were categorized as follows:

- *Physical abuse*—the intentional use of physical force causing pain or bodily harm
- *Psychological abuse*—the intentional infliction of mental anguish by threat, intimidation, humiliation, or other abusive conduct
- *Financial exploitation*—the conversion or unauthorized use of an elderly person's money, property, or other resources
- *Caregiver neglect*—a caregiver's intentional or unintentional failure to fulfill a caregiving obligation needed to maintain an elderly person's well-being
- *Self-neglect*—an older person's failure to provide himself or herself with the necessities of life, such as food, clothing, shelter, adequate medication, and reasonable financial management (AARP, 1993).

An important form of elder abuse is parent abuse. According to Kock and Kock (1980, p. 14), elder abuse refers to "elderly parents who are abused by their children with whom they live or on whom they depend." From their research, the authors have found four main types of parent abuse:

- *Physical abuse*—direct beating and the withholding of personal care, food, medicine, and necessary supervision
- *Psychological abuse*—verbal assaults and threats provoking fear
- *Material abuse*—theft of money or personal property
- *Violation of rights*—forcing a parent out of his or her own dwelling, usually into a nursing home

Profile of an Elder Abuse Victim   With the advent of recent studies that have more thoroughly examined the nature of elder abuse, profiles of both the elderly abuse victim and the perpetrator are beginning to emerge.

The victim: The typical victim is most often a female, over age 75, economically impoverished, (living at or below the poverty level), and mentally or physically disabled. These individuals are usually isolated and dependent on others for care and protection. Many fit the criteria of the *frail elderly* as they are physically and financially unable to remain alone in their own homes and must rely on others. Contrary to common belief, a majority of the elderly live alone or with family or other caretakers. These elderly may be even more prone to some form of maltreatment than those living in institutions where legal protections limit such abuse. In private settings, legal protections are more difficult to enforce because of civil rights and family privacy issues.

The perpetrator: Family members who are caretakers of their elderly relative (parent, older sibling, aunt, uncle) make up more than two-thirds of those reported as abusers. More specifically, this individual is often a female child of the victim who is in some way dependent on the victim (as for money or housing). The perpetrator may often have to contend with various stressors such as long-term medical complaints, marital and financial difficulties, and an alcohol or drug addiction. The existence of such stressors coupled with the demands of caring for the elderly person, may often be too much for the individual. Likely reactions of exhaustion, frustration, and anger are then displaced on the most vulnerable—the older person.

Causal Factors   The literature (Sommers & Shields, 1987; Steinmetz, 1978) cites the following possible reasons for elder abuse that tend to parallel those commonly noted in battered children:

- They are dependent on the abuser for survival,
- They are presumed to be protected by love, gentleness, and caring.
- They are a source of emotional, physical, and financial stress for the caretaker, especially when the elderly person is also physically or mentally disabled (e.g., by stroke or Alzheimer's disease).

Additional contributors to elder abuse have also been cited by gerontological authorities. Hoff (1989) has grouped these factors into five general categories:

1. *Social*—Socially, some elderly, especially the most frail, have no place in the nuclear family and as such are outcasts, relegated to a limited status.

2. *Cultural*—Our youth-oriented society places major emphasis on the young and productive and is less supportive of the elderly who may be seen as a burden. They are often devalued instead of praised for their wisdom, life experience, and status. Ageism, despite the elders' increasing political influence, is still prevalent, further negatively affecting the older individual.

3. *Economic*—Many elderly live at or below the poverty level. Even if they held a well-paying, stable job for most of their lives, the impact of spiraling inflation and mounting health care costs is likely to affect their financial resources adversely.

4. *Psychological*—Many elderly may feel frustrated with their plight and find it difficult to adjust to the many losses they have had to endure. Often, those who are abused may fear retaliation and therefore keep the abuse hidden.

They may also feel deep shame and guilt, especially if there is a history of family violence. A cycle of violence often results. Unresolved conflict between parents and children often is perpetuated throughout the life cycle. Hurt feelings left over from childhood sometimes resurface. The motivation to care for the aging parent may be a sense of guilt, responsibility or duty rather than love and concern. Often, the dependent elder may become the target for the frustrations of the caregiver/adult child.

5. *Legal*—Legal mandates to report elder abuse differ from state to state, with some less stringent than others in their requirements. This factor, combined with the abused elder's shame and fear of retaliation, set up formidable barriers to dealing effectively with elder abuse. (Hoff, 1989, p. 276)

Implications of Elder Abuse   Failing to address elder abuse can have only negative consequences in terms of both human and monetary costs. Some of the possible outcomes are these:

- Decline in health of the elderly person, requiring more costly medical care, hospitalization, or early nursing home placement.
- Increased psychological problems that can lead to maladaptive addictive behaviors (e.g., alcoholism) and ultimately even suicide. (Recall that the suicide rate for the elderly is quite high.)
- Reinforcement of a negative social message about how our nation cares for the elderly. Once a pattern develops, it is more easily reinforced. This is a concern for those who work with the elderly and witness the negative societal attitudes toward them that seem to be mounting.
- Perpetuation of the stigma and fear attached to the aging process.

Treatment Considerations

- Treatment of the abused elderly typically necessitates crisis intervention strategies, knowledge of the legal and ethical issues involved, and professional case management skills.
- Often, being able to intervene with the family is also imperative for proper assessment and follow-up.
- Advocacy of the elderly in social and political arenas can indirectly be helpful.
- Preretirement planning including education regarding social, economic, and physical realities can also help reduce the risk of elder abuse.
- As helpers, we may need to examine our own values, attitudes, and beliefs about growing old. If we can see that with every moment that passes we too are aging, perhaps we can be more empathic and sensitive to those who are older now.

In summary, guidelines for working with elder abuse include these

- Knowledge of elder abuse laws in your state
- Ability to utilize crisis intervention techniques
- Utilization of case management skills
- Family assessment and intervention
- Advocacy for the elderly
- Preretirement planning to help prepare seniors for the many issues they will face on retirement
- Value awareness regarding ageism

SCENARIO

A 65-year-old woman in a wheelchair enters your agency for an intake because of her increasing marital problems. She describes feeling depressed and apathetic about life. In the course of the intake, she tells you of the multiple physical problems she faces as well as the wrath of her husband who is angry that she is not able to provide for him as she once did. Most recently, he has become physically violent with her when she asks him to help with her care. She feels scared and guilty for having to need his help. How should you proceed? What legal and ethical issues are present?

Dependent Adult Care Abuse   The reporting of dependent care abuse is also a mandated disclosure in many states. Many adults, who for various reasons require ongoing care to survive, are also susceptible to abuse. Reporting procedures and treatment considerations similar to those for elder abuse apply here.

Nancy is a 52-year-old, married, Caucasian female who is suffering from the debilitating effects of multiple sclerosis that have left her unable to walk or perform many of the tasks she did in earlier years. Her illness was diagnosed 5 years ago and she has gradually become more and more dependent on John, her spouse of 10 years. Their marriage was never a very good one, and now there seems to be increased tension and a great deal of resentment on John's part for having to care for Nancy. Nancy, who is being seen by visiting nurses, is referred to a social worker for counseling because of the nurses' concerns about her increased depression and possible dependent adult care abuse by John. If you were the social worker assigned to the case what issues might you explore to determine whether there is abuse? How might you proceed in helping Nancy with the various problems she is facing? What are your priorities?

Leslie states (1989) that all licensees, interns, and trainees should be aware of their duty to report child abuse, elder abuse, and dependent adult abuse. Failure to make reports required by the sections of law relating to these kinds of abuse can result in the conviction of a crime, liability in a civil lawsuit for damages, and loss of license or registration. These obligations to breach confidentiality came about because legislators decided that certain public safety interests took precedence over the principle of confidentiality.

Harm to Self and Others

The ultimate harm to self (suicide) and to others (homicide) represent two of the most difficult and anxiety-producing topics human service professionals must confront. As with child abuse, cases involving suicide and/or homicide are often crises and are bound to evoke much anxiety and concern in those involved. In addition, suicide and homicide also have legal and ethical implications that must be taken into consideration by workers in the helping field. If you work with people, regardless of the agency setting or the specific job description you have, you are likely to encounter clients who are suicidal, homicidal, or both. The following sections discuss the multidimensional aspects of both suicide and homicide. They also have guidelines to help you in your future work with those most prone to harming themselves or others.

Suicide

Suicide is not a new phenomenon. Throughout time there have been those who for various reasons have opted to take their own lives. Suicide prevention services began only in 1906 in London. Recognizing a need, the Salvation Army opened a bureau aimed at helping persons who attempted suicide. Almost simultaneously, the National Save-A-Life-League was opened in New York City by the Reverend Harry M. Warren (Roberts, 1990). In the United States, efforts toward suicide prevention were evidenced by the establishment of almost 200 suicide prevention centers through federal funding in the 1960s and early 1970s (Roberts, 1990, p. 398).

In the 1990s, with budget cuts and limited resources, the future of many of these suicide prevention centers is uncertain. What is certain, however, is the unfortunate reality that suicide attempts will not cease nor are likely to decrease.

Statistics   Suicide is among the top 5 to 10 leading causes of death in the United States. Guilliland and James (1990, p. 129) state, "Suicide can strike any family, it cuts across all segments of population, and is prevalent in all age and racial/ethnic groups." As evidenced in a review of the literature, the two major groups where suicide rates have skyrocketed are the elderly and adolescents.

TABLE 7.6   Freud's and Durkheim's Explanations for Suicide

Psychodynamic Approach	Sociological Approach
An intrapsychic conflict emerges when a person is experiencing great psychological stress. This conflict triggers suicide. One's hostility toward other persons or society is turned inward toward oneself. Suicide is the extreme in which self-destruction or self-punishment is chosen over urges to lash out at others. SOURCE: Allen (1977)	Societal pressures and influences are a major determinant of suicidal behavior. Three types of suicides have been identified by Durkheim (1951): 1. Egoistic—One's lack of integration or identification with a group 2. Anomic—Perceived or real breakdown in the norms of society 3. Altruistic—Perceived or real social solidarity Fujihara et al. (1985) cite a fourth: 4. Dying with dignity

Adolescent suicides have shown a dramatic increase, with some estimates suggesting that suicide among those age 15 to 24 is the third leading cause of death for this group, following accidents and homicide, respectively (Berman & Jobes, 1991), some suggest even higher figures. In our experience in working with teens, these statistics demonstrate the stark reality that adolescence is indeed a high-risk age and one in which suicide attempts are climbing. Teens with these tendencies are suffering and clearly require our involvement.

Theoretical Viewpoints   There are numerous theoretical viewpoints regarding suicide. According to Fujimura et al. (1985), two main ones explain suicidal behavior: Freud's psychodynamic approach (Allen, 1977) and Durkheim's (1951) sociological approach. These are summarized in Table 7.6.

Dixon (1987, p. 151), like many others, classifies three major categories for suicidal behaviors: psychosocial, psychopathological, and biological causes. These are summarized in Table 7.7.

TABLE 7.7   Causes of Suicide Identified by Dixon

Psychosocial	Psychopathological	Biological
Suicidal behavior is precipitated by the disappointment related to the loss of a love relationship, loss of self-esteem, or some other psychosocial need perceived to have life or death importance. Past and present relationships must be considered as well as family history that predisposes one to consider suicide as a viable alternative. Psychosocial causes are the most common reason for suicide.	A secondary condition in some psychopathological conditions triggers the suicide. These conditions include: • Major depression • Bipolar conditions (manic depression) • Borderline personality disorder • Schizophrenia	Biological factors known to contribute to suicidal behavior are • Alcohol • Drugs • Physical illness • Organic brain damage • Gender issues: Women attempt more often, yet men more often complete the act. • Age: The incidence of suicide rises with age.

SOURCE: Dixon (1987)

Common Characteristics of Suicidal Individuals   Shneidman (1985, 1987) has identified the 10 most common characteristics of suicide which he grouped into the following six categories:

> *Situational Characteristics:* (1) The common stimulus in suicide is unbearable psychological pain; (2) The common stressor in suicide is frustrated psychological needs.
>
> *Conative Characteristics:* The common purpose of suicide is to seek a solution; (2) The common goal of suicide is cessation of consciousness.
>
> *Affective Characteristics:* (1) The common emotion in suicide is hopelessness-helplessness; (2) The common internal attitude toward suicide is ambivalence.
>
> *Cognitive Characteristics:* The common cognitive state in suicide is constriction.
>
> *Relational Characteristics:* (1) The common interpersonal act in suicide is communication of intention; (2) The common action in suicide is regression.
>
> *Serial Characteristics:* The common consistency in suicide is with lifelong coping patterns. (pp. 124–147)

We believe that Shneidman's list offers a comprehensive yet succinct overview of features most commonly seen in suicidal persons. Be careful, however, in generalizing these traits to all suicidal individuals; suicide is specific and unique to each individual. Therefore, each client must be evaluated separately, taking into consideration his or her unique set of circumstances.

No one is immune to suicide and anyone contemplating it should be taken seriously. Unfortunately, there are many negative stereotypes regarding suicide and many myths and fables that need to be dispelled. Following are some of these potentially hazardous misconceptions with corresponding factual data. We have taken two sets of myths/fables that are identified in the literature, one from Guilliland and James (1993), and one from Shneidman and Farberow (1961). Even though these were written 30 years apart there is still much similarity and overlap between them as you will see in Table 7.8.

Additional stereotypic views of suicide also written by these authors are as follows:

Facts and Fables on Suicide (Shneidman & Farberow, 1961)

Fable: Suicide happens without warning.
Fact: Often many clues and warnings are given regarding suicidal intentions.

Fable: Suicidal people are fully intent on dying.
Fact: Many are undecided and ambivalent about dying. Many times they may "gamble with death," yet leave it to others to save them.

Fable: Suicide is related to socioeconomic status; either it strikes the rich more often or occurs exclusively among the poor.
Fact: There is no specific socioeconomic class that is most affected. It is very "democratic" and is represented proportionately among all levels of society.

Myths and Realities about Suicide (Guilliland & James, 1993)

Myth: If you discuss suicide with a client this will move him or her toward doing it.
Reality: More than often, the client will welcome a discussion of suicide by a supportive empathetic person as it allows him or her the chance to reconsider and the encouragement to explore more appropriate options.

Myth: Suicide is an irrational act.
Reality: At the time, the suicidal person views the suicide as making perfect sense and thus it seems quite rational.

TABLE 7.8

Myths and Realities of Suicide	Facts and Fables on Suicide
1. Clients who threaten suicide don't do it. • A large percentage of people who kill themselves have previously threatened it or disclosed their intent to others.	1. People who talk about suicide don't commit suicide. • Of any 10 persons who kill themselves, eight have given definite warnings of their suicidal intentions.
2. Persons who commit suicide are insane. • Only a small percentage are psychotic or "crazy." Most are just depressed, lonely, helpless, newly aggrieved, or shocked.	2. All suicidal individuals are mentally ill, and suicide is always the act of a psychotic person. • Studies of hundreds of genuine suicide notes indicate that although the suicidal person is extremely unhappy, he or she is not necessarily mentally ill.
3. Suicide runs in families; it is an inherited tendency. • Suicidal tendency is not inherited; it is either learned or situational.	3. Suicide is inherited or "runs in the family." • Suicide does not run in families. It is an individual pattern.
4. Once suicidal, always suicidal. • Many will contemplate suicide yet overcome this desire and live long, productive lives.	4. Once a person is suicidal, he is suicidal forever. • Individuals who wish to kill themselves are "suicidal" only for a limited period of time.
5. When a person attempts suicide and pulls out of it, the danger is over. • In many cases the most dangerous time is during this "upswing period" when the suicidal person feels energized following a period of severe depression. A danger signal is a period of "euphoria" following a period of deep depression.	5. Improvement following a suicidal crisis means that the suicidal risk is over. • Most suicides occur within about three months following the beginning of "improvement," when the individual has the energy to put his or her morbid thoughts and feelings into effect.
SOURCE: Guilliland & James (1993)	SOURCE: Shneidman & Farberow (1961)

Myth: Signs of renewal and recovery are evidenced when a suicidal person begins to show generosity, share belongings, and/or give away personal possessions.

Reality: Often clients who are seriously intent on killing themselves begin to give away their belongings as a way to begin bringing closure to their lives. Their behavior may be indicative of another form of presuicidal euphoria.

Myth: Suicide is always an impulsive act.

Reality: Sometimes this may be true; at others times, suicide is very deliberately planned and carried out.

Suicide Assessment Guidelines "Suicidal behavior is an anxiety-producing subject because it defies the human instinct to survive. While most people are trying to live longer, suicidal people are either thinking about, attempting to, or succeeding at killing themselves" (Dixon, 1987, p. 151). Human service professionals have real anxiety when working with a suicidal client because of the legal, ethical, and often clinical responsibilities. Acquiring specific knowledge regarding appropriate assessment and treatment of suicidal clients will not likely eliminate all your fears, but it can help reduce your level of anxiety.

Guilliland and James (1993) consider three areas of assessment when evaluating a potentially suicidal client. These include risk factors, suicidal clues, and cries for help (p. 133). We follow their format in discussing assessment.

Risk Factors   Dixon highlights the importance of suicidal risk assessment: "Perhaps the greatest concern of the crisis therapist is assessing the degree of suicidal risk; that is, attempting to reasonably predict the likelihood that a particular person will take his or her own life. All suicidal behavior, regardless of apparent significance, should be taken seriously" (Dixon, 1987, p. 152).

Hack (Martin & Moore, 1995) agrees that no matter how you "become aware of the client's intent to end his or her life, there are…procedures to which you should adhere to prevent suicide" (p. 131). A first and most important step prior to exploring risk factors is to address the issue of suicide directly to your client. The fear that talking about suicide will push the client toward it is often a reason for our hesitation. Furthermore, we may feel angry at the client and believe we are responsible for his or her feelings and behaviors. Dixon (1987, p. 151) asserts that helpers often "feel responsible for their clients, but the client's unpredictability causes the helper to feel anxious, helpless, and angry. There is a general tendency among helpers to blame oneself if a client succeeds in the suicide attempt. Consequently, therapists are too eager to dispose of the suicidal client." Hack (1995, p. 130) adds that "although it may seem desirable at times to divert a suicidal client away from discussing the issues surrounding potential suicide, these issues must be dealt with if the suicide is to be prevented." Only when you are willing to explore the latent suicidal ideation can you take precautions to ensure your client's safety. If you are not in a position to conduct a thorough assessment, then seek appropriate assistance from a supervisor, clinician, or crisis team and/or make the needed referral.

Before assessing suicidal potential, carefully observe and listen to your client. Be willing to ask certain pertinent questions in regard to the client's suicidal ideation. As a general guideline, people of any age are dangers to themselves when any one of the following occurs:

- The person has indicated by words or actions an intent to commit suicide or inflict great bodily harm upon himself or herself.
- The person has indicated by words or actions a specific plan and the means and ability to harm or kill himself or herself.
- A third party has reported credible information that indicates the person does indeed have a plan and the means and ability to harm or kill himself or herself.

Questions to ask the person, a family member, or someone who is very familiar with the person's behavior include these:

- Does such person intend to kill or harm herself or himself?
- How does the person intend to kill or harm herself or himself? Look for weapons, pills, the gas left on, or other evidence of a plan and the means and ability to carry out the plan.
- Has the person ever attempted to kill or harm himself or herself in the past? What was the attempt at that time?

- Is there a past history of psychiatric evaluations or psychiatric hospitalizations? When and where?

Hack (1995) encourages assessing some of the same factors and asking similar questions. He indicates that the following pieces of information should be acquired, using direct questions when necessary:

- Frequency, duration, and content of suicidal thoughts
- Antecedents and consequences of these thoughts
- Extent of third-person involvement in suicidal thoughts and the extent of harm to these individuals
- Development of a concrete suicide plan including when, where, and how the suicide will happen
- Acquisition of materials for completing the suicide
- Expected impact of the client's death on family, loved ones, friends, and enemies
- Reasons for not having previously committed suicide and the aspects of life that require attention and modification to prevent suicide
- Preparation for death including farewells to friends and family members and drawing up of a will

Once you have ascertained that the client is indeed contemplating suicide, your next step is to conduct a risk assessment. The literature offers an abundance of valuable information regarding suicidal risk factors; we have compiled a guide from these to help minimize your difficulties in the assessment process.

Classic Suicidal Risk Factors

1. Family history of suicide (this increases the risk)
2. History of previous suicide attempts (80% of suicides were preceded by a prior attempt)
3. Direct verbalization about the desire, intent and plan to commit suicide (one of the most useful single predictors of a suicide)
4. Having a specific plan (the more definite the plan, the more serious the situation; do not be afraid to ask directly about the person's intent)
5. Experiencing serious depression. Sleep disturbance, which can intensify depression and is a symptom of depression as well, is a key sign. Statistics indicate that suicide rates increase when the depression begins to lift, not during the period of time when the person is extremely depressed
6. Pervasive feelings of hopelessness and helplessness, which are often symptoms of depression
7. Recent loss of a loved one through death, divorce, or separation
8. History of drug/alcohol abuse (rational thinking and impulse control can become seriously impaired when under the influence; between one-fourth and one-third of all suicides are associated with alcohol as a contributing factor)
9. Living alone and cut off from contact with others (single individuals are twice as likely as married people to commit suicide)
10. Giving away prized possessions, putting personal affairs in order, or revising wills
11. History of emotional and/or psychiatric problems resulting in previous hospitalizations
12. Preoccupation with anniversary of a particularly traumatic loss
13. Recent family destabilization as a result of loss, personal abuse, violence, or other problems
14. Recent physical trauma with unsuccessful and/or painful medical treatment (chronic or terminal illness has been shown to increase the risk)

15. Radical mood swings or shifts in behavior (apathy, withdrawal, isolation, irritability, panic, anxiety) or changed social habits, eating habits, or school or work habits
16. Preoccupation with past physical, emotional, or sexual abuse
17. Profound and uncharacteristic exhibition of one or more emotions such as anger, aggression, loneliness, guilt, hostility, grief, or disappointment
18. Sudden calm and sense of relief as if having found a "solution"
19. Lack of support systems or reluctance to use existing support networks
20. Unemployment or recent loss of employment
21. Psychotic behavior (secondary to mental illness, emotional crisis, drugs, organic problems)

Clearly, not all these factors indicate a serious suicidal risk. It is when several risk factors occur together that the risk is likely to increase. Trust your intuition about how you view clients and their potential for harming themselves.

A very simplistic view requires identification of three main risk factors:

- *Lethality:* Includes choice or method, circumstances, and how close the individual comes to death. A gun is more lethal than aspirin, for example.
- *Intent:* Determined by motivation. That is, the suicidal behavior has a definite purpose (e.g. revenge, control).
- *Attitude:* Manifested by how the person's ego reacts to the suicidal behavior. It is ego-alien if the person feels ashamed, sick, scared, or uncomfortable.

Risk assessment must be based on demographic, clinical, and historical information obtained from the client or family during the initial encounter. Hack (1995) further summarizes these risk factor categories as follows:

Demographic Risk Factors for Suicide

Age: Increased risk for higher age groups

Employment status: The unemployed at a higher risk

Geographic location: Higher risk for urban dwellers

Marital discord: Higher risk for those who are separated, divorced, or widowed

Race: Higher risk among whites in United States

Sex: Higher attempts by females, yet males succeed more often

Sexual orientation: Increased risk for both male and female homosexuals

Social class: Risk is somewhat higher when there is a drop in socioeconomic status

Clinical Risk Factors for Suicide

Affective: Highest risk for unipolar depression, then bipolar, and finally manic disorder

Alcohol abuse: Many different statistics exists, ranging from 15% to 70% associated with suicide

Antisocial and borderline: Higher risk because of frequent attempts by those affected with either condition.

Biochemical dysfunction: Suicide is associated with a biochemical process accompanied by dysfunction

Depression: Associated with 53% to 70% of attempted suicides

Panic disorder: Lower risk

Schizoaffective disorder: Moderate risk

Schizophrenia: Increased risk for both attempted and actual suicides

Suicidal Clues Suicidal clients are often ambivalent regarding their plans to kill themselves. According to experts in the field, almost everyone who seriously intends suicide leaves clues to this imminent action. Sometimes there are broad hints; sometimes there are only subtle changes in behavior. Shneidman et al. (1976, p. 429) state that "most suicidal clients, feeling high levels of ambivalence or inner conflict, either emit some clues or hints about their serious troubles or call for help in some way." Although often an impulsive act, usually there has been some premeditation. Therefore, being aware of possible clues can help in suicide prevention. Such clues can be broken down into four areas as follows:

1. *Verbal clues:* Spoken or written statements, either direct or indirect
2. *Behavioral clues:* May range from purchasing a grave marker for oneself to slashing one's wrists as "practice run" or suicidal gesture
3. *Situational clues:* Concerns over a wide array of conditions such as the death of a spouse, divorce, a painful physical injury or terminal illness, sudden bankruptcy, preoccupation with the anniversary of a loved one's death, or other drastic changes in one's life situation
4. *Syndromatic clues:* Constellation of suicidal symptoms such as severe depression, loneliness, hopelessness, dependence, and dissatisfaction with life (Shneidman et al., 1976, pp. 431–434).

Cries for Help   As with suicidal clues, cries for help can be clearly evident or at times neatly disguised. Shneidman et al. (1976) say that no person is 100% suicidal, as those with the strongest death wishes can also be ambivalent, confused, and grasping for life (p. 128). Examples of cries for help range from direct verbalization of suicide, calls to suicide hot lines, actual suicidal attempts, to more subtle warnings seen in self-destructive behavior, apathetic view of life, and what may seem "insignificant" suicidal gestures.

We end our discussion on assessment by presenting a clinical instrument designed by Patterson, Dohn, Bird, and Patterson (1993). It is the **SAD PERSONS** scale, an acronym that helps human service professionals determine suicide potential. The high-risk factors include these:

Sex (males)
Age (older clients)
Depression

Previous attempt
Ethanol (alcohol abuse)
Rational thinking loss
Social support system lacking (lonely, isolated clients)
Organized plan
No spouse
Sickness (particularly chronic or terminal illness)

Ethical and Legal Issues Pertaining to Suicide   As human service professionals, we are ethically, legally, and morally bound to protect clients from harming themselves. It is our duty to protect suicidal clients and we therefore can breach confidentiality when there is sufficient cause to suspect suicidal behavior. Hack (1995) states:

> Your clients have the legal right to kill themselves (as suicide has been decriminalized) but it is illegal for you to help them do so. As well as being held legally accountable for a client's death, a therapist (human service professional) may be considered ethically responsible. If anyone believes that a therapist's unprofessional conduct contributed to a client's suicide, that person may face grievance with the organization governing the practice of therapy where [he or she lives]. (p. 150)

According to Szasz (1980), the failure to prevent suicide is now one of the leading reasons for successful malpractice suits against mental health professionals and institutions. These stark realities point further to the value of proper assessment and interventions.

Treatment Considerations

Numerous treatment considerations exist in the area of suicidal clients. Here is a comprehensive list of treatment recommendations we obtained from the literature as well as from our work with suicidal clients.

- Be supportive and empathetic. Working with your clients to create a supportive environment will enhance the treatment outcome. Establishing a sense of rapport and trust right away can help create a working relationship and provide the client with anchor to life.
- Be willing to communicate your caring by attentive listening, active involvement, and firm limit setting.
- Increase your time availability for your suicidal clients. You may need to increase the frequency of the counseling sessions or contacts with these clients. Specify your availability to your clients and offer them other contact persons in case you are not available.
- Delay the suicide by encouraging clients to consider the permanency of suicide and view their situation as a temporary setback. Written or oral agreements not to follow through on their impulses can also be made in an effort to buy time.
- Instill hope by working with client's strengths and desires to remain alive. Attempt to communicate realistic outcomes. It is important to reestablish in clients a sense of hope.
- Contact others to reduce the threat of harm. Enlist the help and support of significant others. You may need to help clients develop a supportive network of family and friends.
- Establish a contract with suicidal clients. It is often very useful to develop a therapeutic contract with clients that will provide reassurance that they will not attempt suicide. Both Hack (1995) and Faiver et al. (1995) indicate the importance of this type of intervention. Hack states "that to be effective, the contract should be time-limited, provide reinforcement for the client in terms of therapist support, be negotiable at the end of the contract period, and be acceptable to the clients.... A key component of suicide contracts, regardless of whether they are written or oral, is the client's agreement to contact the therapist or another qualified professional if he or she feels at high risk for suicide in the future" (p. 146).
- Along the same lines of a contract, establishing check-in times, usually by phone contact, is important during the initial crisis period. Also insist that the client have significant others dispose of all potentially dangerous means (e.g., weapons, pills). Be clear and firm with clients and discourage their manipulation by suicidal threats. By giving clear direct messages, you are avoiding dangerous paradoxical intention strategies.
- Consider hospitalization if clients are imminently suicidal and unwilling or unable to sign a "no-suicide" contract. Corey, Corey, and Callanan (1993, p. 125) feel that if a client is "assessed to be suicidal and unable to control self-destructive impulses, psychiatric hospitalization becomes the most logical course of action." Further they indicate that "a commitment procedure may be called for as a way of protecting the client." Others caution us to consider the drawbacks and weigh the benefits of hospitalization. Especially important is paying particular attention to the increased suicidal risk on being discharged from the hospital.

Encouraging a client to seek voluntary hospitalization is preferable, yet there may be instances when involuntary hospitalization is necessary. In these situations, laws may exist (such as the 5150 California law) that allow involuntary psychiatric hospitalization as they are a threat to themselves and hence pose a danger if not committed.

- Cognitive interventions are quite effective in crisis intervention (Dixon, 1987) and can be especially helpful with suicidal clients. Guilliland and James (1995, p. 50) offer the following cognitive strategies to help suicidal clients:

1. Separate thought from action
2. Reinforce expression of affect
3. Anticipate consequences of action
4. Focus on precipitating events and constructive alternatives

- Collaborate with colleagues and other staff to assess the level of lethality in the client continually.
- Recognize the possibility that you may need to refer clients whose concerns are beyond the scope of your training or competence.
- Documentation is crucial. In order to avoid future problems, document in writing all the steps you take to ensure your clients' well-being. Documentation may be necessary to demonstrate that you used sound professional judgment and acted within acceptable legal and ethical parameters.
- In sum, recommendations for working with suicidal clients include these actions:

  - Recognize the high level of stress inherent in working with suicidal clients. Be aware of your limits and cognizant of the toll this type of client can have on you personally.
  - Continue learning about clinical and assessment issues of suicidal behavior patterns.
  - Maintain ongoing training for suicide prevention and crisis intervention. Stay up to date with current research, theory, and practice.
  - Be cognizant of the possible need to refer some clients; realize that some client issues are beyond the boundary of your competence.
  - Familiarize yourself with legal standards that relate to your practice and be aware of the specific requirements of your agency setting.
  - Be open to consulting with an attorney who has expertise in this area. Most agencies have access to counsel for such consultation purposes.
  - Encourage clients, and perhaps their support networks, to be equally involved in the treatment process. Avoid taking on full responsibility for clients' welfare. Keep in mind that you can only guide them and it is they who must want to get better.
  - Last, keep good documentation to protect yourself from liability.

SCENARIO

Julie is a 17-year-old Caucasian female who is brought into the psychiatry department where you work because of her mother's concerns that she is suicidal. You are the crisis worker on duty, so you are assigned to evaluate her suicidality. Julie appears timid and clearly has a depressed mood. She is poorly groomed and disheveled in appearance, yet her parents seem well groomed and appropriately dressed. She wears her hair pulled back in a pony tail, she has on no makeup, and she seems not to care how she looks. She is soft-spoken and initially quite reluctant to talk openly with you. When you ask her parents to step outside, she seems a little more relaxed and eventually opens up to you. Julie tells you that she has felt very blue since the breakup of her first major relationship. She begins to cry as she tells you how her ex-boyfriend "dumped" her for a new girl in school. Julie feels that "life isn't worth living" since their breakup. When you ask whether she thinks about suicide, she gives a vague response of "sometimes." When you probe deeper about any specific plan, she becomes more agitated and refuses to answer in any definite manner. In the course of the interview, Julie tells you she is upset with her parents, who don't seem to understand her and make light of how difficult this breakup has been for her. She expresses resentment toward them and describes feeling more and more alone and isolated, stating "Nobody seems to understand, not even my best friends. . . . They all tell me he was no good or that I will find someone better." As Julie talks about her feelings of being alone, she begins to cry and eventually sobs openly in the session. At this point, how would you proceed? What further questions would you ask to assess her suicidal risk? How might you approach working with Julie based on the limited information she has provided you?

Homicide

Perhaps even more tragic than taking one's own life is the act of taking the life of another, as in homicidal behavior. Dixon (1987) uses the phrase *homicidal behavior* to mean "potential for or threat to commit physical harm to another individual" (p. 158). Morales and Sheafor (1995) define homicide in a more explicit manner indicating that it is "death due to injuries purposefully inflicted by another person or persons, not including death caused by law enforcement officers or legal execution by federal or state government" (p. 434). They validate our view that as human service workers—whether social workers, counselors, psychologists, or case managers—we are apt to encounter some form of homicidal behavior in our work with others. "As in the case with suicide clientele, many [helpers] deal with people suffering the effects of social, psychological, economic, and political oppression and dehumanization, who eventually become homicide perpetrators or victims" (Morales & Sheafor, 1995, p. 434).

For the human service professional, this presents a set of moral, ethical, and legal ramifications, and also triggers undue distress. Faiver et al. (1995) state: "Homicidal ideation needs to be explored immediately and thoroughly, because there are legal as well as moral and ethical implications if your client actually intends to kill someone" (p. 102). In the case of homicide, it is important to begin with a discussion of the legal and ethical issues as these are germane to an overall understanding of homicidal behavior and relevant treatment issues. Thereafter we will elaborate on the predictive factors, risk assessment, and guidelines for appropriate strategies and interventions.

Ethical and Legal Implications of Homicidal Behavior   As with suicidal clients, ethical, legal, and moral principles apply to clients who threaten homicide. Similarly, confidentiality must be upheld with certain exceptions. Specifically, when and if clients inform human service professionals about their intention to harm others, confidentiality may be breached. "The professional is then essentially faced with making a decision about whether to inform the authorities, significant others, or a potential victim of such threats and taking action to ensure the client does not carry them out" (Guilliland & James, 1993, p. 118).

The underlying basis of this *duty to warn* or *duty to protect* originated from the landmark court decision of *Tarasoff v. Regents of University of California*, 1976. This now well-known Tarasoff case has been extensively discussed in the literature pertaining to ethics and legal issues in general, and to homicide, in particular. Basically, the case was brought against a therapist who was counseling a client, Prosenjit Podder, in the University of California, Berkeley, Counseling Center. During the course of treatment, Mr. Podder, a university graduate student, indicated to his therapist that he intended to kill his girlfriend, Tatiana Tarasoff. On learning that this client had purchased a gun, the therapist informed the campus police. When the police spoke to Mr. Podder, he assured them that he would stay away from Ms. Tarasoff, yet 2 months later, he followed through with his initial intent to harm and stabbed Ms. Tarasoff to death. Ms. Tarasoff's parents brought suit against the university and the therapist. After several appeals, the state Supreme Court of California found for the plaintiff. Negligence was determined because the therapist, his supervisor, and the university police all failed to warn the intended victim or her family of Podder's threats.

The court ruled that three conditions must exist that mandate this *duty to warn:*

1. A "special relationship" must exist between the therapist and the client, implying that in this therapeutic milieu, the therapist has a duty to control the conduct of the client.
2. There must be a clear determination of need to control the conduct of the other person: a clear threat. In this case, the client not only directly verbalized his intent to kill Tatiana Tarasoff, but he also had a gun with which to fulfill his threat.

3. There must be a foreseeable victim, even if not specifically named. In this case, Ms. Tarasoff was specifically named.

As Kopels and Kagle (1993, pp. 101–126) summarized the law, "When a doctor or psychotherapist, in the exercise of his professional skill and knowledge, determines, or should determine, that a warning is essential to avert danger arising from the medical or psychological condition of his patient, he incurs a legal obligation to give that warning."

The court ruling itself states the following:

> Once a therapist does in fact determine, or under applicable professional standards reasonably should have determined, that a patient poses a serious danger of violence to others, he bears a duty to protect the foreseeable victim of that danger...the discharge of this duty may require the therapist...to warn the intended victim or others...to notify police, or to take whatever steps are reasonably necessary. (*Tarasoff v. University of California*, 1976)

Although initially a California decision, Tarasoff became a source of controversy nationwide. Following this case, many Tarasoff-like cases have ensued, creating further landmark court decisions: the Bradley Case, the Jablonski Case, the Hedlund Case, all summarized in Corey, Corey, and Callanan (1993, p. 120).

According to the American Psychological Association (1985), practitioners have a responsibility to protect the public from violent acts inflicted by their clients. Liability for civil damages is high when there is negligence by practitioners. Such negligence is evidenced when (1) there is failure to diagnose or predict dangerousness, (2) there is failure to warn potential victims of violent behavior, (3) there is failure to commit dangerous individuals, and (4) there is a premature discharge of a dangerous client from a hospital.

Implications for human service professionals who are not therapists or clinicians are not as clearly spelled out and would appear to be less severe. Nonetheless, all workers should take the precautions listed and always involve a supervisory staff person and/or clinician when engaging with violent clients.

Predictive Factors   Accurate assessment and prediction of homicidal potential is imperative, yet according to the literature, there is no one efficient way to assess dangerousness or predict homicidality. "No study has indicated an effective way of accurately predicting homicide. Generally, the factors considered most worrisome [are] a history of violence, poor impulse control, or attempted homicide" (Shea, 1988, p. 437). Shea (p. 440) further denotes the following as contributing to the serious possibility of a homicidal situation:

- Psychosis
- Interpersonal conflict
- Need for money and other practical concerns
- Revenge
- Political concerns
- Organized crime
- Pathological murder for pleasure

In the literature, various aspects of a potentially homicidal client have been identified. These may be helpful in clarifying possible causes or reasons for homicidal ideation.

- The threat of violence may be an indication that the individual is struggling against losing control of himself or herself and is very frightened. Sometimes this fear may be hidden by a preoccupation with aggression or with boisterousness.
- When a client is unable to cope with a crisis, this can precipitate a homicidal situation.
- It is important to examine thoughts and feelings. Finding causes is extremely important for ego building, cognitive restoration, understanding the client's psychic structure, and determining the root problem.

- Often when the individual's ego needs, involving self-esteem and feelings of worth and value, are undermined, the individual may react defensively by wanting to hurt those who have hurt him or her. For instance, individuals may resort to violence when they have suffered infidelity, humiliation, or insult.
- Those who have experienced severe emotional deprivation and rejection in early childhood are more likely than others to resort to violence in a crisis situation.
- Those with a childhood history of enuresis, fire setting, and cruelty to animals are more likely to be homicidal risks.
- The more illogical or unjustified the reason for the homicide threat, the greater is the risk. Conversely, the more logical and justified the reason for the threat, the less is the risk.
- When a homicidal impulse is diffused, the risk is lessened. When the threat is directed at a specific person or object, especially if the client has made concrete plans to carry out the threat, the risk is greater.
- If a specific person is the target for the violence, the worker needs to explore the client's relationship with that person to understand better the underlying dynamics.
- The client's capacity for satisfying interpersonal relationships is an indicator of his or her ability to relate to others.
- The client's culture needs to be considered as well. If a client is a member of a culture in which violence is considered acceptable, there is a higher likelihood that the person will act out violently.

(Bellman & Blackman, 1966; Dixon, 1987; MacDonald, 1967)

Risk Assessment   In conducting an assessment to determine whether a client is at risk of becoming homicidal, the following questions should be asked of the client, a family member, or someone who is very familiar with the person's behavior.

- Is the person actively engaged in violent or dangerous behavior?
- Is the person placing others in peril?
- Does the person state intentions to carry out violent or dangerous behavior?
- How does the person intend to harm others? (Look for weapons or other evidence of a plan and the means and ability to carry out the plan.)
- Does the person have a background of violent or dangerous behavior?
- Is there a past history of psychiatric evaluations or psychiatric hospitalization? When and where?

Monahan (1993, p. 242) writes about limiting therapist exposure to Tarasoff liability. In his "Guidelines for Risk Containment," he recommends that four areas be incorporated in an appropriate risk assessment.

> The clinician (helper) must be educated about what information to gather regarding risk, must gather it, must use this information to estimate risk, and, if the clinician is not the ultimate decision maker, must communicate the information and estimate to those who are responsible for making clinical decisions.

These steps highlight the importance of being knowledgeable about risk, even if you will not be directly working with the homicidal individual. We briefly describe the four areas necessary for professional risk assessment.

Education: The clinician must have adequate education regarding the multifaceted issue of homicide. This includes clinical education and legal education.

Information: "Once a clinician knows what information, in general, may be relevant to assessing risk, he or she must take efforts to gather that information in a given case" (Monahan, 1993, p. 243). This includes gathering information from past records, current records, inquiries of the patient, and inquiries of significant others.

Estimation: Helpers must estimate a patient's risk of violent behavior to the best of their ability.

Communication: Workers must communicate relevant information accurately to other mental health professionals. According to Klein, "Information must be transferred between or among clinicians, and significant information must be made salient to the person responsible for making the ultimate decisions regarding the patient" (Monahan, 1993, p. 245).

Workers should document in writing any pertinent information about the case. Documentation is another way to relay important data to others about the situation; it is also necessary to limit potential liability for the worker. Monohan (1993, p. 245) reminds workers that "documenting information received and actions taken, or 'building the record,' is an essential exposure-limitation technique."

In reviewing the literature regarding Tarasoff-type liability (Kopels & Kagle, 1993; Costa & Altekruse, 1994; Mills, Sullivan, & Eth, 1987; Monahan, 1993), we have found that all the authors stress the conditions noted by the Tarasoff judges to be necessary when determining homicidal risk. These conditions include "special relationship," "reasonable prediction of violence," and "foreseeable victim."

Various authors have delineated recommendations or strategies for dealing with Tarasoff-like situations (Ogloff, 1995; Guilliland & James, 1993; Corey, Corey, & Callanan, 1993). We have adapted the following list from these and other authors who have offered their recommendations regarding this case.

Recommendations for Working with Tarasoff-like Cases

• Ascertain whether Tarasoff issues are applicable in your jurisdiction by inquiring about the specific legalities that apply where you practice.

• Provide clients with information regarding informed consent. Also, inform them of the limits of confidentiality. Wilson (1981) notes that a good practice is to convey very clearly what you can and cannot hold in confidence and tell the client this before you begin an intervention.

• On a continual basis and as the need arises, remind clients of the limits of confidentiality. Clients may not recall the specifics of confidentiality and therefore may require that you remind them, especially if they open a discussion on suicide, homicide, or abuse.

• Be alert to the emergence of Tarasoff-like situations. Wilson (1981) says that if clients state a concrete threat, you are bound morally, legally, and ethically to take action. Depending on the statutes of your particular state, you may have to notify the intended victim or person's family.

• Inform your clients of your responsibilities to protect them and the intended victim. Explore options with your clients about warning the victim and possibly having to hospitalize the clients for their protection and the safety of the potential victim. You must first determine when to report/protect, and then what to report and to whom (Guilliland & James, 1993).

• If the potential for harm is unclear, consult with your peers and/or supervisors. Thompson (1983) states, "Consultation is always advisable and is substantiating protection from legal and ethical problems that may arise, particularly when the threat is not clear" (p. 170).

• Take care of yourself! Don't be dominated by guilt or threats of legal reprisal by the client. Seek needed support and validation for your interventions.

Treatment Considerations

If you will be working with homicidal clients in some capacity, consider the following treatment issues. Note that there might be some overlap with recommendations presented earlier.

- Establish a constructive relationship in which clients are encouraged to engage in an open dialogue with you.
- Be respectful of clients' feelings and circumstances. You may not condone their proposed actions or aggressive and irrational behaviors, yet you can be empathetic and cognizant of the underlying dynamics that may exist.
- In your encounter with potentially violent clients, you will benefit by acting decisively and directly and by remaining in charge. (Trying to stay calm will also help!)
- Working with violent clients is often a volatile experience, so the worker must be adept at swift information gathering and rapid assessment. Hesitating to ask pertinent questions could lead to serious consequences.
- As with suicide, take all threats of homicide seriously and conduct a thorough risk assessment.
- Based on your evaluation, you may decide that protective custody or hospitalization is necessary for the client. Approach this issue with care and concern. (In California, clients can be hospitalized against their will if they are homicidal, suicidal, or gravely disabled, based on the 5150 statute that allows involuntary hospitalization when an individual is a threat to himself or others.)
- Assist the client in ventilating pent-up emotions as a way to reduce aggressive impulses.
- Help clients recognize the reasons for their desperation (cognitive grasp) and work toward increasing their cognitive functioning, especially perception, judgment, and reality testing.
- Consider the need for environmental manipulation to reduce potentially explosive situations (e.g., encourage a violent and potentially brutal spouse to stay with friends until his impulse control has improved and his rage is diminished).
- Protect yourself from violence that could be directed your way. By ensuring your safety, you are also helping your clients minimize their liability.

Dealing with homicidal ideation is almost always a stressful and crisis-oriented experience. Understanding the underlying dynamics as well as being aware of appropriate interventions can help you during this tense and anxious time.

SCENARIO

Juan is a 23-year-old Latino male who enters treatment with you after a court orders him to engage in mandated anger management sessions. On this first interview with Juan, you are screening his appropriateness for your upcoming anger management group. He tells you that he was arrested for battering his wife of 2 years who has now left him, taking their 3-year-old son and 6-month-old daughter. Juan becomes angry as he talks about her leaving and seems extremely upset that he does not know where she is. The last thing she told him was that she would not return to him if he didn't get help. He resents having to seek help and believes that he has done nothing wrong. As you inquire more about their relationship, he becomes increasingly agitated and states openly that when she returns, she better never order him around or "that'll be the end." When you inquire into this statement, he tells you bluntly that he would not hesitate to kill her now for leaving him if he only knew where she was. When he notices your visible concern, he quickly changes his tone and states that he is only joking and that he would never hurt her that bad. You are concerned anyway. How would you proceed? What homicidal risk factors are present? What other risk factors would you assess for?

Spousal Abuse

Spousal abuse is another form of domestic violence that is quite prevalent in our society. As with child and elder abuse, reported cases are estimated to be much lower than actual incidence. In fact, the FBI estimates that domestic violence is the most underreported crime in

America (Jacobs, 1989). According to Dieham and Ross (1998), one-third to one-half of all marriages involve physical abuse to the wife. Men are also abused by their wives sometimes, but this is not so common or so frequently reported as wife abuse. The focus here centers primarily on abused women.

Spousal Abuse Defined

Such behaviors as "pushing, slapping, punching, choking, sexual assault, or assault with weapons" are all considered forms of spousal abuse and domestic violence. In addition to physical abuse, verbal and emotional abuse are also forms of spousal abuse. "Contrary to popular opinion, battering is not a black and blue issue. Emotional, mental and psychological abuse typically inflict deeper and more lasting scars than skin-deep or bone breaking battering" (Jacobs, 1989, p. 130). Lenore Walker, a well-known authority in the field of domestic violence, agrees that psychological humiliation and verbal harassment seem more traumatic than the actual physical abuse or the threat of physical violence. Morales and Sheafor (1995) have observed that "most women are reluctant to report their abusers due to shame, guilt, and obstacles created by the legal, health, and social service systems" (p. 307).

As with other forms of abuse, spousal battery does not occur in only one racial, ethnic, or socioeconomic group. Some facts, cited below, will help dispel this and other misconceptions of spousal abuse.

Myths and Facts

Myth: A small percentage of the overall population are battered women.
Fact: Battery occurs in at least 28% of American families. (At least 1.8% million women are battered each year in the United States.)

Myth: Poorer women are more frequently abused than middle-class women.
Fact: There is not a marked difference between upper- or middle-class women and those of lower socioeconomic status. However, the visibility of lower class women is higher. Middle- and upper-class women may fear disclosure or social embarrassment and have ways to hide their abuse.

Myth: Minority women are battered more often than whites.
Fact: Battering crosses all ethnic and racial lines; no one group has been shown to suffer more than another.

Myth: Battered women are crazy.
Fact: Because of the women's survival behaviors, some are misdiagnosed as being insane, unstable, or crazy. In reality, these women are quite emotionally strong and courageous.

Myth: Women with little education are battered more frequently than better educated women.
Fact: Abused women range from those with little or no education to those with graduate and doctoral degrees. Educated or not, all women are susceptible.

Myth: Battered women deserve to get beaten: "She asked for it."
Fact: It is not the women's behavior that leads to the abuse; rather, it is the batterer's loss of control and poor coping skills that do so. In some cases, women may play a part in the conflict, but they are not responsible for the abuse itself.

Myth: Drinking causes the battering.
Fact: Although about half of battered women state that drinking was involved, batterers were prone to abuse regardless of whether they were drinking.

This list is only a partial one; there are many more misconceptions that surround the issue of battering. Hoff (1989), among many others, addressed the multifaceted aspect of spousal abuse:

> Violence occurs not merely as a stress response, but as a complex interplay between conditions of biologic reproduction and economic, political-legal, belief and knowledge systems of particular historical communities. These interacting systems seem to produce a context in which cultural values, the division of labor, power allocation, and beliefs operate to sustain a climate of oppression and conflict. In such a climate, violence against women flourishes, suggesting a link between personal trouble of individual battered women and the public issues of women's status. (1989, p. 267)

The Cycle of Violence

A now renowned text on domestic violence, *The Battered Women* by Lenore Walker (1979), describes the classic cycle of violence that clearly explains the "why" of most spousal abuse cases. According to Walker (1989),

> battered women are not constantly being abused, nor is their abuse inflicted at totally random times. Rather there is a definite battering cycle that entraps them, leads to learned "helplessness behavior," rationalizing and futile hope. The cycle has three distinct phases: the tension-building stage, or the "build-up"; the explosive or acute battering incident, or the "blow-up"; and the calm, loving response, or the "honeymoon." The length of each phase and cycle varies with couples and circumstances. The build-up phase tends to be the longest, while the blowup state rarely lasts longer than a day or two. (p. 37)

A thorough understanding of the three different stages in this cycle of violence is crucial if you expect to comprehend the many aspects of abusive relationships. In our society, it is often easier to criticize or blame the victims of such abuse than to try to understand the underlying dynamics. As human service workers, we, more than any others, need to be sensitive and understanding of the plight of these women. If we are not careful, our pressure to have them leave the abusive partner may further alienate some women, who will then feel further misunderstood and alone. Because of this, many women's shelters now provide intensive training on domestic violence and crisis intervention counseling to their staff. A typical emphasis of these types of workshops is to enhance the helper's understanding of the cycle of violence and the empowerment of the survivor (Jacobs, 1989).

Why Women Stay in Abusive Relationships

According to the research, there are numerous reasons for women/wives tending to stay with the abuser beyond the first complete cycle of abuse. Many have speculated about these reasons. Typically, women are thought to remain in the home if the violence is infrequent, if they were abused by their parents when they were children, or if they are financially dependent on their husbands. Other reasons posited are that certain women may stay in an abusive relationship perhaps to "conquer" their mate. Some may stay because unconsciously they wish to recreate their own parents' violent marital relationship in an effort to make it better. Also, some women who lack positive self-esteem may feel they are undeserving of anyone else who might treat them with respect and dignity; hence, they continue to choose abusive partners despite their history of being abused. According to Zastrow and Kirst-Ashman (1994), "Reasons include lack of self-confidence, traditional beliefs, guilt, economic dependence, fear of the abuser, fear of isolation, fear for their children, and love" (p. 379).

Predictive Factors

It would be untrue to state that there is an exact profile of both a batterer and his spouse, the battered woman. There are, however, various factors that when found together may increase a person's risk of becoming a perpetrator or a victim. Research results regarding these factors are shown in Table 7.9.

Many of these factors are similar for both the perpetrator and the victim. This may not be a coincidence; actually, it may say something about our society and how men and women are socialized and about the violent nature of our world. Sadly, "the United States has been called the most violent society in the Western world" (Hoff, 1989, p. 244). Further, "the historical neglect of victimization, especially in the domestic arena, underscores the value placed on family privacy and the myth of the family as a haven of love and security" (Hoff, 1989, p. 245).

Many authorities believe that spousal abuse is connected to a norm whereby violence is tolerated within the American family. In many families a cultural norm exists which legitimizes hitting, especially in respect to husbands and wives.

Legalities: Reporting Responsibilities

There has been much controversy in the field regarding reporting responsibilities. The most recent law in California, effective January 1996, states that a reporting requirement will be imposed "only when a health practitioner *provides medical services for a physical condition* to a patient who he or she reasonably suspects is suffering from any wound or other injury by means of a firearm, or the patient is seeking treatment for a wound or other physical injury inflicted on the person as a result of assaultive or abusive conduct." (Leslie, 1994, p. 25). Some states uphold similar laws whereas others have no actual legislation regarding this issue. Ethically, the issues are much more difficult and ones you will surely

TABLE 7.9   Predictive Risk Factors for Batterers and Victims

The Batterer-Perpetrator	The Battered Woman-Victim
1. History of family violence	1. History of family violence
2. Aggressive coping style—use of force or violence	2. Passive coping style
3. Abuse of alcohol or drugs	3. Use or abuse of alcohol or drugs, or co-dependency
4. Low self-esteem	4. Low self-esteem, feelings of inadequacy, which seem to mount as the cycle of violence is repeated
5. Strong traditional view of male/female roles	5. Belief in traditional family values—male/female roles
6. Excessive jealousy or potential for this	6. Feels deserving of abuse, may feel guilty for not doing "enough" or for believing she is responsible for battery
7. Use of or access to weapons (guns, knives, or other)	7. May feel trapped in predicament and have no hope
8. Authoritarian toward wife	8. Typically passive and/or submissive when with spouse; may be quite assertive or even aggressive with others
9. Presence of mood swings	9. Highs and lows often depending where she is in the cycle of violence
10. Poor impulse control; uncontrolled anger	10. May have difficulty expressing anger
11. Early or during early relationship, treats girlfriend/spouse roughly	11. May be "blind" to the early signs of abuse
12. Threatening demeanor at times	12. Feels weak and easily dominated
13. In denial; doesn't believe he has a problem	13. Does not believe anyone can or will want to help, especially if she reached out before only to return to the batterer

struggle with. There is a temptation to encourage a woman to get out of the battering situation without recognition of her particular situation. Is it ethical to encourage her to leave? This, too, is a difficult question. The following interventions may help you deal more appropriately with spousal abuse.

Treatment Considerations

Working with spousal abuse and domestic violence can be quite challenging. There are various levels of intervention that are most commonly used: the emergency response, crisis intervention, education, advocacy and referrals, and short- and long-term counseling. We have delineated these as follows:

• *Hot line counseling*—The goal is to listen and ensure the safety of those at risk from further physical, verbal, and psychological harm. Both the woman and her children are protected. Jacobs (1989) considers this "arming her with physical surviving tactics, referring her to a shelter, hotel, motel or private home where she can assess her situation" (p. 133).

• *Crisis intervention*—This is aimed at "empowering the survivor" whereby the woman is helped to deal with her crisis and begin to develop more adaptive coping. Children can also be severely and profoundly affected by the violence that permeates the home. They too should receive some crisis treatment and much needed support and reassurance.

• *Advocacy services*—This involves information and referral to shelters, tips on police protection, assistance in obtaining legal aid, and referral to therapists who specialize in abuse counseling.

• *Educational Services*—Women are taught about the cycle of violence and the role they play. They are helped to pinpoint the precipitators to abuse in their situation and to assess their contribution (e.g., being co-dependent). Their self-esteem is enhanced as is their repertoire of coping skills. Assertiveness training is crucial in helping them begin to stand up for themselves.

• *Short- or long-term counseling, individual or group*—This is often an important component to successful intervention. Women are encouraged to take a closer look at the abuse relationship they have been in and to make plans for their lives. Emphasis is on helping the survivors break through the isolation and work on rebuilding their shattered self-esteem. At times, psychotropic medication may be required to assist the woman with extreme anxiety or depression. Appropriate referrals and coordination with psychiatrists and others in the medical profession is at times essential. Remaining supportive, nonjudgmental, and concerned is necessary to foster trust and hope.

SCENARIO

Farrah is a 30-year-old, Middle Eastern, married female who has four young children ranging from 2 to 10 in age. She comes to the local mental health center where you work because she fears her husband who has become increasingly abusive since their arrival in the United States several years ago. Although he has always been abusive, physically, emotionally, and verbally, his abuse has intensified. He will not allow her out of the home and becomes enraged when she talks to anyone outside their family. He resents her recent relationship with a neighbor and does not want her even to leave home to take their school-age children to or from school. She is beginning to feel smothered and feels little or no support from his family, who agree that her place is in the home. She has few relatives here and those she has talked with also tell her to obey him, for he is the head of the family. The only friend she feels she has is her neighbor, whom she has to sneak out to see. She has told the neighbor of her predicament and was urged to seek help when the neighbor learned she had been beaten severely by her husband when she failed to have dinner ready on his arrival home from work one evening. Farrah enters the session visibly afraid and is very hesitant to say very much until she learns about the confidentiality of your session with her. Based on the information you have, how would you proceed in working with Farrah? What are some of the special ethical issues? How about the cultural factors that are at play?

AIDS

As with emotional and spousal abuse, the issues presented by the AIDS crisis are both numerous and complex.

> *AIDS.* The very word makes us apprehensive, uneasy, even fearful. A new disease, mysterious, insidious, uncontrolled, and as yet, incurable, has appeared. *AIDS* is a most terrifying disease—like something conjured up for a science fiction novel—because it attacks the body's very ability to defend itself against other diseases. Consequently, people with AIDS can become infected with exotic, bizarre, sometimes deadly diseases—diseases that are not a threat at all to people with normal immune systems, but can be deadly to someone with AIDS. (Martelli et al., 1987, p. xiii).

This excerpt illustrates the magnitude of emotion and tragedy of this epidemic disease. Human service workers are called on once again to help those with special needs—in this case, those inflicted with AIDS and their families. Those suffering from AIDS are considered one of three "target populations" that will be a focus in the 1990s. The others include welfare recipients and the homeless (Schmolling et al., 1993). As helpers, we will no doubt be at the forefront of this challenging work and will therefore need to be well educated about this disease.

AIDS Defined

AIDS, correctly known as *acquired immune deficiency syndrome*, is a "contagious, presently incurable disease that destroys the body's immune system. AIDS is caused by the human immunodeficiency virus (HIV), which can be transmitted from one person to another primarily during sexual contact or through sharing of intravenous needles and syringes" (Zastrow, 1990, p. 481).

The seriousness of this disease is evident. Victims are left feeling helpless when the HIV virus attacks their immune systems, leaving them incapacitated and unable to fight off other diseases. They are in an extremely vulnerable position physically, one that also has major emotional and psychological ramifications. Once a person is infected with HIV, there are several potential outcomes:

- Some will remain well and show no signs of the virus, yet will be *HIV carriers*, capable of transmitting the virus to others, primarily through unprotected sex and use of infected needles or syringes.
- Others will develop *AIDS-related complex (ARC)*, which is less severe than the symptoms of AIDS. ARC symptoms include appetite loss, weight loss, night sweats, fever, diarrhea, skin rashes, swollen lymph nodes, tiredness, and low resistance to infection. Some authorities predict that nearly all HIV carriers or those with ARC will eventually develop AIDS (Zastrow, 1990).
- Those who develop *AIDS* will not be able to fight off opportunistic diseases such as pneumocystis carinii pneumonia and Kaposi's sarcoma (a rare form of cancer) (Martelli et al., 1987; Zastrow, 1990). With a heightened ability to acquire infections and other illnesses coupled with an inability to fight off these infections, those with AIDS eventually die. It is reported that most of those who die of AIDS are under 40 years of age (Schmolling et al., 1993). Early death is currently inevitable as no cure has yet been found.

The term *epidemic* is synonymous with such words as *extensive, increasing, prevalent, rampant,* and *widespread,* which all exemplify the nature of AIDS in the United States and in the world today. Although varying statistics have been reported, they all reflect a dramatic rise in the death rate secondary to the AIDS disease. Neukrug (1994) states that in 1993, approximately 1 million people in the United States were infected with HIV. Each year there is an increase in both the number of reported HIV cases and those who actually died of AIDS. Women are the fastest growing segment of the population with AIDS. Further, women of childbearing age (ages 13–39) are increasing in number. Just think of how these few statistics could severely impact our population and endanger so many. Others who appear to have a higher rate of reported HIV or AIDS are gay or bisexual men (57%); 30% are also minorities, with African-Americans significantly overrepresented at 42.5% (Morales & Sheafor, 1995).

How Is Aids Transmitted?

Contrary to many misconceptions and fears surrounding this deadly disease, body fluids (such as fresh blood, semen, urine, and vaginal secretions) infected with the virus (HIV) must enter the bloodstream for transmission to occur from one person to another. Naturally, unprotected sex increases one's chances if any of the body fluids identified above were to enter the bloodstream during the sexual act. Anal intercourse, needle sharing, and blood transfusions are all prime ways for the HIV virus to be transmitted. The risk factors are listed below.

High Risk Factors

• *Anal intercourse.* This form of intercourse is considered very high risk as it can tear the lining of the rectum, which allows infected semen to get into the bloodstream.

• *Multiple sex partners without safe sex practices.* Logically, the risk of infection increases according to the number of sexual partners, male or female. Further, it is believed that engaging in sex with prostitutes (male or female) may increase risk of transmission. Research has shown that "prostitutes are at high risk because they have multiple sex partners and are more apt to be intravenous drug users" (Zastrow, 1990, p. 487).

• *Blood transfusions.* In some cases, although a comparatively low percentage, people have acquired the HIV virus during a blood transfusion needed for medical reasons such as surgery. In these cases, the blood that was used was contaminated. Increased precautions and better testing have minimized this route of infection.

• *Re-usage of contaminated needles and syringes.* Sharing of intravenous needles with an individual who is carrying the virus can be quite dangerous because of the transmission of blood. "A small amount of the previous user's blood is often drawn into the needle and then injected directly into the bloodstream of the next user" (Zastrow, 1990, p. 485).

Many fear that HIV or AIDS can be transmitted through other body fluids. However,

although the virus has been found in small amounts of body fluids, such as saliva, there is no evidence that the virus has been transmitted by these fluids. Further, there is no evidence that AIDS can be transmitted by any casual contact, when normal sanitary procedures are observed. Only theoretically can someone get AIDS from kissing. There is no documented case. [In sum] contact of blood to blood, or semen to blood, is the way it is spread. (Martinelli et al., 1987, p. 7)

In an attempt to reduce the spread of AIDS, various controversial practices have been adopted in some cities. These include (1) exchanging dirty needles for clean ones for drug users and (2) distributing condoms to high school students (Schmolling et al., 1993). As you can imagine, the controversy lies in the concern of many that such measures will encourage drug use and sexual promiscuity.

The complexity of AIDS, its poor prognosis, and the ease with which it can be transmitted are all factors that heighten anxiety levels and encourage stigmatization and alienation of the infected individuals.

Living with AIDS or HIV

Almost anyone would feel extremely overwhelmed on being diagnosed with the HIV virus, or even worse, with having AIDS. With the high probability of being susceptible to a host of other illnesses, of progressively becoming quite ill, and of having an increased chance of dying from this illness, it is no wonder that those infected and their significant others feel bewildered. Unfortunately, people with HIV or AIDS often have even more to contend with—stigma, prejudice, discrimination, alienation, rejection, and abandonment.

Although being diagnosed with AIDS is similar to suffering from other terminal illnesses, such as cancer, the reactions of others are markedly different. As noted by Haney, these reactions are usually negative. "The AIDS epidemic, for example, has fanned existing negative attitudes toward homosexuals, strengthening attempts to bar gay men and lesbian women from teaching jobs and to repeal local government ordinances assuring their free access to housing, employment, and public services" (Popple & Leighninger, 1990, p. 622).

The predicament of people suffering from the AIDS virus is therefore compounded by societal views and the "mysterious and epidemic" nature of this disease. Specifically, there are several distinctive factors confronting those with the AIDS virus:

- A *Stigma* attaches to individuals suffering from this disease.
- *Prejudice* is often held against those who are infected.
- *Fear of infection* given the contagious aspects of AIDS is quite prevalent.
- *Revelations* are often necessary when one is diagnosed with AIDS, such as openly acknowledging that one is gay.
- *Alienation and abandonment* are not only common fears but actual possibilities for those who suffer from AIDS.
- *Use of such defense mechanisms as denial and avoidance* is common in family members and significant others of those who have been diagnosed with AIDS.
- *Increased susceptibility to anxiety and depressive disorders and progressive dementias* further plague the AIDS patient who must deal with these additional psychological and organic impairments.
- *Lack of support* from a variety of sources—family, friends, employers, communities, insurance companies, governmental agencies, and the medical profession—further exacerbate the predicament of those afflicted with AIDS.

For people with AIDS and their significant others, "a complex web of psychological issues, mainly related to some of our deepest fears, comes into play" (Martelli et al., 1987, p. 29). These include the fear of mortality, fear of contagion, fear of similar lifestyles, and fear of helplessness.

Certainly you can empathize with the many added stressors of those with AIDS. As human services professionals, we will likely be the ones most instrumental in helping this segment of the population.

Ethical and Legal Concerns

Legal Implications   The area of AIDS, as with many of the other issues we have discussed (e.g., some forms of abuse, suicide, homicide, domestic violence), has no clear legal obligations for the human service professional. There are two related areas, however, that involve ethical dilemmas that could have legal aspects. These include (1) breaching of confidentiality and (2) duty to warn. Helpers should be concerned when an HIV-infected client is involved in unsafe sexual practices with others. Is it our duty to warn the intended partner as indicated by Tarasoff law, and if so would this be permissible under the laws of confidentiality? These are clearly perplexing questions and ones bound to stir up many thoughts and feelings in the helper. Legally, however, there is yet no clear decision to guide helpers caught in this dilemma. Therefore, as there are currently no legal decisions regarding breaching confidentiality or duty to warn in AIDS cases, "it may be necessary to wait for some future landmark court decisions to get clearer directions on the nature of the therapist's [human service professional's] responsibility in protecting sexual partners of HIV-positive clients" (Corey, Corey, & Callanan, 1993, p. 298).

Ethical Concerns   As a human service worker, you will work with some aspect of the AIDS crisis, be it directly with those diagnosed or indirectly by working with their family, friends, and significant others. Corey, Corey, and Callanan (1993) and Neukrug (1994) emphasize the value of being well educated and knowledgeable about AIDS in order to be ethically competent to work with this population.

In addition to this educational component, human service workers should take a look at their own value systems, underlying attitudes, biases, and misperceptions. They should also be candid with themselves about possible fears and concerns connected with working directly with an AIDS client. By remaining open to your own issues, you can minimize the risk that they will negatively spill over into your work with clients. Having any one of these feelings or issues does not make you an uncaring person, unfit to be a helper; they just make you human. Being aware of your behavior and its potential impact on your clients is what is most crucial.

Throughout your career in human services, you will find certain areas you dislike, feel uncomfortable with, or just aren't interested in. Working with AIDS, as with other terminal illnesses, is quite challenging, not to mention emotionally difficult, and you may decide this is not a specific area you wish to concentrate in. By contrast, you may feel especially touched by those who suffer from this disease and feel a strong commitment to help them in some capacity. In either case, continued self-inquiry into your own feelings and views is critical.

Treatment Considerations

Increasing emphasis on the role of the human service professional is being noted in both the literature and community-based agencies designed specifically for those with AIDS or for significant others. Based on his research, Neukrug (1993, p. 268) notes that

> in the 1990s, the human service worker will be an important component in the education and prevention of this disease, the deliverer of counseling services for those who are already infected, and the supervisors of those many volunteers who are giving of their time and humanity to assist in the caretaking of those infected, their families, and their friends.

• Clearly, there are many functions that we as helpers will perform. Educator, counselor, supervisor, advocate, caretaker, and case manager are all possible roles we may play.
• Buckingham and Van Gorp (1988) cite various areas of treatment intervention that have been useful in working with the AIDS population:

1. Problem solving with everyday concerns and difficulties
2. Estate planning
3. Decreasing the level of hypochondriacal preoccupation
4. Guidance in designing adequate structure and limits for activities of daily living
5. Assisting with family conflict (Buckingham & Van Gorp, 1988, pp. 112–115)

To be effective in any one of these roles or interventions,

• Helpers need to be sensitive, caring, and knowledgeable of the AIDS disease.
• They need to be aware of the ethical dilemmas and legal issues involved.
• Being willing to seek consultation, supervision, and support remains a viable option for anyone who encounters moral conflicts, ethical dilemmas, or just an emotionally difficult time in working with AIDS clients.

SCENARIO

Brandon is a 28-year-old single, gay male who enters a men's group you are co-leading with another male therapist. He is at first very reluctant to talk about his homosexuality in the group for fear of the negative reaction he anticipates if he "comes out." After several months in the group, however, he feels safe enough to open up and share his sexual orientation with the group. Most are supportive of him, yet a few have difficulty accepting him as gay. Many of these issues are worked through in the group and you feel that things are going fairly well. In the last few months, you have grown worried about Brandon's frequent absences and illnesses. When he does attend the group, he seems preoccupied and physically more frail. After many failed attempts to encourage Brandon to talk about any unresolved issues within the group, you ask that he come to see you on an individual basis. During this individual session, you share your concerns and ask whether there is anything he has been struggling with. He bursts into tears and tells you that he has been diagnosed with the HIV virus and has AIDS. You had no idea and are caught totally off guard. Brandon tells you that his shame and fear of sharing his HIV status with the group have kept him from discussing it openly. He further tells you that he has even contemplated not returning to the group. How would you proceed? Would you encourage Brandon to share his condition in the group? If so, how? If not, why not? What treatment interventions would you pursue?

ADDITIONAL ETHICAL AND LEGAL ISSUES

Dual Relationships

In helping clients who are experiencing many of the problems/issues we have discussed, the relationship between helper and client is the most important factor in providing effective treatment. However, *dual relationships* often arise when human service workers blend their professional and personal relationship with their clients. In fact, engaging in any other type of relationship with a client with whom you are professionally involved can be considered a dual relationship (Corey, Corey, & Callanan, 1993).

Another perspective views dual relationships as a "particular type of boundary breakdown that offers the client an invitation to act out negative transference." Ethical standards warn helpers about the possibility of such a "boundary breakdown," indicating the need for helping professionals to be cognizant of their potentially influential position with clients. Moreover, they must avoid exploiting the trust and dependency of their clients (California Association of Marriage and Family Therapists [CAMFT], Sec. 1.2). "Dual (or multiple) relationships may be sexual or nonsexual and may occur when counselors (helpers) simultaneously or sequentially assume two (or more) roles with a client" (Herlihy & Corey, 1993, p. 4).

All these perspectives on dual relationships help to explain why it is such an important and complex ethical issue that affects all human service professionals, regardless of their work setting or client population. Because of its broad applicability, however, it is an area of strong controversy. The notion that workers are to avoid dual relationships with their clients may seem quite logical and appropriate. For instance, without exception, helpers must avoid sexual relationships with their clients. This position is certainly justifiable as such behavior is clearly unethical. However, in other areas, dual relationships may seem appropriate or be unavoidable. What, then, is acceptable and what is not?

The ethical codes of the American Counseling Association suggest that dual relationships be avoided when possible. The code states:

> Counselors are aware of their influential positions with respect to clients, and they avoid exploiting the trust and dependency of clients. Counselors make every effort to avoid dual relationships

with clients that could impair professional judgement or increase the risk of harm to clients. (American Counseling Association [ACA] Ethical Committee 1995 A.6a)

The *Ethical Standards and Code of Ethics* of both human service providers and social workers offer similar stipulations: "Human service workers should avoid any dual relationship that might negatively impact the helping relationship with the client. Sexual relationships with clients are unethical and prohibited" ([NOHSE], 1994, p. 1); and "the social worker should avoid relationships or commitments that conflict with the interests of the clients. The social worker should under no circumstance engage in sexual activities with clients" (National Association of Social Workers [NASW], 1983, p. 2).

In all three of these professional ethical codes/standards there is a concern that the helper could potentially harm a client while engaging in a dual-type relationship, especially when there is sexual contact. Specifically, "exploitation of a sexual nature, including sexual intercourse, sexual contact or sexual intimacy with clients or a client's spouse or partner is prohibited" (Leslie, 1993, p. 6).

Other types of dual relationships that are also to be avoided when possible "include, but are not limited to, familial, social financial, business, or close personal relationships with clients" (ACA, 1995, p. 2). Although not as severe as with sexual intimacy, the inherent danger in these types of relationships is the impaired objectivity and judgment of the professional that would likely result from mixing the two types of relationships.

Unlike the clear boundaries of *no sexual contact*, in some of these situations, boundaries are often difficult to distinguish and are therefore easily crossed. Leslie supports this claim by indicating the many "gray areas" that need careful attention (Leslie, 1992, p. 6). Herlihy and Corey (1993) contend that "developing a sense of ethical self-awareness implies that practitioners are able to sort out the issues involved in appropriately dealing with these gray areas" (p. 5). Below are several of the types of relationships that may present difficulties.

Types of Dual Relationships with Unclear Boundaries

- Engaging in Social Relationships: Client and helper engage in some type of social relationship as in both attending the same clubs, social activities, or religious functions.
- Entering Business Relationships: Client and helper barter for services or engage in some joint business venture.
- Combining Roles of Teacher and Therapist: Educator (such as teacher, professor) agrees to engage in a therapeutic relationship with current student. Sometimes this is not even openly agreed to, yet some form of therapeutic counseling is being provided.
- Combining Roles of Supervisor and Therapist: A supervisor crosses the line and becomes a therapist to his student/employee, or an actual therapeutic relationship develops from the current supervisory relationship.
- Providing Therapy to a Relative or a Friend's Relative: A mental health worker agrees to counsel the daughter of a colleague and friend.

How can these innocent types of duality be considered potentially as dangerous as the more flagrant abuse of sexual contact? In reviewing the literature on dual relationships, Herlihy and Corey (1993) noted a great range of perspectives. More conservative practitioners like Pope (1985), and Pope and Vasquez (1991) advocate stricter ethical codes and enforcement, based on these beliefs:

- Such relationships tend to impair the counselor's judgment.
- The potential exists for conflicts of interest.
- There is the danger of exploiting the client because the counselor holds a more powerful position than the client.
- Boundaries usually become blurred and tend to distort the professional nature of the therapeutic relationship.

Others, including Herlihy and Corey (1993), Hedges (1993), and Tomm (1993), feel that some ethical standards are too rigid and inhibiting and that dual relationships cannot always be avoided. For instance, Herlihy and Corey (1993) state: "Despite some clear clinical and legal risks, dual relationships are not always unethical and unprofessional. Along with associated risks there are also potential benefits." (p. 4). Hedges argues that duality is an essential aspect of all human relationships in which favorable development occurs (Hedges, 1993). Tomm further contends that "maintaining a professional distance" in fact creates a situation where the "focus is on the power differential" that promotes a vertical hierarchy in human relationships (pp. 1–4).

Confronting Problems in Dual Relationships

When helper-client boundaries are crossed and/or some form exploitation occurs, whether intentional or not, problems are sure to arise. Dual relationship problems have been identified by the Ethics Committee of the American Association for Marriage and Family Therapy (AAMFT) in its report on dual relationships. These are outlined as follows:

- Sexual intimacy with clients
- Power differential
- Exploitations
- Role confusions
- Transference and countertransference issues
- Diverse goal relationships
- Triangulation for the client
- Displacement into both relationships
- Financial complications
- Risk of civil suits
- Loss of objectivity

These problems illustrate how serious engaging in dual relationships can be. Clearly, this is an area of ethics that is complex, controversial, and confusing. As stressed before, engaging in dual relationships is often unavoidable for the human service worker who plays many roles that can often overlap. For this reason, we would like to end this discussion by delineating suggested treatment considerations to minimize the risks involved in dual relationships.

Treatment Considerations

- Set healthy boundaries from the outset.
- Discuss with clients any potentially problematic relationship.
- Seek consultation and/or supervision.
- Make a referral if necessary.
  (Adapted from St. Germaine 1993)

In addition, Herlihy and Corey (1993) find these actions important:

- Fully inform clients about any potential risks.
- Disclose and clarify areas of concern.
- Consult with other professions periodically if you are engaged in dual relationships.
- Document discussions about any dual relationships.

Perhaps the most valuable feedback we found was offered by Corey, Corey, and Callanan (1993). They encourage helpers to make honest appraisals of their behavior and its effect on clients. They underscore the need for personal and professional maturity in resolving the ethical dilemmas posed by dual relationships. In sum, you must be sure not to allow your own financial, social, or emotional needs to take precedence in any relationship with a client.

SCENARIO

Leslie is a middle-age, married, Caucasian school teacher who enters treatment with you because she feels overwhelmed by work stress, mounting marital problems, and an increasing need to care for her aging parents with whom she has never had a very good relationship. The more you work with Leslie, the more you enjoy your time together. She is bright, articulate, insightful, and very responsive to the treatment interventions you have used. Also, as you get to know her better, you realize that in many ways you and she are very similar. Furthermore, you learn that she lives very close to you and often visits the same shopping establishments and restaurants you do. She has suggested on several occasions that you get together sometime for lunch or coffee—invitations that you have lightly brushed off. Recently, however, you find yourself truly wanting to accept her offer, for you are finding that your time together is satisfying some of your unmet needs as well. As you look inward to understand your feelings better, you see how similar your current issues are. Leslie's struggles with a father who has Alzheimer's and a mother with cancer are quite similar to your recent struggles with your mother who just had a craniotomy because of a brain tumor. When Leslie talks about her overwhelmed state, you relate to her feelings and find that your therapy sessions have become very therapeutic for you as well. You have not found anyone else whom you can identify with like you do with Leslie, and yet you realize that your relationship must remain professional. How do you proceed? What if Leslie pushes further for a friendship? What if you find yourself no longer being able to treat her objectively?

SCENARIO

You have seen a young male client for depression throughout your internship. He has learned much from you and is quite grateful for your help. He invites you to a party he is having at his home as a way of demonstrating his gratitude and appreciation. How would you proceed?

Ethnic-Sensitive Practice

A theme throughout this text has been the importance of being diversity sensitive. Again we stress that to be effective as human services practitioners, workers have an ethical as well as a practical duty to be sensitive, knowledgeable, and skilled in working with culturally diverse population groups. Corey, Corey, and Callanan share similar sentiments: "Because cultural diversity is a fact of life in today's 'global village,' counselors [human service professionals] can no longer afford to ignore the issue of culture" (1993, p. 240). Sue and Sue (1990) have identified three categories that are essential in working with different ethnic populations. These are listed here.

1. Being Aware of Your Own Assumptions, Values, and Biases

- Knowing your own cultural heritage and valuing and respecting differences
- Being aware of your own values and biases and how they might impact minority clients
- Being comfortable with differences regarding race and beliefs between you and your clients
- Having sensitivity for the need to refer a minority client to a helper of the client's own race/culture or to another helper, when appropriate

- Acknowledging your own racist attitudes, beliefs, and feelings when these are present

2. *Understanding the Worldview of the Culturally Different Client*

- Being knowledgeable and informed about the particular group you will be working with
- Possessing a sound understanding of how the sociopolitical system of the United States operates with respect to its treatment of minorities
- Seeking knowledge and understanding of the generic characteristics of counseling and therapy [helping process]
- Learning of the institutional barriers that prevent minorities from seeking assistance

3. *Developing Appropriate Intervention Strategies and Techniques*

- Using a multitude of response modalities
- Communicating skillfully based on a recognition that communication patterns differ from one culture to the next
- Exercising institutional intervention on behalf of your clients when appropriate
- Knowing your helping style, your limitations, and the likely impact of these on your clients

Remaining True to Ethical and Legal Issues as a Practitioner

Having reviewed both common ethical and legal issues as well as those involving special areas of concern, we hope you are now well armed with a beginning knowledge base regarding the fundamentals. We encourage ongoing consultation and continued training and education so you can stay abreast of recent changes, and new laws. Of utmost importance, remain clear and committed to the ethical codes and guidelines of your specific profession.

Ethical Dilemmas

Despite your awareness of the many ethical codes that provide guidelines and principles, you will likely encounter many situations the codes do not specifically address. These difficult situations will present you with ethical dilemmas that are sometimes difficult to resolve. Often it is translating a code's principles into practical directions for conduct that provides helpers with the greatest challenge.

Faiver et al. (1995) recommend utilizing various resources in your attempt to resolve your ethical dilemmas. These include seeking guidance from supervisors and colleagues (peers), the specific ethics committee of your profession, state licensure or certification boards, and professional journals and texts that directly address ethical issues.

The Role of Professional Training

The last area is one we have also stressed throughout—the need for ongoing learning; whether you are a student intern, a beginning professional, or a seasoned practitioner, continuing to learn is both essential and ethical. Helping clients with a limited knowledge base, inadequate skills, and an outdated approach is simply unacceptable. Both your clients and you, the practitioner, deserve better. To minimize ineffective treatment outcomes and the emergence of competency issues, professionals must take their role in ongoing learning seriously.

CONCLUDING EXERCISES

1. Workers may experience some blocks to reporting when it involves issues that elicit very deep feelings such as child abuse or elder abuse. Desensitization in regard to these issues may help you face and cope with the feelings that arise. What feelings do you imagine you would have on encountering an abused child? What if the child told you about being sexually abused? Physically abused? Would the feelings be different or similar in these situations? Discuss the major feelings you believe you would have.

    How might you become more familiar with areas in which you are apt to struggle emotionally or ethically? _____

    _____

    _____

    _____

    _____

2. Ethical decision making involves the worker's ability to use professional relationships for consultation and help with decision making. Identify the individuals within your current professional position that you would turn to for consultation if you needed help: _____

    _____

    _____

    _____

    _____

    Both your immediate supervisor and/or fieldwork instructor have ethical responsibilities to you as a student. They would need to be included in your plan if you are in that type of setting. Are they approachable or amenable to assisting you with any concerns you might have? If not, how might you handle this situation?

3. What do you believe would be the most challenging ethical situation for you to handle? Why would this type of situation be challenging for you? What feelings would this situation set off in you? Without merely thinking about referring this situation to another professional, how might you attempt to work with the individual if you could? Consider presenting this situation as a role play in front of the class with you playing either the client or the therapist. This could be a learning opportunity for all the class and could stimulate discussion on the particular ethical issue presented._____

    _____

    _____

    _____

    _____

Epilogue

Practice within human services can be a very rewarding and challenging career. It can also be overwhelming and emotionally draining. In order for helpers to be effective, it is important for them to be open to constant learning. Continued self-awareness is crucial, as are knowledge development and the cultivation of clinical skills. Without any one of these, successful practice is likely to be compromised. We hope that this text has assisted you in beginning the process of incorporating these elements into practice. By looking at yourself, you will become aware of any unresolved feelings or issues you may have that may interfere with the therapeutic relationship and become more adept at facing the many challenges common in practice. Learning about the many issues and trends relevant to working with clients is key to better understanding them. The ability to apply this knowledge in practice is essential to the actual helping process. Furthermore, information on appropriate treatment considerations is necessary in facilitating this process.

Appendix A

Process Recording Form
Outline #1
Matrix Format

Student's Name _____

Client's Name _____

Date of Interview _____ Session # _____

I. INTERVIEW CONTENT

Using the format presented below, briefly provide a verbatim account of a small portion of your interview with the client at the beginning, middle and ending of your session.

Content of Interview	Client's Feelings	Client's Thoughts	Student's Feelings	Student's Thoughts	Supervisor's Comments
Beginning of Interview					
Middle of Interview					

Content of Interview	Client's Feelings	Client's Thoughts	Student's Feelings	Student's Thoughts	Supervisor's Comments
End of Interview					

II. ASSESSMENT OF INTERVIEW

How would you assess this interview? Did it go well or were there any problems you encountered? Please describe. _____

_____

_____

_____

III. EVALUATION

Evaluate your interactions with the client and the interventions you used. _____

_____

_____

_____

IV. IMPRESSIONS

Consolidate your impressions by providing a summary of the case. _____

_____

_____

_____

V. FUTURE PLANS

What are your plans for further client contact/interactions? Describe a brief treatment plan here. _____

_____

_____

_____

Appendix B

Process Recording Form
Outline #2
Analytical Type

Student's Name _____

Client's Name _____

Date of Interview _____ Session # _____

I. OBJECTIVES

Describe objective of interview. What is the reason for this client contact?

_____

_____

_____

_____

_____

_____

II. NARRATIVE ACCOUNT OF INTERVIEW

Interview Content	Impression of Client's Feelings	Descriptions of Student's Feelings	Comments of Supervisor

III. OVERALL ASSESSMENT AND IMPRESSIONS OF CASE
(Include Mental Status information here):

_____

_____

_____

_____

_____

IV. CENTRAL ISSUES AND MAJOR THEMES
Discuss what you perceive to be the central issue(s) and/or major theme(s) that emerged during the session.

_____

_____

_____

_____

_____

V. SIGNIFICANT INTERVENTIONS MADE:
Describe what you perceive to be your most significant intervention with the client. How does this intervention fit in with the overall plan for change that you have negotiated with the client?

_____

_____

_____

_____

_____

VI. STUDENT FEELINGS/PROFESSIONAL USE OF SELF
Describe the major feelings that you were aware of while working with this client. How might you use those feelings to help this client to reach the goals that he/she has set for him/herself? (Include body language, use of space, voice, helper's own feelings and values and how those hindered or helped the process, and an account of how the helper is dealing with his or her feelings, etc.)

_____

_____

_____

_____

_____

VII. FURTHER PLANS

Will treatment continue? What are short and long term goals of treatment? Describe specific goals for next session as well as for overall case.

_____

_____

_____

_____

_____

VIII. SUPERVISORY DISCUSSION OF PERTINENT QUESTIONS, CONCERNS, OR ISSUES

Describe one significant question or area of concern that you now face with this client. Is there any specific feedback that you want to receive from your supervisor on this issue?

_____

_____

_____

_____

_____

Appendix C

Process Recording Form
Outline #3
Abbreviated Format

Student's Name _____

Client's Name _____

Date of Interview _____ Session # _____

I. PURPOSE OF INTERVIEW

What are objectives and goals of the interview?

_____

_____

_____

_____

II. DESCRIPTION OF ANY NEW DEVELOPMENTS

Provide any new case information that may have a bearing on your assessment and interventions as well as your evaluation.

_____

_____

_____

_____

III. SIGNIFICANT EVENTS OR CLIENT/HELPER INTERVENTIONS

Provide any specific interview questions and/or responses that were significant during the interview. Were any techniques or interventions used that were pivotal? If so, please describe.

_____

_____

_____

_____

_____

IV. IMPRESSIONS OF CURRENT CASE

What are your impressions of the case?

_____

_____

_____

_____

V. OVERALL EVALUATION OF INTERVENTIONS

How would you evaluate the interventions you used? Which were you most satisfied with and why? Which were you least content with and why?

_____

_____

_____

_____

VI. PLANS FOR FURTHER TREATMENT OR FOLLOW-UP

Describe specific plans for your next interview and for overall treatment of this client.

_____

_____

_____

_____

Appendix D
Process Recording Form
Outline #4
Groups

Student's Name _____

Group Name _____

Date of Group Session _____ Session # _____

Place of Group _____ Time _____

I. OVERALL PURPOSE OF GROUP

What is the general purpose of the group? Why is the group meeting?

_____

_____

_____

_____

_____

_____

_____

II. DIAGRAM OF SEATING ARRANGEMENTS

III. SPECIFIC PURPOSE OF GROUP SESSION

Describe the purpose of this specific group meeting and any planned activities.

_____

_____

_____

_____

_____

_____

IV. GROUP LEADERS

Who are the group leaders and what are their positions within the group?

_____

_____

_____

_____

_____

_____

V. GROUP MEMBERS

Who are the designated group members? List them by name and some brief identifying data as well as their primary problem.

_____

_____

_____

_____

_____

_____

VI. GROUP PARTICIPATION

Who was present at the group? If any members were absent, do you know the reason?

_____

_____

_____

_____

_____

_____

_____

VII. GROUP CONTENT AND DYNAMICS

How did the group start? Were there any significant events, behaviors, or changes during the group process? Briefly describe the group content. How did the group end?

_____

_____

_____

_____

_____

_____

_____

VIII. GENERAL THEMES

Were there any themes or patterns that kept emerging within the group? Please describe and offer any insight as to your thoughts on why this might have occurred.

_____

_____

_____

_____

_____

_____

_____

IX. FEELINGS WITHIN GROUP

What were the most prominent feelings expressed or noted within the group? Describe both the positive and negative responses to the group process and note the emotional tone of the group (i.e. cooperative or conflictual or argumentative).

_____

_____

_____

_____

_____

X. GROUP INTERVENTIONS

What interventions did the group leader(s) use during the group? What was the intent and were they planned or spontaneous?

_____

_____

_____

_____

_____

_____

XI. SUMMARY OF INDIVIDUAL PARTICIPATION

Briefly discuss how each member participated within the group. Note any contributions or changes and members' individual roles, needs, defenses, and communication patterns.

_____

_____

_____

_____

_____

_____

XII. OVERALL GROUP ANALYSIS

How would you assess the group process? Were the initial group objectives met? Were there any significant changes in the group that alters the process?

_____

_____

_____

_____

_____

_____

XIII. GROUP CONTINUITY

Was there continuity with previous sessions? If not, what might be some reasons for this? How might continuity be reestablished?

_____

_____

_____

_____

_____

_____

XIV. FUTURE PLANS

What are future plans for the group as a whole as well as for individual group members?

_____

_____

_____

_____

_____

_____

Appendix E
Process Recording Form
Outline #5
Meetings

Student's Name _____

Agency _____ Field Instructor _____

Title of Meeting _____ Date _____

Place of Meeting _____ Time _____

I. PURPOSE OF MEETING

Describe the specific purpose of this meeting (attach agenda if possible). Is this a regularly scheduled meeting or a one-time event?

_____

_____

_____

_____

_____

_____

II. EXPECTATIONS AND GOALS OF MEETING

What, if any, expectations of this meeting did you have? What are some initial goals you had in mind that might be met at this meeting? What role did you expect to play during the meeting?

_____

_____

_____

_____

_____

_____

III. CONTENT OF MEETING

What was the primary content of the meeting (i.e., what was discussed? What actions took place?, etc.). Who was the chairperson and what role did he or she play? Were any hidden agendas revealed or did unexpected events take place?

_____

_____

_____

_____

_____

_____

IV. OUTCOME AND RESULTS

What was the overall outcome of the meeting? What were some specific results? Were any future plans set? How did the meeting close?

_____

_____

_____

_____

_____

_____

V. ANALYSIS

How would you describe the group process that took place during this meeting and the overall accomplishment of goals and objectives that were initially set? If possible, specify the interaction patterns that took place during the meeting among the various participants.

_____

_____

_____

_____

_____

_____

VI. EVALUATION

How would you evaluate the leadership roles? Were they appropriate? Why or why not? Would you suggest any changes? How do you believe you performed? Do you feel the role you played met your expectations? Why or why not? Were the original goals met and was the meeting productive?

_____

_____

_____

_____

_____

_____

A p p e n d i x F
Initial Intake

Patient name _____ PF# _____

Date of intake _____ Clinician _____

Presenting problem _____

_____

_____

_____

_____

_____

_____

Presenting symptoms _____

_____

_____

_____

_____

Sleep: Normal _____ Disturbed _____ Initial _____ Mid _____ Term _____ Comments _____

Appetite: Normal _____ Disturbed _____ Binging _____ Restricting _____ Comments _____

Suicide/Homicide Risk: Suicidal risk   Low _____ Moderate _____ High _____

Plan _____ Means _____ Intent _____

Homicidal risk   Low _____ Moderate _____ High _____

Plan _____ Means _____ Intent _____

Comments _____

_____

Psychiatric History: Personal mental health history _____

_____

_____

Family history of psychiatric illness _____

_____

_____

Social/marital history _____

_____

_____

_____

_____

_____

_____

_____

Family History _____

_____

_____

_____

_____

_____

_____

_____

Sexual/Physical Abuse History _____

_____

_____

_____

Educational History _____

_____

_____

_____

Employment History _____

_____

_____

_____

Substance Abuse History   Alcohol _____

Illicit drugs _____

Caffeine _____

Nicotine _____

Comments: _____

_____

_____

_____

Physical Health History   Ht. _____ Wt. _____ Increase _____ Decrease _____

Current medical problems: _____

_____

_____

Significant past illnesses/hospitalizations _____

_____

Current medications _____

_____

Past medications _____

_____

Mental Status

Appearance and Behavior   Age _____

As Stated _____   Older _____   Younger _____

Race: Caucasian _____   Black _____   Hispanic _____   Oriental _____   Other _____

Sex: Female _____   Male _____

Orientation: Heterosexual _____   Homosexual _____   Bisexual _____

Eye Contact: Appropriate _____   Avoidant _____   Other _____

Gait: Normal _____   Abnormal (describe) _____

Mannerisms: Appropriate _____   Inappropriate (describe) _____

Grooming: Good _____   Average _____   Poor _____

Manner of Relating to Interviewer: Cooperative _____   Belligerent _____   Warm _____

Apathetic _____   Trusting _____   Distrusting _____   Other _____

Affect and Mood   Affect: (Indicate degrees of severity—Mild, Moderate, Severe)

Anxious _____   Suspicious _____   Worried _____

Depressed _____   Apathetic _____   Hostile _____

Apprehensive _____   Euphoric _____   Happy _____

Angry _____   Other _____

Range of Affect: Broad _____   Restricted _____   Blunted _____   Flat _____   Labile _____

Other _____

Mood: Dysphoric _____   Normal Range _____   Euphoric _____   Irritable _____

Speech   English _____   Spanish _____   Other _____

Effectiveness: Coherent _____   Incoherent _____   Organized _____   Unorganized _____

Quantity: _____   Appropriate _____   Verbose _____   Mute _____

Style: Passive _____   Aggressive _____   Assertive _____   Other _____

Quality and Form: Appropriate _____   Stutters _____   Slurred _____   Pressured _____

Obsessive _____   Concrete _____   Sarcastic _____   Circumstantial _____   Confabulating _____

Neologistic _____   Grandiose _____   Other _____

Thought Content and Process   Associations: Appropriate _____   Loose _____

Blocked _____   Flight of Ideas _____   Ideas of Reference _____   Delusions _____

Hallucinations _____   Illusions _____   Obsessions _____   Compulsions _____

Phobias _____ Other (describe) _____ _____

_____

Level of Insight: High _____   Average _____   Low _____

Judgement: Good _____   Average _____   Poor _____

Ability to Abstract: Good _____   Average _____   Poor _____

Sensorium   Alert: Yes _____   No _____

Oriented: Time _____   Place _____   Person _____   Situation _____

Comment _____

Memory: Normal Ability _____   Deficient Ability _____   Recent _____   Remote _____

Comment _____

Estimate of Intelligence: High _____   Average _____   Low _____

Other _____

Client's Stated Goals: _____

_____

_____

_____

DIAGNOSTIC IMPRESSION:

Axis I: Clinical disorders and other conditions that may be a focus of clinical attention

_____

_____

_____

Axis II: Personality disorders and mental retardation

_____

_____

Axis III: General medical conditions

_____

_____

Axis IV: Psychosocial and environmental problems

Problems with primary support group _____

Problems related to the social environment _____

Educational problems _____

Occupational problems _____

Housing problems _____

Economic problems _____

Problems with access to health care services _____

Problems related to interaction with the legal system/crime _____

_____

Other psychosocial and environmental problems _____

_____

Axis V: Global assessment of functioning

Rating: Current _____   Highest level in past years _____

Initial Treatment Plan and Recommendations:

Additional history and information gathering _____   Psychosocial testing _____

Medication consultation _____   Individual psychotherapy _____

Marital/relational therapy _____   Group therapy _____   Family therapy _____

Other recommendations _____

_____

_____

_____

_____

_____

Signature and degree of clinician _____

Date of intake _____

Appendix G
Initial Evaluation (long version)

1. **Identifying Data**

Name _____

Address _____

_____

(Zip Code)

Phone no. _____

Primary language _____

Age _____ Birthdate _____ Sex _____

Birthplace _____ Race _____

Marital Status: S M W D Sep

Religion _____

Date _____

Occupation _____

Father's name _____

Mother's maiden name _____

Length of Residence _____

(County) (State)

Social Security No. _____

Next of kin _____

Spouse's name _____

Referred by _____

PERSON TO CONTACT IN CASE OF EMERGENCY:

_____

(Name)   (Address)   (Phone)   (Relationship)

OTHER AGENCIES INVOLVED WITH YOU:

_____ Probation/Court   San Bernardino County Hospital _____

Why? _____   Welfare Department _____

Others _____ (include name of person and agency)

Previous therapist _____ (include name of therapist, name of agency, and dates seen)

_____

Previous therapist/therapy terminated because _____

Name and phone number of probation officer and/or welfare worker _____

2. **Presenting Problem**   (Patient's statement of problem and immediate need for tx) _____

   _____

   _____

   _____

3. **Current Problem**   (Therapist's behavioral support for diagnosis)

   _____

   _____

   _____

   History of Problem   (Descriptions and manifestations of symptoms, significant events that led to problem, familial reactions, significant others)

   _____

   _____

   _____

4. **Social History**

   A. *Developmental History*

   Pregnancy/Delivery: _____

   Early problems: _____

   Walked _____   Talked _____   Toileted _____   School _____

   Early traumas: _____

   _____

   _____

   B. *Family and Peer Relationships*

   Sibs and ages: _____

   Raised by: _____

   Adequacy of Environment: _____

   Childhood problems: _____

   _____

   _____

Peers—Type: _____

Number: _____

Intensity of relationships: _____

Lg group _____ Sm group _____ 1:1 _____ Withdrawal/Isolation _____

Amount of Time Spent: _____

Types of activities: _____

_____

_____

C. *Arrests (criminal/legal) and military history*

Criminal/Legal—placements, incarcerations: _____

Arrests, illegal activities: _____

Attitudes, then and residual: _____

Probation, attitude: _____

_____

_____

Military History: _____

_____

D. *Living Arrangements*

Description of residence (house/apt), neighborhood, length of residence, who lives there (age, sex, why?):

_____

_____

Recent changes, anticipated changes, socioeconomic: _____

_____

Privacy in home, individual's living space: _____

_____

Client's attitude toward living arrangements: _____

_____

_____

E. *Marital History and Children*

Current status: _____

Age at marriage: _____

Children (ages): _____

_____

Divorce (reasons, adjustment): _____

_____

Relationship w/children: _____

_____

Religion: _____

_____

F.   *Employment*

Occupation—Type of work: _____

Length: _____

Attitude re: _____

Work history—Types of jobs, length of stay, problems: _____

_____

G.   *Education*

Current/Highest grade _____   High School _____   GED _____

College: _____

Current school: _____

School placement/performance: _____

_____

H.   *Source of Finances*

Current household wage earner(s): _____

_____

Attitude toward job and income level: _____

_____

I.   *Strengths and Support Systems*

Strengths—Self-description (strengths and weaknesses): _____

_____

—Therapist's view (assets and liabilities): _____

_____

Support Systems

Family, friends, alone? _____

Other interested agencies/parties: _____

DPSS, Probation, Court: _____

M.D.: _____

Private therapist, attorneys, guardian: _____

5. **Psychiatric History**

A. *Family History of Psychiatric Illness*   (Mental or emotional problems of family members? Current impaired relationships? What impact has it had on you/family?):_____

_____

B. *Previous Personal Mental Health History*

Psychotherapy/Counseling (who, where, why, patient's age, helpful/not helpful?): _____

_____

Psychotropic Medications (who, when, why, what prescribed, helpful/not helpful? Reactions—Negative or positive? Current attitudes?): _____

_____

Hospitalizations (who, when, why, where, reactions, residual attitude?): _____

_____

Psychological Testing (school, MH agency, when, where?): _____

_____

C. *Alcohol and Drug History*

Alcohol use: _____

_____

Drug use: _____

_____

D. *Suicidal and Homicidal History*

Suicidal—Previous attempts: _____

Details/events: _____

Mode, lethality: _____

Motivation, rescue: _____

Current ideas, thoughts, plan: _____

Homicidal—Threats, violence towards others: _____

6. **Medical History/Physical** (Overall health)

A. *Hospitalization and Major Illnesses* _____

_____

_____

B. *Past Problems*

Childhood Illnesses/Accidents: _____

_____

History of: Heart   B/P   Seizures   Diabetes   Allergies   Headaches   VD

Thyroid   Head Injury   Unconsciousness   Other _____

C. *Current Health Problems* _____

_____

_____

D. *Current Medications* _____

_____

E. *Allergies* _____

_____

F. *Name, Address and Phone No. of M.D.* _____

_____

7. **Mental Status**

A. *General Appearance*

Dress: _____

Hygiene and Grooming: _____

General Health: _____

Body type, height, weight, etc.: _____

State of consciousness (alert, sleepy, lethargic): _____

Manner (cooperative, uncooperative, friendly, suspicious, guarded, open, submissive, dominant):

_____

Barriers to communication—Speech: _____   Eye contact: _____

B. *Motor Behavior*

Gait (abnormal movements, coordination, balance, speed): _____

_____

Posture: _____

_____

Rate and rhythm of activity (hypoactive, hyperactive, excited, agitated): _____

_____

Catatonic features (grimacing, mute, posturing, etc.): _____

_____

C. *Mood and Affect*

Range (constricted, variable, labile): _____

Intensity (increased, normal, decreased): _____

Mood (sad, happy, anxious, angry, apathetic): _____

Appropriateness of Mood: Appropriate _____   Not appropriate _____

Relatedness: _____

Positive, Negative: _____

D. *Speech Pattern and Abnormalities*

Rate, quality (tangential, organized, pressured): _____

Articulation (stuttering, accent, foreign language): _____

Clear, logical: _____

Stream of talk interrupted by: crying, silences, outbursts, etc. _____

E. *Psychotic Features*

Deterioration from previous level of functioning: _____

_____

Content of thought

Delusions (multiple, fragmented, bizarre): _____

Persecutory (of reference, thought broadcasting, thought insertion, through withdrawal):

_____

Delusions (of being controlled, somatic, grandiose, religious, nihilistic): _____

_____

Preoccupation: _____

Markedly illogical thinking: _____

Form of Thought

Loosening of associations: _____

Poverty of content of speech: _____

Perception—Hallucinations (auditory, visual, tactile, olfactory, gustatory, somatic): _____

_____

Affect—Blunting (restricted), flattening, appropriate, inappropriate: _____

_____

Sense of Self

Related to external world: _____

Psychomotor behavior (little spontaneity, catatonic stupor, rigidity, excitement, etc.): _____

_____

Associated features (perplexed, disheveled, eccentrically groomed, magical thinking, ritualistic/
stereotyped behavior, dysphoric mood, depersonalization, derealization, ideas of reference, illusions,
disorganized, catatonic, paranoid, undifferentiated, residual): _____

_____

_____

F. *Neurologic Indicators*

Cognitive Impairment: _____

_____

Clouding of consciousness, perceptual disturbance, incoherent speech, deterioration, memory impairment, rapid and fluctuating clinical symptoms: _____

_____

Intellectual functioning disturbed, memory impairment, impaired judgment, impaired impulse control, personality change, gradual and stable symptomatology: _____

_____

Etiology: No reference   Etiology unknown

G. *Cognitive Functioning*

Fund of knowledge: _____

_____

Sensorium: (clear, confused): _____

_____

Orientation (time, place, person, date): _____

_____

Abstraction (proverbs, interpretation, concrete, literal, understanding meaning, sees relationship to a personal situation, bizarre): _____

_____

Memory—Remote (age, birthdate, personal hx): _____

      Recent (recent events, last day of two): _____

Intelligence (estimate average, above, or below): _____

_____

Attention/concentration (serial-3's, serial-7's, simple arithmetic: _____

_____

H. *Suicidal/Homicidal Potential*

Suicide (thoughts, ideation, plans): _____

_____

Homicide (threats, thoughts): _____

_____

I. *Insight* (to present situation, dynamics, awareness of self and others, etc.): _____

_____

J. *Judgment* (past to present comparison, home, school, family, friends, society, future plans, quality of judgment and plans): _____

_____

K. *Motivation* _____

_____

8. **Diagnostic Impression**

Axis I: _____

_____

Axis II: _____

_____

Axis III: _____

_____

Axis IV: _____

_____

Axis V: Current GAF: _____

Highest GAF, past year: _____

9. **Treatment Plan** (within 3 visits or 30 days)

A.   *Recommendation* (if intake worker only, not therapist): _____

_____

_____

B.   *Recommendation* (if intake worker will be therapist):

1.   Short-term goals: _____

_____

_____

2.   Long-term goals: _____

_____

_____

3.   Expected discharge date: _____

_____

_____

4.   Prognosis: _____

_____

_____

10. **Treatment Plan Conference**

Persons present: _____

_____

Plan discussed: _____

_____

Patient/guardian response: _____

_____

CHILD/ADOLESCENT INTAKE: SOCIAL HISTORY

A. *Developmental Data:*

Was the child planned?   Yes _____   No _____

Pregnancy: Easy _____   Difficult _____   Special Problems _____

Length of Pregnancy _____ months

Delivery: Easy _____   Difficult _____   Cesarean section _____   Complications _____

_____

Early Problems With: Eating _____   Sleeping _____   Rashes _____   Breathing _____

The problems were: _____

Walked: _____ months Well _____ Clumsily _____ Other problems with walking _____

Talked: _____ months Clearly _____ Lisped _____ Other problems with talking _____

_____

Toilet-Training: _____ months Easy _____ Difficult _____ Residual problems _____

Went to School: Easy _____   Difficult _____   Mother accompanied: _____ Yes _____ No

B. *Traumatic Events* (such as divorce, death in family, moving, etc. List age of child at time of event): _____

_____

_____

Losses   (Out of home placements; deaths, friends, relatives, pets; change of residence/school; divorce/separations; school problems; abnormal development; body image; sickness/accidents; arrests; grades repeated/advanced: _____

_____

_____

C. *Identification/Dreams*

Three wishes   Why?

1.

2.

3.

Animal most like to be: Why?

Animal least like to be: Why?

Appendix H
Initial Evaluation (Abbreviated Version)

1. **Identifying Data:**

Patient's name _____

Clinician _____ Date _____

2. **Presenting Problem** (Patient's statement of problem) _____

_____

3. **Current Problem** (History of problem) _____

_____

_____

_____

4. **Social History**

   A. *Developmental history* _____

   _____

   B. *Family history* _____

   _____

   C. *Peer relationships* _____

   _____

   D. *Arrests (criminal/legal) & military history* _____

   _____

   E. *Living arrangements* _____

   _____

F.   *Marital history and children* _____

_____

G.   *Employment* _____

_____

H.   *Education* _____

_____

I.   *Sources of finances* _____

_____

5.   **Psychiatric History**

A.   *Family history of psychiatric illness* _____

_____

B.   *Previous personal mental health history* _____

_____

C.   *Alcohol and drug history* _____

_____

D.   *Suicidal/homicidal history* _____

Suicidal: _____

Homicidal: _____

6.   **Medical History/Physical**

A.   *Hospitalizations and major illnesses* _____

_____

B.   *Past problems* _____

_____

C.   *Current problems* _____

_____

D.   *Current medications* _____

_____

E.   *Allergies* _____

_____

7. **Mental Status**

   A. *General appearance* _____

   _____

   B. *Motor behavior* _____

   _____

   C. *Mood and affect* _____

   _____

   D. *Speech pattern and abnormalities* _____

   _____

   E. *Psychotic features* _____

   _____

   F. *Neurologic indicators* _____

   _____

   G. *Cognitive functioning* _____

   _____

   H. *Suicidal/homicidal potential* _____

   _____

   I. *Insight* _____

   _____

   J. *Judgment* _____

   _____

8. **Diagnostic Impression (DSM III)**

   AXIS I: _____

   _____

   AXIS II: _____

   _____

   AXIS III: _____

   _____

   AXIS IV: _____

   _____

AXIS V: _____

_____

9. **Treatment Plan**

_____

_____

_____

_____

_____

_____

_____

_____

_____

Appendix I
Cultural Family Assessment

In a detailed format, Tseng and Hsu (1991) consider many of the same dimensions found in commonplace assessments in what they call the "Scope of Comprehensive Family Assessment." These greatly significant dimensions include culturally relevant material that enhances the overall assessment process and encourages ethnic sensitivity and multicultural perspective. Tseng and Hsu consider the following major categories (which have been paraphrased) as essential to the assessment process.

I. *Basic information* regarding generation factors and composition; background information and medical/psychiatric/legal history of each family member.

II. *Presenting problems:* Identification of current problems that primarily concern the family's description of any precipitating factors or circumstances and the development of problems; the understanding of the nature of the problem by the family; the family's past attempts to solve the problem; and the family's goals and expectations for treatment.

III. *Ethnic/cultural background:* Marriage form, descent system indicating the rule which defines how descent is affiliated with sets of kin; rule of marital choice; method of mate selection; post-marital residence in terms of proximity to family of origin; household composition; family authority; primary axis or prominent dyad on which the loyalty, affection, or bond exists, and family values/beliefs/attitudes that cover an array of topics.

IV. *Family developmental history* from a longitudinal and chronological review beginning with the establishment of the marriage until the present. This includes typical milestones (marital formation, childbearing, childrearing, family maturing, family contracting) and any disruptions, changes or complications in the course of development (separation, divorce, single-parent family, remarriage).

V. *Current family functioning* noted in the current daily life of the family such as financial situation, social network, family socialization, family task allocation, and family functioning (i.e. decision making, communication, and expression of affections).

VI. *Family interactional function (family mental status)* relates to the family's interactional pattern as observed by the clinician. It may contain the following: boundaries, nature of these boundaries, alignment, cohesion, power/control, communication, subsystem relationships, affection, role division, and the organizational structure of the family.

VII. *Family stress/problems* refers to life events that may cause significant psychological distress in a family. These problems include: situations for which the family has had little or no prior preparation, any sharp or decisive changes for which old patterns of adjustment are not adequate, any emotional tension, anxiety, or dysphoria related to intra family matters, and daily life patterns that bring burdens or deprive the family of the pleasures of life.

VIII. *Problem solving pattern:* Types of coping that can be utilized to solve family problems, dysfunctional coping patterns, and adaptability.

IX. *Family strength* can be defined as those qualities that contribute to a successful marriage and family relationships. It also refers to resources and variations that a family is able to use to prevent and resolve stress and crisis situations in the family system.

X. *Family satisfaction* is measured by the level of satisfaction family members have toward their marriages and family life.

XI. *Family diagnosis and dynamic formulation of problems* is a comprehensive and integrated manner of describing the nature of the family problems and summarizing the way in which the problems are dynamically developed by the effects of contributing factors.

XII. *Suggested intervention plan* describes the plan and strategies needed to carry out intervention for the family.

Appendix J

ACA Code of Ethics and Standards of Practice

CODE OF ETHICS

Preamble

The American Counseling Association is an educational, scientific, and professional organization whose members are dedicated to the enhancement of human development throughout the life-span. Association members recognize diversity in our society and embrace a cross-cultural approach in support of the worth, dignity, potential, and uniqueness of each individual.

The specification of a code of ethics enables the association to clarify to current and future members, and to those served by members, the nature of the ethical responsibilities held in common by its members. As the code of ethics of the association, this document establishes principles that define the ethical behavior of association members. All members of the American Counseling Association are required to adhere to the Code of Ethics and the Standards of Practice. The Code of Ethics will serve as the basis for processing ethical complaints initiated against members of the association.

Section A: The Counseling Relationship

A.1. CLIENT WELFARE

    a. *Primary Responsibility.*
        The primary responsibility of counselors is to respect the dignity and to promote the welfare of clients.

    b. *Positive Growth and Development.*
        Counselors encourage client growth and development in ways that foster the clients' interest and welfare; counselors avoid fostering dependent counseling relationships.

    c. *Counseling Plans.*
        Counselors and their clients work jointly in devising integrated, individual counseling plans that offer reasonable promise of success and are consistent with abilities and circumstances of clients. Counselors and clients regularly review counseling plans to ensure their continued viability and effectiveness, respecting clients' freedom of choice. (See A.3.b.)

    d. *Family Involvement.*
        Counselors recognize that families are usually important in clients' lives and strive to enlist family understanding and involvement as a positive resource, when appropriate.

    e. *Career and Employment Needs.*
        Counselors work with their clients in considering employment in jobs and circumstances that are consistent with the clients' overall abilities, vocational limitations, physical restrictions, general temperament, interest and aptitude patterns, social skills, education, general qualifications, and other relevant characteristics and needs. Counselors neither place nor participate in placing clients in positions that will result in damaging the interest and the welfare of clients, employers, or the public.

A.2. RESPECTING DIVERSITY

    a. *Nondiscrimination.*
        Counselors do not condone or engage in discrimination based on age, color, culture, disability, ethnic group, gender, race, religion, sexual orientation, marital status, or socioeconomic status. (See C.5.a., C.5.b., and D.1.i.)

    b. *Respecting Differences.*
        Counselors will actively attempt to understand the diverse cultural backgrounds of the clients with whom they work. This includes, but is not limited to, learning how the counselor's own cultural/ethnic/racial identity impacts her or his values and beliefs about the counseling process. (See E.8. and F.2.i.)

A.3. CLIENT RIGHTS

    a. *Disclosure to Clients.*
        When counseling is initiated, and throughout the counseling process as necessary, counselors inform clients of the purposes, goals, techniques, procedures, limitations, potential risks, and benefits of services to be performed, and other pertinent information. Counselors take steps to ensure that clients understand the implications of diagnosis, the intended use of tests and reports, fees, and billing arrangements.

413

Clients have the right to expect confidentiality and to be provided with an explanation of its limitations, including supervision and/or treatment team professionals; to obtain clear information about their case records; to participate in the ongoing counseling plans; and to refuse any recommended services and be advised of the consequences of such refusal. (See E.5.a. and G.2.)

b. *Freedom of Choice.*

Counselors offer clients the freedom to choose whether to enter into counseling relationship and to determine which professional(s) will provide counseling. Restrictions that limit choices of clients are fully explained. (See A.1.c.)

c. *Inability to Give Consent.*

When counseling minors or persons unable to give voluntary informed consent, counselors act in these clients' best interests. (See B.3.)

A.4. CLIENTS SERVED BY OTHERS

If a client is receiving services from another mental health professional, counselors, with client consent, inform the professional persons already involved and develop clear agreements to avoid confusion and conflict for the client. (See C.6.c.)

A.5. PERSONAL NEEDS AND VALUES

a. *Personal Needs.*

In the counseling relationship, counselors are aware of the intimacy and responsibilities inherent in the counseling relationship, maintain respect for clients, and avoid actions that seek to meet their personal needs at the expense of clients.

b. *Personal Values.*

Counselors are aware of their own values, attitudes, beliefs, and behaviors and how these apply in a diverse society, and avoid imposing their values on clients. (See C.5.a.)

A.6. DUAL RELATIONSHIPS.

a. *Avoid When Possible.*

Counselors are aware of their influential positions with respect to clients, and they avoid exploiting the trust and dependency of clients. Counselors make every effort to avoid dual relationships with clients that could impair professional judgment or increase the risk of harm to clients. (Examples of such relationships include, but are not limited to, familial, social, financial, business, or close personal relationships with clients.) When a dual relationship cannot be avoided, counselors take appropriate professional precautions such as informed consent, consultation, supervision, and documentation to ensure that judgment is not impaired and no exploitation occurs. (See F.1.b.)

b. *Superior/Subordinate Relationships.*

Counselors do not accept as clients superiors or subordinates with whom they have administrative, supervisory, or evaluative relationships

A.7. SEXUAL INTIMACIES WITH CLIENTS

a. *Current Clients.*

Counselors do not have any type of sexual intimacies with clients and do not counsel persons with whom they have had a sexual relationship.

b. *Former Clients.*

Counselors do not engage in sexual intimacies with former clients within a minimum of 2 years after terminating the counseling relationship. Counselors who engage in such relationship after 2 years following termination have the responsibility to examine and document thoroughly that such relations did not have an exploitative nature, based on factors such as duration of counseling, amount of time since counseling, termination circumstances, client's personal history and mental status, adverse impact on the client, and actions by the counselor suggesting a plan to initiate a sexual relationship with the client after termination.

A.8. MULTIPLE CLIENTS

When counselors agree to provide counseling services to two or more persons who have a relationship (such as husband and wife, or parents and children), counselors clarify at the outset which person or persons are clients and the nature of the relationships they will have with each involved person. If it becomes apparent that counselors may be called upon to perform potentially conflicting roles, they clarify, adjust, or withdraw from roles appropriately. (See B.2. and B.4.d.)

A.9. GROUP WORK

a. *Screening.*

Counselors screen prospective group counseling/therapy participants. To the extent possible, counselors select members whose needs and goals are compatible with goals of the group, who will not impede the group process, and whose well-being will not be jeopardized by the group experience.

b. *Protecting Clients.*

In a group setting, counselors take reasonable precautions to protect clients from physical or psychological trauma.

A.10. FEES AND BARTERING (SEE D.3.A. AND D.3.B.)

a. *Advance Understanding.*

Counselors clearly explain to clients, prior to entering the counseling relationship, all financial arrangements related to professional services including the use of collection agencies or legal measures for nonpayment. (A.11.c.)

b. *Establishing Fees.*

In establishing fees for professional counseling services, counselors consider the financial status of clients and locality. In the event that the established fee structure is inappropriate for a client, assistance is provided in attempting to find comparable services of acceptable cost. (See A.10.d., D.3.a., and D.3.b.)

c. *Bartering Discouraged.*

Counselors ordinarily refrain from accepting goods or services from clients in return for counseling services because such arrangements create inherent potential for conflicts, exploitation, and distortion of the professional relationship. Counselors may participate in bartering only if the relationship is not exploitative, if the client requests it, if a clear written contract is established, and if such arrangements are an accepted practice among professionals in the community. (See A.6.a.)

d. *Pro Bono Service.*

Counselors contribute to society by devoting a portion of their professional activity to services for which there is little or no financial return (pro bono)

A.11. TERMINATION AND REFERRAL

a. *Abandonment Prohibited.*

Counselors do not abandon or neglect clients in counseling. Counselors assist in making appropriate arrangements for the continuation of treatment, when necessary, during interruptions such as vacations, and following termination.

b. *Inability to Assist Clients.*

If counselors determine an inability to be of professional assistance to clients, they avoid entering or immediately terminate a counseling relationship. Counselors are knowledgeable about referral resources and suggest appropriate alternatives. If clients decline the suggested referral, counselors should discontinue the relationship.

c. *Appropriate Termination.*

Counselors terminate a counseling relationship, securing client agreement when possible, when it is reasonably clear that the client is no longer benefiting, when services are no longer required, when counseling no longer serves the client's needs or interests, when clients do not pay fees charged, or when agency or institution limits do not allow provision of further counseling services. (See A.10. b. and C.2.g.)

A.12. COMPUTER TECHNOLOGY

a. *Use of Computers.*

When computer applications are used in counseling services, counselors ensure that (1) the client is intellectually, emotionally, and physically capable of using the computer application; (2) the computer application is appropriate for the needs of the client; (3) the client understands the purpose and operation of the computer applications; and (4) a follow-up of client use of a computer application is provided to correct possible misconceptions, discover inappropriate use, and assess subsequent needs.

b. *Explanation of Limitations.*

Counselors ensure that clients are provided information as a part of the counseling relationship that adequately explains the limitations of computer technology.

c. *Access to Computer Applications.*

Counselors provide for equal access to computer applications in counseling services. (See A.2.a.)

Section B: Confidentiality

B.1. RIGHT TO PRIVACY

a. *Respect for Privacy.*

Counselors respect their clients right to privacy and avoid illegal and unwarranted disclosures of confidential information. (See A.3.a. and B.6.a.)

b. *Client Waiver.*

The right to privacy may be waived by the client or his or her legally recognized representative.

c. *Exceptions.*

The general requirement that counselors keep information confidential does not apply when disclosure is required to prevent clear and imminent danger to the client or others or when legal requirements demand that confidential information be revealed. Counselors consult with other professionals when in doubt as to the validity of an exception.

d. *Contagious, Fatal Diseases.*

A counselor who receives information confirming that a client has a disease commonly known to be both communicable and fatal is justified in disclosing information to an identifiable third party, who by his or her relationship with the client is at a high risk of contracting the disease. Prior to making a disclosure the counselor should ascertain that the client has not already informed the third party about his or her disease and that the client is not intending to inform the third party in the immediate future. (See B.1.c and B.1.f.)

e. *Court-Ordered Disclosure.*

When court ordered to release confidential information without a client's permission, counselors request to the court that the disclosure not be required due to potential harm to the client or counseling relationship. (See B.1.c.)

f. *Minimal Disclosure.*

When circumstances require the disclosure of confidential information, only essential information is revealed. To the extent possible, clients are informed before confidential information is disclosed.

g. *Explanation of Limitations.*

When counseling is initiated and throughout the counseling process as necessary, counselors inform clients of the limitations of confidentiality and identify foreseeable situations in which confidentiality must be breached. (See G.2.a.)

h. *Subordinates.*

Counselors make every effort to ensure that privacy and confidentiality of clients are maintained by subordinates including employees, supervisees, clerical assistants, and volunteers. (See B.1.a.)

i. *Treatment Teams.*

If client treatment will involve a continued review by a treatment team, the client will be informed of the team's existence and composition.

B.2. GROUPS AND FAMILIES

a. *Group Work.*

In group work, counselors clearly define confidentiality and the parameters for the specific group being entered, explain its importance, and discuss the difficulties related to confidentiality involved in group work. The fact that confidentiality cannot be guaranteed is clearly communicated to group members.

b. *Family Counseling.*

In family counseling, information about one family member cannot be disclosed to another member without permission. Counselors protect the privacy rights of each family member. (See A.8., B.3., and B.4.d.)

B.3. MINOR OR INCOMPETENT CLIENTS

When counseling clients who are minors or individuals who are unable to give voluntary, informed consent, parents or guardians may be included in the counseling process as appropriate. Counselors act in the best interests of clients and take measures to safeguard confidentiality. (See A.3.c.)

B.4. RECORDS

a. *Requirement of Records.*

Counselors maintain records necessary for rendering professional services to their clients and as required by laws, regulations, or agency or institution procedures.

b. *Confidentiality of Records.*

Counselors are responsible for securing the safety and confidentiality of any counseling records they create, maintain, transfer, or destroy whether the records are written, taped, computerized, or stored in any other medium. (See B.1.a.)

c. *Permission to Record or Observe.*

Counselors obtain permission from clients prior to electronically recording or observing sessions. (See A.3.a.)

d. *Client Access.*

Counselors recognize that counseling records are kept fort he benefit of clients, and therefore provide access to records and copies of records when requested by competent clients, unless the records contain information that may be misleading and detrimental to the client. In situations involving multiple clients, access to records is limited to those parts of records that do not include confidential information related to another client. (See A.8, B.1.a., and B.2.b.)

e. *Disclosure or Transfer.*

Counselors obtain written permission from clients to disclose or transfer records to legitimate third parties unless exceptions to confidentiality exist as listed in Section B.1. Steps are taken to ensure that receivers of counseling records are sensitive to their confidential nature.

B.5. RESEARCH AND TRAINING

a. *Data Disguise Required.*

Use of data derived from counseling relationships for purposes of training, research, or publication is confined to content that is disguised to ensure the anonymity of the individuals involved. (See B.1.g. and G.3.d.)

b. *Agreement for Identification.*

Identification of a client in a presentation or publication is permissible only when the client has reviewed the material and has agreed to its presentation or publication. (See G.3.d.)

B.6. CONSULTATION

a. *Respect for Privacy.*

Information obtained in a consulting relationship is discussed for professional purposes only with persons clearly concerned with the case. Written and oral reports present data germane to the purposes of the consultation, and every effort is made to protect client identity and avoid undue invasion of privacy.

b. *Cooperating Agencies.*

Before sharing information, counselors make efforts to ensure that there are defined policies in other agencies serving the counselor's clients that effectively protect the confidentiality of information.

Section C: Professional Responsibility

C.1. STANDARDS KNOWLEDGE

Counselors have a responsibility to read, understand, and follow the Code of Ethics and the Standards of Practice.

C.2. PROFESSIONAL COMPETENCE

a. *Boundaries of Competence.*

Counselors practice only within the boundaries of their competence, based on their education, training, supervised experience, state and national professional credentials, and appropriate professional experience. Counselors will demonstrate a commitment to gain knowledge, personal awareness, sensitivity, and skills pertinent to working with a diverse client population.

b. *New Specialty Areas of Practice.*

Counselors practice in specialty are as new to them only after appropriate education, training, and supervised experience. While developing skills in new specialty areas, counselors take steps to ensure the competence of their work and to protect others from possible harm.

c. *Qualified for Employment.*

Counselors accept employment only for positions for which they are qualified by education, training, supervised counselors hire for professional counseling positions only individuals who are qualified and competent.

d. *Monitor Effectiveness.*

Counselors continually monitor their effectiveness as professionals and take steps to improve when necessary. Counselors in private practice take reasonable steps to seek out peer supervision to evaluate their efficacy as counselors.

e. *Ethical Issues Consultation.*

Counselors take reasonable steps to consult with other counselors or related professionals when they have questions regarding their ethical obligations or professional practice. (See H.1.)

f. *Continuing Education.*

Counselors recognize the need for continuing education to maintain a reasonable level of awareness of current scientific and professional information in their fields of activity.

They take steps to maintain competence in the skills they use, are open to new procedures, and keep current with the diverse and/or special populations with whom they work.

g. *Impairment.*

Counselors refrain from offering or accepting professional services when their physical, mental, or emotional problems are likely to harm a client or others. They are alert to the signs of impairment, seek assistance for problems, and, if necessary, limit, suspend, or terminate their professional responsibilities. (See A.11.c.)

C.3. ADVERTISING AND SOLICITING CLIENTS.

a. *Accurate Advertising.*

There are no restrictions on advertising by counselors except those that can be specifically justified to protect the public from deceptive practices. Counselors advertise or represent their services to the public by identifying their credentials in an accurate manner that is not false, misleading, deceptive, or fraudulent. Counselors may only advertise the highest degree earned which is in counseling or a closely related field from a college or university that was accredited when the degree was awarded by one of the regional accrediting bodies recognized by the Council on Postsecondary Accreditation.

b. *Testimonials.*

Counselors who use testimonials do not solicit them from clients or other persons who, because of their particular circumstances, may be vulnerable to undue influence.

c. *Statements by Others.*

Counselors make reasonable efforts to ensure that statements made by others about them or the profession of counseling are accurate.

d. *Recruiting through Employment.*

Counselors do not use their places of employment or institutional affiliation to recruit or gain clients, supervisees, or consultees for their private practices. (See C.5.e.)

e. *Products and Training Advertisements.*

Counselors who develop products related to their profession or conduct workshops or training events ensure that the advertisements concerning these products or events are accurate and disclose adequate information for consumers to make informed choices.

f. *Promoting to Those Served.*

Counselors do not use counseling, teaching, training, or supervisory relationships to promote their products or training events in a manner that is deceptive or would exert undue influence on individuals who may be vulnerable. Counselors may adopt textbooks they have authored for instruction purposes.

g. *Professional Association Involvement.*

Counselors actively participate in local, state, and national associations that foster the development and improvement of counseling.

C.4. CREDENTIALS

a. *Credentials Claimed.*

Counselors claim or imply only professional credentials possessed and are responsible for correcting any known misrepresentations of their credentials by others. Professional credentials include graduate degrees in counseling or closely related mental health fields, accreditation of graduate programs, national voluntary certifications, government-issued certifications or licenses, ACA professional membership, or any other credential that might indicate to the public specialized knowledge or expertise in counseling.

b. *ACA Professional Membership.*

ACA professional members may announce to the public their membership status. Regular members may not announce their ACA membership in a manner that might imply they are credentialed counselors.

c. *Credential Guidelines.*

Counselors follow the guidelines for use of credentials that have been established by the entities that issue the credentials.

d. *Misrepresentation of Credentials.*

Counselors do not attribute more to their credentials than the credentials represent, and do not imply that other counselors are not qualified because they do not possess certain credentials.

e. *Doctoral Degrees from Other Fields.*

Counselors who hold a master's degree in counseling or a closely related mental health field, but hold a doctoral degree from other than counseling or a closely related field, do not use the title "Dr." in their practices and do not announce to the public in relation to their practice or status as a counselor that they hold a doctorate.

C.5. PUBLIC RESPONSIBILITY

a. *Nondiscrimination.*

Counselors do not discriminate against clients, students, or supervisees in a manner that has a negative impact based on their age, color, culture, disability, ethnic group, gender, race, religion, sexual orientation, or socioeconomic status, or for any other reason. (See A.2.a.)

b. *Sexual Harassment.*

Counselors do not engage in sexual harassment Sexual harassment is defined as sexual solicitation, physical advances, or verbal or nonverbal conduct that is sexual in nature, that occurs in connection with professional activities or roles, and that either (1) is unwelcome, is offensive, or creates a hostile workplace environment, and counselors know or are told this; or (2) is sufficiently severe or intense to be perceived as harassment to a reasonable person in the context. Sexual harassment can consist of a single intense or severe act or multiple persistent or pervasive acts.

c. *Reports to Third Parties.*

Counselors are accurate, honest, and unbiased in reporting their professional activities and judgments to appropriate third parties including courts, health insurance companies, those who are the recipients of evaluation reports, and others. (See B.1.g.)

d. *Media Presentations.*

When counselors provide advice or comment by means of public lectures, demonstrations, radio or television programs, prerecorded tapes, printed articles, mailed material, or other media, they take reasonable precautions to ensure that (1) the statements are based on appropriate professional counseling literature and practice; (2) the statements are otherwise consistent with the Code of Ethics and the Standards of Practice; and (3) the recipients of the information are not encouraged to infer that a professional counseling relationship has been established. (See C.6.b.)

e. *Unjustified Gains.*

Counselors do not use their professional positions to seek or receive unjustified personal gains, sexual favors, unfair advantage, or unearned goods or services. (See C.3.d.)

C.6. RESPONSIBILITY TO OTHER PROFESSIONALS

a. *Different Approaches.*

Counselors are respectful of approaches to professional counseling that differ from their own. Counselors know and take into account the traditions and practices of other professional groups with which they work.

b. *Personal Public Statements.*

When making personal statements in a public context, counselors clarify that they are speaking from their personal perspectives and that they are not speaking on behalf of all counselors or the profession. (See C.5.d.)

c. *Clients Served by Others.*

When counselors learn that their clients are in a professional relationship with another mental health professional, they request release from clients to inform the other professionals and strive to establish positive and collaborative professional relationships. (See A.4.)

Section D: Relationships with Other Professionals

D.1. RELATIONSHIPS WITH EMPLOYERS AND EMPLOYEES

a. *Role Definition.*

Counselors define and describe for their employers and employees the parameters and levels of their professional roles.

b. *Agreements.*

Counselors establish working agreements with supervisors, colleagues, and subordinates regarding counseling or clinical relationships, confidentiality, adherence to professional standards, distinction between public and private material, maintenance and dissemination of recorded information, work load, and accountability. Working agreements in each instance are specified and made known to those concerned.

c. *Negative Conditions.*

Counselors alert their employers to conditions that may be potentially disruptive or damaging to the counselor's professional responsibilities or that may limit their effectiveness.

d. *Evaluation.*

Counselors submit regularly to professional review and evaluation by their supervisor or the appropriate representative of the employer.

e. *In-Service.*

Counselors are responsible for in-service development of self and staff.

f. *Goals.*

Counselors inform their staff of goals and programs.

g. *Practices.*

Counselors provide personnel and agency practices that respect and enhance the rights and welfare of each employee and recipient of agency services. Counselors strive to maintain the highest levels of professional services.

h. *Personnel Selection and Assignment.*

Counselors select competent staff and assign responsibilities compatible with their skills and experiences.

i. *Discrimination.*

Counselors, as either employers or employees, do not engage in or condone practices that are inhumane, illegal, or unjustifiable(such as considerations based on age, color, culture, disability, ethnic group, gender, race, religion, sexual orientation, or socioeconomic status) in hiring, promotion, or training. (See A.2.a. and C.5.b.)

j. *Professional Conduct.*

Counselors have a responsibility both to clients and to the agency or institution within which services are performed to maintain high standards of professional conduct.

k. *Exploitative Relationships.*

Counselors do not engage in exploitative relationships with individuals over whom they have supervisory, evaluative, or instructional control or authority.

l. *Employer Policies.*

The acceptance of employment in an agency or institution implies that counselors are in agreement with its general policies and principles. Counselors strive to reach agreement with employers as to acceptable standards of conduct that allow for changes in institutional policy conducive to the growth and development of clients.

D.2. CONSULTATION (SEE B.6.)

a. *Consultation as an Option.*

Counselors may choose to consult with any other professionally competent persons about their clients. In choosing consultants, counselors avoid placing the consultant in a conflict of interest situation that would preclude the consultant being a proper party to the counselor's efforts to help the client. Should counselors be engaged in a work setting that compromises this consultation standard, they consult with other professionals whenever possible to consider justifiable alternatives.

b. *Consultant Competency.*

Counselors are reasonably certain that they have or the organization represented has the necessary competencies and resources for giving the kind of consulting services needed and that appropriate referral resources are available.

c. *Understanding with Clients.*

When providing consultation, counselors attempt to develop with their clients a clear understanding of problem definition, goals for change, and predicted consequences of interventions selected.

d. *Consultant Goals.*

The consulting relationship is one in which client adaptability and growth toward self-direction are consistently encouraged and cultivated. (See A.1.b.)

D.3. FEES FOR REFERRAL

a. *Accepting Fees from Agency Clients.*

Counselors refuse a private fee or other remuneration for rendering services to persons who are entitled to such services through the counselor's employing agency or institution. The policies of a particular agency may make explicit provisions for agency clients to receive counseling services from members of its staff in private practice. In such instances, the clients must be informed of other options open to them should they seek private counseling services. (See A.10.a., A.11. b., and C.3.d.)

b. *Referral Fees.*

Counselors do not accept a referral fee from other professionals.

D.4. SUBCONTRACTOR ARRANGEMENTS

When counselors work as subcontractors for counseling services for a third party, they have a duty to inform clients of the limitations of confidentiality that the organization may place on counselors in providing counseling services to clients. The limits of such confidentiality ordinarily are discussed as part of the intake session. (See B.1.e. and B.1.f.)

Section E: Evaluation, Assessment, and Interpretation

E.1. GENERAL

a. *Appraisal Techniques.*

The primary purpose of educational and psychological assessment is to provide measures that are objective and interpretable in either comparative or absolute terms. Counselors recognize the need to interpret the statements in this section as applying to the whole range of appraisal techniques, including test and nontest data.

b. *Client Welfare.*

Counselors promote the welfare and best interests of the client in the development, publication, and utilization of educational and psychological assessment techniques. They do not misuse assessment results and interpretations and take reasonable steps to prevent others from misusing the information these techniques provide. They respect the client's right to know the results, the interpretations made, and the bases for their conclusions and recommendations.

E.2. COMPETENCE TO USE AND INTERPRET TESTS

a. *Limits of Competence.*

Counselors recognize the limits of their competence and perform only those testing and assessment services for which they have been trained. They are familiar with reliability, validity, related standardization, error of measurement, and proper application of any technique utilized. Counselors using computer-based test interpretations are trained in the construct being measured and the specific instrument being used prior to using this type of computer application. Counselors take reasonable measures to ensure the proper use of psychological assessment techniques by persons under their supervision.

b. *Appropriate Use.*

Counselors are responsible for the appropriate application, scoring, interpretation, and use of assessment instruments, whether they score and interpret such tests themselves or use computerized or other services.

c. *Decisions Based on Results.*

Counselors responsible for decisions involving individuals or policies that are based on assessment results have a thorough understanding of educational and psychological measurement, including validation criteria, test research, and guidelines for test development and use.

d. *Accurate Information.*

Counselors provide accurate information and avoid false claims or misconceptions when making statements about assessment instruments or techniques. Special efforts are made to avoid unwarranted connotations of such terms as IQ and grade equivalent scores. (See C.5.c.)

E.3. INFORMED CONSENT

a. *Explanation to Clients.*

Prior to assessment, counselors explain the nature and purposes of assessment and the specific use of results in language the client (or other legally authorized person on behalf of the client) can understand, unless an explicit exception to this right has been agreed upon in advance. Regardless of whether scoring and interpretation are completed by counselors, by assistants, or by computer or other outside services, counselors take reasonable steps to ensure that appropriate explanations are given to the client.

b. *Recipients of Results.*

The examinee's welfare, explicit understanding, and prior agreement determine the recipients of test results. Counselors include accurate and appropriate interpretations with any release of individual or group test results. (See B.1.a. and C.5.c.)

E.4. RELEASE OF INFORMATION TO COMPETENT PROFESSIONALS

a. *Misuse of Results.*

Counselors do not misuse assessment results, including test results, and interpretations, and take reasonable steps to prevent the misuse of such by others. (See C.5.c.)

b. *Release of Raw Data.*

Counselors ordinarily release data (e.g., protocols, counseling or interview notes, or questionnaires) in which the client is identified only with the consent of the client or the client's legal representative. Such data are usually released only to persons recognized by counselors as competent to interpret the data. (See B.1.a.)

E.5. PROPER DIAGNOSIS OF MENTAL DISORDERS

a. *Proper Diagnosis.*

Counselors take special care to provide proper diagnosis of mental disorders. Assessment techniques (including personal interview) used to determine client care (e.g., locus of treatment, type of treatment, or recommended follow-up) are carefully selected and appropriately used. (See A.3.a. and C.5.c.)

b. *Cultural Sensitivity.*

Counselors recognize that culture affects the manner in which clients' problems are defined. Clients' socioeconomic and cultural experience is considered when diagnosing mental disorders.

E.6. TEST SELECTION

a. *Appropriateness of Instruments.*

Counselors carefully consider the validity, reliability, psychometric limitations, and appropriateness of instruments when selecting tests for use in a given situation or with a particular client.

b. *Culturally Diverse Populations.*

Counselors are cautious when selecting tests for culturally diverse populations to avoid inappropriateness of testing that may be outside of socialized behavioral or cognitive patterns.

E.7. CONDITIONS OF TEST ADMINISTRATION

a. *Administration Conditions.*

Counselors administer tests under the same conditions that were established in their standardization. When tests are not administered under standard conditions or when unusual behavior or irregularities occur during the testing session, those conditions are noted in interpretation, and the results may be designated as invalid or of questionable validity.

b. *Computer Administration.*

Counselors are responsible for ensuring that administration programs function properly to provide clients with accurate results when a computer or other electronic methods are used for test administration. (See A.12.b.)

c. *Unsupervised Test Taking.*

Counselors do not permit unsupervised or inadequately supervised use of tests or assessments unless the tests or assessments are designed, intended, and validated for self-administration and/or scoring.

d. *Disclosure of Favorable Conditions.*

Prior to test administration, conditions that produce most favorable test results are made known to the examinee.

E.8. DIVERSITY IN TESTING

Counselors are cautious in using assessment techniques, making evaluations, and interpreting the performance of populations not represented in the norm group on which an instrument was standardized. They recognize the effects of age, color, culture, disability, ethnic group, gender, race, religion, sexual orientation, and socioeconomic status on test administration and interpretation and place test results in proper perspective with other relevant factors. (See A.2.a.)

E.9. TEST SCORING AND INTERPRETATION

a. *Reporting Reservations.*

In reporting assessment results, counselors indicate any reservations that exist regarding validity or reliability because of the circumstances of the assessment or the inappropriateness of the norms for the person tested.

b. *Research Instruments.*

Counselors exercise caution when interpreting the results of research instruments possessing insufficient technical data to support respondent results. The specific purposes for the use of such instruments are stated explicitly to the examinee.

c. *Testing Services.*

Counselors who provide test scoring and test interpretation services to support the assessment process confirm the validity of such interpretations. They accurately describe the purpose, norms, validity, reliability, and applications of the procedures and any special qualifications applicable to their use. The public offering of an automated test interpretations service is considered a professional-to-professional consultation. The formal responsibility of the consultant is to the consulate, but the ultimate and overriding responsibility is to the client.

E.10. TEST SECURITY

Counselors maintain the integrity and security of tests and otherassessment techniques consistent with legal and contractual obligations. Counselors do not appropriate, reproduce, or modify published tests orparts thereof without acknowledgment and permission from the publisher.

E.11. OBSOLETE TESTS AND OUTDATED TEST RESULTS

Counselors do not use data or test results that are obsolete or outdated forthe current purpose. Counselors make every effort to prevent the misuse ofobsolete measures and test data by others.

E.12. TEST CONSTRUCTION

Counselors use established scientific procedures, relevant standards, andcurrent professional knowledge for test design in the development, publication, and utilization of educational and psychological assessment techniques.

Section F: Teaching, Training, and Supervision

F.1. COUNSELOR EDUCATORS AND TRAINERS

a. *Educators as Teachers and Practitioners.*

Counselors who are responsible for developing, implementing, and supervising educational programs are skilled as teachers and practitioners. They are knowledgeable regarding the ethical, legal, and regulatory aspects of the profession, are skilled in applying that knowledge, and make students and supervisees aware of their responsibilities. Counselors conduct counselor education and training programs in an ethical manner and serve as role models for professional behavior. Counselor educators should make an effort to infuse material related to human diversity into all courses and/or workshops that are designed to promote the development of professional counselors.

b. *Relationship Boundaries with Students and Supervisees.*

Counselors clearly define and maintain ethical, professional, and social relationship boundaries with their students and supervisees. They are aware of the differential in power that exists and the student's or supervisee's possible incomprehension of that power differential. Counselors explain to students and supervisees the potential for the relationship to become exploitive.

c. *Sexual Relationships.*

Counselors do not engage in sexual relationships with students or supervisees and do not subject them to sexual harassment. (See A.6. and C.5.b)

d. *Contributions to Research.*

Counselors give credit to students or supervisees for their contributions to research and scholarly projects. Credit is given through coauthorship, acknowledgment, footnote statement, or other appropriate means, in accordance with such contributions. (See G.4.b. and G.4.c.)

e. *Close Relatives.*

Counselors do not accept close relatives as students or supervisees.

f. *Supervision Preparation.*

Counselors who offer clinical supervision services are adequately prepared in supervision methods and techniques. Counselors who are doctoral students serving as practicum or internship supervisors to master's level students are adequately prepared and supervised by the training program.

g. *Responsibility for Services to Clients.*

Counselors who supervise the counseling services of others take reasonable measures to ensure that counseling services provided to clients are professional.

h. *Endorsement.*

Counselors do not endorse students or supervisees for certification, licensure, employment, or completion of an academic or training program if they believe students or supervisees are not qualified for the endorsement. Counselors take reasonable steps to assist students or supervisees who are not qualified for endorsement to become qualified.

F.2. COUNSELOR EDUCATION AND TRAINING PROGRAMS

a. *Orientation.*

Prior to admission, counselors orient prospective students to the counselor education or training program's expectations, including but not limited to the following: (1) the type and level of skill acquisition required for successful completion of the training, (2) subject matter to be covered, (3) basis for evaluation, (4) training components that encourage self-growth or self-disclosure as part of the training process,

(5) the type of supervision settings and requirements of the sites for required clinical field experiences, (6) student and supervisee evaluation and dismissal policies and procedures, and (7) up-to-date employment prospects for graduates.

b. *Integration of Study and Practice.*

Counselors establish counselor education and training programs that integrate academic study and supervised practice.

c. *Evaluation.*

Counselors clearly state to students and supervisees, in advance of training, the levels of competency expected, appraisal methods, and timing of evaluations for both didactic and experiential components. Counselors provide students and supervisees with periodic performance appraisal and evaluation feedback throughout the training program.

d. *Teaching Ethics.*

Counselors make students and supervisees aware of the ethical responsibilities and standards of the profession and the students' and supervisees' ethical responsibilities to the profession. (See C.1. and F.3.e.)

e. *Peer Relationships.*

When students or supervisees are assigned to lead counseling groups or provide clinical supervision for their peers, counselors take steps to ensure that students and supervisees placed in these roles do not have personal or adverse relationships with peers and that they understand they have the same ethical obligations as counselor educators, trainers, and supervisors. Counselors make every effort to ensure that the rights of peers are not compromised when students or supervisees are assigned to lead counseling groups or provide clinical supervision.

f. *Varied Theoretical Positions.*

Counselors present varied theoretical positions so that students and supervisees may make comparisons and have opportunities to develop their own positions. Counselors provide information concerning the scientific bases of professional practice. (See C.6.a.)

g. *Field Placements.*

Counselors develop clear policies within their training program regarding field placement and other clinical experiences. Counselors provide clearly stated roles and responsibilities for the student or supervisee, the site supervisor, and the program supervisor. They confirm that site supervisors are qualified to provide supervision and are informed of their professional and ethical responsibilities in this role.

h. *Dual Relationships as Supervisors.*

Counselors avoid dual relationships such as performing the role of site supervisor and training program supervisor in the student's or supervisee's training program. Counselors do not accept any form of professional services, fees, commissions, reimbursement, or remuneration from a site for student or supervisee placement.

i. *Diversity in Programs.*

Counselors are responsive to their institution's and program's recruitment and retention needs for training program administrators, faculty, and students with diverse backgrounds and special needs. (See A.2.a.)

F.3. STUDENTS AND SUPERVISEES

a. *Limitations.*

Counselors, through ongoing evaluation and appraisal, are aware of the academic and personal limitations of students and supervisees that might impede performance. Counselors assist students and supervisees in securing remedial assistance when needed, and dismiss from the training program supervisees who are unable to provide competent service due to academic or personal limitations. Counselors seek professional consultation and document their decision to dismiss or refer students or supervisees for assistance. Counselors ensure that students and supervisees have recourse to address decisions made to require them to seek assistance or to dismiss them.

b. *Self-Growth Experiences.*

Counselors use professional judgment when designing training experiences conducted by the counselors themselves that require student and supervisee self-growth or self-disclosure. Safeguards are provided so that students and supervisees are aware of the ramifications their self-disclosure may have on counselors whose primary role as teacher, trainer, or supervisor requires acting on ethical obligations to the profession. Evaluative components of experiential training experiences explicitly delineate predetermined academic standards that are separate and do not depend on the student's level of self-disclosure. (See A.6.)

c. *Counseling for Students and Supervisees.*

If students or supervisees request counseling, supervisors or counselor educators provide them with acceptable referrals. Supervisors or counselor educators do not serve as counselor to students or supervisees over whom they hold administrative, teaching, or evaluative roles unless this is a brief role associated with a training experience. (See A.6.b.)

d. *Clients of Students and Supervisees.*

Counselors make every effort to ensure that the clients at field placements are aware of the services rendered and the qualifications of the students and supervisees rendering those services. Clients receive professional disclosure information and are informed of the limits of confidentiality. Client permission is obtained in order for the students and supervisees to use any information concerning the counseling relationship in the training process. (See B.1.e.)

e. *Standards for Students and Supervisees.*

Students and supervisees preparing to become counselors adhere to the Code of Ethics and the Standards of Practice. Students and supervisees have the same obligations to clients as those required of counselors. (See H.1.)

Section G: Research and Publication

G.1. RESEARCH RESPONSIBILITIES

a. *Use of Human Subjects.*

Counselors plan, design, conduct, and report research in a manner consistent with pertinent ethical principles, federal and state laws, host institutional regulations, and scientific standards governing research with human subjects. Counselors design and conduct research that reflects cultural sensitivity appropriateness.

b. *Deviation from Standard Practices.*

Counselors seek consultation and observe stringent safeguards to protect the rights of research participants when a research problem suggests a deviation from standard acceptable practices. (See B.6.)

c. *Precautions to Avoid Injury.*

Counselors who conduct research with human subjects are responsible for the subjects' welfare throughout the experiment and take reasonable precautions to avoid causing injurious psychological, physical, or social effects to their subjects.

d. *Principal Researcher Responsibility.*

The ultimate responsibility for ethical research practice lies with the principal researcher. All others involved in the research activities share ethical obligations and full responsibility for their own actions.

e. *Minimal Interference.*

Counselors take reasonable precautions to avoid causing disruptions in subjects' lives due to participation in research.

f. *Diversity.*

Counselors are sensitive to diversity and research issues with special populations. They seek consultation when appropriate. (See A.2.a. and B.6.)

G.2. INFORMED CONSENT

a. *Topics Disclosed.*

In obtaining informed consent for research, counselors use language that is understandable to research participants and that (1) accurately explains the purpose and procedures to be followed; (2) identifies any procedures that are experimental or relatively untried; (3) describes the attendant discomforts and risks; (4) describes the benefits or changes in individuals or organizations that might be reasonably expected; (5) discloses appropriate alternative procedures that would be advantageous for subjects; (6) offers to answer any inquiries concerning the procedures; (7) describes any limitations on confidentiality; and (8) instructs that subjects are free to withdraw their consent and to discontinue participation in the project at any time. (See B.1.f.)

b. *Deception.*

Counselors do not conduct research involving deception unless alternative procedures are not feasible and the prospective value of the research justifies the deception. When the methodological requirements of a study necessitate concealment or deception, the investigator is required to explain clearly the reasons for this action as soon as possible.

c. *Voluntary Participation.*

Participation in research is typically voluntary and without any penalty for refusal to participate. Involuntary participation is appropriate only when it can be demonstrated that participation will have no harmful effects on subjects and is essential to the investigation.

d. *Confidentiality of Information.*

Information obtained about research participants during the course of an investigation is confidential. When the possibility exists that others may obtain access to such information, ethical research practice requires that the possibility, together with the plans for protecting confidentiality, be explained to participants as a part of the procedure for obtaining informed consent. (See B.1.e.)

e. *Persons Incapable of Giving Informed Consent.*

When a person is incapable of giving informed consent, counselors provide an appropriate explanation, obtain agreement for participation, and obtain appropriate consent from a legally authorized person.

f. *Commitments to Participants.*

Counselors take reasonable measures to honor all commitments to research participants.

g. *Explanations after Data Collection.*

After data are collected, counselors provide participants with full clarification of the nature of the study to remove any misconceptions. Where scientific or human values justify delaying or withholding information, counselors take reasonable measures to avoid causing harm.

h. *Agreements to Cooperate.*

Counselors who agree to cooperate with another individual in research or publication incur an obligation to cooperate as promised in terms of punctuality of performance and with regard to the completeness and accuracy of the information required.

i. *Informed Consent for Sponsors.*

In the pursuit of research, counselors give sponsors, institutions, and publication channels the same respect and opportunity for giving informed consent that they accord to individual research participants. Counselors are aware of their obligation to future research workers and ensure that host institutions are given feedback information and proper acknowledgment.

G.3. REPORTING RESULTS

a. *Information Affecting Outcome.*

When reporting research results, counselors explicitly mention all variables and conditions known to the investigator that may have affected the outcome of a study or the interpretation of data.

b. *Accurate Results.*

Counselors plan, conduct, and report research accurately and in a manner that minimizes the possibility that results will be misleading. They provide thorough discussions of the limitations of their data and alternative hypotheses. Counselors do not engage in fraudulent research, distort data, misrepresent data, or deliberately bias their results.

c. *Obligation to Report Unfavorable Results.*

Counselors communicate to other counselors the results of any research judged to be of professional value. Results that reflect unfavorably on institutions, programs, services, prevailing opinions, or vested interests are not withheld.

d. *Identity of Subjects.*

Counselors who supply data, aid in the research of another person, report research results, or make original data available take due care to disguise the identity of respective subjects in the absence of specific authorization from the subjects to do otherwise. (See B.1.g. and B.5.a.)

e. *Replication Studies.*

Counselors are obligated to make available sufficient original research data to qualified professionals who may wish to replicate the study.

G.4. PUBLICATION

a. *Recognition of Others.*

When conducting and reporting research, counselors are familiar with and give recognition to previous work on the topic, observe copyright laws, and give full credit to those to whom credit is due. (See F.1.d. and G.4.c.)

b. *Contributors.*

Counselors give credit through joint authorship, acknowledgment, footnote statements, or other appropriate means to those who have contributed significantly to research or concept development in accordance with such contributions. The principal contributor is listed first and minor technical or professional contributions are acknowledged in notes or introductory statements.

c. *Student Research.*

For an article that is substantially based on a student's dissertation or thesis, the student is listed as the principal author. (See F.1.d. and G.4.a.)

d. *Duplicate Submission.*

Counselors submit manuscripts for consideration to only one journal at a time. Manuscripts that are published in whole or insubstantial part in another journal or published work are not submitted for publication without acknowledgment and permission from the previous publication.

e. *Professional Review.*

Counselors who review material submitted for publication, research, or other scholarly purposes respect the confidentiality and proprietary rights of those who submitted it.

Section H: Resolving Ethical Issues

H.1. KNOWLEDGE OF STANDARDS

Counselors are familiar with the Code of Ethics and the Standards of Practice and other applicable ethics codes from other professional organizations of which they are member, or from certification and licensure bodies. Lack of knowledge or misunderstanding of an ethical responsibility is not a defense against a charge of unethical conduct. (See F.3.e.)

H.2. SUSPECTED VIOLATIONS

a. *Ethical Behavior Expected.*

Counselors expect professional associates to adhere to the Code of Ethics. When counselors possess reasonable cause that raises doubts as to whether a counselor is acting in an ethical manner, they take appropriate action. (See H.2.d. and H.2.e.)

b. *Consultation.*

When uncertain as to whether a particular situation or course of action may be in violation of the Code of Ethics, counselors consult with other counselors who are knowledgeable about ethics, with colleagues, or with appropriate authorities.

c. *Organization Conflicts.*

If the demands of an organization with which counselors are affiliated pose a conflict with the Code of Ethics, counselors specify the nature of such conflicts and express to their supervisors or other responsible officials their commitment to the Code of Ethics. When possible, counselors work toward change within the organization to allow full adherence to the Code of Ethics.

d. *Informal Resolution.*

When counselors have reasonable cause to believe that another counselor is violating an ethical standard, they attempt to first resolve the issue informally with the other counselor if feasible, providing that such action does not violate confidentiality rights that may be involved.

e. *Reporting Suspected Violations.*

When an informal resolution is not appropriate or feasible, counselors, upon reasonable cause, take action such as reporting the suspected ethical violation to state or national ethics committees, unless this action conflicts with confidentiality rights that cannot be resolved.

f. *Unwarranted Complaints.*

Counselors do not initiate, participate in, or encourage the filing of ethics complaints that are unwarranted or intend to harm a counselor rather than to protect clients or the public.

H.3. COOPERATION WITH ETHICS COMMITTEES

Counselors assist in the process of enforcing the Code of Ethics. Counselors cooperate with investigations, proceedings, and requirements of the ACA Ethics Committee or ethics committees of other duly constituted associations or boards having jurisdiction over those charged with a violation. Counselors are familiar with the ACA Policies and Procedures and use it as a reference in assisting the enforcement of the Code of Ethics.

STANDARDS OF PRACTICE

All members of the American Counseling Association (ACA) are required to adhere to the Standards of Practice and the Code of Ethics. The Standards of Practice represent minimal behavioral statements of the Code of Ethics. Members should refer to the applicable section of the Code of Ethics for further interpretation and amplification of the applicable Standard of Practice.

Section A: The Counseling Relationship

Standard of Practice One (SP-1): Nondiscrimination. Counselors respect diversity and must not discriminate against clients because of age, color, culture, disability, ethnic group, gender, race, religion, sexual orientation, marital status, or socioeconomic status. (See A.2.a.)

Standard of Practice Two (SP-2): Disclosure to Clients. Counselors must adequately inform clients, preferably in writing, regarding the counseling process and counseling relationship at or before the time it begins and throughout the relationship. (See A.3.a.)

Standard of Practice Three (SP-3): Dual Relationships. Counselors must make every effort to avoid dual relationships with clients that could impair their professional judgment or increase the risk of harm to clients. When a dual relationship cannot be avoided, counselors must take appropriate steps to ensure that judgment is not impaired and that no exploitation occurs. (See A.6.a. and A.6.b.)

Standard of Practice Four (SP-4): Sexual Intimacies With Clients. Counselors must not engage in any type of sexual intimacies with current clients and must not engage in sexual intimacies with former clients within a minimum of 2 years after terminating the counseling relationship. Counselors who engage in such relationship after 2 years following termination have the responsibility to examine and document thoroughly that such relations did not have an exploitative nature.

Standard of Practice Five (SP-5): Protecting Clients During Group Work. Counselors must take steps to protect clients from physical or psychological trauma resulting from interactions during group work. (See A.9.b.)

Standard of Practice Six (SP-6): Advance Understanding of Fees. Counselors must explain to clients, prior to their entering the counseling relationship, financial arrangements related to professional services. (See A.10.a.–d. and A.11.c.)

Standard of Practice Seven (SP-7): Termination. Counselors must assist in making appropriate arrangements for the continuation of treatment of clients, when necessary, following termination of counseling relationships. (See A.11.a.)

Standard of Practice Eight (SP-8): Inability to Assist Clients. Counselors must avoid entering or immediately terminate a counseling relationship if it is determined that they are unable to be of professional assistance to a client. The counselor may assist in making an appropriate referral for the client. (See A.11.b.)

Section B: Confidentiality

Standard of Practice Nine (SP-9): Confidentiality Requirement.
Counselors must keep information related to counseling services confidential unless disclosure is in the best interest of clients, is required for the welfare of others, or is required by law. When disclosure is required, only information that is essential is revealed and the client is informed of such disclosure. (See B.1.a.f.)

Standard of Practice Ten (SP-10): Confidentiality Requirements for Subordinates.
Counselors must take measures to ensure that privacy and confidentiality of clients are maintained by subordinates. (See B.1.h.)

Standard of Practice Eleven (SP-11): Confidentiality in Group Work.
Counselors must clearly communicate to group members that confidentiality cannot be guaranteed in group work. (See B.2.a.)

Standard of Practice Twelve (SP-12): Confidentiality in Family Counseling.
Counselors must not disclose information about one family member in counseling to another family member without prior consent. (See B.2.b.)

Standard of Practice Thirteen (SP-13): Confidentiality of Records.
Counselors must maintain appropriate confidentiality in creating, storing, accessing, transferring, and disposing of counseling records. (See B.4.b.)

Standard of Practice Fourteen (SP-14): Permission to Record or Observe.
Counselors must obtain prior consent from clients in order to record electronically or observe sessions. (See B.4.c.)

Standard of Practice Fifteen (SP-15): Disclosure or Transfer of Records.
Counselors must obtain client consent to disclose or transfer records to third parties, unless exceptions listed in SP-9 exist. (See B.4.e.)

Standard of Practice Sixteen (SP-16): Data Disguise Required.
Counselors must disguise the identity of the client when using data for training, research, or publication. (See B.5.a.)

Section C: Professional Responsibility

Standard of Practice Seventeen (SP-17): Boundaries of Competence.
Counselors must practice only within the boundaries of their competence. (See C.2.a.)

Standard of Practice Eighteen (SP-18): Continuing Education.
Counselors must engage in continuing education to maintain their professional competence. (See C.2.f.)

Standard of Practice Nineteen (SP-19): Impairment of Professionals.
Counselors must refrain from offering professional services when their personal problems or conflicts may cause harm to a client or others. (See C.2.g.)

Standard of Practice Twenty (SP-20): Accurate Advertising.
Counselors must accurately represent their credentials and services when advertising. (See C.3.a.)

Standard of Practice Twenty-One (SP-21): Recruiting Through Employment.
Counselors must not use their place of employment or institutional affiliation to recruit clients for their private practices. (See C.3.d.)

Standard of Practice Twenty-Two (SP-22): Credentials Claimed.
Counselors must claim or imply only professional credentials possessed and must correct any known misrepresentations of their credentials by others. (See C.4.a.)

Standard of Practice Twenty-Three (SP-23): Sexual Harassment.
Counselors must not engage in sexual harassment. (See C.5.b.)

Standard of Practice Twenty-Four (SP-24): Unjustified Gains.
Counselors must not use their professional positions to seek or receive unjustified personal gains, sexual favors, unfair advantage, or unearned goods or services. (See C.5.e.)

Standard of Practice Twenty-Five (SP-25): Clients Served by Others.
With the consent of the client, counselors must inform other mental health professionals serving the same client that a counseling relationship between the counselor and client exists. (See C.6.c.)

Standard of Practice Twenty-Six (SP-26): Negative Employment Conditions.
Counselors must alert their employers to institutional policy or conditions that may be potentially disruptive or damaging to the counselor's professional responsibilities, or that may limit their effectiveness or deny clients' rights. (See D.1.c.)

Standard of Practice Twenty-Seven (SP-27): Personnel Selection and Assignment.
Counselors must select competent staff and must assign responsibilities compatible with staff skills and experiences. (See D.1.h.)

Standard of Practice Twenty-Eight (SP-28): Exploitative Relationships with Subordinates.
Counselors must not engage in exploitative relationships with individuals over whom they have supervisory, evaluative, or instructional control or authority. (See D.1.k.)

Section D: Relationship with Other Professionals

Standard of Practice Twenty-Nine (SP-29): Accepting Fees from Agency Clients.
Counselors must not accept fees or other remuneration for consultation with

persons entitled to such services through the counselor's employing agency or institution. (See D.3.a.)

Standard of Practice Thirty (SP-30): Referral Fees.
Counselors must not accept referral fees. (See D.3.b.)

Section E: Evaluation, Assessment and Interpretation

Standard of Practice Thirty-One (SP-31): Limits of Competence.
Counselors must perform only testing and assessment services for which they are competent. Counselors must not allow the use of psychological assessment techniques by unqualified persons under their supervision. (See E.2.a.)

Standard of Practice Thirty-Two (SP-32): Appropriate Use of Assessment Instruments.
Counselors must use assessment instruments in the manner for which they were intended. (See E.2.b.)

Standard of Practice Thirty-Three (SP-33): Assessment Explanations to Clients.
Counselors must provide explanations to clients prior to assessment about the nature and purposes of assessment and the specific uses of results. (See E.3.a.)

Standard of Practice Thirty-Four (SP-34): Recipients of Test Results.
Counselors must ensure that accurate and appropriate interpretations accompany any release of testing and assessment information. (See E.3.b.)

Standard of Practice Thirty-Five (SP-35): Obsolete Tests and Outdated Test Results.
Counselors must not base their assessment or intervention decisions or recommendations on data or test results that are obsolete or outdated for the current purpose. (See E.11.)

Section F: Teaching, Training, and Supervision

Standard of Practice Thirty-Six (SP-36): Sexual Relationships with Students or Supervisees.
Counselors must not engage in sexual relationships with their students and supervisees. (See F.1.c.)

Standard of Practice Thirty-Seven (SP-37): Credit for Contributions to Research.
Counselors must give credit to students or supervisees for their contributions to research and scholarly projects. (See F.1.d.)

Standard of Practice Thirty-Eight (SP-38): Supervision Preparation.
Counselors who offer clinical supervision services must be trained and prepared in supervision methods and techniques. (See F.1.f.)

Standard of Practice Thirty-Nine (SP-39): Evaluation Information.
Counselors must clearly state to students and supervisees in advance of training the levels of competency expected, appraisal methods, and timing of evaluations. Counselors must provide students and supervisees with periodic performance appraisal and evaluation feedback throughout the training program. (See F.2.c.)

Standard of Practice Forty (SP-40): Peer Relationships in Training.
Counselors must make every effort to ensure that the rights of peers are not violated when students and supervisees are assigned to lead counseling groups or provide clinical supervision. (See F.2.e.)

Standard of Practice Forty-One (SP-41): Limitations of Students and Supervisees.
Counselors must assist students and supervisees in securing remedial assistance, when needed, and must dismiss from the training program students and supervisees who are unable to provide competent service due to academic or personal limitations. (See F.3.a.)

Standard of Practice Forty-Two (SP-42): Self-Growth Experiences.
Counselors who conduct experiences for students or supervisees that include self-growth or self-disclosure must inform participants of counselors' ethical obligations to the profession and must not grade participants based on their nonacademic performance. (See F.3.b.)

Standard of Practice Forty-Three (SP-43): Standards for Students and Supervisees.
Students and supervisees preparing to become counselors must adhere to the Code of Ethics and the Standards of Practice of counselors. (See F.3.e.)

Section G: Research and Publication Standard of Practice Forty-Four (SP-44): Precautions to Avoid Injury in Research.

Counselors must avoid causing physical, social, or psychological harm or injury to subjects in research. (See G.1.c.)

Standard of Practice Forty-Five (SP-45): Confidentiality of Research Information.
Counselors must keep confidential information obtained about research participants. (See G.2.d.)

Standard of Practice Forty-Six (SP-46): Information Affecting Research Outcome.
Counselors must report all variables and conditions known to the investigator that may have affected research data or outcomes. (See G.3.a.)

Standard of Practice Forty-Seven (SP-47): Accurate Research Results.
Counselors must not distort or misrepresent research data, nor fabricate or intentionally bias research results. (See G.3.b.)

Standard of Practice Forty-Eight (SP-48): Publication Contributors.
Counselors must give appropriate credit to those who have contributed to research. (See G.4.a. and G.4.b.)

Section H: Resolving Ethical Issues

Standard of Practice Forty-Nine (SP-49): Ethical Behavior Expected.
Counselors must take appropriate action when they possess reasonable cause that

raises doubts as to whether counselors or other mental health professionals are acting in an ethical manner. (See H.2.a.)

Standard of Practice Fifty (SP-50): Unwarranted Complaints.

Counselors must not initiate, participate in, or encourage the filing of ethics complaints that are unwarranted or intended to harm a mental health professional rather than to protect clients or the public. (See H.2.f.)

Standard of Practice Fifty-One (SP-51): Cooperation With Ethics Committees.

Counselors must cooperate with investigations, proceedings, and requirements of the ACA Ethics Committee or ethics committees of other duly constituted associations or boards having jurisdiction over those charged with a violation. (See H.3.)

Appendix K
NASW Code of Ethics

OVERVIEW

The National Association of Social Workers Code of Ethics is intended to serve as a guide to the everyday professional conduct or social workers. This code includes four sections. Section one, "Preamble," summarizes the social work profession's mission and core values. Section two, "Purpose of the Code of Ethics," provides an overview of the Code's main functions and a brief guide for dealing with ethical issues or dilemmas in social work practice. Section three, "Ethical Principles," presents broad ethical principles, based on social work's core values, that inform social work practice. The final section, "Ethical Standards," includes specific ethical standards to guide social workers' conduct and to provide a basis for adjudication.

PREAMBLE

The primary mission of the social work profession is to enhance human well-being and help meet the basic human needs of all people, with particular attention to the needs and empowerment of people who are vulnerable, oppressed, and living in poverty. A historic and defining feature of social work is the profession's focus on individual well-being in a social context and the well-being of society. Fundamental to social work is attention to the environmental forces that create, contribute to, and address problems in living.

Social workers promote social justice and social change with and on behalf of clients. "Clients" is used inclusively to refer to individuals, families, groups, organizations, and communities. Social workers are sensitive to cultural and ethnic diversity and strive to end discrimination, oppression, poverty, and other forms of social injustice. These activities may be in the form of direct practice, community organizing, supervision, consultation, administration, advocacy, social and political action, policy development and implementation, education, and research and evaluation. Social workers seek to enhance the capacity of people to address their own needs. Social workers also seek to promote the responsiveness of organizations, communities, and other social institutions to individuals' needs and social problems.

The mission of the social work profession is rooted in a set of core values. These core values, embraced by social workers throughout the profession's history, are the foundation of social work's unique purpose and perspective:

- service
- social justice
- dignity and worth of the person
- importance of human relationships
- integrity
- competence.

This constellation of core values reflects what is unique to the social work profession. Core values, and the principles that flow from them, must be balanced within the context and complexity of the human experience.

PURPOSE OF THE NASW CODE OF ETHICS

Professional ethics are at the core of social work. The profession has an obligation to articulate its basic values, ethical principles, and ethical standards. The NASW Code of Ethics sets forth these values, principles, and standards to guide social workers' conduct. The Code is relevant to all social workers and social work students, regardless of their professional functions, the settings in which they work, or the populations they serve.

The NASW Code of Ethics serves six purposes:

1. The Code identifies core values on which social work's mission is based.
2. The Code summarizes broad ethical principles that reflect the profession's core values and establishes a set of specific ethical standards that should be used to guide social work practice.
3. The Code is designed to help social workers identify relevant considerations when professional obligations conflict or ethical uncertainties arise.
4. The Code provides ethical standards to which the general public can hold the social work profession accountable.

429

5. The Code socializes practitioners new to the field to social work's mission, values, ethical principles, and ethical standards.

6. The Code articulates standards that the social work profession itself can use to assess whether social workers have engaged in unethical conduct. NASW has formal procedures to adjudicate ethics complaints filed against its members.[1] In subscribing to this Code, social workers are required to cooperate in its implementation, participate in NASW adjudication proceedings, and abide by any NASW disciplinary rulings or sanctions based on it.

The Code offers a set of values, principles, and standards to guide decision making and conduct when ethical issues arise. It does not provide a set of rules that prescribe how social workers should act in all situations. Specific applications of the Code must take into account the context in which it is being considered and the possibility of conflicts among the Code's values, principles, and standards. Ethical responsibilities flow from all human relationships, from the personal and familial to the social and professional.

Further, the NASW Code of Ethics does not specify which values, principles, and standards are most important and ought to outweigh others in instances when they conflict. Reasonable differences of opinion can and do exist among social workers with respect to the ways in which values, ethical principles, and ethical standards should be rank ordered when they conflict. Ethical decision making in a given situation must apply the informed judgment of the individual social worker and should also consider how the issues would be judged in a peer review process where the ethical standards of the profession would be applied.

Ethical decision making is a process. There are many instances in social work where simple answers are not available to resolve complex ethical issues. Social workers should take into consideration all the values, principles, and standards in this Code that are relevant to any situation in which ethical judgment is warranted. Social workers' decisions and actions should be consistent with the spirit as well as the letter of this Code.

In addition to this Code, there are many other sources of information about ethical thinking that may be useful. Social workers should consider ethical theory and principles generally, social work theory and research, laws, regulations, agency policies, and other relevant codes of ethics, recognizing that among codes of ethics, social workers should consider the NASW Code of Ethics as their primary source. Social workers also should be aware of the impact on ethical decision making of their clients' and their own personal values and cultural and religious beliefs and practices. They should be aware of any conflicts between personal and professional values and deal with them responsibly. For additional guidance social workers should consult the relevant literature on professional ethics and ethical decision making and seek appropriate consultation when faced with ethical dilemmas. This may involve consultation with an agency-based or social work organization's ethics committee, a regulatory body, knowledgeable colleagues, supervisors, or legal counsel.

Instances may arise when social workers' ethical obligations conflict with agency policies or relevant laws or regulations. When such conflicts occur, social workers must make a responsible effort to resolve the conflict in a manner that is consistent with the values, principles, and standards expressed in this Code. If a reasonable resolution of the conflict does not appear possible, social workers should seek proper consultation before making a decision.

The NASW Code of Ethics is to be used by NASW and by individuals, agencies, organizations, and bodies (such as licensing and regulatory boards, professional liability insurance providers, courts of law, agency boards of directors, government agencies, and other professional groups) that choose to adopt it or use it as a frame of reference. Violation of standards in this Code does not automatically imply legal liability or violation of the law. Such determination can only be made in the context of legal and judicial proceedings. Alleged violations of the Code would be subject to a peer review process. Such processes are generally separate from legal or administrative procedures and insulated from legal review or proceedings to allow the profession to counsel and discipline its own members.

A code of ethics can not guarantee ethical behavior. Moreover, a code of ethics can not resolve all ethical issues or disputes or capture the richness and complexity involved in striving to make responsible choices within a moral community. Rather, a code of ethics sets forth values, ethical principles, and ethical standards to which professionals aspire and by which their actions can be judged. Social workers' ethical behavior should result from their personal commitment to engage in ethical practice. The NASW Code of Ethics reflects the commitment of all social workers to uphold the profession's values and to act ethically. Principles and standards must be applied by individuals of good character who discern moral questions and, in good faith, seek to make reliable ethical judgments.

Ethical Principles

The following broad ethical principles are based on social work's core values of service, social justice, dignity and worth of the person, importance of human relationships, integrity, and competence. These principles set forth ideals to which all social workers should aspire.

VALUE: *Service*

Ethical Principle: *Social workers' primary goal is to help people in need and to address social problems.*

Social workers elevate service to others above self-interest. Social workers draw on their knowledge, values, and skills to help people in need and to address social problems. Social workers are encouraged to volunteer some portion of their professional skills with no expectation of significant financial return (pro bono service).

VALUE: *Social Justice*

Ethical Principle: *Social workers challenge social injustice.*

Social workers pursue social change, particularly with and on behalf of vulnerable and oppressed individu-

[1] For information on NASW adjudication procedures, see NASW Procedures for the Adjudication of Grievances.

als and groups of people. Social workers' social change efforts are focused primarily on issues of poverty, unemployment, discrimination, and other forms of social injustice. These activities seek to promote sensitivity to and knowledge about oppression and cultural and ethnic diversity. Social workers strive to ensure access to needed information, services, and resources; equality of opportunity; and meaningful participation in decision making for all people.

VALUE: *Dignity and Worth of the Person*

Ethical Principle: *Social workers respect the inherent dignity and worth of the person.*

Social workers treat each person in a caring and respectful fashion, mindful of individual differences and cultural and ethnic diversity. Social workers promote clients' socially responsible self-determination. Social workers seek to enhance clients' capacity and opportunity to change and to address their own needs. Social workers are cognizant of their dual responsibility to clients and to the broader society. They seek to resolve conflicts between clients' interests and the broader society's interests in a socially responsible manner consistent with the values, ethical principles, and ethical standards of the profession.

VALUE: *Importance of Human Relationships*

Ethical Principle: *Social workers recognize the central importance of human relationships.*

Social workers understand that relationships between and among people are an important vehicle for change. Social workers engage people as partners in the helping process. Social workers seek to strengthen relationships among people in a purposeful effort to promote, restore, maintain, and enhance the well-being of individuals, families, social groups, organizations, and communities.

VALUE: *Integrity*

Ethical Principle: *Social workers behave in a trustworthy manner.*

Social workers are continually aware of the profession's mission, values, ethical principles, and ethical standards and practice in a manner consistent with them. Social workers act honestly and responsibly and promote ethical practices on the part of the organizations with which they are affiliated.

VALUE: *Competence*

Ethical Principle: *Social workers practice within their areas of competence and develop and enhance their professional expertise.*

Social workers continually strive to increase their professional knowledge and skills and to apply them in practice. Social workers should aspire to contribute to the knowledge base of the profession.

Ethical Standards

1. SOCIAL WORKERS' ETHICAL RESPONSIBILITIES TO CLIENTS
   1.01 *Commitment to Clients*
   Social workers' primary responsibility is to promote the well-being of clients. In general, clients' interests are primary. However, social workers' responsibility to the larger society or specific legal obligations may on limited occasions supersede the loyalty owed clients, and clients should be so advised. (Examples include when a social worker is required by law to report that a client has abused a child or has threatened to harm self or others.)

   1.02 *Self-Determination*
   Social workers respect and promote the right of clients to self-determination and assist clients in their efforts to identify and clarify their goals. Social workers may limit clients' right to self-determination when, in the social workers' professional judgment, clients' actions or potential actions pose a serious, foreseeable, and imminent risk to themselves or others.

   1.03 *Informed Consent*
   a. Social workers should provide services to clients only in the context of a professional relationship based, when appropriate, on valid informed consent Social workers should use clear and understandable language to inform clients of the purpose of the services, risks related to the services, limits to services because of the requirements of a third-party payer, relevant costs, reasonable alternatives, clients' right to refuse or withdraw consent, and the time frame covered by the consent. Social workers should provide clients with an opportunity to ask questions.
   b. In instances when clients are not literate or have difficulty understanding the primary language used in the practice setting, social workers should take steps to ensure clients' comprehension. This may include providing clients with a detailed verbal explanation or arranging for a qualified interpreter or translator whenever possible.
   c. In instances when clients lack the capacity to provide informed consent, social workers should protect clients' interests by seeking permission from an appropriate third party, informing clients consistent with the clients' level of understanding. In such instances, social workers should seek to ensure that the third party acts in a manner consistent with clients' wishes and interests. Social workers should take reasonable steps to enhance such clients' ability to give informed consent.
   d. In instances when clients are receiving services involuntarily, social workers should provide information about the nature and extent of services and about the extent of clients' right to refuse service.
   e. Social workers who provide services via electronic media (such as computer, telephone, radio, and television) should inform recipients of the limitations and risks associated with such services.
   f. Social workers should obtain clients' informed consent before audiotaping or

videotaping clients or permitting observation of services to clients by a third party.

1.04 *Competence*

a. Social workers should provide services and represent themselves as competent only within the boundaries of their education, training, license, certification, consultation received, supervised experience, or other relevant professional experience.

b. Social workers should provide services in substantive areas or use intervention techniques or approaches that are new to them only after engaging in appropriate study, training, consultation, and supervision from people who are competent in those interventions or techniques.

c. When generally recognized standards do not exist with respect to an emerging area of practice, social workers should exercise careful judgment and take responsible steps (including appropriate education, research, training, consultation, and supervision) to ensure the competence of their work and to protect clients from harm.

1.05 *Cultural Competence and Social Diversity*

a. Social workers should understand culture and its function in human behavior and society, recognizing the strengths that exist in all cultures.

b. Social workers should have a knowledge base of their clients' cultures and be able to demonstrate competence in the provision of services that are sensitive to clients' cultures and to differences among people and cultural groups.

c. Social workers should obtain education about and seek to understand the nature of social diversity and oppression with respect to race, ethnicity, national origin, color, sex, sexual orientation, age, marital status, political belief, religion, and mental or physical disability.

1.06 *Conflicts of Interest*

a. Social workers should be alert to and avoid conflicts of interest that interfere with the exercise of professional discretion and impartial judgment. Social workers should inform clients when a real or potential conflict of interest arises and take reasonable steps to resolve the issue in a manner that makes the clients' interests primary and protects clients' interests to the greatest extent possible. In some cases, protecting clients' interests may require termination of the professional relationship with proper referral of the client.

b. Social workers should not take unfair advantage of any professional relationship or exploit others to further their personal, religious, political, or business interests.

c. Social workers should not engage in dual or multiple relationships with clients or former clients in which there is a risk of exploitation or potential harm to the client. In instances when dual or multiple relationships are unavoidable, social workers should take steps to protect clients and are responsible for setting clear, appropriate, and culturally sensitive boundaries. (Dual or multiple relationships occur when social workers relate to clients in more than one relationship, whether professional, social, or business. Dual or multiple relationships can occur simultaneously or consecutively.)

d. When social workers provide services to two or more people who have a relationship with each other (for example, couples, family members), social workers should clarify with all parties which individuals will be considered clients and the nature of social workers' professional obligations to the various individuals who are receiving services. Social workers who anticipate a conflict of interest among the individuals receiving services or who anticipate having to perform in potentially conflicting roles (for example, when asocial worker is asked to testify in a child custody dispute or divorce proceedings involving clients) should clarify their role with the parties involved and take appropriate action to minimize any conflict of interest.

1.07 *Privacy and Confidentiality*

a. Social workers should respect clients' right to privacy. Social workers should not solicit private information from clients unless it is essential to providing services or conducting social work evaluation or research. Once private information is shared, standards of confidentiality apply.

b. Social workers may disclose confidential information when appropriate with valid consent from a client or a person legally authorized to consent on behalf of a client.

c. Social workers should protect the confidentiality of all information obtained in the course of professional service, except for compelling professional reasons. The general expectation that social workers will keep information confidential does not apply when disclosure is necessary to prevent serious, foreseeable, and imminent harm to a client or other identifiable person or when laws or regulations require disclosure without a client's consent. In all instances, social workers should disclose the least amount of confidential information necessary to achieve the desired purpose; only information that is directly relevant to the purpose for which the disclosure is made should be revealed.

d. Social workers should inform clients, to the extent possible, about the disclosure of confidential information and the potential consequences, when feasible before the disclosure is made. This applies whether social workers disclose confidential information on the basis of a legal requirement or client consent.

e. Social workers should discuss with clients and other interested parties the nature of confidentiality and limitations of clients' right to confidentiality. Social workers should review with clients circumstances where confidential information may be requested and where disclosure of confidential information may be legally required. This discussion should occur as soon as possible in the social worker—client relationship and as needed throughout the course of the relationship.

f. When social workers provide counseling services to families, couples, or groups, social workers should seek agreement among the parties involved concerning each individual's right to confidentiality and obligation to preserve the confidentiality of information shared by others. Social workers should inform participants in family, couples, or group counseling that social workers can not guarantee that all participants will honor such agreements.

g. Social workers should inform clients involved in family, couples, marital, or group counseling of the social worker's, employer's, and agency's policy concerning the social worker's disclosure of confidential information among the parties involved in the counseling.

h. Social workers should not disclose confidential information to third-party payers unless clients have authorized such.

i. Social workers should not discuss confidential information in any setting unless privacy can be ensured. Social workers should not discuss confidential information in public or semipublic areas such as hallways, waiting rooms, elevators, and restaurants.

j. Social workers should protect the confidentiality of clients during legal proceedings to the extent permitted by law. When a court of law or other legally authorized body orders social workers to disclose confidential or privileged information without a client's consent and such disclosure could cause harm to the client, social workers should request that the court withdraw the order or limit the order as narrowly as possible or maintain the records under seal, unavailable for public inspection.

k. Social workers should protect the confidentiality of clients when responding to requests from members of the media.

l. Social workers should protect the confidentiality of clients' written and electronic records and other sensitive information. Social workers should take reasonable steps to ensure that clients' records are stored in a secure location and that clients' records are not available to others who are not authorized to have access.

m. Social workers should take precautions to ensure and maintain the confidentiality of information transmitted to other parties through the use of computers, electronic mail, facsimile machines, telephones and telephone answering machines, and other electronic or computer technology. Disclosure of identifying information should be avoided whenever possible.

n. Social workers should transfer or dispose of clients' records in a manner that protects clients' confidentiality and is consistent with state statutes governing records and social work licensure.

o. Social workers should take reasonable precautions to protect client confidentiality in the event of the social worker's termination of practice, incapacitation, or death.

p. Social workers should not disclose identifying information when discussing clients for teaching or training purposes unless the client has consented to disclosure of confidential information.

q. Social workers should not disclose identifying information when discussing clients with consultants unless the client has consented to disclosure of confidential information or there is a compelling need for such disclosure.

r. Social workers should protect the confidentiality of deceased clients consistent with the preceding standards.

1.08 *Access to Records*

a. Social workers should provide clients with reasonable access to records concerning the clients. Social workers who are concerned that clients' access to their records could cause serious misunderstanding or harm to the client should provide assistance in interpreting the records and consultation with the client regarding the records. Social workers should limit clients' access to their records, or portions of their records, only in exceptional circumstances when there is compelling evidence that such access would cause serious harm to the client. Both clients' requests and the rationale for withholding some or all of the record should be documented in clients' files.

b. When providing clients with access to their records, social workers should take steps to protect the confidentiality of other individuals identified or discussed in such records.

1.09 *Sexual Relationships*

a. Social workers should under no circumstances engage in sexual activities or sexual contact with current clients, whether such contact is consensual or forced.

b. Social workers should not engage in sexual activities or sexual contact with clients' relatives or other individuals with whom clients maintain a close personal relationship when there is a risk of exploitation or potential harm to the client Sexual activity or sexual contact with clients' relatives or other individuals with whom clients maintain a personal relationship has the potential to be harmful to the client and may make it difficult for the social worker and client to maintain appropriate professional boundaries. Social workers—not their clients, their clients' relatives, or other individuals with whom the client maintains a personal relationship—assume the full burden for setting clear, appropriate, and culturally sensitive boundaries.

c. Social workers should not engage in sexual activities or sexual contact with former clients because of the potential for harm to the client. If social workers engage in conduct contrary to this prohibition or claim that an exception to this prohibition is warranted because of extraordinary circumstances, it is social workers—not their clients—who assume the full burden of demonstrating that the former client has not been exploited, coerced, or manipulated, intentionally or unintentionally.

d. Social workers should not provide clinical services to individuals with whom they have had a prior sexual relationship. Providing clinical services to a former sexual partner has the potential to be harmful to the individual and is likely to make it difficult for the social worker and individual to maintain appropriate professional boundaries.

1.10 *Physical Contact*
Social workers should not engage in physical contact with clients when there is a possibility of psychological harm to the client as a result of the contact (such as cradling or caressing clients). Social workers who engage in appropriate physical contact with clients are responsible for setting clear, appropriate, and culturally sensitive boundaries that govern such physical contact.

1.11 *Sexual Harassment*
Social workers should not sexually harass clients. Sexual harassment includes sexual advances, sexual solicitation, requests for sexual favors, and other verbal or physical conduct of a sexual nature.

1.12 *Derogatory Language*
Social workers should not use derogatory language in their written or verbal communications to or about clients. Social workers should use accurate and respectful language in all communications to and about clients.

1.13 *Payment for Services*
a. When setting fees, social workers should ensure that the fees are fair, reasonable, and commensurate with the services performed. Consideration should be given to clients' ability to pay.

b. Social workers should avoid accepting goods or services from clients as payment for professional services. Bartering arrangements, particularly involving services, create the potential for conflicts of interest, exploitation, and inappropriate boundaries

in social workers' relationships with clients. Social workers should explore and may participate in bartering only in very limited circumstances when it can be demonstrated that such arrangements are an accepted practice among professionals in the local community, considered to be essential for the provision of services, negotiated without coercion, and entered into at the client's initiative and with the client's informed consent. Social workers who accept goods or services from clients as payment for professional services assume the full burden of demonstrating that this arrangement will not be detrimental to the client or the professional relationship.

c. Social workers should not solicit a private fee or other remuneration for providing services to clients who are entitled to such available services through the social workers' employer or agency.

1.14 *Clients Who Lack Decision-Making Capacity*
When social workers act on behalf of clients who lack the capacity to make informed decisions, social workers should take reasonable steps to safeguard the interests and rights of those clients.

1.15 *Interruption of Services*
Social workers should make reasonable efforts to ensure continuity of services in the event that services are interrupted by factors such as unavailability, relocation, illness, disability, or death.

1.16 *Termination of Services*
a. Social workers should terminate services to clients and professional relationships with them when such services and relationships are no longer required or no longer serve the clients' needs or interests.

b. Social workers should take reasonable steps to avoid abandoning clients who are still in need of services. Social workers should withdraw services precipitously only under unusual circumstances, giving careful consideration to all factors in the situation and taking care to minimize possible adverse effects. Social workers should assist in making appropriate arrangements for continuation of services when necessary.

c. Social workers in fee-for-service settings may terminate services to clients who are not paying an overdue balance if the financial contractual arrangements have been made clear to the client, if the client does not pose an imminent danger to self or others, and if the clinical and other consequences of the current nonpayment have been addressed and discussed with the client.

d. Social workers should not terminate services to pursue a social, financial, or sexual relationship with a client.

e. Social workers who anticipate the termination or interruption of services to clients should notify clients promptly and seek the transfer, referral, or continuation of services in relation to the clients' needs and preferences.

   f. Social workers who are leaving an employment setting should inform clients of appropriate options for the continuation of services and of the benefits and risks of the options.

2. Social Workers' Ethical Responsibilities to Colleagues

2.01 *Respect*

   a. Social workers should treat colleagues with respect and should represent accurately and fairly the qualifications, views, and obligations of colleagues.

   b. Social workers should avoid unwarranted negative criticism of colleagues in communications with clients or with other professionals. Unwarranted negative criticism may include demeaning comments that refer to colleagues' level of competence or to individuals' attributes such as race, ethnicity, national origin, color, sex, sexual orientation, age, marital status, political belief, religion, and mental or physical disability.

   c. Social workers should cooperate with social work colleagues and with colleagues of other professions when such cooperation serves the well-being of clients.

2.02 *Confidentiality*

Social workers should respect confidential information shared by colleagues in the course of their professional relationships and transactions. Social workers should ensure that such colleagues understand social workers' obligation to respect confidentiality and any exceptions related to it.

2.03 *Interdisciplinary Collaboration*

   a. Social workers who are members of an interdisciplinary team should participate in and contribute to decisions that affect the well-being of clients by drawing on the perspectives, values, and experiences of the social work profession. Professional and ethical obligations of the interdisciplinary team as a whole and of its individual members should be clearly established.

   b. Social workers for whom a team decision raises ethical concerns should attempt to resolve the disagreement through appropriate channels. If the disagreement cannot be resolved, social workers should pursue other avenues to address their concerns consistent with client well-being.

2.04 *Disputes Involving Colleagues*

   a. Social workers should not take advantage of a dispute between a colleague and an employer to obtain a position or otherwise advance the social workers' own interests.

   b. Social workers should not exploit clients in disputes with colleagues or engage clients in any inappropriate discussion of conflicts between social workers and their colleagues.

2.05 *Consultation*

   a. Social workers should seek the advice and counsel of colleagues whenever such consultation is in the best interests of clients.

   b. Social workers should keep themselves informed about colleagues' areas of expertise and competencies. Social workers should seek consultation only from colleagues who have demonstrated knowledge, expertise, and competence related to the subject of the consultation.

   c. When consulting with colleagues about clients, social workers should disclose the least amount of information necessary to achieve the purposes of the consultation.

2.06 *Referral for Services*

   a. Social workers should refer clients to other professionals when the other professionals' specialized knowledge or expertise is needed to serve clients fully or when social workers believe that they are not being effective or making reasonable progress with clients and that additional service is required.

   b. Social workers who refer clients to other professionals should take appropriate steps to facilitate an orderly transfer of responsibility. Social workers who refer clients to other professionals should disclose, with clients' consent, all pertinent information to the new service providers.

   c. Social workers are prohibited from giving or receiving payment for a referral when no professional service is provided by the referring social worker.

2.07 *Sexual Relationships*

   a. Social workers who function as supervisors or educators should not engage in sexual activities or contact with supervisees, students, trainees, or other colleagues over whom they exercise professional authority.

   b. Social workers should avoid engaging in sexual relationships with colleagues when there is potential for a conflict of interest. Social workers who become involved in, or anticipate becoming involved in, a sexual relationship with a colleague have a duty to transfer professional responsibilities, when necessary, to avoid a conflict of interest.

2.08 *Sexual Harassment*

Social workers should not sexually harass supervisees, students, trainees, or colleagues. Sexual harassment includes sexual advances, sexual solicitation, requests for sexual favors, and other verbal or physical conduct of a sexual nature.

2.09 *Impairment of Colleagues*

   a. Social workers who have direct knowledge of a social work colleague's impairment that is due to personal problems, psychosocial distress, substance abuse, or mental health difficulties and that interferes with practice effectiveness should consult with that colleague when feasible and assist the colleague in taking remedial action.

    b. Social workers who believe that a social work colleague's impairment interferes with practice effectiveness and that the colleague has not taken adequate steps to address the impairment should take action through appropriate channels established by employers, agencies, NASW, licensing and regulatory bodies, and other professional organizations.

2.10 *Incompetence of Colleagues*

    a. Social workers who have direct knowledge of a social work colleague's incompetence should consult with that colleague when feasible and assist the colleague in taking remedial action.

    b. Social workers who believe that a social work colleague is incompetent and has not taken adequate steps to address the incompetence should take action through appropriate channels established by employers, agencies, NASW, licensing and regulatory bodies, and other professional organizations.

2.11 *Unethical Conduct of Colleagues*

    a. Social workers should take adequate measures to discourage, prevent, expose, and correct the unethical conduct of colleagues.

    b. Social workers should be knowledgeable about established policies and procedures for handling concerns about colleagues' unethical behavior. Social workers should be familiar with national, state, and local procedures for handling ethics complaints. These include policies and procedures created by NASW, licensing and regulatory bodies, employers, agencies, and other professional organizations.

    c. Social workers who believe that a colleague has acted unethically should seek resolution by discussing their concerns with the colleague when feasible and when such discussion is likely to be productive.

    d. When necessary, social workers who believe that a colleague has acted unethically should take action through appropriate formal channels (such as contacting a state licensing board or regulatory body, an NASW committee on inquiry, or other professional ethics committees).

    e. Social workers should defend and assist colleagues who are unjustly charged with unethical conduct.

3. SOCIAL WORKERS' ETHICAL RESPONSIBILITIES IN PRACTICE SETTINGS

3.01 *Supervision and Consultation*

    a. Social workers who provide supervision or consultation should have the necessary knowledge and skill to supervise or consult appropriately and should do so only within their areas of knowledge and competence.

    b. Social workers who provide supervision or consultation are responsible for setting clear, appropriate, and culturally sensitive boundaries.

    c. Social workers should not engage in any dual or multiple relationships with supervisees in which there is a risk of exploitation of or potential harm to the supervisee.

    d. Social workers who provide supervision should evaluate supervisees' performance in a manner that is fair and respectful.

3.02 *Education and Training*

    a. Social workers who function as educators, field instructors for students, or trainers should provide instruction only within their areas of knowledge and competence and should provide instruction based on the most current information and knowledge available in the profession.

    b. Social workers who function as educators or field instructors for students should evaluate students' performance in a manner that is fair and respectful.

    c. Social workers who function as educators or field instructors for students should take reasonable steps to ensure that clients are routinely informed when services are being provided by students.

    d. Social workers who function as educators or field instructors for students should not engage in any dual or multiple relationships with students in which there is a risk of exploitation or potential harm to the student. Social work educators and field instructors are responsible for setting clear, appropriate, and culturally sensitive boundaries.

3.03 *Performance Evaluation*

Social workers who have responsibility for evaluating the performance of others should fulfill such responsibility in a fair and considerate manner and on the basis of clearly stated criteria.

3.04 *Client Records*

    a. Social workers should take reasonable steps to ensure that documentation in records is accurate and reflects the services provided.

    b. Social workers should include sufficient and timely documentation in records to facilitate the delivery of services and to ensure continuity of services provided to clients in the future.

    c. Social workers' documentation should protect clients' privacy to the extent that is possible and appropriate and should include only information that is directly relevant to the delivery of services.

    d. Social workers should store records following the termination of services to ensure reasonable future access. Records should be maintained for the number of years required by state statutes or relevant contracts.

3.05 *Billing*

Social workers should establish and maintain billing practices that accurately reflect the nature and extent of services provided and that identify who provided the service in the practice setting.

3.06 *Client Transfer*

a. When an individual who is receiving services from another agency or colleague contacts a social worker for services, the social worker should carefully consider the client's needs before agreeing to provide services. To minimize possible confusion and conflict, social workers should discuss with potential clients the nature of the clients' current relationship with other service providers and the implications, including possible benefits or risks, of entering into a relationship with a new service provider.

b. If a new client has been served by another agency or colleague, social workers should discuss with the client whether consultation with the previous service provider is in the client's best interest.

3.07 *Administration*

a. Social work administrators should advocate within and outside their agencies for adequate resources to meet clients' needs.

b. Social workers should advocate for resource allocation procedures that are open and fair. When not all clients' needs can be met, an allocation procedure should be developed that is nondiscriminatory and based on appropriate and consistently applied principles.

c. Social workers who are administrators should take reasonable steps to ensure that adequate agency or organizational resources are available to provide appropriate staff supervision.

d. Social work administrators should take reasonable steps to ensure that the working environment for which they are responsible is consistent with and encourages compliance with the NASW Code of Ethics. Social work administrators should take reasonable steps to eliminate any conditions in their organizations that violate, interfere with, or discourage compliance with the Code.

3.08 *Continuing Education and Staff Development*

Social work administrators and supervisors should take reasonable steps to provide or arrange for continuing education and staff development for all staff for whom they are responsible. Continuing education and staff development should address current knowledge and emerging developments related to social work practice and ethics.

3.09 *Commitments to Employers*

a. Social workers generally should adhere to commitments made to employers and employing organizations.

b. Social workers should work to improve employing agencies' policies and procedures and the efficiency and effectiveness of their services.

c. Social workers should take reasonable steps to ensure that employers are aware of social workers' ethical obligations as set forth in the NASW Code of Ethics and of the implications of those obligations for social work practice.

d. Social workers should not allow an employing organization's policies, procedures, regulations, or administrative orders to interfere with their ethical practice of social work. Social workers should take reasonable steps to ensure that their employing organizations' practices are consistent with the NASW Code of Ethics.

e. Social workers should act to prevent and eliminate discrimination in the employing organization's work assignments and in its employment policies and practices.

f. Social workers should accept employment or arrange student field placements only in organizations that exercise fair personnel practices.

g. Social workers should be diligent stewards of the resources of their employing organizations, wisely conserving funds where appropriate and never misappropriating funds or using them for unintended purposes.

3.10 *Labor—Management Disputes*

a. Social workers may engage in organized action, including the formation of and participation in labor unions, to improve services to clients and working conditions.

b. The actions of social workers who are involved in labor—management disputes, job actions, or labor strikes should be guided by the profession's values, ethical principles, and ethical standards. Reasonable differences of opinion exist among social workers concerning their primary obligation as professionals during an actual or threatened labor strike or job action. Social workers should carefully examine relevant issues and their possible impact on clients before deciding on a course of action

4. SOCIAL WORKERS' ETHICAL RESPONSIBILITIES AS PROFESSIONALS

4.01 *Competence*

a. Social workers should accept responsibility or employment only on the basis of existing competence or the intention to acquire the necessary competence.

b. Social workers should strive to become and remain proficient in professional practice and the performance of professional functions. Social workers should critically examine and keep current with emerging knowledge relevant to social work. Social workers should routinely review the professional literature and participate in continuing education relevant to social work practice and social work ethics.

c. Social workers should base practice on recognized knowledge, including empirically based

4.02 *Discrimination*

Social workers should not practice, condone, facilitate, or collaborate with any form of discrimination on the basis of race, ethnicity, national origin, color, sex, sexual orientation, age, marital status, political belief, religion, or mental or physical disability.

4.03 *Private Conduct*

Social workers should not permit their private conduct to interfere with their ability to fulfill their professional responsibilities.

4.04 *Dishonesty, Fraud, and Deception*

Social workers should not participate in, condone, or be associated with dishonesty, fraud, or deception.

4.05 *Impairment*

a. Social workers should not allow their own personal problems, psychosocial distress, legal problems, substance abuse, or mental health difficulties to interfere with their professional judgment and performance or to jeopardize the best interests of people for whom they have a professional responsibility.

b. Social workers whose personal problems, psychosocial distress, legal problems, substance abuse, or mental health difficulties interfere with their professional judgment and performance should immediately seek consultation and take appropriate remedial action by seeking professional help, making adjustments in workload, terminating practice, or taking any other steps necessary to protect clients and others.

4.06 *Misrepresentation*

a. Social workers should make clear distinctions between statements made and actions engaged in as a private individual and as a representative of the social work profession, a professional social work organization, or the social worker's employing agency.

b. Social workers who speak on behalf of professional social work organizations should accurately represent the official and authorized positions of the organizations.

c. Social workers should ensure that their representations to clients, agencies, and the public of professional qualifications, credentials, education, competence, affiliations, services provided, or results to be achieved are accurate. Social workers should claim only those relevant professional credentials they actually possess and take steps to correct any inaccuracies or misrepresentations of their credentials by others.

4.07 *Solicitations*

a. Social workers should not engage in uninvited solicitation of potential clients who, because of their circumstances, are vulnerable to undue influence, manipulation, or coercion.

b. Social workers should not engage in solicitation of testimonial endorsements (including solicitation of consent to use a client's prior statement as a testimonial endorsement) from current clients or from other people who, because of their particular circumstances, are vulnerable to undue influence.

4.08 *Acknowledging Credit*

a. Social workers should take responsibility and credit, including authorship credit, only for work they have actually performed and to which they have contributed.

b. Social workers should honestly acknowledge the work of and the contributions made by others.

5. SOCIAL WORKERS' ETHICAL RESPONSIBILITIES TO THE SOCIAL WORK PROFESSION

5.01 *Integrity of the Profession*

a. Social workers should work toward the maintenance and promotion of high standards of practice.

b. Social workers should uphold and advance the values, ethics, knowledge, and mission of the profession. Social workers should protect, enhance, and improve the integrity of the profession through appropriate study and research, active discussion, and responsible criticism of the profession.

c. Social workers should contribute time and professional expertise to activities that promote respect for the value, integrity, and competence of the social work profession. These activities may include teaching, research, consultation, service, legislative testimony, presentations in the community, and participation in their professional organizations.

d. Social workers should contribute to the knowledge base of social work and share with colleagues their knowledge related to practice, research, and ethics. Social workers should seek to contribute to the profession's literature and to share their knowledge at professional meetings and conferences.

e. Social workers should act to prevent the unauthorized and unqualified practice of social work.

5.02 *Evaluation and Research*

a. Social workers should monitor and evaluate policies, the implementation of programs, and practice interventions.

b. Social workers should promote and facilitate evaluation and research to contribute to the development of knowledge.

c. Social workers should critically examine and keep current with emerging knowledge relevant to social work and fully use evaluation and research evidence in their professional practice.

d. Social workers engaged in evaluation or research should carefully consider possible consequences and should follow guidelines developed for the protection of evaluation and research participants. Appropriate institutional review boards should be consulted.

e. Social workers engaged in evaluation or research should obtain voluntary and written informed consent from participants, when appropriate, without any implied or actual deprivation or penalty for refusal to participate; without undue inducement to participate; and with due regard for participants' well-being, privacy, and dignity. Informed consent should include information about the nature, extent, and duration of the participation requested and disclosure of the risks and benefits of participation in the research.

f. When evaluation or research participants are incapable of giving informed consent, social workers should provide an appropriate explanation to the participants, obtain the participants' assent to the extent they are able, and obtain written consent from an appropriate proxy.

g. Social workers should never design or conduct evaluation or research that does not use consent procedures, such as certain forms of naturalistic observation and archival research, unless rigorous and responsible review of the research has found it to be justified because of its prospective scientific, educational, or applied value and unless equally effective alternative procedures that do not involve waiver of consent are not feasible

h. Social workers should inform participants of their right to withdraw from evaluation and research at anytime without penalty.

i. Social workers should take appropriate steps to ensure that participants in evaluation and research have access to appropriate supportive services.

j. Social workers engaged in evaluation or research should protect participants from unwarranted physical or mental distress, harm, danger, or deprivation.

k. Social workers engaged in the evaluation of services should discuss collected information only for professional purposes and only with people professionally concerned with this information.

l. Social workers engaged in evaluation or research should ensure the anonymity or confidentiality of participants and of the data obtained from them. Social workers should inform participants of any limits of confidentiality, the measures that will be taken to ensure confidentiality, and when any records containing research data will be destroyed.

m. Social workers who report evaluation and research results should protect participants' confidentiality by omitting identifying information unless proper consent has been obtained authorizing disclosure.

n. Social workers should report evaluation and research findings accurately. They should not fabricate or falsify results and should take steps to correct any errors later found in published data using standard publication methods.

o. Social workers engaged in evaluation or research should be alert to and avoid con-

flicts of interest and dual relationships with participants, should inform participants when a real or potential conflict of interest arises, and should take steps to resolve the issue in a manner that makes participants' interests primary.

p. Social workers should educate themselves, their students, and their colleagues about responsible research practices.

6. SOCIAL WORKERS' ETHICAL RESPONSIBILITIES TO THE BROADER SOCIETY

6.01 *Social Welfare*

Social workers should promote the general welfare of society, from local to global levels, and the development of people, their communities, and their environments. Social workers should advocate for living conditions conducive to the fulfillment of basic human needs and should promote social, economic, political, and cultural values and institutions that are compatible with the realization of social justice.

6.02 *Public Participation*

Social workers should facilitate informed participation by the public in shaping social policies and institutions.

6.03 *Public Emergencies*

Social workers should provide appropriate professional services in public emergencies to greatest extent possible.

6.04 *Social and Political Action*

a. Social workers should engage in social and political action that seeks to ensure that all people have equal access to the resources, employment, services, and opportunities they require to meet their basic human needs and to develop fully. Social workers should be aware of the impact of the political arena on practice and should advocate for changes in policy and legislation to improve social conditions in order to meet basic human needs and promote social justice.

b. Social workers should act to expand choice and opportunity for all people, with special regard for vulnerable, disadvantaged, oppressed, and exploited people and groups.

c. Social workers should promote conditions that encourage respect for cultural and social diversity within the United States and globally. Social workers should promote policies and practices that demonstrate respect for difference, support the expansion of cultural knowledge and resources, advocate for programs and institutions that demonstrate cultural competence, and promote policies that safeguard the rights of and confirm equity and social justice for all people.

d. Social workers should act to prevent and eliminate domination of, exploitation of, and discrimination against any person, group, or class on basis of race, ethnicity, national origin, color, sex, sexual orientation, age, marital status, political belief, religion, or mental or physical disability.

Appendix L

Ethical Standards of Human Service Professionals
National Organization for Human Service Education Council for Standards in Human Service Education

---

PREAMBLE

Human services is a profession developing in response to and in anticipation of the direction of human needs and human problems in the late twentieth century. Characterized particularly by an appreciation of human beings in all of their diversity, human services offers assistance to its clients within the context of their community and environment. Human service professionals, regardless of whether they are students, faculty, or practitioners, promote and encourage the unique values and characteristics of human services. In so doing human service professionals uphold the integrity and ethics of the profession, partake in constructive criticism of the profession, promote client and community well-being, and enhance their own professional growth.

The ethical guidelines presented are a set of standards of conduct which the human service professional considers in ethical and professional decision making. It is hoped that these guidelines will be of assistance when the human service professional is challenged by difficult ethical dilemmas. Although ethical codes are not legal documents, they may be used to assist in the adjudication of issues related to ethical human service behavior.

Human service professionals function in many ways and carry out many roles. They enter into professional-client relationships with individuals; families, groups, and communities who are all referred to as "clients" in these standards. Among their roles are caregiver, case manager, broker, teacher/educator, behavior changer, consultant, outreach professional, mobilizer, advocate, community planner, community change organizer, evaluator and administrator. The following standards are written with these multifaceted roles in mind.

THE HUMAN SERVICE PROFESSIONAL'S RESPONSIBILITY TO CLIENTS

STATEMENT 1•Human service professionals negotiate with clients the purpose, goals, and nature of the helping relationship prior to its onset as well as inform clients of the limitations of the proposed relationship.

STATEMENT 2•Human service professionals respect the integrity and welfare of the client at all times. Each client is treated with respect, acceptance, and dignity.

STATEMENT 3•Human service professionals protect the client's right to privacy and confidentiality except when such confidentiality would cause harm to the client or others, when agency guidelines state otherwise, or under other stated conditions (e.g., local, state, or federal laws). Professionals inform clients of the limits of confidentiality prior to the onset of the helping relationship.

STATEMENT 4•If it is suspected that danger or harm may occur to the client or to others as a result of a client's behavior, the human service professional acts in an appropriate and professional manner to protect the safety of those individuals. This may involve seeking consultation, supervision, and/or breaking the confidentiality of the relationship.

STATEMENT 5•Human service professionals protect the integrity, safety, and security of client records. All written client information that is shared with other professionals, except in the course of professional supervision, must have the client's prior written consent.

STATEMENT 6•Human service professionals are aware that in their relationships with clients power and status are unequal. Therefore, they recognize that dual or multiple relationships may increase the risk of harm to, or exploitation of, clients, and may impair their professional judgment. However, in some communities and situations it may not be feasible to avoid social or other nonprofessional contact with clients. Human service professionals support the trust implicit in the helping relationship by avoiding dual relationships that may impair professional judgment, increase the risk of harm to clients, or lead to exploitation.

STATEMENT 7 Sexual relationships with current clients are not considered to be in the best interest of the client and are prohibited. Sexual relationships with previous clients are considered dual relationships and are addressed in STATEMENT 6 (above).

STATEMENT 8 The client's right to self-determination is protected by human service professionals. They recognize the client's right to receive or refuse services.

STATEMENT 9 Human service professionals recognize and build on client strengths.

THE HUMAN SERVICE PROFESSIONAL'S RESPONSIBILITY TO THE COMMUNITY AND SOCIETY

STATEMENT 10 Human service professionals are aware of local, state, and federal laws. They advocate for change in regulations and statutes when such legislation conflicts with ethical guidelines and/or client rights. Where laws are harmful to individuals, groups, or communities, human service professionals consider the conflict between the values of obeying the law and the values of serving people and may decide to initiate social action.

STATEMENT 11 Human service professionals keep informed about current social issues as they affect the client and the community. They share that information with clients, groups, and community as part of their work.

STATEMENT 12 Human service professionals understand the complex interaction between individuals, their families, the communities in which they live, and society.

STATEMENT 13 Human service professionals act as advocates in addressing unmet client and community needs. Human service professionals provide a mechanism for identifying unmet client needs, calling attention to these needs, and assisting in planning and mobilizing to advocate for those needs at the local community level.

STATEMENT 14 Human service professionals represent their qualifications to the public accurately.

STATEMENT 15 Human service professionals describe the effectiveness of programs, treatments, and/or techniques accurately.

STATEMENT 16 Human service professionals advocate for the rights of all members of society, particularly those who are members of minorities and groups at which discriminatory practices have historically been directed.

STATEMENT 17 Human service professionals provide services without discrimination or preference based on age, ethnicity, culture, race, disability, gender, religion, sexual orientation, or socioeconomic status.

STATEMENT 18 Human service professionals are knowledgeable about the cultures and communities within which they practice. They are aware of multiculturalism in society and its impact on the community as well as individuals within the community. They respect individuals and groups, their cultures and beliefs.

STATEMENT 19 Human service professionals are aware of their own cultural backgrounds, beliefs, and values, recognizing the potential for impact on their relationships with others.

STATEMENT 20 Human service professionals are aware of sociopolitical issues that differentially affect clients from diverse backgrounds.

STATEMENT 21 Human service professionals seek the training, experience, education and supervision necessary to ensure their effectiveness in working with culturally diverse client populations.

THE HUMAN SERVICE PROFESSIONAL'S RESPONSIBILITY TO COLLEAGUES

STATEMENT 22 Human service professionals avoid duplicating another professional's helping relationship with a client They consult with other professionals who are assisting the client in a different type of relationship when it is in the best interest of the client to do so.

STATEMENT 23 When a human service professional has a conflict with a colleague, he or she first seeks out the colleague in an attempt to manage the problem. If necessary, the professional then seeks the assistance of supervisors, consultants, or other professionals in efforts to manage the problem.

STATEMENT 24 Human service professionals respond appropriately to unethical behavior of colleagues. Usually this means initially talking directly with the colleague and, if no resolution is forthcoming, reporting the colleague's behavior to supervisory or administrative staff and/or to the professional organization(s) to which the colleague belongs.

STATEMENT 25 All consultations between human service professionals are kept confidential unless to do so would result in harm to clients or communities.

THE HUMAN SERVICE PROFESSIONAL'S RESPONSIBILITY TO THE PROFESSION

STATEMENT 26 Human service professionals know the limit and rope of their professional knowledge and offer services only within their knowledge and skill base.

STATEMENT 27 Human service professionals seek appropriate consultation and supervision to assist indecision-making when there are legal, ethical, or other dilemmas.

STATEMENT 28 Human service professionals act with integrity, honesty, genuineness, and objectivity.

STATEMENT 29 Human service professionals promote cooperation among related disciplines (e.g., psychology, counseling, social work, nursing, family and consumer sciences, medicine, education) to foster professional growth and interests within the various fields.

STATEMENT 30 Human service professionals promote the continuing development of their profession. They encourage membership in professional associations, support research endeavors, foster educational advancement, advocate for appropriate legislative actions, and participate in other related professional activities.

STATEMENT 31 Human service professionals continually seek out new and effective approaches to enhance their professional abilities.

THE HUMAN SERVICE PROFESSIONAL'S RESPONSIBILITY TO EMPLOYERS

STATEMENT 32 Human service professionals adhere to commitments made to their employers.

STATEMENT 33 Human service professionals participate in efforts to establish and maintain employment conditions which are conducive to high quality client services. They assist in evaluating the effectiveness of the agency through reliable and valid assessment measures.

STATEMENT 34 When a conflict arises between fulfilling the responsibility to the employer and the responsibility to the client, human service professionals advise both of the conflict and work conjointly with all involved to manage the conflict.

THE HUMAN SERVICE PROFESSIONAL'S RESPONSIBILITY TO SELF

STATEMENT 35 Human service professionals strive to personify those characteristics typically associated with the profession (e.g., accountability, respect for others, genuineness, empathy, pragmatism).

STATEMENT 36 Human service professionals foster self-awareness and personal growth in themselves. They recognize that when professionals are aware of their own values, attitudes, cultural background, and personal needs, the process of helping others is less likely to be negatively impacted by those factors.

STATEMENT 37 Human service professionals recognize a commitment to lifelong learning and continually upgrade knowledge and skills to serve the populations better.

References

Adams, P. L. (1982). *A primer of child psychotherapy* (2nd ed.). Boston: Little, Brown.

Adler, A. (1964). *Social interest: A challenge to mankind.* New York: Capricorn.

Aguilera, D. C., & Messick, J. M. (1982). *Crisis intervention: Theory and methodology* (4th ed.). St. Louis, MO: CV Mosby.

Alberti, R., & Emmons, M. (1995). *Your perfect right: A guide to assertive living* (7th ed.). San Luis Obispo, CA: Impact.

Alinsky, S. D. (1972). *Rules for radicals.* New York: Random House.

Alinsky, S. D. (1974). *Rules for radicals: A pragmatic primer for realistic radicals.* New York: Vintage Books.

Alle-Corliss, L. (1982). *Working with the elderly.* Paper presented to clinical staff at Verdugo Mental Health Clinic, Glendale, CA.

Alle-Corliss, L., & Alle-Corliss, R. (1998). *Human service agencies: An orientation to fieldwork.* Pacific Grove, CA: Brooks/Cole.

American Association of Retired Persons (AARP). (1993). *Public Policy Institute FACT SHEET.*

American Counseling Association (ACA). (1995). *Code of Ethics and Standards of Practice.* Alexandria, VA: Author.

American Psychiatric Association (1994). *Diagnostic and statistical manual of mental disorders* (4th ed.). Washington, DC: Author.

Anderson, H., & Goolishian, H. (1992). The client is the expert: A not-knowing approach to therapy. In S. McNamee & K. J. Gergin (Eds.), *Therapy as social construction* (pp. 25–39). Newbury Park, CA: Sage.

Anderson, M. L., & Collins, P. H. (Eds.). (1995). *Race, class, and gender: An anthology* (2nd ed.). Belmont, CA: Wadsworth.

Aponte, H. (1994). *Bread and spirit: Therapy with the new poor: Diversity and races, culture, and values.* New York: Norton.

Ashford, J. B., Lecroy, C. W., & Lortie, K. L. (1997). *Human behavior in the social environment: A multidimensional perspective.* Pacific Grove, CA: Brooks/Cole.

Association for Specialists in Group Work. (1991). Professional standards for the training of group workers. *Together: Association for Specialists in Group Work Newsletter, 20*(1), 9–14.

Atkinson, D., Morten, G., & Sue, D. W. (1993). *Counseling American minorities: A cross-cultural perspective* (4th ed.). Madison, WI: WCB Brown and Benchmark.

Austin, C. D. (1990). Case management: Myths and realities. *Families in Society, 2,* 398–405.

Axelson, L. J., & Dail, P. W. (1988). The changing character of homelessness in the U.S. *Family Relations, 37,* 463–469.

Axism, J., & Levin, H. (1982). *Social welfare: A history of the American response to need* (2nd ed.). New York: Longman.

Baars, B. J. (1986). *The cognitive revolution in psychology.* New York: Guilford Press.

Baca Zina, M., & Eitzen, D. S. (1990). *Diversity in families.* New York: Harper & Row.

Bagarozzi, D., & Kurtz, L. F. (1983). Administrator's perspectives in case management. *Arete, 8,* 13–21.

Ball, S. (1994). A group model for gay and lesbian clients with chronic mental illness. *Social Work 39*(1), 109–115.

Baird, B. N. (1996). *The internship, practicum, and field placement handbook: A guide for helping professions.* Upper Saddle River, NJ: Prentice Hall.

Baltes, P. B. (1987). Theoretical propositions of life span developmental psychology: On the dynamics between growth and decline. *Developmental Psychology, 23,* 611–626.

Banks, W. M. (1972). The black client and the helping professionals. In R. L. Jones (Ed.), *Black psychology* (pp. 205–224). New York: Harper & Row.

Baptiste, D. A. (1987). Psychotherapy with gay lesbian couples and their children in "step-families": A challenge for marriage and family therapists. *Journal of Homosexuality, 14,*(½), 223–238.

Barker, R. L. (1987). *The social work dictionary.* Silver Spring, MD: National Association of Social Workers.

Barker, R. L. (1991). *The social work dictionary* (2nd ed.). Silver Spring, MD: National Association of Social Workers.

Barringer, F. (1992, July 24). White-black disparity in income narrowed in the 80's, census shows. *The New York Times,* pp. Al, A10.

Barusch, A. S. (1995). Programming for family care of elderly dependents: mandates, incentives, and service rationing. *Social Work 40*(3), 315–322.

Basse, D. T., & Greenstreet, K. L. (1991). On counseling men. *Journal of the Colorado Association for Counseling and Development, 19,* 3–6.

Beck, A. (1976). *Cognitive therapy and emotional disorders.* New York: International Universities Press.

Beck, A. T. (1993). Static communication. In V. Cyrus (Ed.), *Experiencing race, class, and gender in the United States.* Mountain View, CA: Mayfield.

Beckhard, R. (1969). *Organization development: Strategies and models.* Reading, MA: Addison-Wesley.

Beklin, G. S. (1984). *Introduction to counseling* (2nd ed.). Dubuque, IA: William C. Brown.

Bell, A. P., & Weinberg, M. S. (1978). *Homosexualities: A study of diversity among men and women.* New York: Simon and Schuster.

Bellah, R. N., Madsen, R., Sullivan, W. M., Swidler, A., & Tipton, S. M. (1985). *Habits of the heart.* New York: Harper & Row.

Berengarten, S. (1957). Identifying learning patterns of individual students: An exploratory study. *Social Service Review, 31*(4): 407–417.

Berg, I. K., & Miller, S. D. (1992). *Working with the problem drinker: A solution-focused approach.* New York: Norton.

Berman, A. L., & Jobes, D. A. (1991). *Adolescent suicide: Assessment and intervention.* Washington, DC: American Psychological Association.

Bernard, J. (1975). *Women, wives, and mothers: Values and options.* New York: Aldene.

Bernstein, A. G. (1981). *Case managers: Who are they and are they making any difference in mental health?* Unpublished doctoral dissertation, University of Georgia, Athens.

Berry, B. (1965). *Race and ethnic relations* (3rd ed.). Boston, MA: Houghton Mifflin.

Bertalanffy, L. (1934). *Modern theories of development: An introduction to theoretical biology.* London: Oxford University Press.

Block, J. H. (1976). Issues, problems and pitfalls in assessing sex differences. *Merrill-Palmer Quarterly, 22,* 282–308.

Blood, P., Tuttle, A., & Lackey, G. (1992). Understanding and fighting sexism: A call to men. In M. L. Anderson and P. H. Collins (Eds.), *Race, class, and gender: An anthology* (2nd ed., pp. 154–161). Belmont, CA: Wadsworth.

Bly, R. (1990). *Iron John: A book about men.* Reading, MA: Addison-Wesley.

Bole, T. J. (1971). Systematic observation of behavior change with older children in group therapy. *Psychological Reports, 28,* 26.

Bolton, R. (1979). *People skills.* New York: Simon and Schuster.

Bowlby, J. (1969). *Attachment.* New York: Basic Books.

Bowlby, J. (1973). *Separation.* New York: Basic Books.

Bowen, M. (1960). A family concept of schizophrenia. In D. Jackson (Ed.), *The etiology of schizophrenia.* New York: Basic Books.

Bowen, M. (1961). Family psychotherapy. *American Journal of Orthopsychiatry, 31,* 40–60.

Bowen, M. (1971). Family therapy and family group therapy. In D. Jackson (Ed.), *The etiology of schizophrenia.* New York: Basic Books.

Boyd, N. (1938). Play as a means of social adjustment. In J. Lieberman (Ed.), *New trends in group work.* New York: Association Press.

Brackett, J. (1895). *The charity organization movement: Its tendency and its duty.* New Haven, CT: Proceedings of the 22nd National Conference of Charities and Corrections.

Bragger, G., & Holloway, S. (1978). *Changing human service organizations.* New York: Free Press.

Brammer, L. M. (1988). *The helping relationship: Process and skills* (4th ed.). Englewood Cliffs, NJ: Prentice Hall.

Brammer, L. M., & Shostrom, E. L. (1982). *Therapeutic psychology: Fundamentals of counseling and psychotherapy* (4th ed.). Englewood Cliffs, NJ: Prentice Hall.

Brill, N. (1990). *Working with people* (4th ed.). White Plains, NY: Longman.

Brill, N. (1995). *Working with people* (5th ed.). White Plains, NY: Longman.

Brockett, D. R., & Gleckman, A. D. (1991). Countertransference with the older client: The importance of mental health counselor awareness and strategies for effective management. *Journal of Mental Health Counseling, 13*(3), 343–315.

Brodley, B. T. (1986, Spetember 3–7). *Client centered therapy—What is it? What is it not?.* Presentation at First Annual Meeting of the Association for the Development of the Person-Centered Approach, Chicago, IL.

Brown, L. S. (1995). Therapy with same-sex couples: An introduction. In N. S. Jacobsen & A. S. Gurman (Eds.), *Clinical handbook of couple therapy* (pp. 295–316). New York: Guilford.

Bruggemann, W. G. (1996). *The practice of macro social work.* Chicago: Nelson-Hall.

Brughardt, S. (1987). Community-based social action. In A. Minihan (Editor-in-Chief), *Encyclopedia of social work* (18th ed., vol. 1, pp. 292–299). Silver Spring, MD: National Association of Social Workers.

Buckingham, S. L., & Van Gorp, W. G. (1988). Essential knowledge about AIDS dementia. *Social Work, 33,*(2), 112–115.

Buhler, C. (1968). The developmental structure of goal setting in group and individual studies. In C. Buhler & F. Massarik (Eds.), *The course of human life: A study of goals in the humanistic perspective.* New York: Springer.

California Department of Justice. (1976). Information Pamphlet #8, *Child abuse: The problem of the abused and neglected child.* Los Angeles, CA: California Department of Justice, Information Services.

Calopinto, J. (1991). Structural family therapy. In A. Gurman & D. Knuskern (Eds.), *Handbook of family therapy* (vol. 2). New York: Brunner/Mazer.

Caplan, G. (1961). *An approach to community mental health.* New York: Grune & Stratton.

Carlson, J. & Lewis, J. (1988). (Eds.) *Counseling the adolescent.* Denver, CO: Love.

Carter, E. A., & McGoldrick, M. (Eds.). (1980). *The family life cycle: A framework for family therapy.* New York: Gardner Press.

Carter, E. A., & McGoldrick, M. (Eds.). (1988). *The changing life cycle. A framework for family therapy* (2nd ed.). New York: Gardner Press.

Cartwright, D. (1968). The nature of group cohesiveness. In D. Cartwright & A. Zonder (Eds.), *Group dynamics: Research and theory* (3rd ed.). New York: Harper & Row.

Castex, G. M. (1997). Providing services to Hispanic/Latino populations: Profiles on diversity. *Social Work, 39*(3), 288–296.

Cecchin, G. (1987). Hypothesizing, circularity and neutrality revisited: An invitation to curiosity. *Family Process, 26*(4), 405–414.

Chaidez, L. (Ed.). (1991). *The California child abuse and neglect reporting law: Issues and answers for health practitioners.* Sacramento, CA: State Department of Social Services, Office of Child Abuse Prevention.

Cherlin, A. (1983). Changing family and household: Contemporary lessons from historical research. *Annual Review of Sociology, 9,* 51–66.

Cherniss, C., & Krantz, D. L. (1983). The ideological community as an antidote to burnout in human services. In B. A. Farber (Ed.), *Stress and burnout in the human service professionals* (pp. 198–212). New York: Pergamon Press.

Chess, W. A., & Norlin, J. M. (1988). *Behavior and the social environment.* Boston: Allyn & Bacon.

Chiaferi, R., & Griffin, M. (1997). *Developing fieldwork skills: A guide for human services, counseling, and social work students.* Pacific Grove, CA: Brooks/Cole.

Code, B., & O'Hanlon, W. H. (1993). *A brief guide to brief therapy.* New York: Norton.

Cohler, B. J., & Lieberman, M. A. (1980). Social relations and mental health: Middle aged and older men and women from three European ethnic groups. *Research on Aging, 2,* 445–469.

Compton, B. R., & Galaway, B. (1984). *Social work processes* (3rd ed.). Pacific Grove, CA: Brooks/Cole.

Compton, B. R., & Galaway, B. (1989). *Social work processes* (4th ed.). Pacific Grove, CA: Brooks/Cole.

Compton, B. R., & Galaway, B. (1994). *Social work processes* (5th ed.). Pacific Grove, CA: Brooks/Cole.

Congress, E. P. (1994). The use of culturagrams to assess and empower culturally diverse families. *Families in Society, 75,* 531–540.

Conyne, R. K., Wilson, R. F., Kline, W. B., Morran, D. K., & Ward, D. E. (1993). Training group workers: Implications for the new ASGW training standards for training and practice. *Journal for Specialist in Group Work, 18*(1), 11–23.

Cooper, M. (1991). The rich in America. *U.S. News and World Report,* Nov. 18, 1991, 34–40.

Cormier, S., & Cormier, B. (1998). *Interviewing strategies for helpers: Fundamental skills and cognitive behavioral interventions.* Pacific Grove, CA: Brooks/Cole.

Corey, G. (1996). *Theory and practice of counseling and psychotherapy* (5th ed.). Pacific Grove, CA: Brooks/Cole.

Corey, G., & Corey, M. S. (1993). *Becoming a helper* (2nd ed.). Pacific Grove, CA: Brooks/Cole.

Corey, G., & Corey, M. S. (1998). *Becoming a helper* (3rd ed.). Pacific Grove, CA: Brooks/Cole.

Corey, G., Corey, M. S., & Callanan, P. (1993). *Issues and ethics in the helping professions* (4th ed.). Pacific Grove, CA: Brooks/Cole.

Corey, G., Corey, M. S., & Callanan, P. (1998). *Issues and ethics in the helping professions* (5th ed.). Pacific Grove, CA: Brooks/Cole.

Corey, M. S., & Corey, G. (1997). *Groups: Process and practice* (5th ed.). Pacific Grove, CA: Brooks/Cole.

Corsini, R., & Rosenberg, B. (1955). Mechanisms of group psychotherapy. *Journal of Abnormal and Social Psychiatry, 51,* 406–411.

Costa, L., & Alterkruse, M. (1994). Duty-to-warn guidelines for mental health counselors. *Journal of Counseling and Development, 72,* 346–350.

Coulton, C. J. (1996). Poverty, work, and community: A research agenda for an era of diminishing federal responsibility. *Social Work, 41*(5), 509–520.

Cournoyer, B. (1996). *The social work skills workbook* (2nd ed., pp. 53–68). Pacific Grove, CA: Brooks/Cole.

Cowan, G., & Avants, S. K. (1988). Children's influence on strategies: Structure, sex differences, and bilateral mother/child influences. *Child Development, 53,* 984–990.

Cowger, C. D. (1992). Assessment of client strengths. In Saleebey, D. (Ed.), *The strengths perspective in social work practice* (pp. 139–147). New York: Longman.

Coyle, G. L. (1930). *Social process in organized groups.* New York: Richard R. Smith.

Cozby, P. C. (1973). Self disclosure: A literature review. *Psychological Bulletin, 79,* 73–91.

Crooks, R., & Baur, K. (1993). *Our sexuality* (5th ed.). Redwood City, CA: Benjamin/Cummings.

Child Welfare League of America (CWLA). (1990). *Standards for service for abused or neglected children and their families.* Washington, DC: Author.

Cyrus, V. (Ed). (1993). *Experiencing race, class, and gender in the United States.* Mountain View, CA: Mayfield.

Daniels, R., & Kitano, H. H. L. (1970). *American racism: Exploitation of the nature of prejudice.* Englewood Cliffs, NJ: Prentice Hall.

Danish, J., D'Augelli, A., & Hauer, A. (1980). *Helping skills: A basic training program.* New York: Human Services Press.

Davis, L. E., & Gelsamino, J. (1994). An assessment of practitioner cross-racial treatment experiences. *Social Work, 39*(1), 116–123.

Davis, M. (1995, Winter). Critical incident stress debriefing: The case of corrections. In *The keeper's voice* (http://www.assp.uic.edu/IACO/ku160145.htm).

Davitt, J. K., & Kaye, L. W. (1996). Supporting patient autonomy: Decision making in home health care. *Social Work, 41*(1), 41–50.

DeJong, P., & Miller, S. D. (1995). How to interview for client's strengths. *Social Work, 40*(6), 729–736.

Dember, W. N. (1974). Motivation and the cognitive revolution. *American Psychologist, 29,* 161–168.

deShazer, S. (1982). *Patterns of brief family therapy: An ecosystemic approach.* New York: Guilford.

deShazer, S. (1985). *Keys to solution in brief therapy.* New York: Norton.

deShazer, S., Kinberg, I., Lipchik, E., Numnally, E., Molnar, A., Gingerich, W., & Weiner-Davison, M. (1986). Brief therapy: Focused solution development. *Family Practice, (2nd ed. 25,* 207–220).

DeVoe, D. (1990). Feminist and nonsexist counseling: Implications for the male counselor. *Journal of Counseling and Development, 69*(1), 33–38.

Devore, W., & Schlesinger, E. G. (1981). *Ethnic sensitive social work practice.* St. Louis, MO: C. V. Mosby.

Devore, W., & Schlesinger, E. G. (1995). *Ethnic sensitive social work practice.* St. Louis, MO: C. V. Mosby.

Dinkmeyer, D. C., Dinkmeyer, D. C., Jr., & Sperry, L. (1987). *Adlerian counseling and psychotherapy* (2nd ed.). Columbus, OH: Merrill.

Dixon, S. M. (1987). *Working with people in crisis* (2nd ed.). Columbus, OH: Merrill.

Doster, J., Surrat, F., & Webster, T. (1995) *Interpersonal variables affecting psychological communications of hospitalized psychiatric patients.* Paper presented at meeting of Southeastern Psychological Association, Atlanta.

Douglass, R. (1980). *A study of maltreatment of the elderly and other vulnerable adults.* Ann Arbor, MI: University of Michigan, Institute of Gerontology and School of Public Health.

Downs, S. W., Costin, L. B., & McFadden, E. (1996). *Child welfare and family services: Policies and practices.* White Plains, NY: Longman.

Dower, J. W. (1986). *War without mercy: Race and power in the pacific war.* New York: Pantheon.

Doyle, R. E. (1992). *Essential skills and strategies in the helping process.* Pacific Grove, CA: Brooks/Cole.

Drachman, D. (1995). Immigration statuses and their influence on service provision, access, and use. *Social Work, 40*(2), 195.

Drachman, D., & Halberstadt, A. (1992). A stage of migration framework as applied to recent soviet emigres. *Journal of Multicultural Social Work, 2,* 63–78.

Drakeford, J. (1967). *The awesome power of the listening ear.* Waco, TX: Word.

Duvall, E. M. (1990). *Family development.* Philadelphia: J. B. Lippincott.

Dye, T. R. (1975). *Understanding public policy* (2nd ed.). Englewood Cliffs, NJ: Prentice Hall.

D'Zurilla, T. J. (1988). Problem-solving therapies. In K. S. Dobson (Ed.), *Handbook of cognitive-behavioral therapies.* New York: Guilford Press.

Eaton, W. O., & Evans, L. R. (1986). Sex differences in human motor activity level. *Psychological Bulletin, 100,* 19–28.

Edelwich, J., & Brodsky, A. (1982). Training guidelines: Linking the workshop experience to needs on and off the job. In W. S. Paine (Ed.), *Job stress and burnout* (pp. 133–154). Beverly Hills, CA: Sage.

Edlefsen M., & Baird, M. (1994). Making it work: Preventive mental health care for disadvantaged preschoolers. *Social Work, 39*(5), 566–573.

Egan, G. (1994). *The skilled helper* (5th ed.). Pacific Grove, CA: Brooks/Cole.

Eggebeen, D. J., & Lichter, D. T. (1991). Race, family structure, and changing poverty among american children. *American Sociological Review, 56,* 801–817.

Ellis, A. E. (1962). *Reason and emotion of psychotherapy.* New York: Lyle Stuart.

Encyclopedia of Social Work. (1974). Guilford, CT: Duskin.

Erickson, E. (1950). *Childhood and society.* New York: Norton.

Erickson, E. (1968). *Identity: Youth and crisis.* New York: Norton.

Erickson, E. H. (1982). *The life cycle completed.* New York: Norton.

Erickson, M. H. (1954). Special techniques of brief hypnotherapy. *Journal of Clinical and Experimental Hypnosis, 2,* 109–129.

Etzioni, A. (1964). *Modern organizations.* Englewood Cliffs, NJ: Prentice Hall.

Eubank, H. (1932). *Sociologist, 2,* pp. 160–173.

Evans, D. R., Hearn, M. T., Uhlemann, M. R., & Ivey, A. E. (1984). *Essential interviewing: A programmed approach to effective communication* (2nd ed.). Monterey, CA: Brooks/Cole.

Fables, R. A., Einsenberg, N., & Miller, P. A. (1990). Maternal correlates of children's vicarious emotional responsiveness. *Developmental Psychology, 26,* 639–648.

Fairchild, H. P. (1964). *Dictionary of sociology.* Paterson, NJ: Littlefield, Adams.

Faiver, C., Eisengart, S., & Colonna, R. (1995). *The counselor intern's handbook.* Pacific Grove, CA: Brooks/Cole.

Fanning, P., & McKay, M. (1993). *Being a man: A guide to the new masculinity.* Oakland, CA: New Harbinger Publications.

Farber, B. A. (Ed.). (1983). *Stress and burnout in human service professions.* New York: Pergamon Press.

Fellin, P. (1987). *The community and the social worker.* Itasca, IL: F. E. Peacock.

Fellin, P. (1995). *The community and the social worker.* Itasca, IL: F. E. Peacock.

Fennel, D. L., & Weinhold, B. K. (1989). *Counseling families: An introduction to marriage and family therapy.* Denver, CO: Love.

Festinger, L. (1953). An analysis of compliant behavior. In M. Sheif (Ed.), *Group relationships at the crossroads.* New York: Harper & Row.

Filip, J., Schene, P., & McDaniel, N. (Eds.). (1991). *Helping in child protective services: A casework handbook* (rev. ed.). Englewood, CO: American Association for Protecting Children.

First, R. J., Rife, J. C., & Toomey, B. (1994). Homelessness in rural areas: Causes, patterns, and trends. *Social Work 39*(1), 97–108.

Fisch, R., Weakland, J. H., & Segel, L. (1982). *Tactics of change: Doing therapy briefly.* San Francisco: Jossey-Bass.

Fleuridas, C., Nelson, T. S., & Rosenthal, D. M. (1986). The evolution of circular questions: Training family therapists. *Journal of Marital and Family Therapy, 12*(2), 113–127.

Forney, D. S., Wallace-Schutzman, F., & Wiggers, T. T. (1982). Burnout among career development professionals: Preliminary findings and implications. *Personnel and Guidance Journal, 60,* 435–439.

Franklin, C., & Jordan, C. (1999). *Family practice: Brief systems methods for social work.* Pacific Grove, CA: Brooks/Cole.

Freeman, E. M. (1996). Welfare reforms and services for children and families: Setting a new practice, research, and policy agenda. *Social Work, 41*(2), 521–532.

Freud, S. (1947). *The ego and the id.* London: Hogarth Press.

Fujimara, L. E., Weis, D. M., & Cochran, J. R. (1985). Suicide: Dynamics and implications for counseling. *Journal of Counseling and Development, 63*(6), 212–215.

Gailbraith, J. R. (1973). *Designing complex organizations.* Reading, MA: Addison-Wesley.

Galigor, J. (1977). Perceptions of the group therapist and the dropout from group. In A. R. Wolberg & M. L. Aronson (Eds.), *Group therapy 1997: An overview.* New York: Stratton Inter Continental Medical Book Corp.

Gallup, G. (1989). *The Gallup Poll: Public opinion.* Wilmington, DE: Scholarly Resources.

Garbarino, J., & Abramowitz, R. H. (1992). Sociocultural risk and opportunity. In J. Barbarino (Ed.), *Children and family in the social environment* (2nd ed.). New York: Aldene de Gruyter.

Garwin, C. D., & Cox, F. M. (1987). A history of community organizing since the Civil War with special reference to oppressed communities. In F. M. Cox, J. L. Erlich, J. Rothman, & J. Tropman (Eds.), *Strategies of community organizing.* Itasca, IL: F. E. Peacock.

Gelso, C. J., & Carter, J. A. (1985). The relationship in counseling and psychotherapy: Components, consequences and theoretical antecedents. *The Counseling Psychologist, 13,* 155–243.

Gerdes, E. P. (1981). The effects of sex and sex-role concept on self-disclosure. *Sex-Roles, 7,* 789–798.

Gerhardt, U. C. (1990). Principles of case management. *Caring for the chronically mentally ill.* Itasca, IL: Peacock.

Germain, C. (1985). The place of community work within an ecological approach to social work practice. In S. A. Taylor & R. W. Roberts (Eds.), *Theory and practice of community social work.* New York: Columbia University Press.

Germain, C. B. (1991). *Human behavior in the social environment: An ecological view.* New York: Columbia University Press.

Gharajedaghi, J., & Ackoff, R. L. (1986). *A prologue to national development planning.* New York: Greenwood Press.

Gilman, D., & Koverola, C. (1995). Cross-cultural counseling. In D. G. Martin & A. D. Moore (Eds.), *First steps in the art of interventions: A guidebook for trainers in the helping professions* (pp. 378–391). Pacific Grove, CA: Brooks/Cole.

Goldenberg, I., & Goldenberg, H. (1991). *Family therapy: An overview.* Pacific Grove, CA: Brooks/Cole.

Goldfried, M. R., & Castonguary, L. G. (1992). The future of psychotherapy integration. *Psychotherapy, 29*(1), 4–10.

Goldman, D. (1995). *Emotional intelligence.* New York: Bantam.

Goldstein, H. (1993). Starting where the client is. *Social Casework, 64*(5), 267–275.

Goodman, M., Brown, J., & Dietz, P. (1992). *Managing managed care: A mental health practitioner's survival guide.* Washington, DC: American Psychiatric Press.

Goolishian, H., & Anderson, H. (1987). Language systems and therapy: An evolving idea. *Psychotherapy, 24*(35), 529–538.

Gooren, B., Fliers, E., & Courtney, K. (1990). Biological determinants of sexual orientation. *Annual Review of Sex Research, 1,* 175–196.

Gordon, M. (1964). *Assimilation in American life: The role of race, religion, and national origins* (pp. 27–28). New York: Oxford University Press.

Gravold, D. K. (Ed.). (1996). *Cognitive and behavioral treatment methods and applications.* Pacific Grove, CA: Brooks/Cole.

Green, B. (1982). *Cultural awareness in the human services.* Englewood Cliffs, NJ: Prentice Hall.

Greene, R. R. (1987). Case management. *Helping the homeless, mentally ill, and persons with AIDS and their families.* Silver Spring, MD: National Association of Social Workers.

Grotevant, H. D., & Carlson, C. I. (1989). *Family assessment: A guide to methods and measures.* New York: Guilford.

Guilliland, B., & James, R. (1990). *Crisis intervention strategies.* Pacific Grove, CA: Brooks/Cole.

Gulliland, B. E., & James, R. K. (1993). *Crisis intervention strategies* (2nd ed.). Pacific Grove, CA: Brooks/Cole.

Gulick, L. (1937). Notes on the theory of organizations. In L. Guther & L. Irwick (Eds.), *Papers on the science of organizations.* New York: Institute of Public Administration.

Gustafson, J. (1998). *Brief versus long psychotherapy: When, why, and how?* (http://www.aronson.com/ppp/gusstaf.html).

Guttierez, L. (1995). Working with women of color: An empowerment perspective. In J. Rothman, J. L. Erlich, & J. E. Tropman (Eds.), *Strategies of community intervention* (5th ed., pp. 204–211). Itasca, IL: F. E. Peacock.

Guttierez, L., Alvarez, A. R., Nemon, H., & Lewis, E. A. (1996). Multicultural community organizing: A strategy for change. *Social Work, 41*(5), 501–508.

Guttierez, L., Parsons, R. J., & Cox, E. O., (1998). *Empowerment.* Pacific Grove, CA: Brooks/Cole.

Hack, T. F. (1995). Suicide risk assessment and intervention. In D. G. Martin & A. Moore (Eds.). (1995). *First steps in the art of intervention: A guidebook for trainers in the helping professions.* Pacific Grove, CA: Brooks/Cole.

Hackney, H., & Cormier, L. S. (1988). *Counseling strategies and interventions* (3rd ed.). Englewood Cliffs, NJ: Prentice Hall.

Haley, J. (1971a). Approach to family therapy. In J. Haley (Ed.), *Changing families: A family therapy reader.* New York: Grune & Stratton.

Haley, J. (1973). *Uncommon therapy.* New York: Norton.

Haley, J. (1990). *Problem-solving therapy.* San Francisco: Jossey-Bass.

Haley, J. (Ed.). (1971b). *Changing families.* New York: Grune & Stratton.

Hansen, J., Stevic, R., & Warner, R. (1986). *Counseling: Theory and process.* Boston: Allyn & Bacon.

Hansenfeld, Y. (1983). *Human service organizations.* Englewood Cliffs, NJ: Prentice Hall.

Hansenfeld, Y. (1995). Program development. In J. Rothman, J. L. Erlick, & J. E. Tropman (Eds.). *Strategies of community intervention.* Itasca, IL: F. E. Peacock.

Hare, A. (1976). *Handbook of small group research* (2nd ed.). New York: Free Press.

Harlow, J. (1996). *Counseling basics for wiccan clergy* (Home page, http://,e,bers,@aol.com/Jehanas/c_basics/c_models. html).

Hartford, M. E. (1972). *Groups in social work.* New York: Columbia University Press.

Hartman, A. (1978). Diagrammatic assessment of family relationships. *Social Casework, 59,* 456–476.

Harvey, D. F., & Brown, D. F. (1992). *An experiential approach to organization development* (4th ed.). Englewood Cliffs, NJ: Prentice Hall.

Hawkins-McNary, I. D. (1979). The effect of institutional racism on the therapeutic relationship. *Perspectives in Psychiatric Care, 17*(11), 25–54.

Haynes, K. S., & Michelson, J. S. (1986). *Affecting change: Social workers in the political arena.* New York: Longman.

Heges, L. E. (1993, May/June). In praise of dual relationships. *The California Therapist,* 46–50.

Henker, B., & Whalen, B. (1989). Hyperactivity and attention deficits. *American Psychologists, 44,* 216–223.

Henry, S. (1992). *Group skills in social work: A four-dimensional approach* (2nd ed.). Pacific Grove, CA: Brooks/Cole.

Hepworth D. H., & Larsen, J. A. (1990). *Direct social work practice: Theory and skills* (3rd ed.). Belmont, CA: Wadsworth.

Hepworth D. H., Rooney, R. H., & Larsen, J. A. (1997). *Direct social work practice: Theory and skills* (5th ed.). Pacific Grove, CA: Brooks/Cole.

Herek, G. M., & Berrill, K. (1990). Violence against lesbians and gay men: Issues for research, practice, and policy. *Journal of Interpersonal Violence, 5.*

Herlihy, B., & Corey, G. (1993). *Dual relationships in counseling.* Alexandria, VA: American Association for Counseling and Development.

Hersey, P., & Blanchard, K. H. (1988). *Management of organizational behavior: Utilizing human resources.* Englewood Cliffs, NJ: Prentice Hall.

Hetterington, E. M. (1989). Coping with family transitions: Winners, losers, and survivors. *Child Development, 60,* 1–14.

Hill, M., Glaser, K., & Harden, J. (1995). A feminist model for ethical decision making. In E. G. Rave & C. C. Larsen (Eds.), *Ethical decision making in therapy: Feminist perspectives* (pp. 18–37). New York: Guilford Press.

Ho, D. Y. F. (1991). Cultural values and professional issues in clinical psychology: Implications from the Hong Kong experience. *American Psychologist, 40*(11), 1212–1218.

Ho, M. K. (1987). *Family therapy and ethnic minorities.* Newbury Park, CA: Sage.

Hoebel, E. A. (1972). *Anthropology: The study of man* (4th ed.). New York: McGraw-Hill.

Hoff, L. A. (1989). *People in crisis: Understanding and helping* (3rd ed.). Redwood City, CA: Addison-Wesley.

Hokenstad, M. C. (1997). Teaching practitioners ethical judgement. *NASW News, 32,* 4.

Holden, W. (1972). Process recording. *Social Work Encyclopedia Reporter, 20,* 67–69.

Holland, T. P., & Pitchers, M. K. (1987). Organizations: Context for social service delivery. *Encyclopedia of social work.* Silver Spring, MD: National Association of Social Workers.

Hollis, F. (1972). *Casework: A psychosocial theory.* New York: Random House.

Homan, M. S. (1994). *Promoting community change: Making it happen in the real world.* Pacific Grove, CA: Brooks/Cole.

Hooyman, N. R., Frederiksen, K. L., & Perlmutter, B. (1995). In J. Rothman, J. L. Erlich, & J. E. Tropman (Eds.), *Strategies of community intervention* (5th ed. pp. 417–426). Itasca, IL: F. E. Peacock.

Hulewat, P. (1996). Resettlement: A cultural and psychological crisis. *Social Work, 41*(2), 129–135.

Hutchins, D. E., & Cole, C. G. (1992). *Helping relationships and strategies* (2nd ed.). Pacific Grove, CA: Brooks/Cole.

Institute for Rehabilitation and Disability Management and National Center for Social Policy and Practice. (1988). *Case management: The present and future tools for improving employment opportunities for people with disabilities.* Unpublished proposal.

Intagliata, J. (1982). Improving the quality of community care for the chronically mentally disabled: The role of case management. *Schizophrenia Bulletin, 8*(4), 655–674.

Internet. *Client centered terms* (http://www.galluudet.edu/~11mgoum/tersms3.html).

Ivey, A. (1986). *Developmental therapy: Theory into practice.* San Francisco: Jossey-Bass.

Ivey, A. E. (1991). *Developmental strategies for helpers: Individual, family, and network interventions.* Pacific Grove, CA: Brooks/Cole.

Ivey, A. E., Ivey, M. B., & Simek-Morgan, L. (1993). *Counseling and psychotherapy: A multicultural perspective.* Boston: Allyn & Bacon.

Jacklin, C. N. (1989). Male and female issues of gender. *American Psychologist, 44,* 127–133.

Jacobs, C. (1989, April). The truth about abuse. *Shape,* 95.

Janosik, E. H. (1984). *Crisis counseling: A contemporary approach.* Belmont, CA: Wadsworth.

Johnson, W. (1986). *The social services: An introduction* (2nd ed.). Itasca, IL: F. E. Peacock.

Jones, D. L. (1979). African American clients: Clinical practice issues. *Social Work, 24,* 112–118.

Jones, R. M. (1995). The price of welfare dependency: Children pay. *Social Work, 40*(4), 496–505.

Jordan, C., & Franklin, C. (1995). *Clinical assessment for social workers: Qualitative and qualitative methods.* Chicago: Lyceum Press.

Jung, C. J. (1971). The stages of life. In J. Campbell (Ed.), *The portable Jung.* New York: Viking Press.

Jung, M. (1998). *Chinese American family therapy: A new model for clinicians.* San Francisco: Jossey-Bass.

Jung, M. (1998). Presentation of roles, functions, and tasks of case management. Skill building in case management symposium, May 27, 1998. Conference center at Riverside County Department of Social Services, CA: Sponsored by California State University, San Bernadino, Department of Social Work.

Justice, R., & Justice, B. (1990). Crisis intervention with child abusing families: Short-term cognitive coercive group therapy using goal attainment scaling. In A. Roberts (Ed.), *Crisis intervention handbook: Assessment, treatment, and research.* Belmont, CA: Wadsworth.

Kadushin, A., & Martin, J. A. (1988). *Child welfare services* (4th ed.). New York: Macmillan.

Kahn, M. (1991). *Between therapist and client.* New York: W. H. Freeman.

Kalter, N., Kloner, A., Schreier, S., & Okla, K. (1989). Predictors of children's post divorce adjustment. *American Journal of Orthopsychiatry, 59,* 605–619.

Kamerman, S. B. (1996). The new politics of child and family policies. *Social Work 41*(5), 453–467.

Kamerman, S. B., & Khan, A. J. (1976). *Social services in the United States: Policies and programs.* Philadelphia, PA: Temple University Press.

Kamya, H. A. (1997). African immigrants in the U.S.: The challenge for research and practice. *Social Work, 42*(2), 154–165.

Kanel, K. (1999). The ABC model of crisis intervention: A formula for understanding and treating crisis. In K. Kanel, *A guide to crisis intervention.* Ohio: ITP Custom Publishing.

Kaplan, K. O. (1990). Recent trends in case management. In A. Minihan (Editor-in-Chief), *Encyclopedia of social work,* (18th ed.). 1990 supplement (pp. 60–77). Silver Spring, MD: National Association of Social Workers.

Kapp, M. B. (1987). Interprofessional relationships in geriatrics: Ethical and legal considerations. *The Gerontologist, 27*(5), 547–552.

Kaul, M. L. (1995). Serving oppressed communities: The self-help approach. In J. Rothman, J. L. Erlich, & J. E. Tropman (Eds.), *Strategies of community intervention* (5th ed., pp. 268–274). Itasca, IL: F. E. Peacock.

Keen, S. (1991). *Fire in the belly: On being a man.* New York: Bantam Books.

Kerr, M. E. (1981). Family systems theory and therapy. In A. Horne, M. Gurnin, & D. P. Knishern (Eds.). *Handbook of family therapy.* New York: Brunner/Mazel.

Kerr, M. E., & Bowen, M. (1988). *Family evaluation: An approach based on Bowen theory.* New York: Norton.

Kettner, P., & Martin, L. (1994). Privatization. In M. Austin & I. J. Lowe (Eds.), *Controversial issues in communities and organizations* (pp. 165–172). Boston, MA: Allyn & Bacon.

Kimmell, D. C. (1974). *Adulthood and aging.* New York: Wiley.

Kinsey, A. C. (1948). *Sexual behavior in the human male.* Philadelphia, PA: Saunders.

Kirst-Ashman, K. K., & Hull, G. H., Jr. (1993). *Understanding generalist practice.* Chicago, IL: Nelson Hall.

Kitano, H. H. L. (1981). Asian Americans, the Chinese, Japanese, Koreans, Filipinos, and Southeast Asians. *The Annuals, 454,* 125–138.

Kock, L., & Kock, J. (1980, January 27). Parent abuse—A new plague. *Parade,* p. 14.

Kohut, H. (1984). *How does analysis cure?* Chicago: University of Chicago Press.

Kopels, S., & Kagle, J. D. (1993, March). Do social workers have a duty to warn? *Social Service Review*, 101–126.

Kottler, J. F. (1986). *On being a therapist.* San Francisco, CA: Jossey-Bass.

Kramer, R. M. (1981). *Voluntary agencies in the welfare state.* Berkeley: University of California Press.

Laird, J., & Hartman, A. (1985). *A handbook of child welfare.* New York: Free Press.

Lambert, J. J. (1992). Psychotherapy outcome research: Implications for integrative and eclectic therapists. In J. C. Norcross & M. R. Goldfried (Eds.), *Handbook of psychotherapy integration* (pp. 3–45). New York: Basic Books.

Lauffer, A. (1981). The practice of social planning. In N. Gilbert & H. Specht (Eds.), *Handbook for social services.* Englewood Cliffs, NJ: Prentice Hall.

Lauffer, A. (1984). *Understanding your social agency.* Beverly Hills, CA: Sage.

Lazzari, J. J., Ford, H. R., & Haughey, K. J. (1996). Making a difference: Women of action in the community. *Social Work, 41*(2), 197–205.

Leacock, E. (1971). *The culture of poverty: A critique.* New York: Simon & Schuster.

Leader, A. (1958). The problem of resistance in social work. *Social Work, 3,* 19–23.

Leitner, L. A. (1974). *Crisis counseling: A contemporary approach.* Monterey, CA: Wadsworth Health Sciences Division.

Leslie, R. S. (1989, July/August). Confidentiality. *The California Therapist.*

Leslie, R. S. (1993, January/February). Dual relationships. *The California Therapist, 6.*

Leslie, R. S. (1994, November/December). Reporting of domestic violence? *California Therapist.*

Levant, R. F. (1990). Men's changing roles. *The Family Psychologist, 6,* 4–6.

Levant, R. F. (1992). Toward the reconstruction of masculinity. *Journal of Family Psychology, 5*(3/4), 379–402.

Levant, R. H, Hirsh, L., Celentano, E., Cozza, T., Hill, S., MacEachern, M., Marty, N., & Schnedeker, J. (1992). The male role: An investigation of norms and stereotypes. *Journal of Mental Health Counseling.*

Lewis, O. (1966). The culture of poverty. *Scientific American, 215,* 19–25.

Lieberman, M., & Borman, L. (Eds.). (1979). *Self-help groups for coping with crisis.* San Francisco: Jossey-Bass.

Lieberman, M. A., Yalom, I. D., & Miles, M. E. (1973). *Encounter groups: First facts.* New York: Basic Books.

Lieberson, S. (1980). *A piece of the pie: Black and white immigrants since 1880.* Berkeley: University of California Press.

Lindemann, E. (1944). Symptomatology and management of acute grief. *American Journal of Psychiatry, 101,* 141–148.

Lipchick, E. (1986). The purposeful interview. *The Journal of Strategic and Systemic Therapies, 5* (1/2), 88–99.

Loeser, L. H. (1957). Some aspects of group dynamics. *Interpersonal Journal of Group Psychotherapy, 7,* 5–16.

Long, V. O. (1996). *Facilitating personal growth in self and others.* Pacific Grove, CA: Brooks/Cole.

Louer, J., Lourie, I., Salus, M., & Broadhurst, D. (1980). *The role of the mental health professional in the prevention and treatment of child maltreatment.* Washington, DC: U.S. Department of Health, Education, and Welfare.

Lum, D. (1996). *Social work practice and people of color* (3rd ed.). Pacific Grove, CA: Brooks/Cole.

Mackelprang, R. W., & Salisgiver, R. O. (1996). People with disabilities and social work: Historical and contemporary issues. *Social Work, 41*(1), 7–14.

Maccoby, E. E. (1990). Gender and relationships: A developmental account. *American Psychologist, 45,* 513–520.

Maccoby, E. E., & Jacklin, C. N. (1974). *The psychology of sex differences.* Stanford, CA: Stanford University Press.

Madanes, C. (1984). *Behind the one-way mirror: Advances in the practice of strategy therapy.* San Francisco: Jossey-Bass.

Mahoney, M. J. (1991). *Human change processes.* New York: Basic Books.

Martelli, L. J. (1987). *When someone you know has AIDS: A practical guide.* New York: Crown.

Martin, D. G., & Moore, A. D. (Eds.). (1995). *First steps in the art of intervention: A guidebook for trainers in the helping professions.* Pacific Grove, CA: Brooks/Cole.

Maslack, C. (1982). Understanding burnout: Definitional issues in analyzing a complex phenomenon. In W. S. Paing (Ed.), *Job stress and burnout* (pp. 29–40). Beverly Hills, CA: Sage.

McEntee, M. K. (1995). Deaf and hard-of-hearing clients: Some legal implications. *Social Work, 40*(2), 183–187.

McGoldrick, M., Pearce, J. K., & Giordano, J. (Eds.). (1982). *Ethnicity and family therapy.* (pp. 3–30). New York: Guilford Press.

McLemore, S. D. (1994). *Racial and ethnic relations in America* (4th ed.). Boston, MA: Allyn & Bacon.

McMahon, S. (1996). *The portable problem solver: Having healthy relationships.* New York: Dell.

McManus, M. C. (1991). Serving lesbian and gay youth. *Focal Point, 5,* 1–4.

Meenaghan, T. M. (1987). Macro practice: Current trends and issues. In A. Minihan (Editor-in-Chief), *Encyclopedia of social work* (18th ed. vol. 2, pp. 82–89). Silver Spring, MD: National Association of Social Workers.

Mehrabian, A. (1971). *Silent messages.* Belmont, CA: Wadsworth.

Meichenbaum, D. (1975). Theoretical and treatment implications of developmental research on verbal control of behavior. *Canadian Psychological Review, 16,* 22–27.

Meier, S. T., & Davis, S. R. (1993). *The elements of counseling* (2nd ed.). Pacific Grove, CA: Brooks/Cole.

Mendez, H. A. (1976). Single fatherhood. *Social Work, 21,* 308–312.

Meyer, C. (1990). *Can social work keep up with the changing family?* The Fifth Annual Robert O'Leary Memorial Lecture (pp. 1–24). Columbus: Ohio State University College of Social Work.

Miles, R. E. (1975). *Theories of management: Implications for organizational behavior and development.* New York: McGraw-Hill.

Miller, S. C. (1969). *The unwelcome immigrant: The American image of the Chinese, 1785–1882.* Berkeley: University of California Press.

Mills, M. J., Sullivan, G., & Eth, S. (1987). Protecting third parties: A decade after Tarasoff. *American Journal of Psychiatry, 144*(1), 68–74.

Minuchin, S. (1974). *Families and family therapy.* Cambridge, MA: Harvard University Press.

Minuchin, S., & Fishman, C. (1981). *Family therapy techniques.* Cambridge, MA: Harvard University Press.

Minuchin, S., Montalvo, B., Guerney, B. G., Rosman, B. L., & Schumer, F. (1967). *Families and their slums: An exploratory study of their structure and treatment.* New York: Basic Books.

Minuchin, S., Rosman, B., & Baker, L. (1978). *Psychosomatic families: Anorexia nervosa in context.* Cambridge, MA: Harvard University Press.

Mizarahi, T., & Rosentahl, B. B. (1993). Managing dynamic tensions in social changes coalitions. In T. Mizarahi & J. D. Morrison (Eds.), *Community organization and social administration: Advances, trends, and emergency principles.* New York: Hawthorn Press.

Mokuau, N., & Matsuoka, J. (1992). The appropriateness of personality theories for social work with Asian Americans. In S. Furato, R. Biswas, D. Chung, K. Munrase, & F. Ross-Scheriff (Eds.), *Social work practice with Asian Americans* (pp. 67–81). Newbury Park, CA: Sage.

Monahan, J. (1993). Limiting therapist exposure to Tarasoff liability. *American Psychologist, 48*(3), 242–250.

Moore, A. D., & Hiebert-Murphy, D. (1995). Paperwork and Writing Reports. In D. G. Martin & A. D. Moore (Eds.), *First steps in the art of intervention: A guidebook for trainers in the helping professions* (pp. 404–448). Pacific Grove, CA: Brooks/Cole.

Moore, S. T. (1990). A social work practice model of case management: The case management grid. *Social Work, 35*(5), 444–448.

Morales, A. T., & Sheafor, B. W. (1995). *Social work: A profession of many faces* (7th ed.). Needham Heights, MA: Simon & Schuster.

Morris, R. (1987). Social welfare policy: Trends and issues. In A. Minihan (Editor-in-Chief), *Encyclopedia of social work* (18th ed., vol. 2, pp. 664–681). Silver Spring, MD: National Association of Social Workers.

Morrow, W. L. (1975). *Public administration: Practices and the political process.* New York: Random House.

Moses, A. E., & Hawkins, R. O. (1982). *Counseling lesbian women and gay men: A lifetime approach.* St. Louis, MO: Mosby.

Moxley, D. P. (1989). *The practice of case management.* Newburg Press, CA: Sage.

Munrase, K. (1977). Minorities and Asian Americans. In J. B. Turner (Ed.), *Encyclopedia of social work* (vol. 2, no. 7, p. 953). Washington, DC: National Association of Social Workers.

NASW Provisional Council on Clinical Social Work. (1987). In A. Minahan et al. (Eds.), *Encyclopedia of social work* (18th ed., vol. 2; Appendix 3, pp. 965–966). Silver Spring, MD: National Association of Social Workers.

National Association of Social Workers (NASW), (1980). *Code of ethics.* Washington, DC: Author.

National Association of Social Workers. (1981). *Guidelines for the selection and use of social workers.* Silver Spring, MD: Author.

National Organization of Human Service Education. (1994a). *Code of Ethics.* Author.

National Organization for Human Service Education. (1995b, October). *Ethical Standards of Human Service Professionals.* Council for Standards in Human Service Education.

Nelson, T. (1994). *Handbook for field instructors in social work.* San Bernardino, San Bernardino, CA: California State University, Department of Social Work.

Netting, E. F., Kettner, P. M., & McMurtry, S. (1993). *Social work macro practice.* New York: Longman.

Neugeboren, B. (1985). *Organization, policy, and practice in human services.* New York: Longman.

Neukrug, E. (1994). *Theory, practice, and trends in human services.* Pacific Grove, CA: Brooks/Cole.

Newman, B., & Newman, P. (1995). *Development through life: A psychosocial approach* (6th ed.). Pacific Grove, CA: Brooks/Cole.

Nieto, D. S. (1982). Aiding the single father. *Social Work, 27,* 473–478.

Nilsen, A. P. (1993). Sexism in English: A 1990's update. In V. Cyrus (Ed.), *Experiencing race, class, and gender in the United States.* Mountain View, CA: Mayfield.

Norcross, J. C., & Newman, C. F. (1992). Psychotherapy integration: Setting the context. In J. C. Norcross & M. R. Goldfried (Eds.), *Handbook of psychotherapy integration* (pp. 3–45). New York: Basic Books.

Norten, H. (1982). *Clinical social work.* New York: Columbia University Press.

Nugent, F. A. (1990). *An introduction to the profession of counseling.* Columbus, OH: Merrill.

O'Brien, C., & Briggen, P. (1985). Our personal and professional lives: Learning positive connotation and circular questioning. *Family Process, 24,* 311–322.

Office of Attorney General of California. (1988). *Child abuse prevention handbook.* Sacramento, CA: Crime Prevention Center.

Ogloff, J. R. P. (1995). Navigating the quagmire: Legal and ethical guidelines. In D. Martin and A. D. Moore (Eds.), *First steps in the art of intervention* (pp. 349–373). Pacific Grove, CA: Brooks/Cole.

O'Hare, T., Williams, C. L., & Ezoviski, A. (1996). Fear of AIDS and homophobia: Implications for direct practice and advocacy. *Social Work, 41*(1), 51–58.

Okun, B. F., & Rappaport, L. J. (1980). *Working with families: An introduction to family therapy.* Pacific Grove, CA: Brooks/Cole.

Olmstead, M. (1959). *The small group.* New York: Random House.

O'Malley, H. (1979). *Elder abuse: A review of recent literature.* Boston: Legal Research and Services for the Elderly.

Ozawa, M. N. (1995). The economic status of vulnerable older women. *Social Work, 40*(3), 323–333.

Paine, W. S. (1982). Overview of burnout stress syndromes and the 1980's. In W. S. Paine (Ed.), *Job stress and burnout* (pp. 11–25). Beverly Hills, CA: Sage.

Pakel, E., Prusoff, B., & Uhlenhuth, E. (1971). Scaling life events. *Archives of General Psychiatry, 25,* 340–347.

Pallozzoli, M. S., Curillo, S., Selvin, M., & Sorrentino, A. M. (1989). *Family games: General model of psychotic processes in the family.* New York: Norton.

Pantoja, A., & Perry, W. (1992). Community development and restoration: A perspective. In T. G. Rivera & J. L. Erlick (Eds.), *Community organizing in a diverse society.* Boston: Allyn & Bacon.

Papell, C., & Rothman, B. (1980). Social group work models: Possession and heritage. In A. Alissi (Ed.), *Perspectives on social group work practice.* New York: Free Press.

Parad, H. J. (1965). *Crisis intervention: Selected readings.* New York: Family Service Association of American.

Park, R. (1950). *Race and culture.* Glencoe, IL: Free Press.

Patterson, J. M. (1988). Families experiencing stress. *Family Systems Medicine, 6,* 202–237.

Patterson, W., Dohn, H., Bird, J., & Patterson, G. (1983). Evaluation of suicidal patients: The SAD PERSONS scale. *Psychosomatics, 24,* 343–349.

Patti, R. J. (1985). In search of purpose for social welfare administration. *Administration in Social Work, 9,* 1–14.

Paul, G. L. (1967). Strategy of outcome research in psychotherapy. *Journal of Counseling Psychology, 31,* 109–118.

Pecora, P. J., Whittaker, J. K., & Maluccio, A. (1992). *The child welfare challenge: Policy, practice, and research.* New York: Aldine de Gruyter.

Pederson, P. B. (1991). Multiculturalism as a generic approach to counseling. *Journal of Multicultural Counseling and Development, 15,* 16–24.

Perry, M. A., & Furukawa, M. J. (1986). Modeling methods. In F. H. Keifer & A. P. Goldstein (Eds.), *Helping people change: A textbook of methods* (3rd ed., pp. 66–110). New York: Pergamon Press.

Piaget, T. (1967). *The child's conception of the world.* New York: Harcourt Brace.

Piercy, F., & Sprenkle, D. (1986). *Family therapy sourcebook.* New York: Guilford.

Pines, A., & Aronson, E., with Krafty, D. (1981). *Burnout: From tedium to personal growth.* New York: Free Press.

Pope, K. S. (1985, April). Dual relationships: A violation of ethical, legal, and clinical standards. *California State Psychologist, 20*(3), 3–5.

Pope, K. S., & Vasquez, M. J. T. (1991). *Ethics in psychotherapy and counseling: A practical guide for psychologists.* San Francisco: Jossey-Bass.

Popple, P. R., & Leighninger, L. H. (1990). *Social work, social welfare, and American society.* Boston, MA: Allyn & Bacon.

Potter-Efron, R., & Potter-Efron, P. (1995). *Letting go of anger.* Oakland, CA: New Harbinger Publications.

Proctor, C. D., & Groze, V. K. (1994). Risk factors for suicide among gay, lesbian, and bisexual youths. *Social Work, 39*(5), 504–514.

Proctor, E. K., & Davis, L. E. (1994). The challenge of racial differences: Skills for clinical practice. *Social Work, 39,* 314–323.

Rabinowitz, F. E. (1991). The male-to-male embrace: Breaking the touch taboo in a men's therapy group. *Journal of Counseling and Development, 69*(6), 574–576.

Rapp, C. A., & Poertner, J. (1995). *A performance model for human service management.* Manuscript. Lawrence: University of Kansas.

Rapport, L. *Crisis intervention goals and steps* (http://childlaw.law.sc.edu/usermanual/crisis/seatonz.htm).

Reed, W. J. (1987). Research in social work. In A. Minihan (Editor-in-Chief), *Encyclopedia of social work* (18th ed., vol 2, pp. 479–487). Silver Spring, MD: National Association of Social Workers.

Reid, K. (1981). *From character building to social treatment: The history of the use of groups in social work.* Westport, CT: Greenwood Press.

Reynolds, B. (1965). *Learning and teaching in the practice of social work.* New York: Russell and Russell.

Ridley, C. R. (1989). Racism and counseling as an adverse behavioral process. In P. Pedersen, J. Draguns, W. Lonner, & U. Trimble. (Eds.), *Counseling across cultures* (3rd ed., pp. 55–77). Honolulu, HI: University of Hawaii Press.

Riggar, T. F. (1985). *Stress burnout: An annotated bibliography.* Carbondale: Southern Illinois University Press.

Rivera, F., & Erlick, J. (1992). *Community organizing in a diverse society.* Boston, MA: Allyn & Bacon.

Robbins, S. (1992). *Essentials of organizational behavior* (3rd ed.). Englewood Cliffs, NJ: Prentice Hall.

Russo, R. J. (1985). *Serving and surviving as a human-service worker.* Prospect Heights, IL: Waveland Press.

Roberts, J. (1990). *Crisis intervention handbook: Assessment, treatment and research.* Belmont, CA: Wadsworth.

Roberts-DeGennaro, M. (1987). Developing case management as a practice model. *Social Casework, 68* (8), 466–470.

Rogers, C. R. (1951). *Client-centered therapy.* Boston: Houghton Mifflin.

Rogers, C. (1959). A theory of therapy, personality, and interpersonal relationships as developed in the client-centered framework. In S. Kock (Ed.), *Psychology: A study of a science* (vol. 3, pp. 210–211). New York: McGraw-Hill.

Rogers, C. (1961). *On becoming a person.* Boston: Houghton Mifflin.

Rooney, R. H. (1992). *Strategies for work with involuntary clients.* New York: Columbia University Press.

Rossi, P. H. (1990). The old homeless and the new homelessness in historical perspective. *American Psychologist, 45*(8), 954–959.

Rothblum, E. D., & Franks, V. (1983). Introduction: Warning: Sex-role stereotypes may be hazardous to your health. In V. Franks & E. D. Rothblum (Eds.), *The stereotyping of women: Its effects on mental health.* New York: Springer.

Rothman, J. (1987). Community theory and research. In A. Minihan (Editor-in-Chief), *Encyclopedia of social work* (18th ed., vol. 1, pp. 308–316). Silver Spring, MD: National Association of Social Workers.

Rothman, J., Erlich, J. L., & Tropman, J. E. (1995). *Strategies of community intervention* (5th ed.). Itasca, IL: F. E. Peacock.

Rothman, J. & Tropman, J. (1987). Models of community organization and macro practice perspectives: Their mixing and phasing. In F. M. Cox, J. L. Erlich, J. Rothman, & J. Tropman (Eds.), *Strategies of community organizing* (pp. 3–26). Itasca, IL: F. E. Peacock.

Rothman, R. (1991). A model of case management: Toward empirically based practice. *Social Work, 36*(6), 520–527.

Rubin, A., & Babbie, E. (1989). *Research methods for social work.* Belmont, CA: Wadsworth.

Rubin, H., & Rubin, I. S. (1992). *Community organizing and development* (2nd ed.). New York: Macmillan.

Ruth, S. (1990). *Issues in feminism.* Mountain View, CA: Mayfield.

Ryan, W. (1976). *Blaming the victim.* (rev. ed.). New York: Vintage.

Sabin, J. E. (1995). Time efficient long term psychotherapy in managed care. *Harvard Review of Psychiatry, 3,* 163–165.

Sarin, R. C. (1987). Administration in social welfare. In *Encyclopedia of social work* (vol. 1). Silver Spring, MD: National Association of Social Workers.

Satir, V. (1967). *Conjoint family therapy.* Palo Alto, CA: Science and Behavior Books.

Satir, V. (1972). *Peoplemaking.* Palo Alto, CA: Science and Behavior Books.

Satir, V. (1971). The family as a treatment unit. In J. Haley (Ed.), *Changing families.* New York: Grune & Stratton.

Saxton, A. (1971). *The indispensable enemy: Labor and the American-Chinese movement.* Berkeley: University of California Press.

Schaefer, C. E., Johnson, L., & Wherry, J. N. (1982). *Group therapies for children and youth.* San Francisco: Jossey-Bass.

Schmolling, P., Jr., Youkeles, M., & Burger, W. R. (1993). *Human services in contemporary America* (3rd ed.). Pacific Grove, CA: Brooks/Cole.

Scott, G. G. (1990). *Resolving conflict with others and within yourself.* Oakland, CA: New Harberger Publications.

Scott, W. R. (1981). *Organizations: Rational, natural, and open systems.* Englewood Cliffs, NJ: Prentice Hall.

Seligman, M. (1996, Spring). Good news for psychotherapy: The Consumer Report study. *The Communicator.*

Selznick, Philip. (1992). The moral commonwealth: Social theory and practice of community social work. *NY: The promise of community.* Berkeley: University of California Press.

Shertzer, B., & Stone, S. (1980). *Fundamentals of counseling* (3rd ed.). Boston, MA: Houghton and Mifflin.

Shimkunas, A. (1972). Demand for intimate self-disclosure and pathological verbalization in schizophrenia. *Journal of Abnormal Psychology, 80,* 197–205.

Shneidman, E. S. (1985). *Definition of suicide.* New York: Wiley.

Shneidman, E. S. (1987, March). At the point of no return: Suicidal thinking follows a predictable path. *Psychology Today,* 54–58.

Shneidman, E. S., Farberow, N. L., & Litman, R. E. (1976). *The psychology of suicide* (p. 128). New York: Aronson.

Skidmore, R. A. (1990). *Social work administration: Dynamic management and human relationships* (2nd ed.). Englewood Cliffs, NJ: Prentice-Hall.

Slaikeu, K. (1990). *Crisis intervention: A handbook for practice and research.* Boston: Allyn & Bacon.

Siegel, J. M. (1974). A brief review of the effects of race in clinical service interactions. *American Journal of Orthopsychiatry, 44,* 555–562.

Sigelman, C. K., & Shaffer, D. R. (1995). *Life-span human development* (2nd ed.). Pacific Grove, CA: Brooks/Cole.

Simons, R. L. (1995). Generic social work skills in social administration: The example of persuasion. In J. Rothman, J. L. Erlick, & J. E. Tropman (Eds.), *Strategies of community intervention.* Itasca, IL: F. E. Peacock.

Simons, R. L., & Aigner, S. M. (1985). *Practice principles: A problem solving approach to social work.* New York: Macmillan.

Single moms view selves as strong and resourceful, *National Association of Social Workers News, 32*(2), 1, 16.

Smith, D. S. (1982). Trends in counseling and psychotherapy. *American Psychologist, 37,* 802–809.

Soloman, B. (1976). *Black empowerment: Social work in oppressed communities.* New York: Columbia University Press.

Sommers, T., & Shields, L. (1987). *Women take care.* Gainesville, FL: Triad.

Speiter, G. (1978a). *Playing their game our way: Using the political process to meet community needs.* Amherst, MA: University of Massachussets, Citizen Involvement Project.

Spenser, M. B., & Dornbusch, S. M. (1990). Challenges in studying minority youth. In S. Feldman & G. R. Elliot (Eds.), *At the threshold: The developing adolescent.* Cambridge, MA: Harvard University Press.

Steinberg, S. (1981). *The ethnic myth: Race, ethnicity, and class in America.* Boston: Beacon Press.

Steinmetz, S. (1978). Battered parents. *Society, 15,* 54–55.

Steinmetz, S. (1994). Dependency, stress, and violence between middle-aged caregivers and their elderly parents. *Social issues resource manual.* New York: Guilford Press.

Stepick, A. (1984). Haitian boat people: A study in the complicating forces shaping U.S. immigration policy. In R. R. Hofstedler (Ed.), *U.S. immigration policy* (pp. 163–196). Durham, NC: Duke University Press.

St. Germaine, J. (1993). Dual relationships: What's wrong with them? *American Counselor, 2*(3), 25–30.

Storms, M. D. (1981). A theory of erotic orientation development. *Psychological Review, 83,* 340–353.

Straus, M. A., Gelles, R. J., & Steinmetz, S. K. (1980). *Behind closed doors: Violence in the American family.* New York: Anchor Press.

Sue, W. D., & Sue, D. (1990). *Counseling the culturally different: Theory and practice* (2nd ed.). New York: Wiley.

Szapocznik, J., & Hernandez, R. (1988). The Cuban American family. In C. H. Mindel, R. Witabenstein, & R. Wright, Jr. (Eds.), *Ethnic families in America* (3rd. ed., pp. 160–172). New York: Elsevier.

Szasz, T. (1986). The case against suicide prevention. *American Psychologist, 41*(7), 806–812.

Tabachnick, N. D., & Farberow, N. L. (1965). The assessment of self-destructive potentiality. In N. L. Farberow & E. S. Shneidman (Eds.), *The cry for help.* New York: McGraw-Hill.

Taeuber, D. M. (1991). *Statistical handbook on women in America.* Phoenix, AZ: Onyx Press.

Tarasoff v. Board of Regents of the University of California (1976), 131 *California Reporter 14,* 551 P.2d 334.

Taylor, R. L. (Ed.). (1994). *Minority families in the United States: A multicultural perspective.* Englewood Cliffs, NJ: Prentice Hall.

Teamer, B. (1977). *The empathetic understanding response process.* Chicago Counseling Center discussion paper.

Thomas, R. R., Jr. (1991). *Beyond race and gender: Unleashing the power of your total work force by managing diversity.* New York: AMACOM.

Thomlison, B., Rogers, G., Collins, D., & Grinnell, R. M., Jr. (1996). *The social work practicum: An access guide* (2nd ed.). Itasca, IL: F. E. Peacock.

Thompson, A. (1983). *Ethical concerns in psychotherapy and their legal ramifications.* Lanham, MD: University Press of America.

Thompson, D. C. (1993). The male role stereotype. In V. Cyrus (Ed.), *Experiencing race, class, and gender in the United States.* Mountain View, CA: Mayfield.

Thornton, S., & Garrett, K. (1995). Ethnography as a bridge to multicultural practice. *Journal of Social Work Education, 31*(1), 67–74.

Toman, W. (1961). *Family constellation.* New York: Springer.

Tomm, K. (1984). One perspective in the Milan Systematic Approach. Part II. Description of session format, interviewing style and interventions. *Journal of Marital and Family Therapy, 10*(3), 253–271.

Tomm, K. (1993, January/February). The ethics of dual relationships. *The California Therapist,* 7–19.

Toseland, R. W., & Rivas, R. E. (1984). *An introduction to group work.* New York: MacMillan.

Toughy, W. (1974). World Health Agency zeros in on suicide. *Los Angeles Times,* October 25, 1974, pp. 1–3.

Trattner, W. I. (1989). *From poor low to welfare state: A history of social welfare in America* (4th ed.). New York: Free Press.

Traux, C., & Carkhuff, R. (1967). *Toward effective counseling and psychotherapy: Training and practice.* Chicago, IL: Aldene.

Travers, A. W. (1993). *Supervision: Techniques and new dimensions* (2nd ed.). Englewood Cliffs, NJ: Regents/Prentice Hall.

Trotzer, J. P. (1981). The centrality of values in families and family therapy. *International Journal of Family Therapy, 3,* 42–55.

Tseng, W. S. (1986). Cultural aspects of family assessment. *International Journal of Family Psychiatry, 7,* 19–31.

Tseng, W. S., & Hsu, J. (1991). *Culture and family: Problems and therapy.* New York: Hawthorn Press.

Tseng, W. S., & Hsu, J. (1986). The family in Micronesia. In W. S. Tseng, W. S. Lee, & C. A. Lee (Eds.), *Culture and mental health in Micronesia.* Honolulu: University of Hawaii, John A. Burns School of Medicine, Department of Psychiatry.

United Nations. (1955). *Social progress through community development* (p. 6). New York: Author.

U.S. Bureau of Census. (1991). *Statistical abstract of the United States 1991.* Washington, DC: US Government Printing Office.

U.S. Bureau of Census. (1992). *Statistical abstract of the United States, 1991.* Washington, DC: U.S. Government Printing Office.

U.S. Department of Health and Human Services. (1984). *The educator's role in the prevention and treatment of child abuse and neglect.* Publication No. (OHDS) 84-30172, p. 13. (Modified). Washington, DC: Author.

Van Hoose, W. H., & Paradise, L. V. (1979). *Ethics in counseling and psychotherapy: Perspectives in issues and decision making.* Cranston, RI: Carroll.

Wagley, C., & Harris, M. (1956). *Minorities in the new world.* New York: Columbia University Press.

Walker, L. E. (1979). *The battered woman.* New York: Harper & Row.

Wallerstein, J. S. (1988). Children of divorce: Stress and developmental tasks. In N. Garmezy & M. Rutter (Eds.), *Stress, coping, and development in children.* Baltimore, MD: Johns Hopkins University Press.

Wallerstein, J. S., & Blakesless, S. (1989). *Second chances.* New York: Ticknor & Fields.

Ward, D. E. (1983). The trend toward eclecticism and the development of comprehensive models to guide counseling and psychotherapy. *Personnel and Guidance Journal, 67,* 154–157.

Warren, P. (1972). *The community in America.* Chicago: Rand McNally.

Warren, R. L. (1975). Types of purposive social change at the community level. In R. M. Kramer & H. Specht (Eds.), *Readings in community organization practice.* (2nd ed., pp. 134–149). Englewood Cliffs, NJ: Prentice Hall.

Watkins, E. C., Jr. (1985, February). Countertransference: Its impact on the counseling situation. *Journal of Counseling and Development, 63,* 356–359.

Watzlawick, P., Weakland, J. H., & Fisch, R. (1974). *Change: Principles of problem formulation and problem resolution.* New York: Norton.

Webster's Ninth Collegiate Dictionary. (1984). Springfield, MA: Merriam-Webster.

Weeks, D. (1992). *The eight essential steps to conflict resolution.* New York: Putnam.

Weick, A., Rapp, C., Sullivan, P., & Kisthardt, W. (1989). A strengths perspective for social work practice. *Social Work, 34,* 350–354.

Weiner, M. E. (1990). *Human services management: Analysis and applications* (2nd ed.). Belmont, CA: Wadsworth.

Weiner-Davis, M., deShazer, S., & Gingerich, W. (1987). Building on pre-treatment change to construct the therapeutic solution: An exploratory study. *Journal of Marital and Family Therapy, 13*(14), 359–363.

Westwood, M. J., & Ishiyama, F. I. (1990). The communication process as a critical intervention for direct change in cross-cultural counseling. *Journal of Multicultural Counseling and Development, 18,* 163–171.

Wildavsky, A. (1977). *A comparative theory of budgetary processes.* Boston, MA: Little, Brown.

Williams, E. E., & Ellison, F. (1996). Culturally informed social work practice with American Indian clients: Guidelines for non-Indian workers. *Social Work, 41*(2), 147–151.

Wilson, S. J. (1978). *Confidentiality in social work: Issues and principles.* New York: Free Press.

Wilson, S. J., (1980). *Recording: Guidelines for social workers.* New York: Free Press.

Woodside, M., & McClam, T. (1994). *An introduction to human services* (2nd ed.). Pacific Grove, CA: Brooks/Cole.

Yalom, I. (1975). *The theory and practice of group psychotherapy* (2nd ed.). New York: Basic Books.

Yamashiro, G., & Matsuoka, K. (1997). Help seeking among Asian and Pacific Americans: A multiperspective analysis. *Social Work, 42*(2), 176–186.

Zastrow, C. (1985a). *The practice of social work.* Chicago: Dorsey Press.

Zastrow, C. (1985b). *Social work with groups.* Chicago: Nelson-Hall.

Zastrow, C. (1995). *The practice of social work* (3rd ed.). Pacific Grove, CA: Brooks/Cole.

Zastrow, C. (1996). *Introduction to social work and social welfare* (6th ed.). Pacific Grove, CA: Brooks/Cole.

Zastrow, C., & Kirst-Ashman, K. (1994). *Understanding human behavior and the social environment* (3rd ed.). Chicago: Nelson-Hall.

Zimring, F. M., & Ruskin, N. J. (1992). Carl Rogers and client/person-centered therapy. In D. K. Freedheim (Ed.), *History of psychotherapy: A century of change* (pp. 629–656). Washington, DC: American Psychological Association.

INDEX

CREDITS

This page constitutes an extension of the copyright page. We have made every effort to trace the ownership of all copyrighted material and to secure permission from copyright holders. In the event of any question arising as to the use of any material, we will be pleased to make the necessary corrections in future printings. Thanks are due to the following authors, publishers, and agents for permission to use the material indicated.

Chapter 3: 97: Table 3.3 from *Student Manual for Theory and Practice of Counseling and Psychotherapy,* 5th Edition, by G. Corey, p. 126. Copyright © 1996 Brooks/Cole Publishing Company. All rights reserved. **130:** Table 3.5 reprinted by permission of Marshall Jung. **133:** Table 3.6 adapted from *Counseling Basics for Wiccan Clergy,* by J. Harrow. Home page: http://members.aol.com/JehanaS/c_basics/c_models.html. Reprinted by the permission of Russell & Volkening as agents for the author. Copyright © 1996 by Judy Harrow.

Chapter 4: 143: Figure 4.1(A) from *The Practice of Social Work,* Sixth Edition, by C. H. Zastrow, p. 182. Copyright © 1999 Brooks/Cole Publishing Company. All rights reserved.

Appendices: 413: Appendix J reprinted from ACA *Codes of Ethics and Standards of Practice.* Copyright © 1995 ACA. Reprinted with permission. No further reproduction authorized without written permission of the American Counseling Association. **429:** Appendix K reprinted from the *Code of Ethics,* National Association of Social Workers. Copyright © 1996 National Association of Social Workers, Inc. Reprinted by permission. **441:** Appendix L from "Ethical Standards for Human Service Professionals," (1996), *Human Service Education,* 16(1), 11–17. Copyright © 1996 National Organization for Human Service Education. Reprinted by permission.

IN-BOOK SURVEY

At Brooks/Cole, we are excited about creating new types of learning materials that are interactive, three-dimensional, and fun to use. To guide us in our publishing/development process, we hope that you'll take just a few moments to fill out the survey below. Your answers can help us make decisions that will allow us to produce a wide variety of videos, CD-ROMs, and Internet-based learning systems to complement standard textbooks. If you're interested in working with us as a student Beta-tester, be sure to fill in your name, telephone number, and address. We look forward to hearing from you!

In addition to books, which of the following learning tools do you currently use in your counseling/human services/social work courses?

_____ **Video** _____ in class _____ school library _____ own VCR

_____ **CD-ROM** _____ in class _____ in lab _____ own computer

_____ **Macintosh disks** _____ in class _____ in lab _____ own computer

_____ **Windows disks** _____ in class _____ in lab _____ own computer

_____ **Internet** _____ in class _____ in lab _____ own computer

How often do you access the Internet? _____

My own home computer is:

_____ Macintosh _____ DOS _____ Windows _____ Windows 95

The computer I use in class for counseling/human services/social work courses is:

_____ Macintosh _____ DOS _____ Windows _____ Windows 95

If you are NOT currently using multimedia materials in your counseling/human services/social work courses, but can see ways that video, CD-ROM, Internet, or other technologies could enhance your learning, please comment below:

_____

_____

_____

Other comments (optional): _____

_____

Name _____ Telephone _____

Address _____

School _____

Professor/Course_____

You can fax this form to us at (831) 375-6414; e-mail to: info@brookscole.com; or detach, fold, secure, and mail.

Optional:

Your name: _____ Date: _____

May Brooks/Cole quote you, either in promotion for *Advanced Practice in Human Service Agencies: Issues, Trends, and Treatment Perspectives* or in future publishing ventures?

Yes: _____ No: _____

Sincerely,

Lupe A. Alle-Corliss
Randy M. Alle-Corliss

FOLD HERE

BUSINESS REPLY MAIL
FIRST CLASS   PERMIT NO. 358   PACIFIC GROVE, CA

POSTAGE WILL BE PAID BY ADDRESSEE

ATT: <u>Editor, Assessment and Counseling</u>

Brooks/Cole, Cengage Learning
20 Davis Drive
Belmont, CA 94002

FOLD HERE

TO THE OWNER OF THIS BOOK:

I hope that you have found *Advanced Practice in Human Service Agencies* useful. So that this book can be improved in a future edition, would you take the time to complete this sheet and return it? Thank you.

School and address: _____

_____

Department: _____

Instructor's name: _____

1. What I like most about this book is: _____

_____

_____

2. What I like least about this book is: _____

_____

_____

3. My general reaction to this book is: _____

_____

_____

4. The name of the course in which I used this book is:_____

_____

5. Were all of the chapters of the book assigned for you to read? _____

   If not, which ones weren't? _____

6. In the space below, or on a separate sheet of paper, please write specific suggestions for improving this book and anything else you'd care to share about your experience in using this book.

_____

_____

_____

_____

_____

_____

BROOKS/COLE
CENGAGE Learning

BUSINESS REPLY MAIL
FIRST-CLASS MAIL   PERMIT NO. 34   BELMONT CA

POSTAGE WILL BE PAID BY ADDRESSEE

Attn: Editor, Assessment and Counseling

Brooks/Cole, Cengage Learning
20 Davis Drive
Belmont, CA 94002

OPTIONAL:

Your name: _____ Date: _____

May we quote you, either in promotion for *Advanced Practice in Human Service Agencies*, or in future publishing ventures?

Yes: _____   No: _____

Sincerely yours,

Lupe A. Alle-Corliss
Randy M. Alle-Corliss

CPSIA information can be obtained
at www.ICGtesting.com
Printed in the USA
FFOW01n2223210715
15378FF

9 780534 348113